Diagenesis and Basin Development

Edited by

A. D. Horbury

and

A. G. Robinson

AAPG Studies in Geology #36

Published by
The American Association of Petroleum Geologists
Tulsa, Oklahoma, U.S.A.
Printed in the U.S.A.

Published 1993

ISBN: 0-89181-044-7

Association Editor: Susan A. Longacre
Science Director: Gary D. Howell
Publications Manager: Cathleen P. Williams
Special Projects Editor: Anne H. Thomas

On the cover: cathodoluminescence photo showing sequence stratigraphy and carbonate cements. Cements in the plate show a succession as follows. Alternating bright and nonluminescing calcite are related to 4th-order eustatic lowstands and establishment of shallow meteroic water lenses at the onset of Gondwana glaciation. Orange-brown luminescent calcite is related to basin margin uplift at the onset of post-rift deposition. This ferroan calcite cement was precipitated down dip in the aquifer at relatively high temperatures. The bright yellow luminescent phase precipitated during late diagenesis as a result of hydrocarbon expulsion and basin dewatering during the Hercynian orogeny.

Contents

AAPG
Wishes to thank the following
for their generous contributions
to

Diagenesis and Basin Development

Netherlands Research School
of Sedimentary Geology

Statoil

Contributions are applied against the production costs of the publication, thus directly reducing the book's purchase price and making the volume available to a greater audience.

About the Editors

ANDREW HORBURY

Andrew did his undergraduate degree at Bristol University, England and a PhD in carbonate sedimentology at Manchester University. Following completion of his PhD in 1986, Andrew joined BP Exploration in London as a sedimentologist to work on exploration projects. These were initially all on the sedimentology of redbed clastic fluvial systems in NW Europe, but a chance to work on the sedimentology and stratigraphy of Northern Iraq and Kurdistan for a year brought him back to his carbonate roots. There followed projects on the Tethyan geology of Syria, Jordan, Tunisia and Ethiopia. Andrew moved into the South East Asia Group of BP as the resident sedimentologist in 1990, and immediately worked on and later ran a geological review of BP as the resident sedimentologist in 1990, and immediately worked on and later ran a geological review of the Himalayan foreland basins. He later worked on the sedimentology of Miocene carbonates in Vietnam and China, and also completed a sedimentological and sequence stratigraphic analysis of the lacustrine clastics of the Bohai basin in China. Andrew left BP in 1992 and formed a small consultancy company primarily dedicated to carbonate depositional systems and their diagenesis (Cambridge Carbonates Ltd.). The consultancy has worked on Kurdistan, the Tarim basin of onshore China, SE Asian Miocene carbonate fields and platforms, Eocene to Miocene carbonates of Indonesia, Miocene buildups in Vietnam and the Philippines and Palaeogene carbonates in Libya.

ANDREW ROBINSON

After undergraduate work at Oxford University, England, Andrew completed a MSc degree at the University of Toronto, Canada in minerals exploration. After a year working in gold exploration in Spain, Andrew joined Shell International Petroleum as an exploration geologist working in the Netherlands. In 1984, Andrew joined BP Research in England and spent the following five years developing and carrying out research programs in the fields of reservoir quality prediction and sedimentology. In 1990, he moved into BP Exploration's Frontiers' Exploration group in England where he has been involved in projects in offshore Newfoundland, Yemen, the Black Sea and Eastern Europe. He has written scientific papers in a number of geological disciplines including minerals exploration, clastic diagenesis, inorganic geochemistry and sedimentary basin development and has co-authored a book published by Blackwells Scientific: "Inorganic geochemistry: applications to petroleum geology." He is a member of an International Lithosphere Programme working party on the origin of sedimentary basins. At the time of writing, he still works for BP.

Preface

Diagenesis and Basin Development grew out of a conference of the same name held on May 10, 1991 at the Geological Society of London, England. Most (though not all) of the chapters herein were presented at that conference. The purpose of both conference and volume is to examine links between sediment diagenesis—and consequent porosity and permeability modification—and aspects of the development of sedimentary basins. We began from the premise that most diagenetic phenomena, at least in the relatively coarse-grained sediments that constitute petroleum reservoirs, are not directly related to compaction alone, but rather to subsurface fluid flow. This in turn is determined by large-scale features of basin development such as relative changes in sea level, faulting, or changes in heating or subsidence rates. Diagenetic phenomena ought therefore in principle to be predictable from seismic data. We point out in our introductory chapter that the volume offers little comfort to those who would prefer easy solutions to porosity and permeability prediction, simply because we are still fairly ignorant about the causes of cementation and mineral dissolution and about the often complex relationships between diagenesis and facies. Nonetheless, the chapters do provide some important guidelines and insights that may be useful to the exploration geologist in the development of new play concepts. For the academically minded, the volume also provides many stimulating ideas for further research.

We would point out that the views expressed above and in the introductory chapter do not necessarily represent those of the authors who have kindly contributed to the project.

Andrew Horbury and Andrew Robinson

ACKNOWLEDGMENTS

The conference on which this volume is based was held under the auspices of the British Sedimentological Research Group at the Geological Society of London and was supported by BP Research and BP Exploration. We owe our thanks to all of the authors who have contributed papers and also to friends and colleagues in BP who have contributed over many years to our evolving ideas. Finally, our thanks go to our families and partners for their forebearance.

Chapter 1

Diagenesis, Basin Development, and Porosity Prediction in Exploration—An Introduction

Andrew Horbury
and
Andrew Robinson

Accurate prediction of porosity and permeability of reservoir targets ahead of the bit is one of the most important tasks facing the petroleum geologist. This is as true for a wildcat well in an undrilled basin as it is for a production or injection well in a producing field. The porosity prediction influences the perception of likely reserves while that of permeability—the key control on well flow rate—has a major impact on the Net Present Value ascribed to an opportunity and thus the likelihood that it will be drilled. Porosity and permeability are related both to the original fabric and composition of sediments and to diagenetic processes. The latter are the subject of this volume.

The usual method of predicting porosity and permeability involves the use of correlations between porosity and depth (or sometimes a time-depth integral) and between porosity and permeability (Scherer, 1987; Schmoker and Gautier, 1988; Schmoker and Halley, 1982; Bloch, 1991). This procedure requires an estimate of the depth to the target (which is always available); a set of data relating average porosity to depth for the target formation or another selected on some basis as analogous; and core data relating porosity to permeability. It also requires of course that depth, porosity and permeability should show reasonable correlations. Correlations often work well but there are cases in which they can be very difficult to apply with any confidence. The most obvious is in a frontier exploration setting where local empirical relationships between depth, porosity and permeability are simply not available. One can resort to regional or global data sets but these can mislead as much as they can help. Another instance in which empirical prediction is difficult is when depth, porosity and permeability are poorly correlated. The method has one further important drawback. Regression relationships smooth out anomalies and make it impossible to predict, for example, unusually porous or permeable sands at great depths; yet it is often these that are the most important predictions, particularly in basins which are in a mature phase of exploration.

At the other end of the spectrum of approaches to porosity and permeability prediction lies the use of computer models that calculate water flow through the burial history of a sediment and how the water reacts with the mineral components, either precipitating or dissolving minerals. This approach is exemplified by AAPG Memoir 49 (Meshri and Ortoleva, 1990). The models can calculate mineral volumes as well as growth/dissolution sequences and can therefore be used to make quantitative porosity (though not permeability) predictions. The models are undeniably tremendously powerful as an aid to probing the processes responsible for diagenesis, but they involve a number of assumptions about the causes of mineral precipitation and dissolution that may be incorrect.

When plausible depth-porosity-permeability relationships of known reliability are available, it would be perverse not to use them. When they are not, it is necessary to try to predict the nature and effects of the diagenetic processes to which the target formation is likely to have been subjected. The diagenesis of a coarse-grained sediment does not usually take place in a closed chemical system (e.g., Gluyas and Coleman, 1992). Chemical components are moved in and/or out of the rock either by diffusion or, more importantly, by advection involving moving pore waters. Water moves through sedimentary basins in response to changes in potential caused by features of basin evolution, including for example generation and release of overpressure, eustatic changes in sea level, tectonic elevation of basin margins, faulting, and increased heat flow. The theme of this volume is the relationships between such different aspects of basin development and sediment diagenesis. The chapters it contains do not in the main offer easy solutions to porosity and permeability prediction. Our understanding of diagenetic processes is still too limited. They do, however, contribute to understanding how and why a sediment's burial history—in its broadest sense—is related to its reservoir properties.

This approach is in a sense intermediate between that exemplified by AAPG Memoir 49 and the straightforward use of porosity-depth plots. It presupposes that by looking at their geological history, we can make some semi-quantitative predictions about where, when, and how rocks become cemented. The ultimate goal of such an approach would be prediction of porosity and permeability from basin analysis based on interpreted seismic sections. Perhaps a measure of how far we still have to go along this road is the fact that none of the chapters collected here contain a seismic line.

In the remainder of this introduction, we briefly review the contents and major conclusions of each chapter, attempting at the same time to place each in context. The contributions are grouped into four themes that are not mutually exclusive:

(1) Constraints on diagenetic processes
(2) Diagenesis and basin hydrodynamics
(3) Diagenesis and faults
(4) Diagenesis, sequence stratigraphy and changes in relative sea level.

CONSTRAINTS ON DIAGENETIC PROCESSES

Chapter 2, "Mechanisms of quartz cementation in North Sea reservoir sandstones: constraints from fluid compositions," by Andrew Aplin et al., uses oxygen isotope and fluid inclusion data from quartz cements to characterize the formation waters involved in quartz precipitation. These turn out to be highly saline and shifted to heavy isotopic compositions as a result of significant water-rock interaction involving low water/rock ratio. This is important because the volumes of waters such as these in sedimentary basins are strictly limited. The data seem to point to a cementation mechanism which involves either water recycling or diffusion. This would rule out meteoric water driven by a hydraulic head, pointing to a mechanism such as density-driven convection. Convection does not however appear capable of driving pore water at rates that are high enough (at least centimeters to meters per year; e.g., Robinson and Gluyas, 1992) to precipitate significant amounts of quartz (Wood and Hewett, 1984).

If there remains a problem in identifying what the processes responsible for cementation actually *are*, the next chapter, "Geochemical evidence for a temporal control on sandstone cementation," by Jon Gluyas et al., attempts to describe *how* the mechanisms operate. It interprets a correlation between fluid inclusion homogenization temperature and depth in Haltenbanken (offshore Norway) and parts of the North Sea as reflecting relatively short-lived (of the order of 10 m.y.) cementation episodes during which the clastic reservoirs in question lost most of their porosity and permeability. K-Ar ages of illite cements in the Rotliegend gas reservoirs of the Southern North Sea also suggest a regional cementation event, this time involving illite rather than quartz. The authors tentatively link these phases of cementation to changes in regional fluid flow patterns caused by changes in basin subsidence rates. Changes in subsidence rates can be interpreted from seismic sections so this would offer a means of predicting the timing of cementation phases in undrilled areas.

Both of these chapters draw partly on fluid inclusion data. The chapter by Stuart Haszeldine and Mark Osborne, "Fluid inclusion temperatures in diagenetic quartz reset by burial: implications for oilfield cementation," questions whether fluid inclusions in diagenetic quartz provide reliable information about the temperature of quartz growth, or whether the temperatures have been reset by the high internal pressures built up during burial and consequent heating. This is an important caveat. Inclusion leakage is now widely regarded as a general problem in diagenetic carbonates (Goldstein, 1986; Prezbindowski and Larese, 1987) but there is still a question mark as to whether the same is true of quartz.

DIAGENESIS AND BASIN HYDRODYNAMICS

Five chapters in this section examine the influences of particular fluid-flow mechanisms on sediment diagenesis. In "Meteoric water and sandstone diagenesis in the Western Canada sedimentary basin," Fred Longstaffe reviews a large body of stable isotope data on a wide range of mineral cements. He shows how meteoric water was the dominant influence on cementation throughout the history of the foreland basin, affecting not only non-marine, but also marginal to marine sediments, even during early diagenesis. This hints at the importance of looking at early diagenesis in particular, in a sequence-stratigraphic context (relationship to relative sea-level changes) and points to the next group of chapters. Early diagenesis notwithstanding, Longstaffe shows that most cementation in the basin occurred from meteoric water when the sediments were close to their maximum burial depths. He relates this to uplift of the basin margins during the main phase of the Laramide Orogeny in the early Eocene.

The following chapter, "Diagenetic pathways in sedimentary basins," by Wendy Harrison and Regina Tempel, calculates diagenetic pathways for two sandstones from the Gulf of Mexico that are likely to have had very different hydrologic histories. This is accomplished via a relatively simple method for integrating fluid-flow and burial-history models with chemical equilibrium modeling without a computational link. One of the sandstones—from the onshore Wilcox Group—would have been subjected initially to flushing by meteoric water and later to water expelled from the overpressured part of the basin. The other—offshore Frio Formation—sandstone is at the boundary between normally and overpressured zones. The modeling seems capable of predicting diagenetic mineral growth sequences that correspond closely to those observed. Diagenetic changes are concentrated during periods of greatest water flow and are thus effectively episodic, related to changes in the hydraulic evolution of the basin.

"A discussion of flow mechanisms responsible for alteration and mineralization in the Cambrian aquifers of the Ouachita-Arkoma basin—Ozark system," by Larry Cathles, examines the kinds of fluid flow that could have produced the extensive dolomite cements with consistent cathodoluminescence zoning that are found in the Cambrian aquifers of the U.S. Mid-Continent. These cements and associated zinc-lead mineralization bear witness to extraordinary phases of water flow affecting vast areas: >100 km^3 of Cambrian limestone was dolomitized over an area of >16,000 km^2 (Gregg, 1985). The fact that the aquifers are still filled with brine (even though they are presently being flushed by meteoric recharge) constrains models for the dolomitization. Cathles's mathematical modeling indicates that flow must have been rapid and pulsed in order to explain the homogenization temperatures of fluid inclusions in dolomite. Compaction or gas-driven dewatering seem the most likely causes. One might speculate as to the effect such pulsed flow of hot water might have on the diagenesis of clastic rocks in other sedimentary basins.

Another chapter, "Stress-induced fluid flow in rifted basins," by Ronald van Balen and Sierd Cloetingh, describes an investigation of the possible role of changes in horizontal stress fields in driving fluid flow. Changes in horizontal stresses are generated by plate tectonic movements and transmitted over large distances. They affect relative sea level by raising and lowering basin margins relative to their centers and so could be responsible for triggering phases of fluid flow. Van Balen and Cloetingh present calculations that show that compressive stresses can raise overpressures and associated flow velocities by a few tens of percent and that tensile stresses can reduce both by similar amounts. These effects seem fairly small, but the authors' calculations are conservative; greater effects might be associated with for example, the major Neogene subsidence of the Norwegian continental margin. Compressive stresses can also cause a relative fall in sea level by raising basin margins, which might have the effect of triggering meteoric water flow. Such effects are of interest because of their potential to cause diagenesis through time scales of a few to ten m.y. that correspond with tectonic phenomena (as implied by the chapter by Gluyas et al.).

Much of the important diagenesis in carbonates occurs before significant burial and is related to water circulation systems within the platforms. In "Circulation of saline ground water in carbonate platforms: a review and case study from the Bahamas," Fiona Whitaker and Peter Smart review flow mechanisms in carbonate platforms and ramps and describe a modern example of an isolated carbonate platform where large scale fluid flow is occurring today—the Great Bahamas Bank. Around $10^4 m^3$ of saline water is being discharged through karstic "blue holes" every day per kilometer of coastline along the Tongue of the Ocean. Salinity and temperature measurements made at blue holes suggest that refluxing brines mix with seawater in a major circulation system involving cool oceanic waters. The waters emerging are depleted in Mg, suggesting that they are dolomitizing the platform limestones. This model may be applied to ancient analogs.

DIAGENESIS AND FAULTS

We have sometimes felt that as explanations for diagenetic phenomena, faults are, to misquote Doctor Johnson, "...the last refuge of a scoundrel." They can be invoked as conduits for or barriers to fluid flow, as desired, and their movements may even themselves drive fluid flow. In "The influence of fault zone processes and diagenesis on fluid flow," Rob Knipe demonstrates that he is not a scoundrel by examining aspects of the roles of faults in fluid flow. Fault movements can change fluid pressures and potentials and thereby drive fluid flow: water is drawn into dilatant fault zones prior to movement and is expelled both during and after the main rupture over periods of perhaps three months (the aftershock period). Extensive flow will, however, only be possible if there is an appropriate combination of fault event magnitude, fracture geometry, and permeable fault rock development. These are the factors that need to be evaluated for particular fault systems at particular phases of basin development, if they are to be demonstrated as major influences on diagenesis. The absence of other chapters in this section is a reflection partly of the novelty of Knipe's approach and partly of the difficult nature of the work.

DIAGENESIS, SEQUENCE STRATIGRAPHY, AND CHANGES IN RELATIVE SEA LEVEL

The chapters by Longstaffe and by Whitaker and Smart have already hinted at the importance of considering early diagenesis in particular in a sequence stratigraphic context. In "Eustatic and tectonic controls on porosity evolution beneath sequence-bounding unconformities and parasequence disconformities on carbonate platforms," Fred Read and Andrew Horbury are more specific about the link for carbonates. They look at the magnitude and frequency of sea level changes (greenhouse/icehouse relationships) that influence limestone cylicity through geological time. When this is examined in the context of climate, diagenetic pattern types and porosity development can be predicted in a qualitative sense. These authors conclude that sequence boundaries (sensu strictu) are often responsible for regional calcite cementation as well as cavernous porosity development in the near-surface, particularly under wet climatic conditions. Higher amplitude and frequency sea level changes—such as those caused by Pleistocene glaciations—are more commonly associated with karstification and cementation on a more local scale and favor primary porosity preservation. Arid climates result in less early diagenesis than do humid climates for all types of sequence boundaries, and under arid conditions, primary porosity is preserved whilst dolomitic porosity often develops; these cycles are usually sealed by evaporites. The distribution of giant field hydrocarbon reservoirs is strongly controlled by these cycle types, with the largest fields occurring in arid "greenhouse" cycles, many "icehouse" cycles, and beneath humid-climate 2nd–3rd-order eustatic and tectonic cycle top paleokarst. This chapter shows how a knowledge of the age, sequence stratigraphy, and paleoclimatological setting of a target reservoir would permit pre-drill predictions of porosity distribution.

In "Diagenesis and porosity evolution of Lower Permian *Palaeoaplysina* buildups, Bjørnøya, Barents Sea—an example of diagenetic response to high-frequency sea level fluctuations in an arid climate," Lars Stemmerik and Geir Larssen take a look at the early diagenetic effects associated with high-amplitude (glacio-eustatic) sea level fluctuations in an arid climate and use the data collected from onshore Bjørnøya to predict reservoir quality in time-equivalent, undrilled buildups offshore. The dolomitization models associated with these systems contrast strongly with the diagenetic effects expected under humid climatic conditions (e.g., Meyers, 1974). Evidently, paleoclimatological modeling is an important tool in predicting the early diagenesis—and thus porosity—of carbonates.

The next chapter, by Clyde Moore and Ezat Heydari, "Burial diagenesis and hydrocarbon migration in platform limestones—a conceptual model based on the Upper Jurassic of the Gulf Coast of the USA," also considers carbonate diagenesis in an explicitly sequence-stratigraphic context. The "conceptual model"—based on Moore's detailed work on the Smackover Formation—refers to a carbonate that was buried soon after deposition and that was not subjected to subsequent subaerial exposure, providing a counterpoint and contrast to Hendry's work. Moore shows that in such cases, ultimate diagenetic pathways are established during very early diagenesis under the influence of changes in relative sea level. The model relates to a particular combination of sequence distributions and burial history that can be interpreted from seismic records and thus can be used predictively in undrilled areas.

The chapter by Jim Hendry, "Geologic controls on regional subsurface carbonate cementation: an isotopic-paleohydrologic investigation of Middle Jurassic limestones in central England," provides an excellent case history that illustrates the overprinting effects of a 3rd order tectonic sequence boundary (the Late Cimmerian unconformity) over 4th order sequence boundaries. The resulting aquifer is regionally developed and confined by shales. Hendry describes isotopic analyses and textures from the relatively high-temperature anoxic aquifer. There are notable contrasts with low temperature shallow meteoric aquifers of the type more frequently described in the literature.

Finally, "How important is the Late Cimmerian unconformity in controlling formation of kaolinite in sandstones of the North Sea? (with examples from the Snorre Field)," by Per-Arne Bjørkum et al., continues with the theme of the influence of the Late Cimmerian unconformity on diagenesis, this time on clastics. The authors look at the influence of a relative sea level fall on the diagenesis of feldspathic sandstones, pointing out that diagenetic effects related to unconformities will only be preserved if, on average, the erosion rate is slower that the rate of propagation of the reaction front. Their calculations suggest that this was not true for kaolinization of feldspars in the Lunde Field during Late Jurassic–Early Cretaceous emergence, and bolster the argument by demonstrating that kaolinite distribution is not obviously spatially related to the unconformity.

REFERENCES CITED

Bloch, S., 1991, Empirical prediction of porosity and permeability in sandstones, AAPG Bulletin, v.75, p.1145-1160.

Gluyas, J.G. and M.L. Coleman, 1992, Material flux and porosity change during sediment diagenesis, Nature, v.356, p.52-54.

Goldstein, R.H., 1986, Reequilibration of fluid inclusions in low temperature calcium carbonate cement, Geology, v.14, p.792-795.

Gregg, J.M.. 1985. Regional epigenetic dolomitization in the Boneterre Dolomite (Cambrian), southeastern Missouri. Geology, v. 13, p. 503–506.

Meshri, I.D. and P.J. Ortoleva, 1990, Prediction of reservoir quality through chemical modeling, AAPG Memoir 49, 175pp.

Meyers, W.J., 1974, Carbonate cement stratigraphy of the Lake Valley Formation (Mississippian) Sacramento Mountains, New Mexico, Journal of Sedimentary Petrology, v.44, p.837-861.

Prezbindowski, D.R. and R.E. Larese, 1987, Experimental stretching of fluid inclusions in calcite—implications for diagenetic studies, Geology, v. 15, p.333-336.

Robinson, A.G. and J.G. Gluyas, 1992, Model calculations of loss of porosity in sandstones as a result of compaction and quartz cementation, Marine and Petroleum Geology, v. 9, p.319-323.

Scherer, M., 1987, Parameters influencing porosity in sandstones: a model for sandstone porosity prediction, AAPG Bulletin, v.71, p.485-491.

Schmoker, J.W. and D.L. Gautier, 1988, Sandstone porosity as a function of thermal maturity, Geology, v.16, p.1007-1010.

Schmoker, J.W. and R.B. Halley, 1982, Carbonate porosity versus depth—a predictable relation for south Florida, AAPG Bulletin, v.66, p.2561-2570.

Wood, J.R. and T.A. Hewett, 1984, Reservoir diagenesis and convective fluid flow, *in* Clastic Diagenesis, D.A. McDonald and R.C. Surdam, eds., AAPG Memoir 37, p.99-1109.

PART 1

Constraints on Diagenetic Processes

Chapter 2

Mechanisms of Quartz Cementation in North Sea Reservoir Sandstones: Constraints From Fluid Compositions

Andrew C. Aplin
Newcastle Research Group in Fossil Fuels and Environmental Geochemistry
The University of Newcastle
Newcastle upon Tyne, U.K.

Edward A. Warren and Shona M. Grant[1]
BP Research
Sunbury-on-Thames
Middlesex, U.K.

Andrew G. Robinson
BP Exploration
Stockley Park, Uxbridge
Middlesex, U.K.

ABSTRACT

Quartz occurs widely as a cement in pre-Tertiary North Sea reservoir sands. Fluid inclusion and oxygen isotope data show that much of the cement formed in the Tertiary from ^{18}O-rich waters with variable, but often high salinities. The volume of this type of water is limited and is unlikely to be more than the void space of the sedimentary basin. The mineralizing fluids also show considerable compositional heterogeneity over relatively small areas and depth ranges.

The composition and compositional heterogeneity of the mineralizing fluids limit the mechanisms by which silica transport and quartz cementation could have occurred. Meteoric recharge is ruled out. Mass balance considerations strongly suggest that compaction-driven flow is extremely unlikely to have led to the type of pervasive quartz cementation observed in the pre-Tertiary North Sea section. The compositional heterogeneity of the mineralizing fluids suggests that large-scale convective overturn is also an unlikely transport mechanism. It is probable that the silica required for quartz cementation was supplied locally, and not by large-scale fluid flow. We believe that the fluid data restrict potential silica transport mechanisms to: (1) reservoir-scale convection and (2) diffusion. However, significant mass transport by either of these mechanisms has yet to be demonstrated in sedimentary basins.

[1]Present address:
BP Petroleum Development (Norway) Ltd.
Stavanger, Norway

INTRODUCTION

Quartz is an almost ubiquitous and volumetrically important cement in pre-Tertiary North Sea reservoir sands, comprising up to 24 % and averaging 12 to 15% of the rock volume (e.g. Scotchman et al., 1989; Glasmann et al., 1989a,b; Burley et al., 1989). As the prediction of reservoir quality is a key goal of exploration success and diagenetic research, there are thus good reasons for trying to understand what controls the distribution of diagenetic quartz in actual and potential reservoirs.

Point-counting of thin sections using the optical microscope suggests that the volume of quartz cement in pre-Tertiary North Sea sandstones often exceeds the volume of quartz which could have been supplied locally from intergranular pressure solution (Gluyas, 1985; Ehrenberg, 1990; Gluyas & Coleman, 1992). This implies that silica has been imported into the sand. However, the processes which conspired to cause cementation—the origin of the silica and the mechanisms by which it was transported—are very poorly understood.

Economic geologists have shown repeatedly how the chemical and isotopic composition of mineralizing fluids can be used to understand the processes which led to mineral deposition (e.g. Taylor, 1974; Sverjensky, 1984). Diagenetic systems are amenable to a similar approach since most diagenetic reactions involve the aqueous phase (Curtis, 1977). Information about the fluids from which diagenetic minerals precipitated is retained in fluid inclusions and in the isotopic composition of the mineral. The aim of this chapter is to demonstrate how the number of alternative explanations of mechanisms of silica transport and quartz cementation can be constrained from simple observations about the fluids from which quartz precipitated. We present data for the salinity and oxygen isotopic composition of fluids from which quartz cements formed in several pre-Tertiary North Sea reservoir sands (Figure 1). Like present-day formation waters in the deep North Sea, the mineralizing fluids are generally highly saline, enriched in ^{18}O and compositionally heterogeneous over small areas. These results place strict limits both on the absolute volumes of mineralizing fluids and the extent to which mass transport could have been achieved by large-scale fluid flow.

SAMPLES AND ANALYTICAL TECHNIQUES

The fluid inclusion, oxygen isotopic and water compositional data presented in this paper are all taken from Permian Rotliegend and Mesozoic (mainly Jurassic) sandstones from a large part of the North Sea (Figure 1). We present a compilation of new and published fluid inclusion data (Hogg, 1989; Glasmann et al., 1989a; 1989b). Previously unpublished fluid inclusion analyses were carried out on a Linkam THM600 heating-cooling stage attached to a petrographic microscope and connected to a Linkam TMS90 controller. This was calibrated from -125 to 440°C using synthetic inclusions, pure bulk standards and liquid crystals. Accuracy is around ±1°C for a homogenization temperature measurement and about ±0.2°C for a determination of ice final melting temperature.

The temperature of quartz cementation in the samples was estimated from the homogenization temperatures (T_H) of primary (within cement) and pseudosecondary inclusions (at grain-cement boundaries). Maximum ages for cementation were calculated from the homogenization temperatures and the modeled burial history of the sample. The accuracy of these estimates is difficult to assess and depends on several factors including the age of cementation itself, the accuracy of the bottom-hole temperature determination, the time since a rifting event and other factors intrinsic to the model used. We believe that in general, the accuracy of the age of cementation is of the order of one geological stage.

Oxygen isotopic analyses were made on quartz overgrowths which had been separated from detrital quartz using the procedure of Lee and Savin (1986). Separates were size-fractionated and each size-fraction analyzed. Purity checks using cathodoluminescence microscopy showed that authigenic quartz was concentrated in the finer size fractions. Plots of $\delta^{18}O$ versus grain size have a hyperbolic form which tend towards a constant $\delta^{18}O$ value in the finer grain sizes. This is assumed to give the isotopic composition of authigenic quartz, although errors of up to 1‰ are possible (a complete data set for the size fractions is available from the authors on request).

MINERALIZING FLUIDS

Isotopic And Chemical Composition

Oxygen isotope data for authigenic quartz are given in Table 1. Table 1 also displays estimates of precipitation temperatures, which were taken to be the same as the fluid inclusion homogenization temperatures. This procedure assumes that homogenization temperatures are equal to true trapping temperatures, which is reasonable if the fluids contain small amounts of methane (Hanor, 1980). The assumption may be justifiable here because all the samples are from oil and gas reservoirs and contain authigenic quartz that formed at temperatures at which oil and gas were being generated in adjacent or down flank mudstones.

Combining the isotopic composition of the quartz with a precipitation temperature allows the oxygen

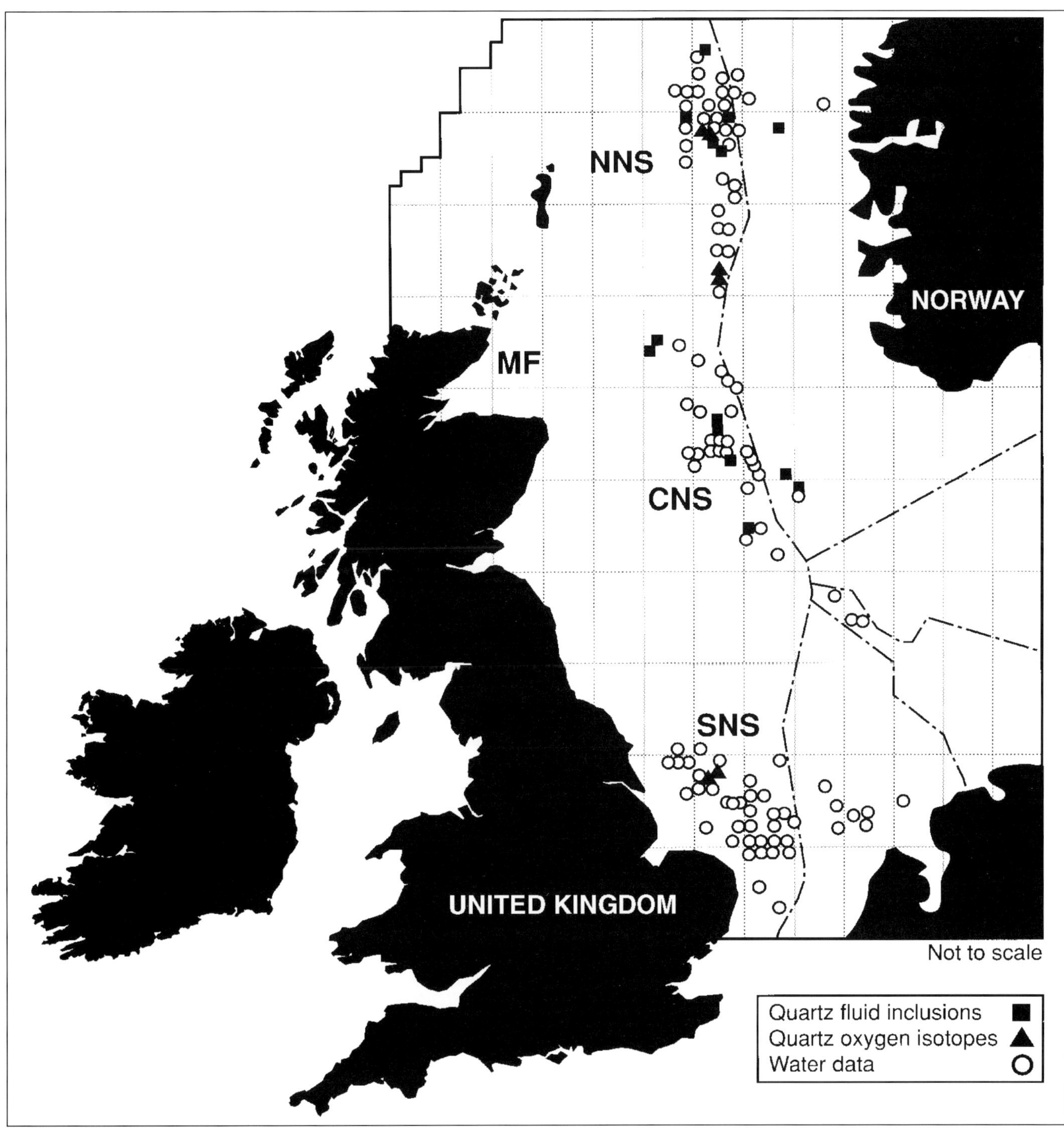

Figure 1. Location map of datasets. NNS = Northern North Sea; CNS = Central North Sea; SNS = Southern North Sea; MF = Moray Firth.

isotopic composition of the mineralizing fluid to be calculated. These are, in every case, enriched in ^{18}O relative to meteoric water present in the region at the time of cementation (Figure 2). The oxygen isotopic composition of the putative paleometeoric water is an estimate based on the strong present-day relation between latitude and $\delta^{18}O$ (Dansgaard, 1964). A similar relation is thought to have prevailed during the Tertiary (Lawrence and Taylor, 1971; Sheppard, 1986).

The salinities (equivalent weight % NaCl) of the mineralizing fluids are plotted as a function of present-day depth in Figure 3. Quartz precipitated from fluids with an extremely diverse compositional range. However, whilst many of the cements precipitated from fluids which were more saline than sea water (3.5 wt.% NaCl), less than five per cent precipitated from fresh water (< 0.5 wt.% NaCl).

An important conclusion of this study is that

quartz precipitated from fluids that were compositionally heterogeneous on a variety of scales. Regional heterogeneity is suggested by the data from sandstones in both the Northern and Central North Sea. In the Northern North Sea (NNS), there is a broad correlation between salinity and present-day depth, suggesting that the cements may have precipitated from a compositionally stratified water column (Figure 3). In the Moray Firth and Central North Sea (CNS) the salinities of mineralizing fluids vary enormously over relatively small areas, both vertically and laterally. Such compositional heterogeneity points towards the local involvement of salt in influencing fluid composition and suggests that the mineralizing fluids were compartmentalized. This observation has important implications for potential mass transport mechanisms, since large-scale fluid flow would tend to homogenize fluid compositions.

The approximate timing of quartz cementation can be estimated by combining fluid inclusion homogenization temperatures with a burial history for the relevant stratigraphic unit. In the NNS and Moray Firth areas this procedure consistently indicates that quartz cementation took place in the Late Cretaceous to early Tertiary (70 to 40 Ma). In the CNS, quartz cementation has occurred between the Miocene and the present-day. Commonly, quartz precipitated when the sediments were undergoing relatively rapid subsidence (see Gluyas et al., this volume).

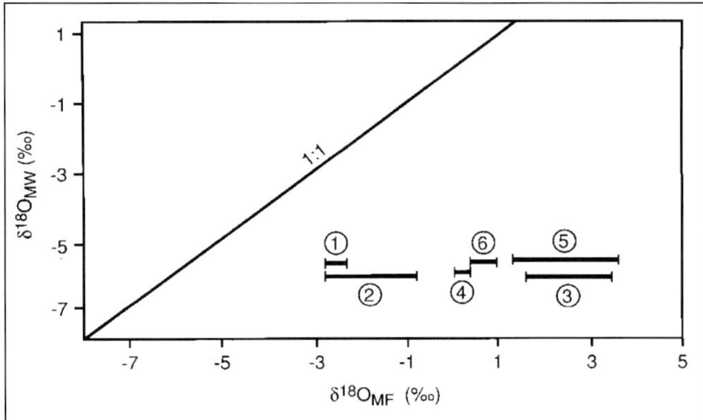

Figure 2. $\delta^{18}O$ of mineralizing fluid ($\delta^{18}O_{MF}$) versus the $\delta^{18}O$ of local meteoric water ($\delta^{18}O_{MW}$). This shows that the mineralizing fluids are universally enriched in ^{18}O compared with a putative meteoric water present in the region at the time of cementation. The isotopic composition of the meteoric water was estimated from the strong correlation between latitude and $\delta^{18}O$ (Dansgaard, 1964). Bars represent the estimated analytical uncertainty; numbers refer to sample numbers in Table 1.

Comparison With Present-Day Formation Waters

The Tertiary tectonic evolution of the North Sea is dominated by subsidence. During the Tertiary, therefore, there has been little opportunity for meteoric water to enter the basin from its margins. The waters now present in the basin may therefore be those from which quartz precipitated during the Tertiary. If these Tertiary waters and present-day formation waters turned out to be chemically similar, we would be more confident in suggesting that quartz formed under conditions similar to those presently existing in the deep parts of the basin—a regime dominated by very sluggish, compaction-driven flow.

There are very few published analyses of formation waters from the North Sea. The only data of regional coverage are salinities (Figure 4), both measured (Egeberg and Aagaard, 1989; Abbotts, 1991) and derived from water resistivity (R_w) data (SPWLA, 1989). Salinities in the Permian Rotliegend sandstone of the Southern North Sea (SNS; Figure 4) are generally high and many are close to that of halite-saturated water (250,000 mg/L) Those for Paleocene and mid-Jurassic sandstones of the CNS are much more variable (50–250,000 mg/L) but also approach halite saturation. In contrast, formation water salinities for the Jurassic sandstones of the NNS are far lower (20–70,000 mg/L), many being less than seawater (37,000 mg/L). The high salinities of formation waters in the SNS and CNS appear to be controlled by the distribution of Permian Zechstein evaporites. High salinity waters occur at high stratigraphic intervals in the CNS where salt diapirs penetrate upward into Eocene sediments, indicating significant water-salt interaction even in young sediments. In the NNS fields where salt is absent, salinities are distinctly lower. These observations suggest significant stratification and compartmentalization of formation waters in the North Sea.

Table 1. Sample details and isotope data.

Sample No.	Area	Age/Unit	$\delta^{18}O_q$[1] (‰)	T_{ppt}[2]	$\delta^{18}O_w$[3] (‰)	Reference
1	North Viking Graben	U. Jurassic (Brae equiv.)	+18.5 to +19.0	112–135	-2.8 to -2.3	This chapter
2	North Viking Graben	U. Jurassic (Brae equiv.)	+16.4 to +16.9	112–135	-2.8 to +0.2	This chapter
3	Southern North Sea	Permian (Rotliegend)	+22.9	95–110	+1.6 to +3.5	Lee et al., 1989
4	Brent	M. Jurassic (Brent Group)	+19	115–120	0	This chapter
5	Hoton	Permian (Lower Leman)	+16.5 to +18.8	155	+1.3 to +3.6	This chapter
6	Hoton	Permian (Lower Leman)	+17.8 to +18.5	130	+0.3 to +1.0	This chapter

[1]q=quartz
[2]Precipitation temperature, estimated using fluid inclusion homogenization temperatures.
[3]w=water from which quartz precipitated

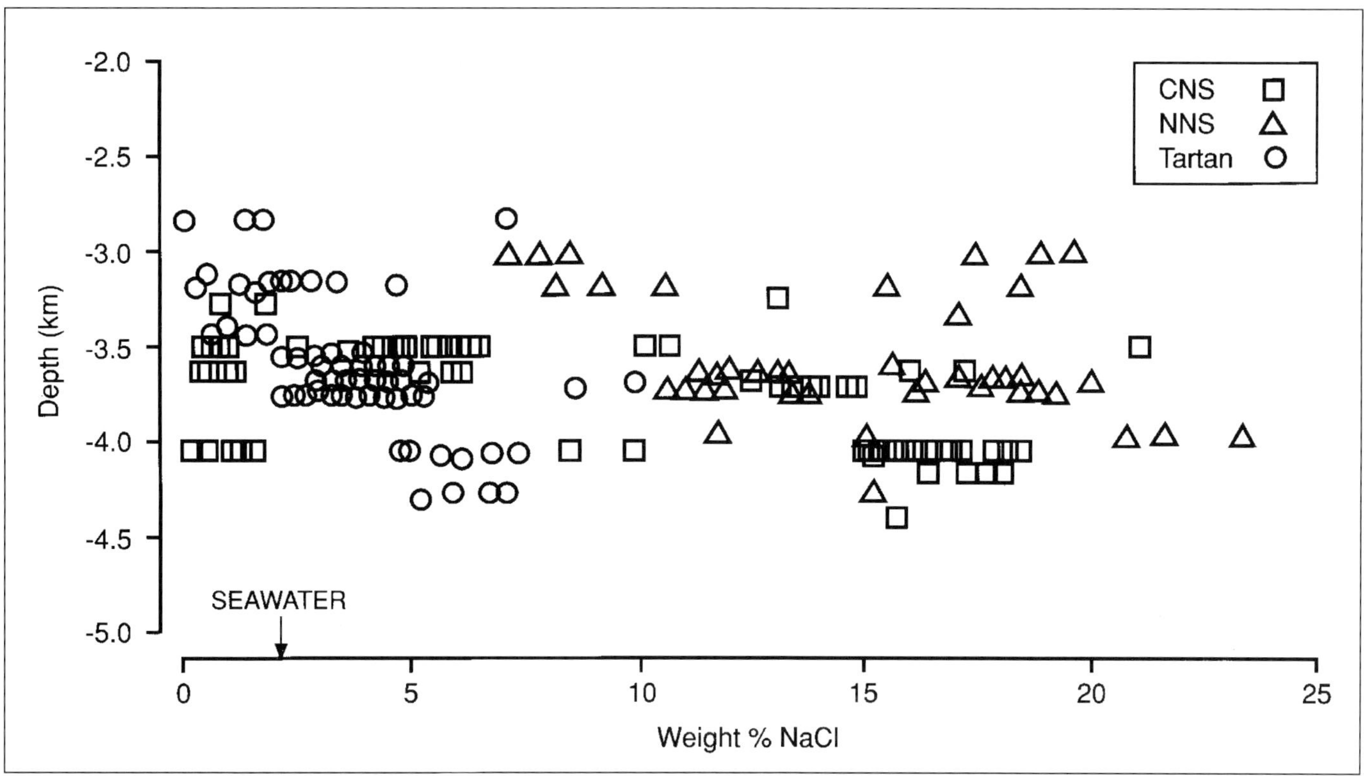

Figure 3. Salinities (expressed as wt % equivalent NaCl) of fluids from which quartz precipitated, plotted as a function of present-day burial depth. Salinities were estimated from freezing experiments on fluid inclusions. CNS=Central North Sea; NNS=Northern North Sea. Tartan field is located about 100 mi east of the northeast tip of Scotland.

The gross, present-day salinity variations between the CNS and NNS mimic the variations in the fluid inclusion salinities found in quartz cements: salinities of the mineralizing fluids are generally lower in the NNS than in the CNS (Figure 3). Such salinity variations suggest that the processes which have the greatest control on salinity, such as salt dissolution, have been active over significant periods of time. Furthermore, the data show that no case can be made for large scale fluid movement between the CNS and NNS, either now or during the Tertiary.

The most complete set of formation water analyses is that of Egeberg and Aagaard (1989) for the Norwegian sector of the North Sea. In reservoir sands buried below 2 km, they found a range of NaCl contents between 1.8 and 24.7 weight %, and a range of $\delta^{18}O$ values between –5.0 and +3.9 ‰. All the waters plot to the right of the meteoric water line (Figure 5). These variably saline, ^{18}O-enriched waters have a compositional range which is broadly similar to that of the waters from which quartz precipitated, and may therefore be genetically related to them.

Egeberg and Aagaard's (1989) data also demonstrate the important fact that, like the mineralizing fluids, present-day waters exhibit significant isotopic and chemical heterogeneity over relatively small areas. Salinity generally increases with depth, and different fields contain compositionally distinct waters. This argues against large-scale mixing or overturn of waters. Another excellent example of vertically stratified waters is the Piper reservoir system, located 100 mi east of the northeast tip of Scotland (Burley et al., 1989) (Figure 6). Because salinity exerts a greater influence on density than increasing temperature, waters like those in Piper are density-stratified and thus gravitationally stable. Similar trends are seen in many sedimentary basins (e.g. Hanor, 1979). The few oxygen isotopic analyses of North Sea formation waters support the implication of the salinity variations—that waters are compositionally heterogeneous and thus poorly mixed (Table 2; Figure 7).

In summary, the waters from which quartz precipitated in North Sea sands during the Tertiary are compositionally similar to those which are present today in the deep parts of the North Sea basin. These waters are saline and are enriched in ^{18}O compared with local meteoric water. Furthermore, both present-day and ancient waters show large compositional variations over relatively small areas. Waters appear to be compartmentalized and, in some cases, vertically stratified. There is no evidence for large-scale mixing or convective overturn.

Origin And Volume Of Mineralizing Fluids

One way to move material in a sedimentary basin is by fluid flow. Since quartz (like most diagenetic

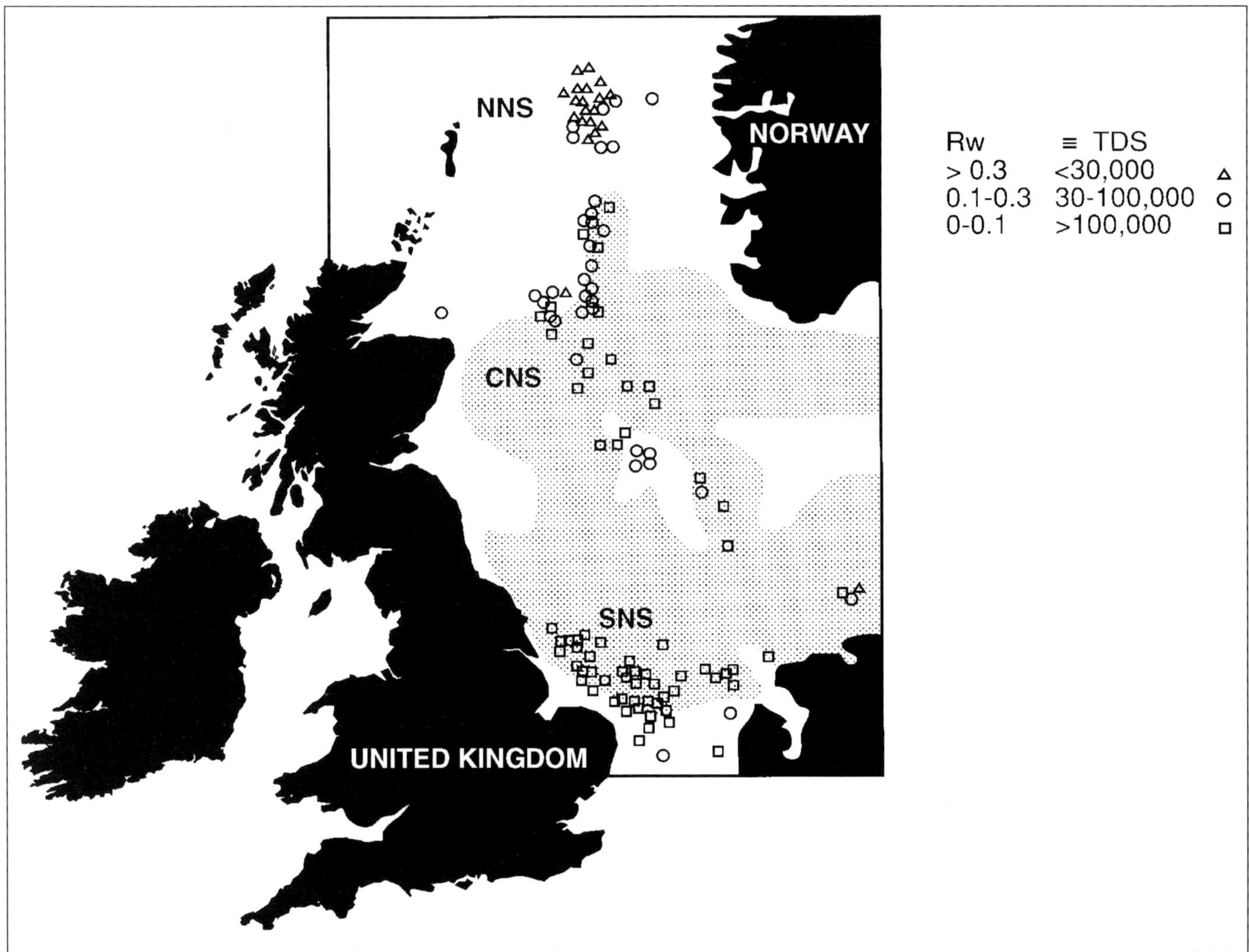

Figure 4. Present-day salinity variations in North Sea formation waters superimposed on the distribution of Late Permian Zechstein evaporites (stippled area). Salinities were derived analytically from formation water samples or from formation resistivities (TDS = Total Dissolved Solids). Data for the Southern North Sea (SNS) are from Early Permian Rottliegend sands that underlie Zechstein evaporites. Data for the Central North Sea (CNS) are from Paleocene and Jurassic sands. Data for the Northern North Sea (NNS) are from Jurasic sands.

minerals) is sparingly soluble, the observation that deeply-buried sandstones often contain large masses of authigenic quartz has been used to infer that large volumes of fluid must have passed through the sandstones in order to account for the observed volumes of cement (Sibley and Blatt, 1976; Leder and Park, 1986; Robinson and Gluyas, 1992). This idea can be tested if constraints can be placed on the absolute volumes of the fluid from which quartz precipitated. The oxygen isotopic and chemical composition of mineralizing fluids provide these constraints by placing limits on their origin. Formation waters in sedimentary basins may contain components from waters of several sources: meteoric water, sea water, the residues of evaporated sea water ("bitterns") and "metamorphic" water (originating from basement rocks). Each of these end-member water types has a distinct and recognizable chemical and/or isotopic composition which can be modified by diagenetic reactions (Figure 5). It is difficult to define uniquely the origin of a formation water because mixtures can explain most of the compositional variations observed in formation waters (Figure 5). There are therefore several possible solutions to the problem of producing a saline and ^{18}O-enriched pore water. Nonetheless, each leads to the important conclusion that the volume of fluid from which quartz precipitated was strictly limited:

1. The waters are mixtures of meteoric water and bitterns. The volume of bitterns is essentially that of the pores in evaporites at the time of deposition. Since evaporites generally form less than 10% of basin fill (Hitchon, 1968; Garrels and Mackenzie, 1971), the total volume of bitterns, and of bittern-meteoric water mixtures, is correspondingly limited.

2. The waters are meteoric in origin but were enriched in ^{18}O by reaction with rock minerals; salts were derived from the dissolution of evaporites. In sedimentary basins, enrichment of water in ^{18}O occurs when minerals which formed at low temperature, such as clay minerals and carbonates, recrystallize at higher temperatures (Clayton et al., 1966; Suchecki and Land, 1983). During this process, oxygen isotopes are repartitioned between mineral and water, resulting in a relative enrichment of water in ^{18}O. The degree of enrichment is highly dependent on the relative amounts of reacting minerals and water (the water-rock ratio). If the mass of water oxygen is high relative to the mass of rock oxygen (high water-rock ratio), reaction and isotopic exchange between water and rock will have little effect on the isotopic composition of the water. Low water-rock ratios are required if the isotopic composition of the water is to be shifted significantly from its initial value.

The extent to which the isotopic composition of water is shifted by reaction with rock minerals can be estimated using Taylor's (1974) isotope mass balance approach to water-rock interaction. In a sediment that is buried and exchanges isotopes with its trapped interstitial water (a "closed system"), the isotope mass balance equation is:

$$W.\delta_{wi}.C_w + R.\delta_{ri}.C_r = W.\delta_{wf}.C_w + R.\delta_{rf}.C_r \qquad (1)$$

where W and R are the masses of water and rock minerals in the reaction system, i = initial value, f = final value after isotopic exchange, C_w = atom per cent of oxygen in water, C_r= atom per cent of oxygen in reacting rock minerals, and $\delta = \delta^{18}O$. If equilibrium occurs between water and rock, then:

$$W/R = \{(\delta_{rf} - \delta_{ri})/[\delta_{wi} - (\delta_{rf} - \Delta)]\} \times C_r/C \qquad (2)$$

where $\Delta = \delta_{rf} - \delta_{wf}$

It is possible to calculate δ_{wf} by (a) taking W/R to be the ratio of porosity to reacting minerals in a

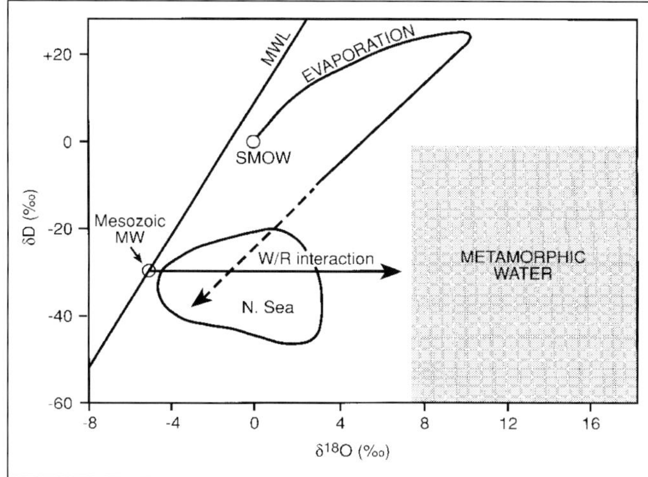

Figure 5. Present-day stable isotopic composition of formation waters from the Norwegian Shelf (data from Egeberg and Aagaard, 1989) superimposed on various end member water compositions. Metamorphic water is from Sheppard (1986). The sea water evaporation curve is from Knauth and Beeunas (1986). The arrow denoting water-rock (W/R) interaction shows the evolutionary trend of a putative Mesozoic meteoric water (MW) trapped within a smectite-bearing mudstone as it recrystallizes to illite. (SMOW = Standard Mean Ocean Water; MWL = Meoteoric Water Line.) The isotopic composition of the meteoric water is based on the relation between latitude, δD and $\delta^{18}O$ (Dansgaard, 1964). See text for further details.

Table 2. Stable isotope data (‰ SMOW) for North Sea and Haltenbanken formation waters.

Area	Age/Unit	Depth (m)	$\delta^{18}O$	δD
North Viking Graben	U. Jurassic (Kimmeridge Clay Formation)	2850	1.9	–28.6
North Viking Graben	M. Jurassic (Brent Group)	4260	1.9	–27.0
North Viking Graben	M. Jurassic (Brent Group)	4260	2.6	–25.0
North Viking Graben	M. Jurassic (Brent Group)	4260	1.1	–23.0
South Viking Graben	U. Jurassic (Brae equiv.)	4910	2.2	–22.0
North Viking Graben(NOCS)	U. Jurassic	1570	-4.2	–27.0
South Viking Graben (NOCS)	Eocene	1950	-1.8	–23.2
Central Graben (NOCS)	U. Jurassic	3600	3.26	–29.4
Central Graben (NOCS)	U. Jurassic	4050	3.34	–29.4
Central Graben	Triassic (Skagerrak equiv.)	3550	2.6	–33.0
South Viking Graben	M. Jurassic (Brent Group)	3515	–0.32	–36.7
South Viking Graben	M. Jurassic (Brent Group)	3515	–0.01	–27.0
South Viking Graben	M. Jurassic (Brent Group)	3856	–0.02	–26.0
South Viking Graben	U. Jurassic (Brae equiv.)	4880	2.2	–22.0
South Viking Graben	U. Jurassic (Brae equiv.)	5099	–1.2	–16.0
Haltenbanken (NOCS)	M. Jurassic	2500	–2.93	–26.0
Haltenbanken (NOCS)	M. Jurassic	4300	1.3	–18.8

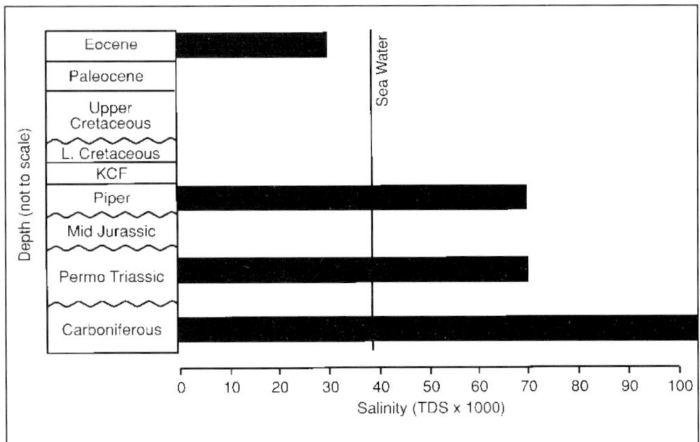

Figure 6. An example of a vertically stratified water column: the Piper Reservoir, 100 miles east of the northeast tip of Scotland (data from Burley et al., 1989) KCF = Kimmeridge Clay Formation.

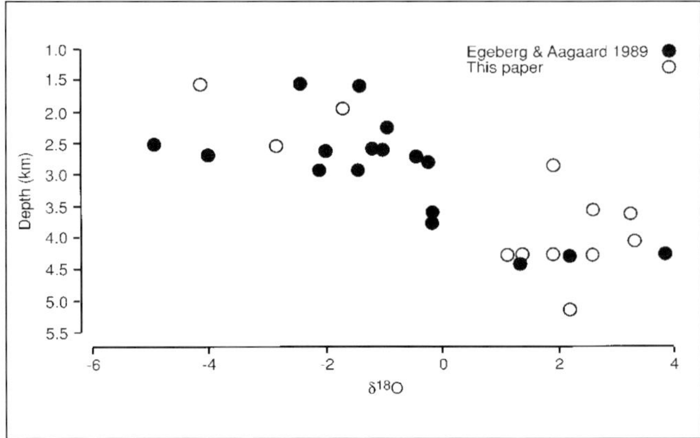

Figure 7. The oxygen isotopic compositions of North Sea formation waters plotted as a function of present-day burial depth. Data from Egeberg and Aagaard (1989) (filled dots) and this chapter (open dots).

defined volume of rock; (b) assuming the initial isotopic compositions of water and minerals; (c) assuming a temperature at which the minerals reacted or recrystallized. In a compacting, clastic sedimentary sequence, most of the water and reactive minerals (clays; especially smectite) are within mudstones. It is therefore likely that diagenetic reactions within mud rocks are a major influence on the isotopic composition of formation waters. For illustrative purposes, we can use equation (1) to calculate the extent of the isotopic shift of water in a mudstone containing 15% water and 30% smectite with an initial $\delta^{18}O$ = 22‰ and an initial δD = –70‰. The assumptions used in this illustrative calculation are that (1) the smectite was converted to illite at 100°C; (2) the mud rock behaved as a closed system and (3) the water in the mud rock at the time that smectite was converted to illite was Mesozoic meteoric water (used because the hydrogen isotopic composition of present-day formation waters suggests that they could contain a meteoric component, probably introduced into the North Sea during the Early Cretaceous). This simple, illustrative calculation shows that waters derived from the closed-system evolution of a typical mudstone sequence can yield waters which have an isotopic composition close to that of the most evolved waters now in reservoir sands (Figure 5). Very similar results are obtained if marine carbonate is used instead of smectite. We do not seek to suggest that this very simple model is necessarily the correct interpretation of the isotopic composition of North Sea formation waters. Rather, the implication is that if the isotopically heavy water is derived from water-mineral reaction, then its absolute volume is strictly limited.

Waters which are less enriched in ^{18}O than the diagenetic end-member described above may be the result of either lesser degrees of mineral recrystallization (higher effective water-rock ratio), or mixing between the end-member and more ^{18}O-depleted waters. Nevertheless, any enrichment of ^{18}O implies low water-rock ratios. Using equation (2) and the same smectite and illite data as above, it is possible to calculate $\delta^{18}O$ of water as a function of water-rock ratio. If water/rock volume ratios exceed about 5, then there is little change in the isotopic composition of the water after its reaction with the rock (Figure 8). If more than a few volumes of water pass through the porespace of a sandstone, its isotopic composition will remain effectively unchanged.

3. "Metamorphic" waters contribute to the formation waters. There are few data with which to estimate the extent to which waters expelled from basement rocks contribute to formation waters. An absolute maximum value can be calculated by assuming that 2 weight % of the basement rock is expelled as water during low- to high-grade metamorphism (Fyfe et al., 1978). A 24 km thick crust can then expel a volume of water twice that contained in a 6 km thick sedimentary basin. This is strictly a maximum figure because the basement will dehydrate over a much longer period than the lifetime of the sedimentary basin.

In summary, neither the present-day formation waters nor the mineralizing fluids are unmodified meteoric waters but are what can be loosely termed "basinal" or "oil field" brines (Hanor, 1987). There are restricted volumes of these ^{18}O-rich, saline waters; probably little more than the interstitial volume of the deeper parts of the basin. This type of fluid distinguishes those parts of many basins which are isolated from the basin margins and which are characterized by compaction-driven or thermobaric flow (Galloway, 1984).

MASS TRANSPORT MECHANISMS

Mass transport in sedimentary basins may occur by four basic processes:

1. Fluid flow in response to the creation of hydraulic

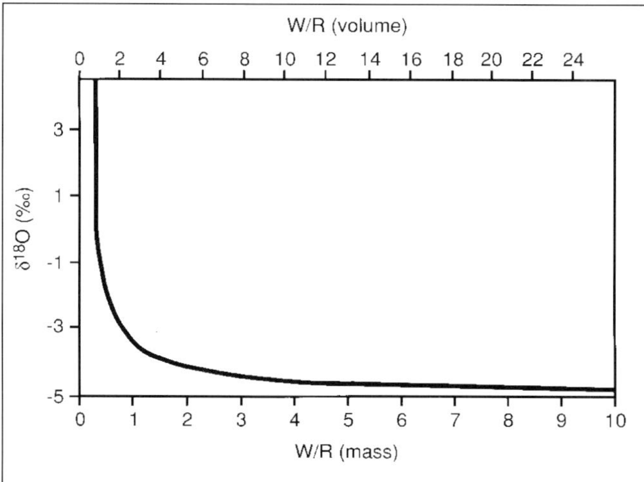

Figure 8. A model for the evolution of the oxygen isotopic composition of water during reaction with smectite. In this example, water with an initial $\delta^{18}O$ = –5‰ reacts with increasing amounts of smectite in a closed system, resulting in illite and isotopically altered water. W/R refers to the water/rock ratio. Further details of the calculation are given in the text.

gradients through the uplift and exposure of basin margins.

2. Compaction-driven fluid flow.
3. Diffusion through an essentially static volume of fluid.
4. Recycling of a single volume of fluid, by convection or through the periodic release of tectonically induced stresses (e.g. seismic pumping).

All these processes have been proposed as mechanisms by which silica transport and quartz cementation may occur (e.g. Heald, 1955; Siever, 1959; Sibley and Blatt, 1976; Haszeldine et al., 1984; Davis et al., 1985; Leder and Park, 1986; Bjorlykke et al., 1988; Burley et al., 1989; McBride, 1989). In this section we explore how the composition of the mineralizing fluids can allow us to differentiate between these possibilities.

1. Flow of meteoric water in response to the exposure of basin margins

Meteoric water can enter an aquifer when a hydraulic head is generated in response to the exposure of basin margins (Garven and Freeze, 1984; Bethke, 1986). Waters entering the basin in this way are fresh and have isotopic compositions which fall on or close to the meteoric water line, even when they penetrate deep into the basin (Airey et al., 1979). The waters from which quartz precipitated in pre-Tertiary North Sea sands were not fresh and did not have oxygen isotopic signatures consistent with unmodified Mesozoic or Tertiary meteoric waters (Figure 2). Influx of meteoric water from periodically exposed basin margins can be firmly ruled out as the mechanism for silica transport and quartz cementation in these sandstones.

2. Compaction-driven flow

The identification of basinal brines as mineralizing fluids does not allow us to differentiate between compaction-driven flow, diffusion and fluid recycling as mass transport mechanisms. In order to constrain further the mechanisms by which quartz cementation occurred, we need to consider the geological and geochemical implications of each process.

The predominant direction of compaction-driven flow is upwards, towards cooler parts of sedimentary basins (Bethke, 1986; Harrison & Summa, 1991). Since the solubility of quartz increases with temperature (Siever, 1959), the migration and cooling of fluid will lead to quartz precipitation. However, if the concentration of dissolved silica is controlled by quartz solubility, huge volumes of water are required to form significant volumes of quartz cement. For example, to form 5 volume percent quartz cement, between 10^4 and 10^6 volumes of water must pass through each pore, depending on the precipitation temperature (Sibley and Blatt, 1976; McBride, 1989). We have already shown that the volume of ^{18}O-rich, saline fluid responsible for quartz cementation is unlikely to be more than the pore volume of the basin. Since most of the basin's porosity is within mud rocks, there is a limited potential for focusing the expelled water through sandstones. If sands contain 15% of the basin's total porespace (Pettijohn et al., 1987), 6 (i.e., 100/15) pore volumes of the ^{18}O-rich mineralizing fluid can be channeled through the sands; this leaves a mere 9994 to 999,994 still to find!

One way to move large volumes of basinal brine through a particular sand is to focus it from a large catchment area. However, simple calculations show that the degree of focusing required is extremely unlikely. As an example, we have calculated the volume of water required to form 5 volume percent quartz cement in a 1 billion barrel oil field, if the water is cooled from 101 to 100° C. The volume required is the entire inventory of water expelled from a block of sediment 500 km × 500 km × 1 km during its burial from 2 to 3 km. Not only does this seem untenable, but neither can such focusing explain the pervasive nature of quartz cement.

The volume of water required to precipitate a given volume of quartz from a cooling fluid decreases if the concentration of dissolved silica is greater than that dictated by quartz solubility. However, the few measurements of dissolved silica in formation waters suggest that concentrations are rarely greater than twice that of quartz solubility (Land and MacPherson, 1992). There is some evidence that organic anions may chelate silica and enhance its solubility by as much as seven times quartz saturation (Bennett and Siegel, 1987). If, for example, the concentration of dissolved silica was raised to that defined by amorphous silica (nine times quartz solubility at 100° C), then

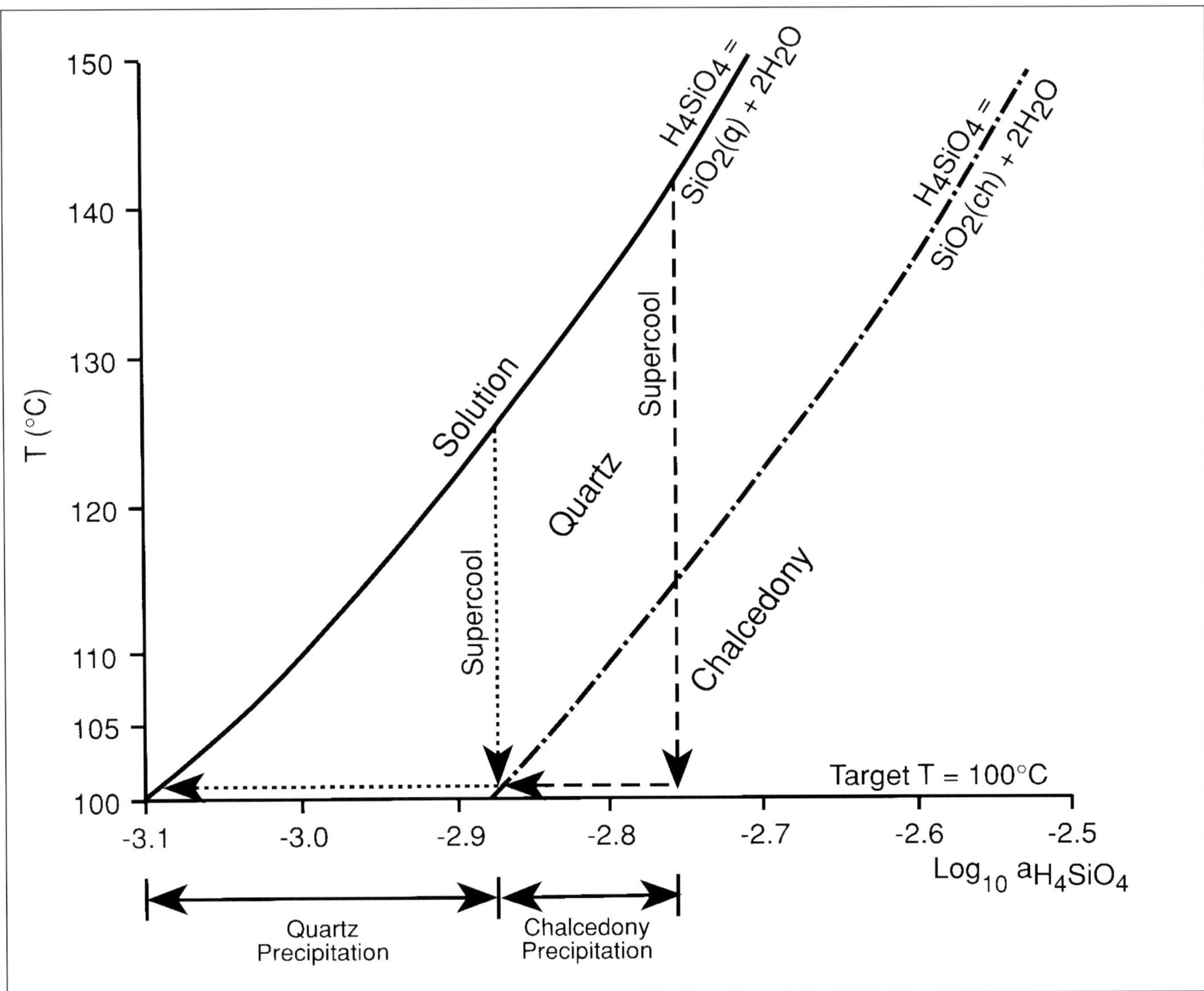

Figure 9. Solubility of quartz and chalcedony as a function of temperature (data from Helgeson et al., 1978). The cooling paths show that quartz or chalcedony will precipitate from a cooling fluid, depending on the extent of cooling.

about 10 times less water would be required to form the same volume of quartz cement. This concentration enhancement can be considered a maximum permissible value for quartz precipitation, because at even higher dissolved silica concentrations, metastable phases such as amorphous silica and opal C-T are more likely to precipitate than quartz. For the 1 billion barrel oil field described above, it would now be necessary to focus the water expelled from a block of sediment 1 km × 160 km × 160 km during its burial from 2 to 3 km. This is still impossible for a cement as widespread as quartz.

Can a sufficiently large volume of water be expelled from basement rocks to account for the observed volumes of quartz cement? We showed previously that the absolute maximum volume of water that can be expelled from basement rocks is equivalent to twice the total interstitial volume of the overlying sedimentary basin. If sands contain 15% of the interstitial volume of the basin, then 13 volumes [(100/15) × 2] of "metamorphic" water can be flushed through the sands. This volume is insignificant in the context of what is apparently required.

3. Diffusion

Diffusive transport occurs along gradients of chemical potential, or, to a reasonable approximation, along concentration gradients. Large fluid volumes are unnecessary for diffusive transport because the fluid simply acts as a medium through which material is transported. Diffusive transport is thus compatible with the relatively small volumes of water from which quartz precipitated in the sands studied here.

Because the solubility of quartz increases with temperature, concentration gradients of dissolved silica should exist wherever there is a thermal gradient. In Appendix 1 we explore the consequences of this by calculating how much silica diffuses down a normal geothermal gradient in a dipping sand body. We have assumed that the concentration of dissolved silica is defined by equilibrium with quartz and that concentration gradients are maintained by quartz precipitation as silica diffuses down temperature. The amount of quartz cement generated in this way is negligible; about 0.3 cm^3 per cubic meter of porespace per million years, or $3 \times 10^{-5}\%$ of porespace per million years.

Diffusive transport of silica can only be significant where enhanced concentration gradients are maintained for long periods of time. If the concentration of dissolved silica is defined by quartz solubility, then concentration gradients should not be expected in quartz-rich sandstones. The situation is more complex in sandstones with chert or amorphous silica (e.g. sponge spicules) or in mudstones with more varied mineral asssemblages. Here, silica could be buffered above quartz saturation by mineral assemblages other than quartz, for example K feldspar-illite-kaolinite, or chalcedony. At 100°C, the concentration of silica in equilibrium with chalcedony is about 50 mg/L greater than that dictated by quartz, whilst that in equilibrium with K feldspar-illite-kaolinite is about 30 mg/L greater (Helgeson et al., 1978).

The possibility that the concentrations of dissolved silica in mudstones are enhanced above quartz saturation is untested because it is extremely difficult to extract the interstitial-water of deeply-buried mudstones. There are, however, many data from shallow (< 1 km), fine-grained sediments cored during the DSDP (Deep Sea Drilling Project) and ODP (Ocean Drilling Program) that show that concentrations of dissolved silica can be much greater than quartz saturation (5 mg/kg): up to 70 mg/kg, with most lying between 10 and 35/kg (Gieskes et al., 1990). These data indicate that in muds, a steady state may exist in which the concentration of dissolved silica is maintained above quartz saturation through a balance of reactions which supply and remove silica from solution. Reactions supplying silica to pore-waters include the dissolution of biogenic silica and amorphous aluminosilicates (Foscolos and Powell, 1980). These phases are unlikely to be abundant in deeply buried muds so that the relevance of observations in shallow-buried muds to deeper sections is questionable. However, in more deeply buried mudstones, silica is a known by-product of several diagenetic reactions including the destruction of smectite and associated production of illite (Towe, 1962; Hower et al., 1976). The concentration of dissolved silica in the interstitial waters of the mudstones will therefore reflect the relative rates of silica generation and precipitation. Although there are experimental data and theoretical expressions for the rates of dissolution and precipitation of various SiO_2 polymorphs (Rimstidt and Barnes, 1980; Lasaga, 1984), the range of these rates is poorly constrained in natural systems. Furthermore, in natural systems silica may be generated and removed by reactions other than those involving SiO_2 polymorphs. The situation is particularly complicated in mudstones, in which quartz precipitation may be inhibited by a lack of nuclei. A similar phenomenon has been observed in sandstones, where grain-coating chlorite appears to inhibit the formation of quartz overgrowths (Heald and Larese, 1974).

4. Intergranular pressure solution

Enhanced silica solubility at grain contacts can result in transport away from highly stressed sections towards less stressed parts (Weyl, 1959; Houseknecht, 1988). This can occur in a fixed body of water and is therefore consistent with the fluid compositional data. Experimentally determined kinetics of the precipitation of silica on clean quartz grains suggest that silica generated by pressure solution and transported by diffusion should not travel more than a few meters before reprecipitating as quartz overgrowths (Rimstidt and Barnes, 1980). Across-formation transport of silica should therefore be rare. However, although it is notoriously difficult to quantify how much material has been lost by a diagenetic reaction, petrographic and some geochemical studies have been used to infer across-formation transport of silica (Houseknecht, 1988; McBride, 1989; Gluyas and Coleman, 1992). These observations contain important implications. Firstly, we must question our understanding of the kinetics of quartz precipitation. If the kinetics of quartz precipitation in natural systems are correctly described by Rimstidt and Barnes' (1980) experimental data, then silica must be transported by movement of water into and out of the sandstones undergoing pressure solution. Secondly, to what extent might pressure solution be occurring in mudstones? Fuchtbauer (1978) observed that pressure solution was better developed in siltstones than in adjacent sandstones. Extrapolations based on grain size alone would suggest even stronger development of pressure solution in mudstones, with associated potential for diffusive transport out of the mudstone. And thirdly, how accurate are our estimates of material loss from sandstones? These are questions which need to be addressed.

5. Fluid recycling: convection and seismic pumping

The apparent water volume problem can be overcome if a finite volume of water can be recycled many times through the same volume of rock. The most often cited mechanisms of fluid recycling are convection (Wood and Hewett, 1982; Davis et al., 1985) and seismic pumping or valving (Sibson et al., 1976; Burley et al., 1989). The compositional heterogeneity of mineralizing fluids over relatively small areas of the North Sea (Figure 3) suggests that large-scale recycling is unlikely to have occurred, though this does not preclude convection within individual sand bodies (but see Bjorlykke et al., 1988). Recycling can also occur in response to changes in pore pressure induced by changes in tectonic stress. The best known example of this is flow associated with earthquakes (Sibson et al., 1976; Knipe, this volume).

The extent to which fluids must be recycled can be estimated by calculating the number of times the fluid must pass through a pore in order to precipitate a given volume of cement. The amount of quartz deposited during each cycle depends on the extent to which the fluid is cooled before precipitation occurs. Consider a sand in which quartz is precipitating at 100°C. Assuming that the silica content of the diagenetic fluid remains in equilibrium with quartz, the amount of quartz deposited is the difference between the solubility of quartz at the initial (hottest) and final precipitation (coolest) temperature. This is shown in Figure 9, which shows the solubility of quartz and chalcedony as a function of temperature. A temperature step of 25°C or more will cause the precipitation of chalcedony. Since chalcedony is rarely seen as a cement in reservoir sands, 25°C is a maximum permissible temperature step.

The solubility data can be used to calculate the volume of water required to precipitate a given volume of quartz cement. Dividing the volume of water by the volume of an average pore gives the required number of pore volumes, or the number of times the water must be recycled. To precipitate 4 % quartz overgrowths, water must be recycled 17,000 times for a temperature step of 25°C, and 520,000 times for a 1°C step. Assuming a geothermal gradient of 30°C/km, a temperature step of 25°C requires water to be moved (17,000 times) almost 1 km vertically, compared with 30 m for a 1°C temperature step. It is unclear whether recycling on the required scale is possible in seismically quiet regions of sedimentary basins.

SUMMARY

Quartz is the most widespread and volumetrically important cement in Mesozoic North Sea sands. An understanding of the mechanisms of quartz cementation is therefore of great relevance to the accurate prediction of reservoir quality, both at the exploration and early appraisal stages. In this chapter we have demonstrated how data on fluid compositions can be used to place some important constraints on the considerations of the mechanisms by which silica was transported and caused quartz cementation in these particular sandstones.

Quartz cementation occurred during the Tertiary from waters (1) enriched in ^{18}O relative to local Tertiary meteoric water and (2) having high and variable salinities. The composition of the mineralizing fluids is broadly similar to that of the waters presently in the pre-Tertiary section of the North Sea. Since the Tertiary, evolution of the North Sea has been dominated by subsidence; there has been limited scope for the introduction of new fluids during the Tertiary. It is therefore possible that the present-day fluids are essentially those from which quartz precipitated.

The source(s) of the mineralizing fluids and present-day formation waters cannot be specified precisely. Their isotopic and chemical composition suggest that the fluids could be mixtures of evaporated sea water and meteoric water, with a possible contribution from basement-derived fluid. However, the composition of these original end-member waters may well have been modified by diagenetic reactions with rock minerals. Whatever the exact origin(s) of the fluids, their oxygen isotopic compositions place strict limits to their possible absolute volume; it is unlikely that the total volume of isotopically heavy, mineralizing fluid exceeded the total void space of the sedimentary basin.

Another vital characteristic of both present-day formation waters and Tertiary mineralizing fluids is that they are compositionally heterogeneous on both local and regional scales. This implies that the waters are, and were at the time of cementation, poorly mixed. Since large-scale flow would tend to reduce compositional variability, we use this observation to argue against large-scale fluid flow or convection as important mass transport mechanisms in these sandstones.

The composition and compositional heterogeneity of the mineralizing fluids constrain the mechanisms by which silica could have been transported and caused quartz cementation in these North Sea sandstones:

1. Transport by meteoric recharge is ruled out by the isotopic and chemical composition of the mineralizing fluids.
2. Compaction-driven flow, although broadly consistent with the fluid data, is very unlikely to have caused widespread quartz cementation. Mass balance calculations show that the relatively small volumes of water expelled during compaction contain several orders of magnitude less silica than the observed mass of quartz cement. It is also impossible to appeal to the focusing of compactional waters to explain quartz cementation, partly because of the immense degree of focusing required and partly because quartz is a widespread, not a local cement.
3. Quartz appears to have formed from stratified and/or compositionally heterogeneous waters, suggesting that large-scale convection or seismic pumping are unlikely to be relevant transport mechanisms.

The fluid data suggest that silica was sourced locally and was transported either by reservoir-scale convection or diffusion. However, significant mass transport by either of these mechanisms remains to be

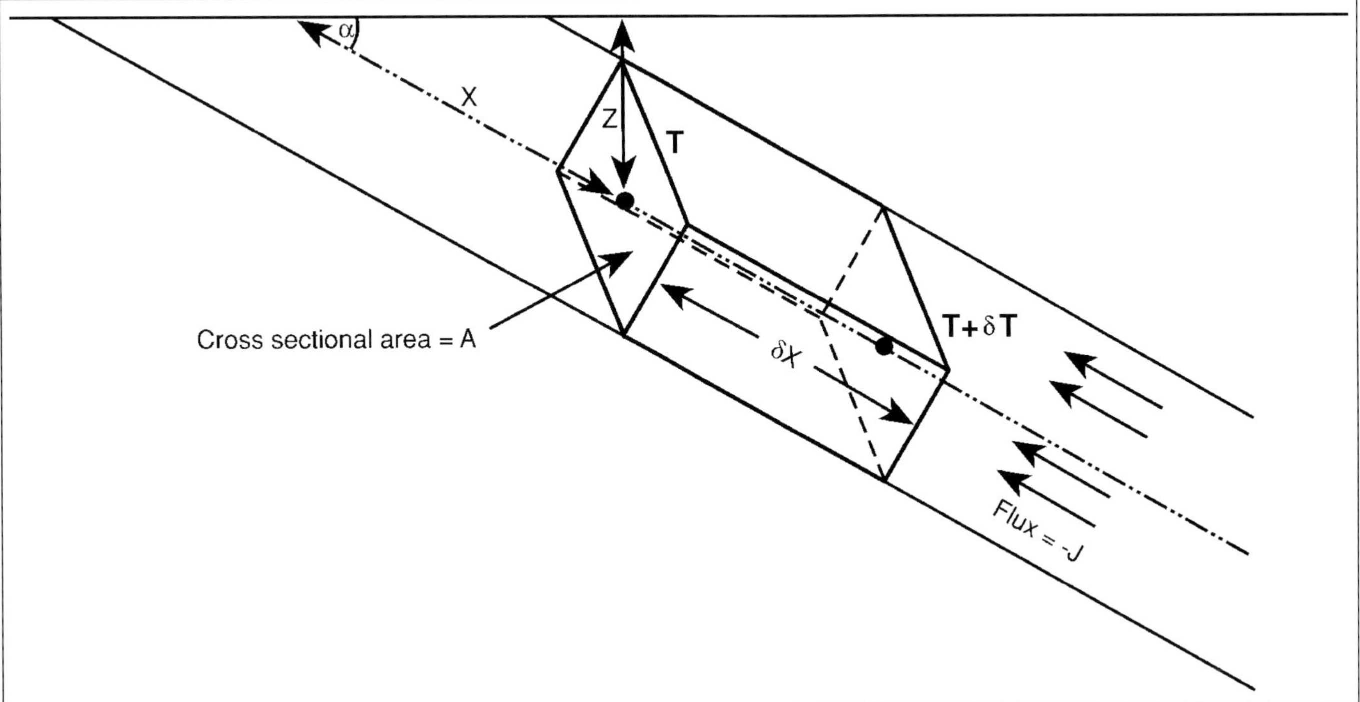

Figure A1. Schematic representation of diffusion through a dipping body of rock.

demonstrated in sedimentary basins. The extent of fluid convection in reservoir sands remains unclear (Wood and Hewett, 1982; Bjorlykke et al., 1988) and it is not certain that convection proceeds at a rate sufficient to precipitate significant amounts of quartz. Significant diffusive transport requires that high silica concentration gradients be maintained for several million years, either within the reservoir, or between mudstones and adjacent sandstones. This is impossible if, as is commonly assumed, the concentration of silica is everywhere controlled by quartz solubility. High concentration gradients could, however, occur where a disparity existed between the rates at which silica was generated and precipitated. For example, silica is a by-product of several diagenetic reactions occurring in mud rocks. If the rate of generation of silica in a mud rock exceeded its rate of precipitation, then a concentration gradient between mud and sand might result. Further progress in this area requires a clearer understanding of the kinetics of silica generation and precipitation in clastic sediments.

ACKNOWLEDGMENTS

We thank Gordon Macleod and Sam Savin, Tony Fallick and Charles Curtis for making constructive and valuable comments on the manuscript. Thanks also to BP for allowing us to publish the work and to Yvonne Hall and John Ager for producing the final text and diagrams. Peter Corvi, the barking boffin, is courteously thanked for his mathematical assistance.

APPENDIX 1

Derivation of the equation for calculating the mass (M) of quartz precipitated from a saturated fluid as a result of diffusion down a temperature gradient.

Step 1 (geometry; see Figure A1):

$$z = x \times \sin\alpha \rightarrow d/dx = dz/dx \times d/dz = \sin\alpha \times (d/dz) \quad \text{(A1)}$$

where z=vertical distance, x=distance along sand body, and α=dip.

Step 2 (mass conservation):

$$\delta M/\delta t = -A \times \delta x \times (dJ(x)/dx)$$

where t = time, M = mass of quartz, A = cross sectional area of sand body and J(x) = flux of silica in the x direction.

Step 3 (Fick's 1st Law):

$$J(x) = -D \times dc/dx$$

where D = the diffusion coefficient and c=concentration.

Step 4 (Equilibrium thermodynamics):

$$\ln C = 0.348 - 2676/T$$

$$\rightarrow 1/C \times (dc/dx) = 2676/T^2 \times (dT/dx) = [(2676.\sin\alpha)/T^2] \times dT/dZ \quad \text{(A2)}$$

$$\rightarrow J(x) = (-2676 \times D \times \sin\alpha) \times (C/T^2) \times (dT/dz)$$

$$\rightarrow \delta M/\delta t = 2676 \times D \times A \times \delta x \times (\sin^2\alpha) \times (d/dz) \times [(C/T^2) \times (dT/dz)]$$

where C = the concentration of dissolved silica and T = temperature.

Thus

$\delta M/\delta t = 2676 \times D \times A \times \delta x \times \sin^2\alpha \times [(C/T^2) \times (d^2T/dz^2)+\{(T^{-2}(dC/dz)- 2 \times C \times T \times (dT/dz))/T^4\} \times (dT/dz)]$

But $T^2 (dC/dz) = 2676 \times C \times (dT/dz)$ from eqns A1 and A2

$\rightarrow \delta M/\delta t = 2676 \times D \times A \times \delta x \times \sin^2\alpha \times \{(C/T^2) \times (d^2T/dz^2) + [(2676-2T) \times C(dT/dz)^2]/T^4\}$

Provided the temperature gradient is uniform: $d^2T/dz^2 = 0$ (i.e., dT/dz = constant)

$$\delta M/\delta t = [2676 \times D \times A \times \delta x \times \sin^2\alpha \times (2676-2T)/T^4] \times C(dT/dz)^2 \qquad \text{(eqn A3)}$$

The final step is to substitute $C = \exp[0.348 - (2676/T)]$

Equation (A3) may be used to predict the rate of increasing mass of precipitated quartz, due to diffusion, in a pore space volume V ($V = A \times \delta x$) at any given temperature (T, in Kelvin), with geothermal gradient (dT/dz) and bed dip (α).

For example:

For 1 m^3 of pore space, vertical flow ($\alpha = 90°$), a geothermal gradient of 0.03°C/m, a temperature of 100°C (373°K), and a diffusion coefficient for silica of $1.3 \times 10^{-10} m^2s^{-1}$, the rate of increase of quartz due to diffusion alone is given by:

$$\delta M/\delta t = 3.4 \times 10^{-20} \text{ kg/s} = 1 \text{ mg/m.y.}$$

REFERENCES CITED

Abbotts, I.L., (editor), 1991, United Kingdom Oil and Gas Fields 25 years commemorative volume: The Geological Society Memoir 14, 565 p.

Airey, P.L., G.E. Calf, B.L. Campbell, P.E. Hartley, and M.A. Habermehl, 1979, Aspects of the isotope hydrology of the Great Artesian Basin: IAEA-SM-228/12, p. 205- 219.

Bennett, P.C. and D.I. Siegel, 1987, Increased solubility of quartz in water due to complexing by organic compounds: Nature, v. 326, p. 684-686.

Bethke, C.M., 1986, Hydrologic constraints on the genesis of Upper Mississippi Valley mineral district from Illinois Basin brines: Economic Geology, v. 81, p. 233-249.

Bjorlykke, K., A. Mo, and E. Palm, 1988, Modelling of thermal convection and its relevance to diagenetic reactions: Marine and Petroleum Geology, v. 5, p. 338- 351.

Burley, S.M., J. Mullis, and A. Matter, 1989, Timing diagenesis in the Tartan reservoir (UK North Sea): constraints from combined cathodoluminescence and fluid inclusion studies: Marine and Petroleum Geology, v. 6, p. 98-120.

Clayton, R.N., I. Friedman, D.L. Graf, T.K. Mayeda, Meents, W.F. and N.F. Shimp, 1966, The origin of saline formation waters. Journal of Geophysical Research, v. 71, p. 3869-3882.

Curtis, C.D., 1977, Sedimentary geochemistry: environments and processes dominated by involvement of an aqueous phase: Philosophical Transactions of the Royal Society of London, A, v. 286, p. 353-372.

Dansgaard, W., 1964, Stable isotopes in precipitation: Tellus, v. 16, p. 436-468.

Davis, S.H., S. Rosenblad, J.R. Wood and T.A. Hewitt, 1985, Convective fluid flow and diagenetic patterns in domed sheets. American Journal of Science, v.285, p. 207- 223.

Egeberg, P.K., and P. Aagaard, 1989, Origin and evolution of formation waters from oil fields on the Norwegian shelf: Applied Geochemistry, v. 4, p. 131-142.

Ehrenberg, S.N., 1990, Relationship between diagenesis and reservoir quality in sandstones of the Garn Formation, Haltenbanken, Mid-Norwegian Continental Shelf. AAPG Bulletin, v. 74, p. 1538-1558.

Foscolos, A.E. and T.G. Powell, 1980, Mineralogical and geochemical transformation of clays during catagenesis and their relation to oil generation. *in* A.D. Miall, ed., Facts and principles of world petroleum occurrence, Canadian Society of Petroleum Geologists Memoir, v. 6, p. 153-172.

Fuchtbauer, H., 1978, Zur Herkunft des Quarzzements. Abschätzung der Quarzauflösung in Silt- und Sandsteinen: Geologische Rundschau, v. 67, p. 991-1008.

Fyfe, W.S., N.J. Price, and A.B. Thompson, 1978, Fluids in the Earth's Crust: Elsevier, 383 p.

Galloway, W.E., 1984, Hydrologic regimes of sandstone diagenesis: *in* D.A. MacDonald and R.C. Surdam, eds. Clastic diagenesis: American Association of Petroleum Geologists Memoir 37, p. 3-14.

Garrels, R.N., and F.T. Mackenzie, 1971, Evolution of Sedimentary Rocks: New York, W.W. Norton.

Garven, G., and R.A. Freeze, 1984, Theoretical analysis of the role of groundwater flow in the genesis of stratabound ore deposits, 1. Mathematical and numerical model: American Journal of Science, v. 284, p. 1085-1124.

Gieskes, J.M., G. Blanc, P. Vrolijk, H. Elderfield, and R. Barnes, 1990, Interstitial water chemistry - major constituents: Proceedings of the Ocean Drilling Program, Scientific Results, v. 110, p.155-177.

Glasmann, J.R., P.D. Lundegard, R.A. Clark, B.K. Penny and I.D. Collins, 1989a, Geochemical evidence for the history of diagenesis and fluid migration: Brent Sandstone, Heather Field, North Sea: Clay Minerals, v. 24, p. 255-284.

Glasmann, J.R., R.A. Clark, S. Larter, N.A. Briedis and P.D. Lundegard, 1989b, Diagenesis and hydrocarbon accumulation, Brent Sandstone (Jurassic), Bergen High Area, North Sea: AAPG Bulletin, v. 73, p. 1341-1360.

Gluyas, J.G., 1985, Reduction and prediction of sandstone reservoir potential, Jurassic, North Sea:

Philosophical Transactions of the Royal Society, A, v. 315, p. 187- 202.

Gluyas, J.G. and M.L. Coleman, 1992, Material flux during sediment diagenesis: a contradiction to fluid flow prediction. Nature, v. 356, p. 22-23

Hanor, J.S., 1979, The sedimentary genesis of hydrothermal fluids: *in*, H.L. Barnes, ed., Geochemistry of hydrothermal ore deposits, New York, Wiley, p. 137-172.

Hanor, J.S., 1980, Dissolved methane in sedimentary brines: potential effect on the PVT properties of fluid inclusions: Economic Geology, v. 75, p. 603-609.

Hanor, J.S., 1987, Origin and migration of subsurface sedimentary brines: SEPM Short Course Notes no. 21, p. 247

Harrison, W.J. and L.L. Summa, 1991, Paleohydrology of the Gulf of Mexico Basin: American Journal of Science, v. 291, p. 109-176.

Haszeldine, R.S., I.M. Samson, and C. Cornford, 1984, Quartz diagenesis and convective fluid movement: Beatrice Oilfield, UK North Sea: Clay Minerals, v. 19, p. 391-402

Heald, M.T., 1955, Stylolites in sandstones: Journal of Geology, v. 63, p.101-114.

Heald, M.T., and R.E. Larese, 1974, Influence of coatings on quartz cementation: Journal of Sedimentary Petrology, v. 44, p. 1269-1274.

Helgeson, H.C., J.M. Delany, H.W. Nesbitt and D.K. Bird, 1978, Summary and critique of the thermodynamic properties of rock-forming minerals. American Journal of Science, v. 278A, 229p.

Hitchon, B., 1968, Rock volume and pore volume data for Plains Region of Western Canada Sedimentary Basin between 49° and 60°N: American Association of Petroleum Geologists Bulletin, v. 52, p. 2318-2323.

Hogg, A.J.C., 1989, Petrographic and isotopic constraints on the diagenesis and reservoir properties of the Brent Group Sandstones, Alwyn South, Northern UK North Sea: Ph.D.thesis, Aberdeen University, Microfilms International No. DX89573, 414 p.

Houseknecht, D.W., 1988, Intergranular pressure solution in four quartzose sandstones: Journal of Sedimentary Petrology, v. 58, p. 228-246.

Hower, J., E.V. Eslinger, M.E. Hower and E.A. Perry, 1976, Mechanism of burial metamorphism of argillaceous sediment: 1. Mineralogical and chemical evidence: Geological Society of America Bulletin, v. 87, p. 725- 737.

Knauth, L.P., and M.A. Beeunas, 1986, Isotope geochemistry of fluid inclusions in Permian halite with implications for the history of ocean water and the origin of saline formation waters: Geochimica et Cosmochimica Acta, v. 50, p. 419-433.

Land, L.S., and G.L. MacPherson, 1992, Geothermometry from brine analysis: lessons from the Gulf Coast, USA: Applied Geochemistry, v. 7, p. 333-340.

Lasaga, A.C., 1984, Chemical kinetics of water-rock interactions: Journal of Geophysical Research, v. 89, p. 9009-9025.

Lawrence, J.R. and H.P. Taylor, 1971, Deuterium and oxygen-18 correlation: clay minerals and hydroxides in Quaternary soils compared to meteoric waters: Geochim. Cosmochim. Acta, v. 35, p.993-1003.

Leder, F., and W.C. Park, 1986, Porosity reduction in sandstone by quartz overgrowth: American Association of Petroleum Geologists Bulletin, v. 70, p. 1713-1728.

Lee, M., and S. Savin, 1986, Isolation of diagenetic overgrowths on quartz sand grains for oxygen isotopic analysis: Geochimica et Cosmochimica Acta, v. 49, p. 497-501.

McBride, E.F., 1989, Quartz cement in sandstones: A review: Earth Science Reviews, v. 26, p. 69-112.

Pettijohn, E.J., P.E. Potter, and R. Siever, 1987, Sands and Sandstone, 2nd edition: Heidelberg, Springer Verlag, 553 p.

Rimstidt, J.D. and H.L. Barnes, 1980, The kinetics of silica-water reactions: Geochimica et Cosmochimica Acta, v. 44, p. 1683-1699.

Robinson, A. and J. Gluyas, 1992, Model calculations of loss of porosity as a result of compaction and quartz cementation: Marine and Petroleum Geology, v. 9, p. 319-323.

SPWLA, 1989, North Sea R_w catalogue: Society of Professional Well Log Analysts, London Chapter.

Scotchman, I.C, L.H. Johnes and R.S. Miller, 1989, Clay diagenesis and oil migration in Brent Group sandstones of NW Hutton Field, UK North Sea: Clay Minerals, v. 24, p. 339-374.

Sheppard, S.M.F., 1986, Characterization and isotopic variations in natural waters, *in* Valley, J.W., Taylor, H.P., and O'Neil, J.R., eds., Mineralogical Society of America Reviews in Mineralogy, v. 16, p. 165-183.

Sibley, D.F., and H. Blatt, 1976, Intergranular pressure solution and cementation of the Tuscarora orthoquartzite: Journal of Sedimentary Petrology, v. 46, p. 881-896.

Sibson, R.H., J.McM. Moore, and A.H. Rankin, 1976, Seismic pumping—a hydrothermal fluid transport mechanism: Journal of the Geological Society of London, v. 131, p. 653-659.

Siever, R., 1959, Petrology and geochemistry of silica cementation in some Pennsylvanian sandstones, in Ireland, H.A., ed., Silica in Sediments: Society of Economic Paleontologists and Mineralogists Special Publication, v. 7, p. 55-79.

Suchecki, R.N., and L.S. Land, 1983, Isotopic geochemistry of burial-metamorphosed volcanogenic sediments, Great Valley sequence, California: Geochimica et Cosmochimica Acta, v. 47, 1487-1499.

Sverjensky, D.A., 1984, Oil field brines as ore-forming solutions: Economic Geology, v. 79, p. 23-37.

Taylor, H.P., 1974, The application of oxygen and

hydrogen isotope studies to problems of hydrothermal alteration and ore deposition: Economic Geology, v. 69, p. 843- 883.

Towe, K.M., 1962, Diagenesis of clay minerals as a possible source of silica cement in sedimentary rocks: Journal of Sedimentary Petrology, v. 32, p.26-28.

Weyl, P.K., 1959, Pressure solution and the force of crystallisation—a phenomenological theory. Journal of Geophysical Research, v. 64, p. 2001-2125.

Wood, J.R., and T.A. Hewett, 1982, Fluid convection and mass transfer in porous sandstones - a theoretical model: Geochimica et Cosmochimica Acta, v. 46, p. 1707- 1713.

Chapter 3

Geochemical Evidence for a Temporal Control on Sandstone Cementation

J.G. Gluyas, S.M. Grant,[1] and A.G. Robinson[2]

BP Research
Sunbury on Thames
Middlesex, U.K.

ABSTRACT

Porosity-depth plots for sandstones commonly show a decrease in porosity with depth. In more deeply buried sandstones, this reflects an increase in the amount of mineral cement. This correlation is often thought to indicate a causal relationship; however, in this paper, we examine data that argue against a direct depth control on cementation. In sandstones from the Garn Formation, Haltenbanken, the homogenization temperatures of primary fluid inclusions in quartz cement are correlated with depth over the entire basin. The inclusions do not appear to have stretched or leaked so the correlation suggests a widespread cementation event affecting large parts of the basin at about the same time. K-Ar ages of illite cements from eolian sandstones of the Rotliegend Formation (Village Fields area, southern North Sea) all fall within the Middle to Late Jurassic and show no correlation with depth at time of cementation. There is, however, a weak correlation between illite oxygen isotope ratio and depth at time of illite growth. Both ages and isotope ratios again suggest a widespread cementation phase taking place within a restricted time period. Published data from other sandstones, including the Brent sandstone reservoirs of the North Sea, tell a similar story, indicating cementation at a particular time rather than temperature or depth. In the case of quartz, cementation seems to coincide with rapid subsidence and heating, while in the Southern North Sea, illite growth is approximately synchronous with rifting and consequent rapid subsidence and heating. These effects may have influenced fluid flow and/or precipitation mechanisms in such a way that regional cementation occurred.

[1]Present address:
BP Petroleum Development Norway, Ltd.
Stavanger, Norway

[2]Present address:
BP Exploration Operating Company Ltd.,
Uxbridge, Middlesex, England

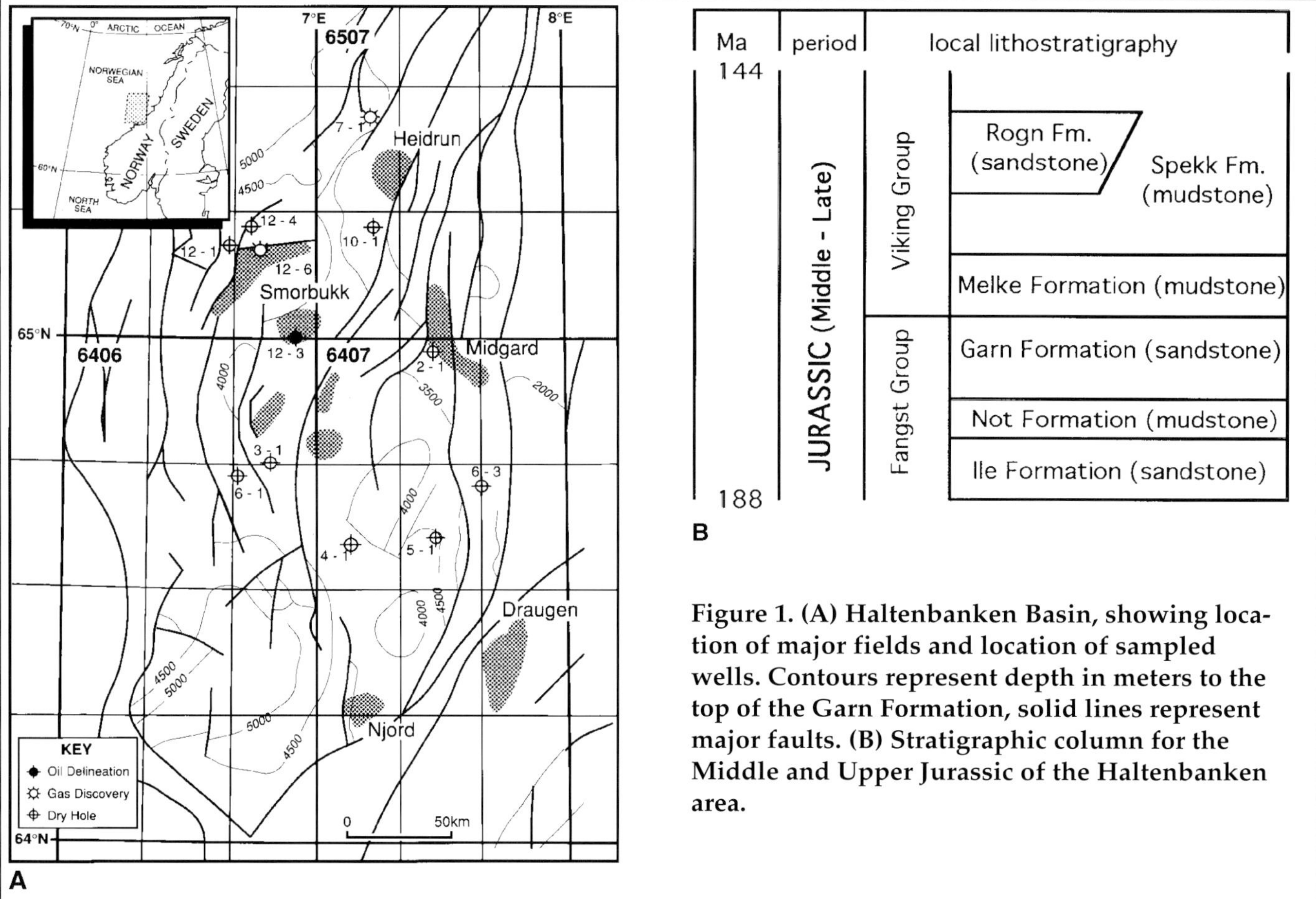

Figure 1. (A) Haltenbanken Basin, showing location of major fields and location of sampled wells. Contours represent depth in meters to the top of the Garn Formation, solid lines represent major faults. (B) Stratigraphic column for the Middle and Upper Jurassic of the Haltenbanken area.

INTRODUCTION

Some studies specifically link sandstone cementation to progressive burial (e.g., Leder and Park, 1986; Schmoker and Gautier, 1988). The volumes of some mineral cements—for example, quartz—often increase with increasing depth and reduce porosity; such relationships are useful for predicting the porosity of undrilled petroleum prospects (Scherer, 1987; Bloch, 1991). However, the existence of a correlation does not mean that burial is the *cause* of mineral cement growth (Robinson and Gluyas, 1992a). Indeed, the occurrence of uncemented sandstones at considerable depth suggests that the correlation reflects some other process or processes that may only have an indirect link with burial.

In this chapter, we compare the results of two regional studies of sandstone diagenesis: quartz cement growth in Haltenbanken (offshore Mid-Norway) and illite cement growth in the Village Fields area of the Southern North Sea. Both involve the use of geochemical data to describe and identify patterns in the cementation process. The geochemical methods used in the two cases are different: fluid inclusion analysis in quartz cement, and K-Ar dating and oxygen isotope analysis of illite cement. The information that each provides about cementation is therefore independent. The common picture that emerges has implications for the cause of cementation in sandstones generally and for reservoir quality prediction in petroleum exploration.

QUARTZ CEMENTATION (HALTENBANKEN): EVIDENCE FROM FLUID INCLUSIONS

Geological Setting

The Haltenbanken Basin is situated about 150 km north and west of the coast of Norway and is one of a number of basins formed by rifting of the Norwegian continental margin between the late Paleozoic and Eocene (Figure 1A) (Bukovics and Ziegler, 1985). The Middle Jurassic Garn Formation (Figure 1B) is the most important petroleum reservoir in the basin. These sandstones were deposited in a shallow marine to fluvial environment and are subarkosic, generally well sorted and very fine to coarse grained. The Garn was deposited just before a phase of Late Jurassic to Early Cretaceous rifting, and soon after deposition it was eroded from local structural highs. During and immediately after rifting, mudstones belonging to the Melke and Spekk Formations (Figure 1B) were deposited and now form the regional seal. The Spekk

Figure 2. SEM-based, backscattered electron image (upper), cathodoluminescence image (lower) of quartz overgrowth in the Garn Formation from well 6407/5-1 at 4218 m sub-sea level.

Formation is also the main petroleum source rock in the basin. Sedimentation rates were low between the end of rifting and the late Tertiary, and a deep water basin developed. This basin was filled during the Pliocene-Pleistocene when large volumes of sediment were generated by uplift of the Norwegian land mass. This period of rapid Pliocene-Pleistocene subsidence has resulted in two broadly different hydrodynamic regimes. In the heavily faulted area to the west, moderate to severe (> 40 MPa) overpressures have developed, while the more open structures in the east and central parts have remained normally pressured (Vik et al., 1991).

Quartz is the most abundant cement in the Garn Formation and increases in volume with depth. Jurassic sandstones buried to less than about 2 km as in, for example, the shallow (1.6 to 1.7 km) Draugen field, contain virtually no diagenetic quartz and are poorly consolidated. At 2.5 km, the quartz cement content is around 4-5% and at 4 km, about 15-17% (Ehrenberg, 1990). At depths greater than about 2.5 km, the volume of quartz cement therefore controls the observed decline in porosity (of approximately 9 porosity units per km) for the Garn Formation.

Samples And Methods

The Garn Formation in 13 wells was sampled for fluid inclusion microthermometry, covering a depth range of approximately 2.5 to 4.3 km. All samples were cemented by authigenic quartz, and examination of the overgrowths by SEM-based cathodoluminescence reveals complex zoning. This comprises a thin, either non- or moderately luminescent layer (not always developed), followed by alternating low and highly luminescent sector-zoned layers that are often succeeded by a poorly luminescent outer layer (Figure 2).

Analyses were performed on a Linkam THM600 heating-cooling stage attached to a petrographic microscope and connected to a Linkam TMS90 controller. This was calibrated from –125 to +440°C using synthetic fluid inclusions, pure bulk standards and liquid crystals. The accuracy of a homogenization temperature measurement is estimated to be around ±1°C. Measurements were carried out on inclusions isolated within quartz cement (with no relationship to fractures) or located at grain-cement boundaries.

Results

Homogenization Temperatures

Figure 3 shows homogenization temperatures plotted against depth (Grant and Oxtoby, 1992). The range of T_h for the entire data set is about 75°C. Individual wells have ranges of T_h less than 30°C and usually less than 15°–20°C. All *mean* homogenization temperatures for the wells are less than present day formation temperatures but a very small number of individual measurements are slightly higher. The wells in the west of Haltenbanken (blocks 6506 and 6406) generally record slightly lower temperatures than those in the east, particularly in well 6406/3-1. Figure 3 shows two prominent features:

(1) T_h is correlated with depth; and

(2) The *range* of T_h in any well (over a small depth range) is limited.

It was not possible to measure and compare for

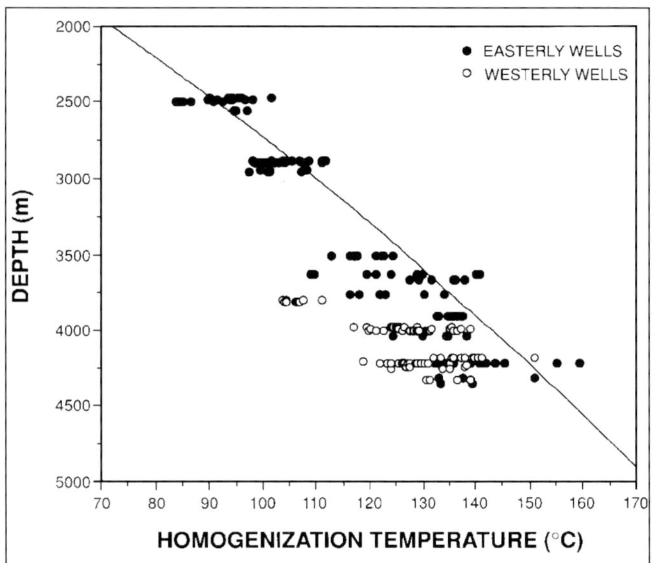

Figure 3. Homogenization temperatures of primary fluid inclusions in quartz cement. Data are for the Garn Formation, from Grant and Oxtoby (1992). Primary petroleum inclusions were found in one well (6506/12-4) and are not included. T_h measurements reported by Konnerup-Madsen and Dypvik (1988), Walderhaug (1990) and Ehrenberg (1990) fall within two standard deviations of the means reported here with the exception of those from well 6507/7-1. The solid line shows an approximate relationship between present-day (maximum) temperature and depth in the Garn Formation.

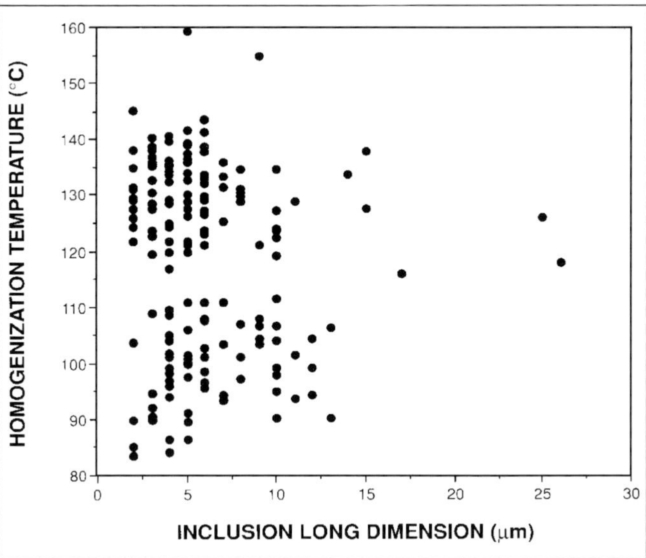

Figure 4. Homogenization temperature versus inclusion size (from Robinson et al., 1992). By permission of the publishers Butterworth Heinemann Ltd.

any one overgrowth the T_h of inclusions isolated within the overgrowth to those from the grain-cement boundary (cf. Burley et al., 1989). However, measurements of both from the same sample completely overlap (Grant and Oxtoby, 1992). Thus, despite the complex zoning observed under cathodoluminescence, there was no corresponding systematic variation in T_h, implying that at least on the scale of an individual sample, all quartz cement precipitated within a very narrow temperature interval.

Homogenization temperatures represent minimum values for the temperature of quartz cement growth provided that:

(1) The inclusions are primary; and

(2) They have not undergone non-elastic deformation (stretching or leakage) since trapping.

Fluid inclusions in many diagenetic minerals have a tendency to leak because of the increase in temperature to which they are subjected as the sediment continues to subside after the inclusions were trapped. In most minerals, leakage tends to produce a correlation between T_h and depth (equivalent to temperature), but fluid inclusions in quartz are resistant to leakage and usually faithfully record the conditions under which they were trapped (Robinson et al., 1992). Furthermore, the Garn inclusions show no relationship between size and homogenization temperature as might be expected in a population of leaked inclusions (Figure 4). No pressure correction has been applied to the homogenization temperatures because of the uncertainties inherent in the calculation (see Grant and Oxtoby, 1992). The pressure correction will be small to negligible if the inclusions contain methane (Hanor, 1980).

Ice Final Melting Temperatures

Salinity (calculated from ice final melting temperature) is plotted against T_h in Figure 5. At T_h >115°C, corresponding to a present depth of about 3 km, the range of fluid salinity is small, with an average of about 5%. At lesser depths, salinity is much more variable, from >10% to almost zero (fresh water). The restricted distribution of ice final melting temperatures suggests that only one fluid was involved in quartz precipitation at depths greater than 3 km. At shallower depths, it seems likely that more than one fluid was involved.

Implications

The fluid inclusion data suggest that at any particular depth (paleo or present day), quartz cement grew over a limited temperature range and that the actual temperatures involved were higher at greater depths. The range and absolute values of T_h measurements can be used to estimate the duration and age of cementation by modeling the subsidence and thermal history of a basin (Robinson and Gluyas, 1992b). In the Garn Formation, this approach indicates that quartz precipitated over a wide depth range (Figure 6), in both the west and east of Haltenbanken, within a period of a few million years, mainly during rapid Pliocene-Pleistocene subsidence (Grant and Oxtoby, 1992). Importantly there is no evidence that quartz cement precipi-

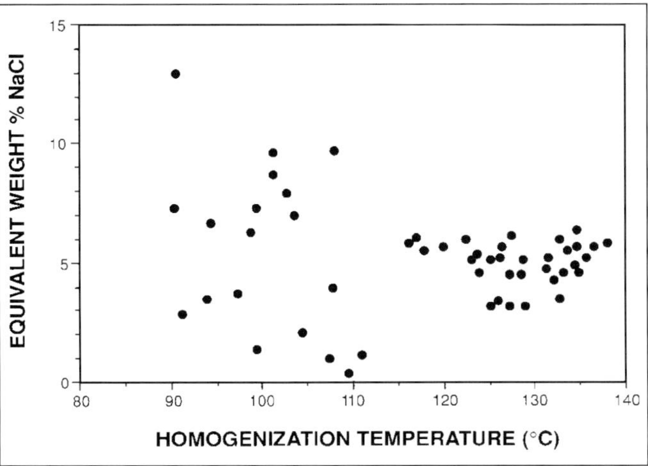

Figure 5. Inclusion salinity versus homogenization temperature. Salinity is calculated from ice final melting temperature and the equation of Potter et al. (1977). Data from the Garn Formation.

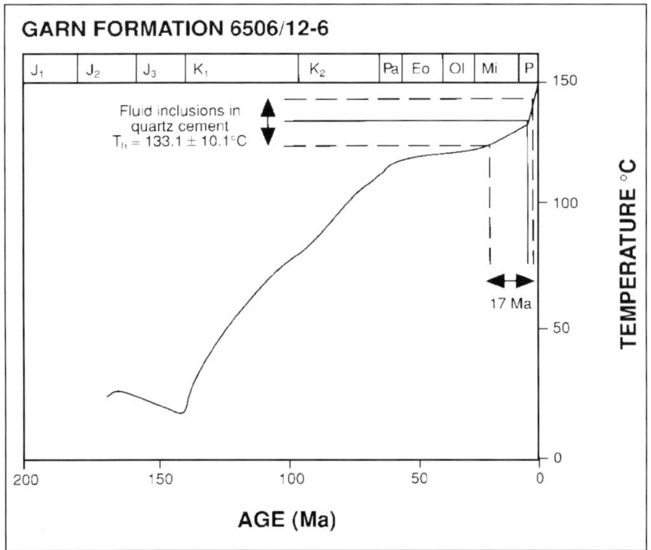

Figure 6. Modeled burial and thermal history for 6506/12-6 (from Robinson and Gluyas, 1992b). Modeled temperature history for the Garn Formation samples from well 6506/12-6. The methodology is as described in Allen and Allen (1990). All heat flow is assumed to be conductive. Input required comprises stratigraphic thicknesses, sediment types and thermal conductivities and surface temperature and basal heat-flow history. By permission of the publishers Butterworth Heinemann Ltd.

tated significantly earlier than this despite the fact that, in the case of the deeper samples, the 100°C isotherm (the precipitation temperature of the shallow samples) was passed as long ago as 100 Ma.

A similar precipitation age was determined by Walderhaug (1990) from a fluid inclusion study on one well from a depth of 2.5 km. Walderhaug used the presence of petroleum fluid inclusions to propose that quartz cementation was synchronous with petroleum filling of the reservoir. In the present study primary petroleum inclusions were only found in one well (6506/12-4), and it appears that either quartz cementation largely predated petroleum filling of the various reservoirs sampled here, or for some as yet unknown reason, petroleum was not trapped in the quartz overgrowths in these wells. The latter is less likely because secondary petroleum inclusions, although rare, are present in other wells. These, along with secondary aqueous inclusions, occur along healed microfractures which cross cut both detrital quartz and overgrowth and clearly postdate precipitation of the overgrowth.

ILLITE CEMENTATION (SOUTHERN NORTH SEA): EVIDENCE FROM K-AR AGES AND $^{18}O/^{16}O$ RATIOS

Geological Background

The Southern North Sea is the western part of the Southern Permian Basin, a major petroleum province extending from eastern England to Poland (Figure 7A). The main gas reservoir in the western part of the Southern North Sea is the lower Leman Sandstone Formation (Figure 7B), which rests unconformably on Carboniferous sedimentary rocks and comprises aeolian dune sandstones, interdune sabkha siltstones and fluvial, fine to coarse, even pebbly sandstones with associated siltstones and intraformational conglomerates (Glennie et al., 1978; Glennie and Boegner, 1981; Arthur et al., 1986; Glennie, 1990). Eastward, toward the basin center, the time equivalent of the Leman Sandstone is finer grained lacustrine deposits known collectively as the Silverpit Formation (Figure 7B). Both are overlain by the metalliferous black mudstone of the Kupferschiefer (Figure 7B) and then by thick marine evaporites of the Zechstein Group (Figure 7B), which constitute an effective regional seal.

The geological history of the area is complex. Subsidence and sedimentation continued from the Permian, through the Triassic and into the Early Jurassic. This was followed by regional uplift due to thermal doming and then by Middle- to Late Jurassic rifting. Erosion continued in the Early Cretaceous and was followed by subsidence to the middle to late Tertiary when there was major Alpine inversion along the Sole Pit fault system (Glennie and Boegner, 1981).

The aeolian sandstones of the lower Leman Sandstone constitute the best reservoir rocks, but their permeability is strongly influenced by the presence of diagenetic illite cement. There is no smectite in the aeolian sandstones and the authigenic illite observed in the core precipitated directly from solution (Robinson et al., 1993b). As such, and in contrast to the subarkosic sandstones from the Garn Formation in Haltenbanken (Ehrenberg, 1990), there is no possibility of progressive "illitization" of a smectitic precursor.

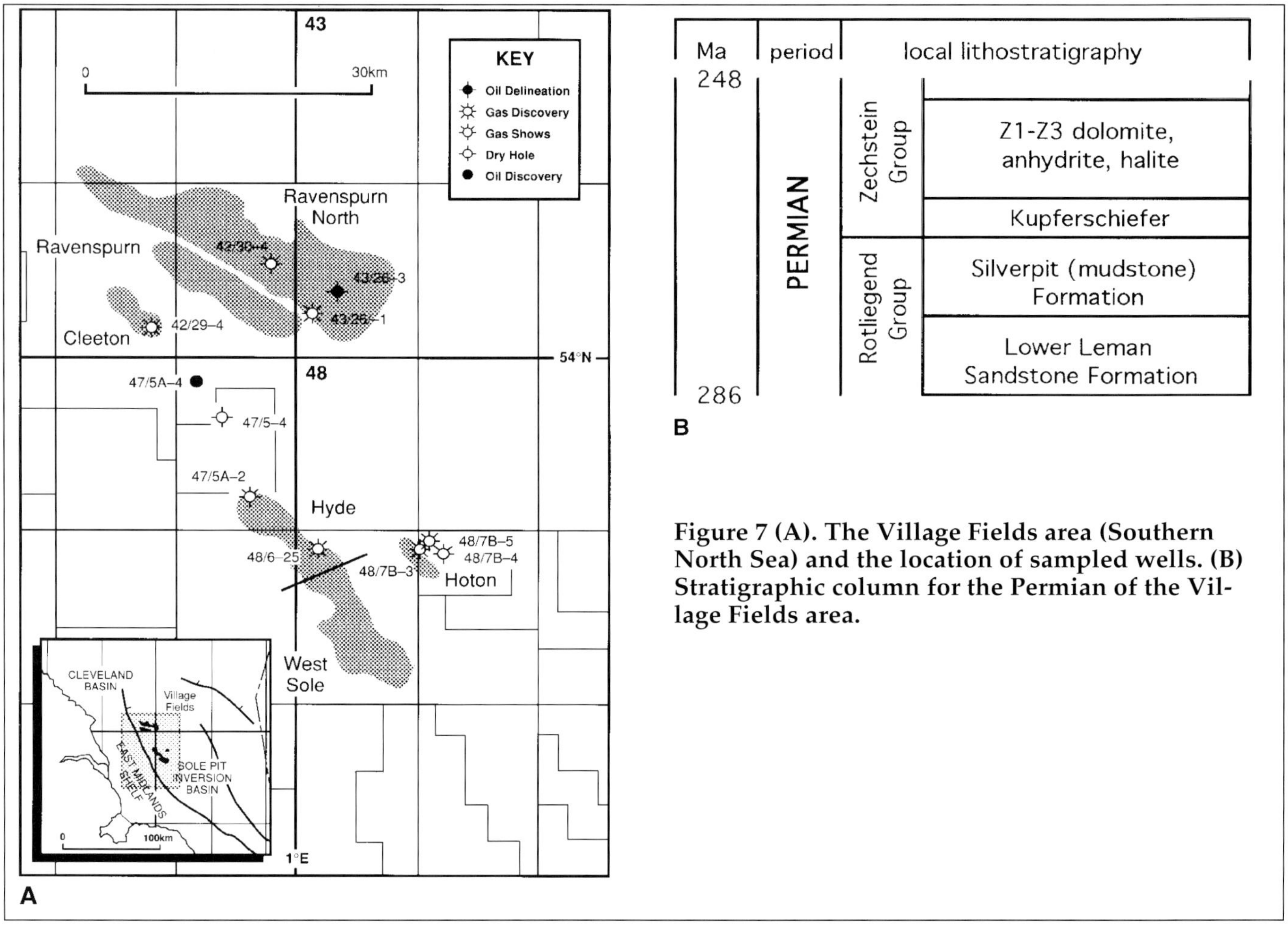

Figure 7 (A). The Village Fields area (Southern North Sea) and the location of sampled wells. (B) Stratigraphic column for the Permian of the Village Fields area.

Results

Illite K-Ar Ages

Figure 8 shows distributions of 39 illite K-Ar age determinations from four gas fields (9 wells) and two further wells located away from major structures (Figure 7A), all from an area in the west of the basin known as the Village Fields area (Robinson et al., 1993). Samples were selected from both gas bearing and water bearing sandstones. All but two fall within the Middle to Late Jurassic: the average for all measurements is 158 million years before the present (158 Ma) ± 13.6 and falls within the Callovian. Both the average date and range of dates obtained for the time of illite precipitation in the Village Fields area are comparable with those data obtained by Lee et al. (1989) for the Leman and Indefatigable fields some 75 km further east. Figure 9 shows K-Ar age plotted against depth at 160 Ma (Figure 9B)—the approximate time of illite cementation—and, for reference, against present depth (Figure 9A). In neither case do age and depth appear to be correlated.

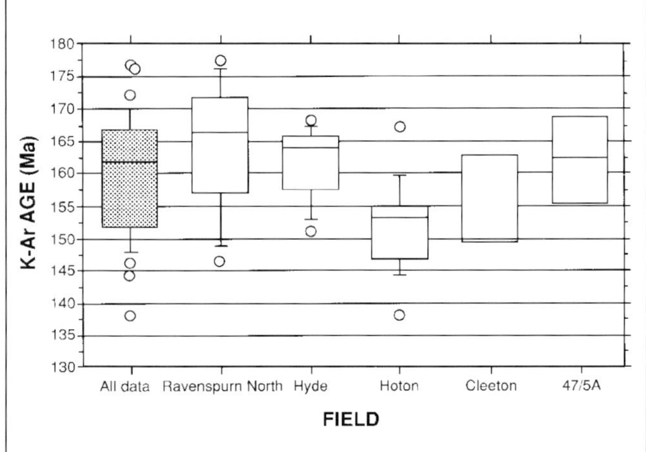

Figure 8. Illite K-Ar age distributions for the sampled fields. The box plots show the modes [not the means as quoted in the text] in the center of the box, and the 10th, 25th, 75th, and 90th percentiles of the distributions. Data points outside the 10th and 90th percentiles are shown individually as open circles.

The measured K-Ar ages will reflect the true age of illite growth only if certain assumptions are met (Dalrymple and Lanphere, 1969; Faure, 1986):

(1) The dated mineral separates must be free of contamination by K-bearing phases;

(2) Illite must have remained a closed system to K and Ar since it grew; and

(3) All non-radiogenic Ar in the samples must have the isotopic composition of the present day atmosphere (no "excess argon"; Shafiqullah and Damon, 1974).

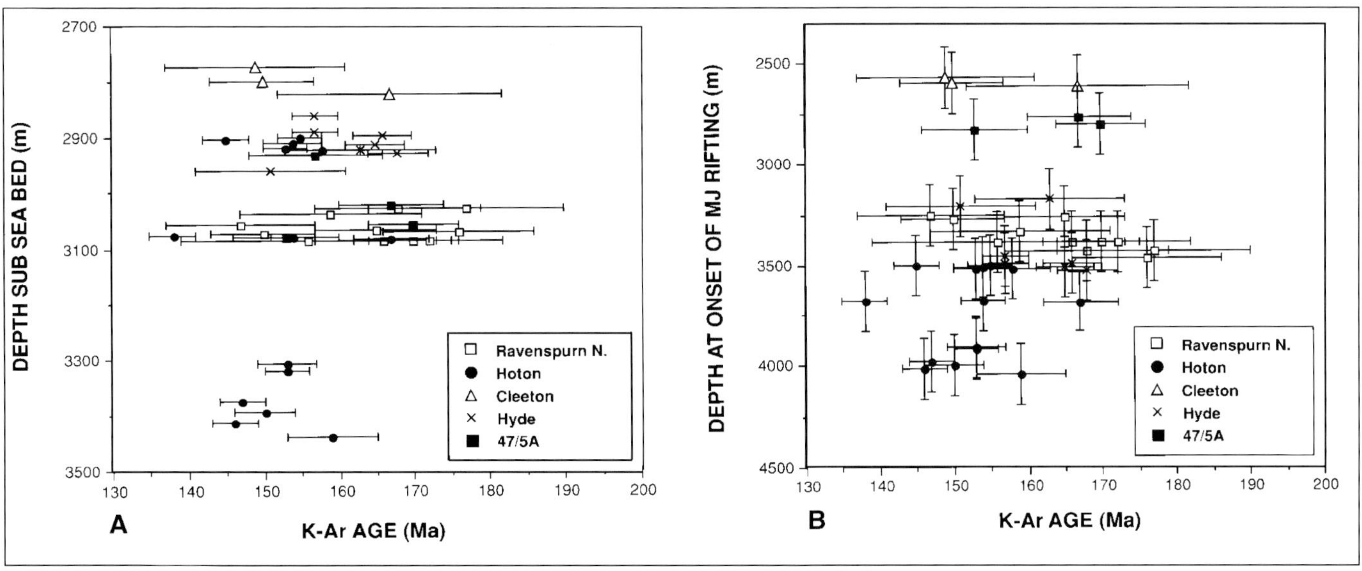

Figure 9. Illite K-Ar age plotted against (A) current sub-sea floor depth and (B) depth restored to 160 Ma. Vertical error bars of ±150 m represent an estimate of the accuracy of the paleodepth estimate obtained by back-stripping. Horizontal error bars represent precision expressed as ±2σ. Depths are plotted sub-sea floor.

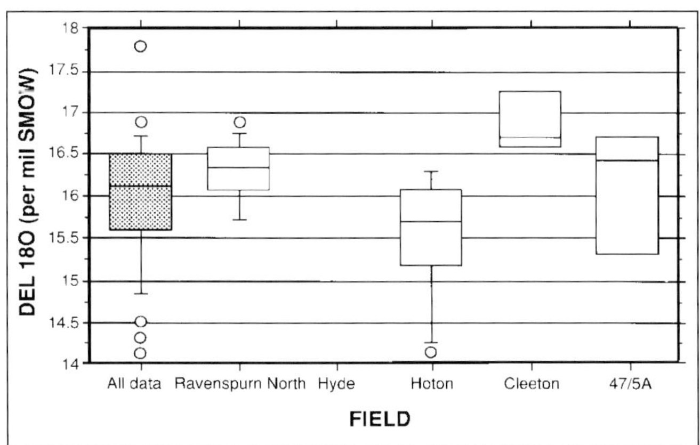

Figure 10. Illite oxygen isotope ratios for the sampled fields. SMOW=Standard Mean Ocean Water.

All samples were from aeolian sandstones that tend to have little or no detrital illite (Lee, 1984; Lee et al., 1985). Furthermore, clay separates were examined using a TEM, and samples containing what appeared to be detrital illite, or K-feldspar, were not dated. Estimates of Ar diffusion kinetics suggest that illites can undergo minor Ar loss that may reduce ages by a few percent (Hamilton et al., 1989). However, the absence of a correlation between age and depth at 160 Ma (which would show an approximately linear relationship to maximum burial temperature) suggests that the ages have not been substantially reduced by Ar diffusion. A plot of the age data on a ^{40}Ar against K "isochron diagram" suggests that there *may* be excess Ar in the samples (see Robinson et al., 1993). The best-fit isochron age is 138 Ma (Early Cretaceous).

Oxygen Isotope Ratios

Figure 10 shows the distribution of illite oxygen isotope ratio measurements across the study area. In Figure 11, δ^{18}O is plotted against both the present day depth (Figure 11A) and depth at 160 Ma (Figure 11B). There is a crude correlation between δ^{18}O and depth at the time of illite growth in the middle Jurassic. The broken lines on Figure 11 represent the δ^{18}O values that illite would have if it had precipitated from a pore water of constant δ^{18}O at temperatures determined by depth. The trend of the data is consistent with precipitation of illite at higher temperatures at greater depths, from pore waters with a restricted range of oxygen isotope composition.

Implications

The uniformity of illite K-Ar ages and the lack of a relationship between age and depth at time of illite growth suggest that illite grew over the entire study area (> 50×50 km) within a restricted period, regardless of the depth of the lower Leman Sandstone at the time. Illite-oxygen isotope ratios show a general pattern of decreasing δ^{18}O with depth at time of cementation. This is compatible with precipitation of illite during a restricted time interval and over a range of temperatures determined by depth.

DISCUSSION

The geochemical data reviewed in this chapter suggest that:

(1) cementation can occur in discrete, geologically relatively short events; and

(2) cementation does not proceed simply because a certain depth or temperature is reached.

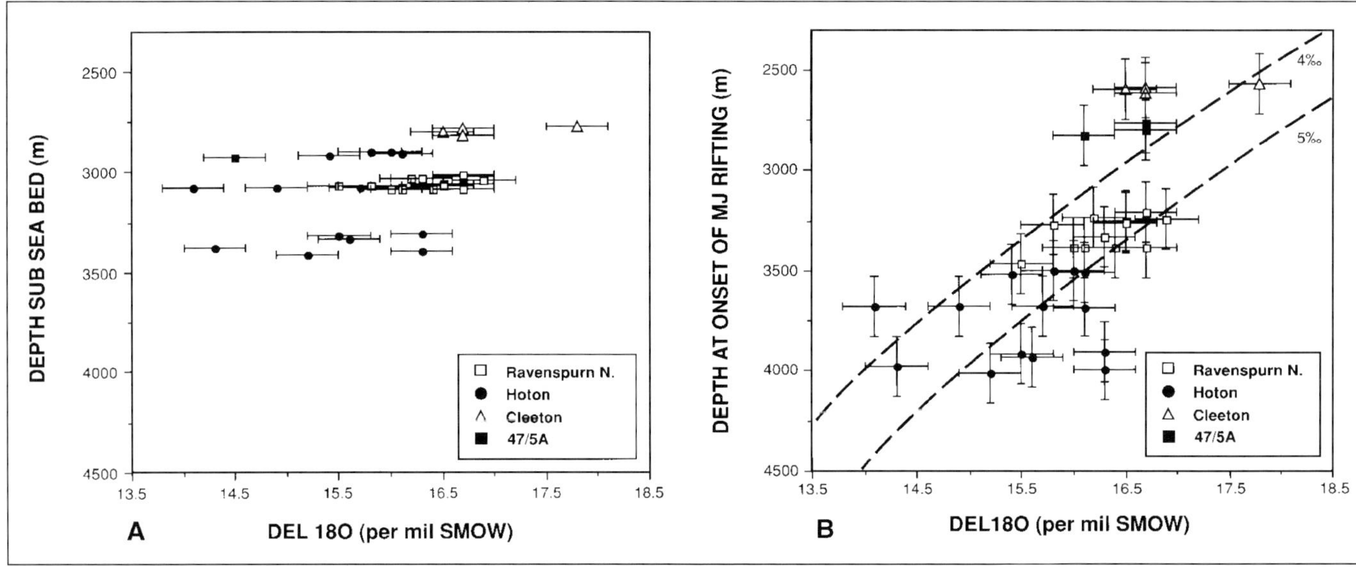

Figure 11. Illite $\delta^{18}O$ plotted against (A) current sub-sea floor depth and (B) depth restored to 160 Ma. Vertical error bars of ±150 m represent an estimate of the accuracy of the restoration. Horizontal error bars represent precision expressed as ±2σ. The two broken lines show the isotope ratios that illite would have if it had precipitated from a water of constant $\delta^{18}O$. Water ratios of +4 and +5‰ are used in this example and are calculated using the fractionation equation of Eslinger and Savin (1973) for a sea-floor temperature of 10°C and a thermal gradient of 30°C.km^{-1}.

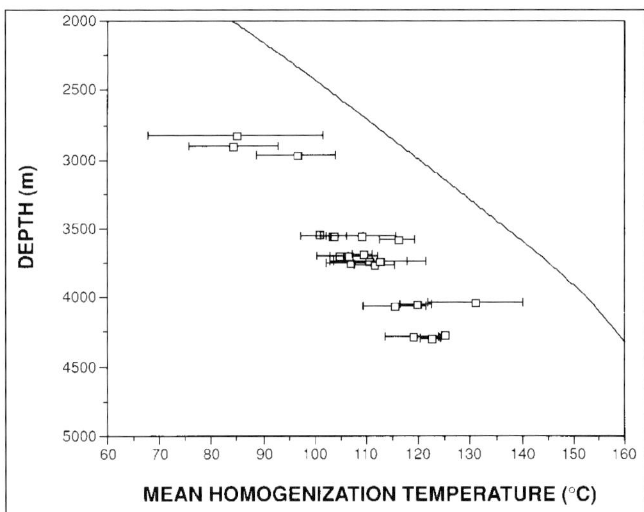

Figure 12. Homogenization temperatures of primary fluid inclusions in quartz cement from Brent Group sandstones from the Northern North Sea (Glasmann et al., 1989 a and b; Hogg, 1989). The solid lines show the approximate relationship between present-day (maximum) temperatures and depth in the Brent Group sandstones. Bars represent ±1σ either side of the mean.

The data presented for the Garn and Rotliegend sandstones are not unique. Fluid inclusion homogenization temperatures from quartz cements in the Brent Group sandstones of the Northern North Sea show a correlation with depth similar to that presented for Haltenbanken (Figure 12). When converted to time using burial histories, these data indicate a Paleocene to Eocene period of cementation (Glasmann et al., 1989a,b; Hogg, 1989; Robinson and Gluyas, 1992b). Furthermore a compilation of homogenization temperatures from different areas around the world demonstrates that quartz cementation *generally* takes place over a narrow temperature interval (usually < 20°C for any particular sample) but at different temperatures in different areas, irrespective of the age of the sediments (Figure 13; see also McBride, 1989). These data suggest that quartz cementation is not initiated at any one temperature threshold. We do not dispute that temperature may influence the *volumes* of cement precipitated; only that it is not the trigger for cementation. The data do suggest that there may be a minimum temperature requirement for quartz precipitation: approximately 70°C in the case of Figure 13 and approximately 50°C for the data in McBride (1989). The attainment of such temperatures by a sediment does not, however, appear to guarantee quartz precipitation.

There are at least three potential and possibly related candidates that may cause cementation:

(1) Tectonically induced changes in fluid transport rates and/or directions

(2) Changes in the thermal regime

(3) Changes in pore fluid chemistry (possibly related to the maturation of organic matter and/or other reactions in mud rocks).

At present, we are unable to provide a mechanism that explains the observations. We can, however, place some constraints on possible solutions both directly, from geochemistry, and indirectly, by relat-

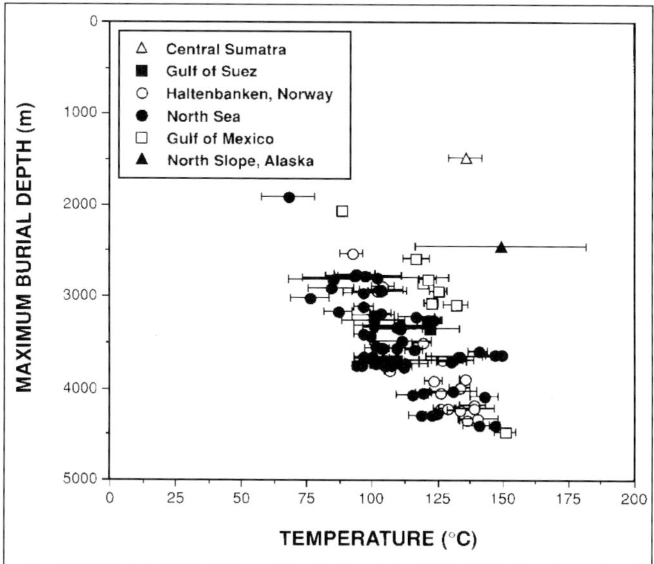

Figure 13. Homogenization temperatures of primary fluid inclusions in quartz cement from sandstones ranging in age from Jurassic to Miocene from several areas around the world (Haszeldine et al., 1984; Glasmann et al., 1989a and b; Burley et al, 1989; Hogg, 1989; Grant and Oxtoby, 1992; and unpublished in-house studies). The solid lines show the approximate relationship between present day (maximum) temperatures and depth in the Brent Group sandstones. Bars represent ±1σ either side of the mean. T_h is plotted against present-day depth for reference only.

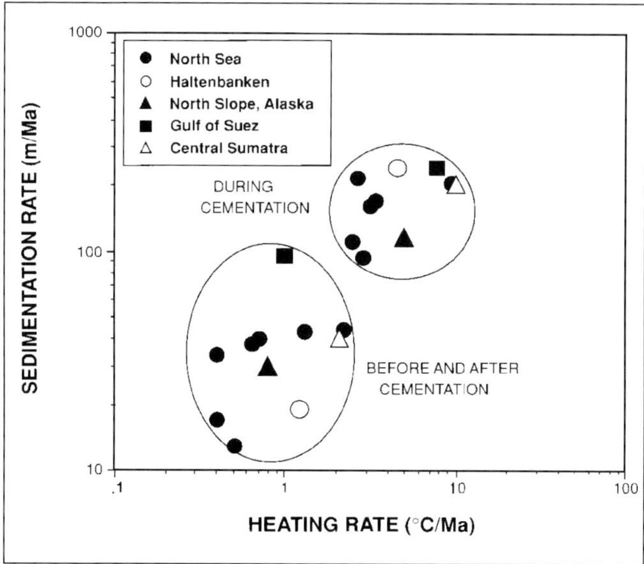

Figure 14. Estimated rates of sedimentation and heating for sandstones in extensional basins before, during and after periods of quartz cementation (Table 1). This figure was constructed by converting the temperature of quartz cementation to time, using burial and thermal histories, and then reading off both the sedimentation and heating rates before, during, and after this time. The significant feature of the graph is not that sedimentation and heating rate are correlated, but that measurements of both during cementation are distinctly high relative to those that characterize periods when no cementation was occurring.

ing the timing of cementation in basins to their geologic history. Aplin et al. (this volume) demonstrate that the pore waters responsible for quartz precipitation in the North Sea are generally stratified, although locally there is considerable spatial and temporal heterogeneity, particularly in the Moray Firth (Burley et al., 1989) and Central North Sea. This indicates that large scale, net vertical movement of compositionally homogeneous cementing fluids from basin margin to basin center (or vice versa) is unlikely, at least in the North Sea. Instead it suggests that the waters involved were locally derived. The salinities of the Haltenbanken fluid inclusions also point to the involvement of waters with different salinities at different depths.

In studies on the timing of quartz cementation in extensional basins worldwide, quartz appears to precipitate during and/or immediately after periods of rapid sedimentation (> 100 m/m.y.) and high heating rates (>2°C/m.y.; Figure 14, Table 1). The illite in the Village Fields area of the Southern North Sea apparently precipitated at about the time of rifting when heat flow would have been high. A similar conclusion was reached by Lee et al. (1989), who linked illite precipitation with tectonic activity. Exactly how changes in heating rate might affect reactions in, for example, mud rocks, or precipitation in sandstones, is unclear at present. Periods of rapid sedimentation correspond to the time when formation waters are released most rapidly for any given depth of burial in hydrostatically pressured sandstones. As a result, the waters will cool much more quickly as they move laterally and vertically upwards. If, on the other hand, the pore waters could not escape, rapid subsidence would promote a rapid buildup of overpressures and hence the possibility for rapid release of pore waters if and when the sandstones regained compaction equilibrium. Both scenarios may apply in the case of Haltenbanken, where the wells in the west developed overpressures while those in the east were normally pressured during quartz cementation.

CONCLUSIONS

Independent geochemical techniques—K-Ar ages, oxygen isotope ratios, and fluid inclusions analyses—suggest that sandstone cementation by quartz and illite tends to be restricted in time and tends to affect large areas and volumes of rock. The sandstones studied were not progressively cemented during burial. On the contrary, it appears that they suffered little diagenetic modification (other than

Table 1. Burial rate and heating rate data for sandstones before, during, and after cementation events. Periods of cementation commonly correspond with phases of rapid burial and heating (Figure 14).

	Basin	Field/Well	Burial rate (m/m.y.)	Heating rate (°C/m.y.)	Relationship to cementation
1	North Sea	Brent	170	3.40	During
2	North Sea	Brent	43	1.30	Before
3	Haltenbanken		240	4.50	During
4	Haltenbanken		19	1.20	Before
5	North Sea	Brae	160	3.20	During
6	North Sea	Brae	38	0.65	Before
7	North Sea	Ula	94	2.90	During
8	North Sea	Ula	40	0.70	Before
9	North Sea	Magnus	202	9.40	During
10	North Sea	Magnus	44	2.20	Before
11	North Sea	Magnus	13	0.50	After
12	North Sea	25/7-2	110	2.50	During
13	North Sea	25/7-2	34	0.40	Before
14	North Sea	13/29-1	217	2.70	During
15	North Sea	13/29-1	17	0.40	Before
16	North Slope, Alaska		115	5.00	During
17	North Slope, Alaska		30	0.80	Before
18	Central Sumatra		200	10.0	During
19	Central Sumatra		40	2.10	Before

compaction) for most of the basin history, with most mineral precipitation taking place rather quickly in geologic terms. Although the *amount* of cement precipitated in an area at any one time may have been influenced by depth (porosity and depth are, after all, often correlated), burial is not of itself sufficient to *cause* cementation.

Geochemical techniques have not determined the cause of cementation but they have placed some constraints on possible mechanisms. They must affect large parts of basins at approximately the same time over a limited period (substantially less than the burial history), effectively switching on and off. In our studies on the timing of quartz cement growth in extensional basins such as Haltenbanken, major cementation seems frequently to occur during or immediately after periods of rapid subsidence (>100 m/m.y.) and high heating rates (> 2°C/m.y.). In the Southern North Sea, this also seems to be true of illite growth. Significant cementation thus appears to require a major perturbation to the status quo, either in the form of loading and/or thermal energy. It is likely that rapid increases in subsidence affect the rates of both precipitation and water flow (perhaps also flow paths) in such a way as to cause widespread mineral precipitation. Exactly how this might occur is the subject of future investigation. If the link is made, we will be within sight of being able to predict the timing and possibly extent of major episodes of porosity reduction from an analysis of basin evolution.

ACKNOWLEDGMENTS

We thank BP Exploration, Statoil and partners for permission to publish this material. This chapter has also benefited greatly from discussions with our Sunbury colleagues, Norman Oxtoby, Andrew Hogg, Tim Primmer, and Edward Warren.

REFERENCES CITED

Allen, P.A. and J.R.Allen, 1990, Basin Analysis: Blackwell Scientific Publications.

Arthur, T.J., D.Pilling, D.Bush and L.Macchi, 1986, The Leman Sandstone Formation in UK Block 49/28. Sedimentation, diagenesis and burial history: *in* Brooks, J., J.C.Goff and B.van Hoorn, Habitat of Palaeozoic gas in N.W.Europe, Geological Society Special Publication 23, p.251-266.

Bloch, S., 1991, Empirical prediction of porosity and permeability in sandstones: AAPG Bulletin, v.75, p.1145-1160.

Bukovics, E. and P.A.Ziegler, 1985, Tectonic development of the mid-Norway continental margin: Marine and Petroleum Geology, v.2, p.2-22.

Burley, S.D., K.Mullis, and A.Matter, 1989, Timing diagenesis in the Tartan reservoir (UK North Sea): constraints from combined cathodoluminescence microscopy and fluid inclusion studies. Marine and Petroleum Geology, v.6, p.98-120.

Dalrymple, G.B. and M.A.Lanphere, 1969, Potassium-Argon Dating: W.H.Freeman and Co.

Ehrenberg, S.N., 1990, Relationship between diagenesis and reservoir quality in sandstones of the Garn Formation, Haltenbanken, Mid-Norwegian continental shelf: AAPG Bulletin, v.74, p.1538-1558.

Eslinger, E.V. and S.M.Savin, 1973, Mineralogy and oxygen isotope geochemistry of the hydrothermally altered rocks of the Ohaki-Broadlands, New Zealand geothermal area: American Journal of Science, v.273, p.240-270.

Faure, G., 1986, Principles of Isotope Geology: John Wiley and Sons.

Glasmann, J.R., P.D. Lundegard, R.A Clark, B.K. Penny, and I.D. Collins, 1989a, Geochemical evidence for the history of diagenesis and fluid migration: Brent sandstone, Heather Field, North Sea. Clay Minerals, v.24, 255-284.

Glasmann, J.R., R.A. Clark, S. Larter, N.A. Briedis, P.D. Lundegard, 1989b, Diagenesis and hydrocarbon accumulation, Brent Sandstone (Jurassic), Bergen High area, North Sea. AAPG Bulletin, v.73, p.1341-1360.

Glennie, K.W., 1990, Early Permian: Rotliegend: *in* K.W.Glennie ed., Introduction to the Petroleum Geology of the North Sea, Blackwell Scientific Publications, p.120-152.

Glennie, K.W. and P.L.E.Boegner, 1981, Sole Pit inversion tectonics: *in* Petroleum Geology of the Continental Shelf of N.W.Europe, Institute of Petroleum, London, p.110-120.

Glennie, K.W., G.C.Mudd and P.J.C.Nagtegaal, 1978, Depositional environment and diagenesis of Permian Rotliegendes sandstone in Leman Bank and Sole Pit areas of the UK southern North Sea: Journal of the Geological Society of London, v.135, p.25-34.

Grant, S.M. and N.H.Oxtoby, 1992, The timing of quartz cementation in Mesozoic sandstones from Haltenbanken, offshore mid-Norway: fluid inclusion evidence (Journal of the Geological Society of London, in press).

Hamilton, P.J., S.Kelley, and A.E.Fallick, 1989, K-Ar dating of illite in hydrocarbon reservoirs: Clay Minerals, v.24, p.215-231.

Hanor, J.S., 1980, Dissolved methane in sedimentary basins: potential effect on PVT properties of fluid inclusions: Economic Geology, v.75, p.603-609.

Haszeldine, R.S., I.M. Samson, and C. Cornford, 1984, Quartz diagenesis and convective fluid movement: Beatrice oilfield, UK North Sea. Clay Minerals, v.19, p.391-402.

Hogg, A.G.C., 1989, Petrographic and isotopic constraints on the diagenesis and reservoir properties of the Brent Group sandstones, Alwyn South, Northern UK, North Sea. Unpublished PhD thesis, University of Aberdeen, UMI reference DX89573.

Konnerup-Madsen, J. and H.Dypvik, 1988, Fluid inclusions and quartz cementation in Jurassic sandstones from Haltenbanken, offshore mid-Norway. Bulletin de Mineralogie, v.111, p.401-411.

Leder, F. and W.C.Park, 1986, Porosity reduction in sandstones by quartz overgrowth: AAPG Bulletin, v.70, p.1713-1728.

Lee, M.C., 1984, Diagenesis of the Permian Rotliegendes sandstone, North Sea: K/Ar, O^{18}/O^{16} and petrologic evidence: PhD thesis, Case Western University, Cleveland, 346pp.

Lee, M.C., J.L.Aronson and S.M.Savin, 1985, K/Ar dating of time of gas emplacement in Rotliegendes sandstone, Netherlands: AAPG Bulletin, v.69, p.1381-1385.

Lee, M.C., J.L.Aronson and S.M.Savin, 1989, Timing and conditions of Permian Rotliegende sandstone diagenesis, Southern North Sea: K/Ar and oxygen isotope data: AAPG Bulletin, v.73, p.195-215.

McBride, E.F. 1989, Quartz cement in sandstones: A review. Earth Science Reviews 26, 69-112.

Potter, R.W., M.A.Clynne and D.A.Brown, 1977, Freezing point depression of aqueous sodium chloride solutions: Economic Geology, v.73, p.284-285.

Robinson, A.G. and J.G.Gluyas, 1992a, Model calculations of sandstone porosity loss due to compaction and quartz cementation: Marine and Petroleum Geology, v.9, p.319-323.

Robinson, A.G. and J.G.Gluyas, 1992b, Duration of quartz cementation in sandstones, North Sea and Haltenbanken basins: Marine and Petroleum Geology, v.9, p.324-327.

Robinson, A.G, S.M.Grant and N.H.Oxtoby, 1992, Evidence against natural deformation of fluid inclusions in diagenetic quartz: in press in Marine and Petroleum Geology.

Robinson, A.G., M.L.Coleman and J.G.Gluyas, 1993, The age of illite cement growth, Village Fields area, Southern North Sea: evidence from K-Ar ages and 18O/16O ratios: in press in AAPG Bulletin.

Scherer, M., 1987, Parameters influencing porosity in sandstones: a model for sandstone porosity prediction: AAPG Bulletin, v.71, p.485-491.

Schmoker, J.W. and D.L.Gautier, 1988, Sandstone porosity as a function of thermal maturity: Geology, v.16, p.1007-1010.

Shafiqullah, M. and P.E.Damon, 1974, Evaluation of K-Ar isochron methods: Geochimica et Cosmochimica Acta, v.38, p.1341-1358.

Vik, E., O.R. Heum, and K.G. Amaliksen, 1991, Leakage from deep reservoirs: possible mechanisms and relationship to shallow gas in the Haltenbanken area, mid-Norwegian shelf. In England, W.A. and Fleet, A.J. (eds), Petroleum Migration. Geological Society Spec. Public. No. 59, p. 273.

Walderhaug, O., 1990, Fluid inclusion study of quartz cemented sandstones from offshore mid-Norway—possible evidence for continued quartz cementation during oil emplacement. Journal of Sedimentary Petrology, v.60, p.203-210.

Chapter 4

Fluid Inclusion Temperatures in Diagenetic Quartz Reset by Burial: Implications for Oil Field Cementation

R. Stuart Haszeldine and Mark Osborne
Department of Geology & Applied Geology
University of Glasgow
Glasgow, U.K.

ABSTRACT

Fluid inclusion (FI) homogenization temperatures in diagenetic quartz are often used as crucial evidence to infer abnormally hot paleotemperatures and precipitation of pore-filling quartz cements following advection of large volumes of hot waters from depth. However, we consider that FI temperatures do not record their original paleotemperatures.

North Sea and Norwegian Jurassic oil field sandstones show volumes of diagenetic quartz cement which today increase monotonically with depth from 1.8 km to at least 4.2 km. Minimum, modal, and maximum FI temperatures within this diagenetic quartz all show a progressive temperature increase with depth. If subsidence of an oil field has been slow during the past 5 m.y., FI modal temperatures coincide exactly with rock temperature, but if subsidence has been rapid (>10 m/m.y.), then they systematically lag beneath rock temperatures. Additional geological evidence favoring resetting includes: high temperature quartz FIs in texturally and isotopically shallow veins; and high temperature quartz FIs inside and outside shallow-formed carbonate cement. Laboratory experiments also show some resetting of quartz FIs within one year of overheating. The resetting mechanism is unlikely to be by fracturing and leakage, but more likely to be via stretching and local shape change. Inclusion compositions are probably unaltered. An FI temperature distribution from any single depth records the pauses in subsidence and maximum burial temperature, but does not simply record the temperature of quartz growth. Resetting of FIs negates temperature evidence for massive circulation of hot basinal fluids transporting quartz. We conclude that diagenetic quartz in these rift basin sandstones may be supplied by local diffusion, so that cement volumes are inherently predictable.

INTRODUCTION

Quartz cement is an important cause of porosity reduction during deep burial diagenesis of sandstones. Such cements can form 5–15% of the rock volume and so can reduce porosity by 50% (e.g., Giles et al., 1992). The processes by which such quartz cements originate are thus crucially important both industrially—to understand why sandstone reservoirs contain only half the hydrocarbons they might—and intellectually, to understand how ions forming 15% of a rock can be moved in the subsurface. Two rival hypotheses are current to explain quartz cements. Neither diagenetic cement textures nor isotopic evidence are conclusive in testing between these two hypotheses (Haszeldine et al., 1992), so that FI evidence has a pivotal role, and its reliability must be examined closely. The two hypotheses are:

(1) Quartz grew as a mineralizing event transported by hot fluids that circulated in the deep basin (Haszeldine et al., 1984a,b, Jourdan et al., 1987, Burley et al., 1989, Glasmann et al., 1989b, Gluyas et al., this volume); and

(2) Quartz grew during a geologically long time span at equilibrium temperature; ions were locally supplied (Blanche and Whitaker, 1978, Walderhaug, 1990, Bjørlykke et al., 1992; Aplin et al., this volume).

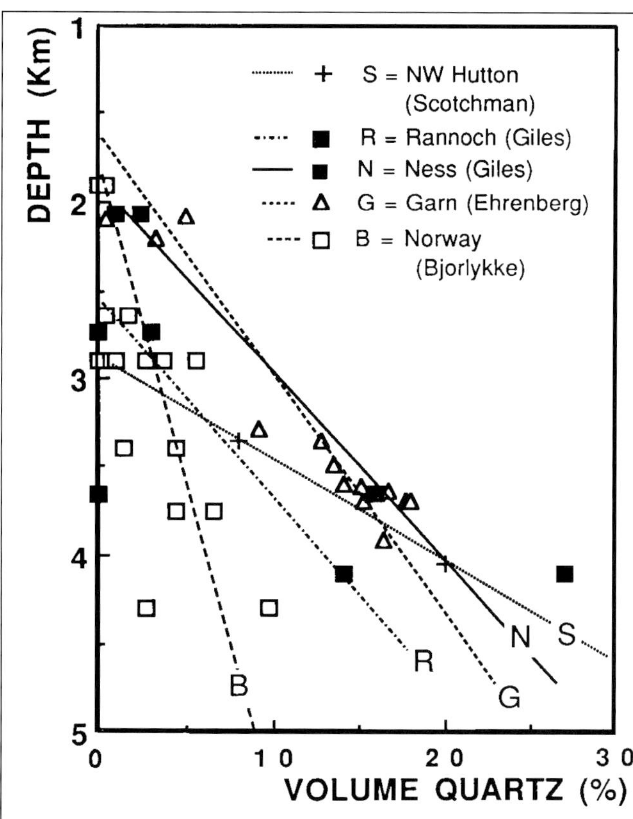

Figure 1. Quartz overgrowth volume plotted against present-day burial depth in the Middle Jurassic Brent Group and Garn Formation. The amount of quartz cement tends to increase with depth. Data replotted from Scotchman et al., 1989, Ehrenberg 1990, Giles et al., 1992, Bjørlykke et al., 1992).

Evidence of elevated temperatures from fluid inclusions (FIs) (Haszeldine et al., 1984b, Burley et al., 1989) is frequently used as critical data to support the first hypothesis. In this chapter, we have compiled a large FI data set from published and unpublished North Sea and Norwegian sources to test the accuracy of the FI paleotemperature record. We find that FI homogenization temperatures in diagenetic quartz show a correlation with present-day burial depth and an excellent correlation with burial temperature, when corrected for subsidence rate. We conclude that FIs in diagenetic quartz are reset during burial, and do not record the paleotemperatures of quartz growth. Thus the key evidence supporting hypothesis 1 is probably wrong. It is very probable that diagenetic quartz grew over a long time span, supplied by local diffusive processes rather than by large scale fluid advection. Consequently, porosity reduction by quartz cementation can potentially be modeled as a closed system.

FLUID INCLUSIONS IN DIAGENETIC QUARTZ

Of critical importance in the following discussion is the observation that quartz cement increases monotonically with depth, especially as this can be documented within a single field (Scotchman et al., 1989) by sampling in a pattern that moves 1 km vertically over an 8-km lateral distance. Regional compilations of Middle Jurassic data show the same effect (Glasmann et al., 1989a, Ehrenberg 1990, Giles et al., 1992, Bjørlykke et al., 1992). These data imply that quartz cement has formed, and is still forming, gradually with depth (Figure 1), commencing around 70°C at depths of approximately 2 km. If quartz had formed from a single short-term mineralizing event in ancient times (Robinson and Gluyas, 1992), then sandstones would have been cemented in relation to their paleodepth, and would not show a relationship of quartz cement to their present depth. If quartz had formed from a recent open system mineralizing event, then overpressures would not be extensively developed today in basins with very different subsidence histories.

During burial of sandstones, diagenetic quartz nucleates on detrital grains to form overgrowths that may reach 100 µm thick and gradually fill primary porosity. The great majority of FIs form at the boundary between the detrital sand grain and the overgrowth or at boundaries of cathodoluminescence growth zones, suggesting diastems in the growth history (Figure 2). Burley et al. (1989) and Hogg et al. (1992) show that growth zones within these overgrowths are, in general, concentric, so that FIs at a grain/overgrowth boundary would have been

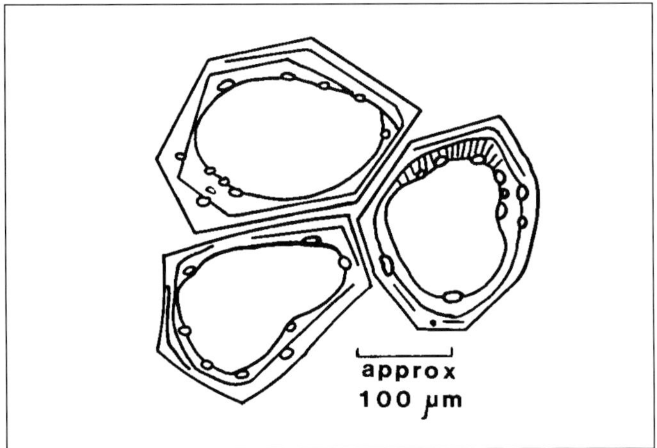

Figure 2. Schematic Cathode Luminescence (CL) thin section view of quartz overgrowth. FIs occur preferentially at grain-overgrowth boundaries (Burley et al., 1989, Hogg et al., 1992). Fewer FI occur at CL boundaries within the overgrowth, and still fewer as isolated FIs within one CL zone.

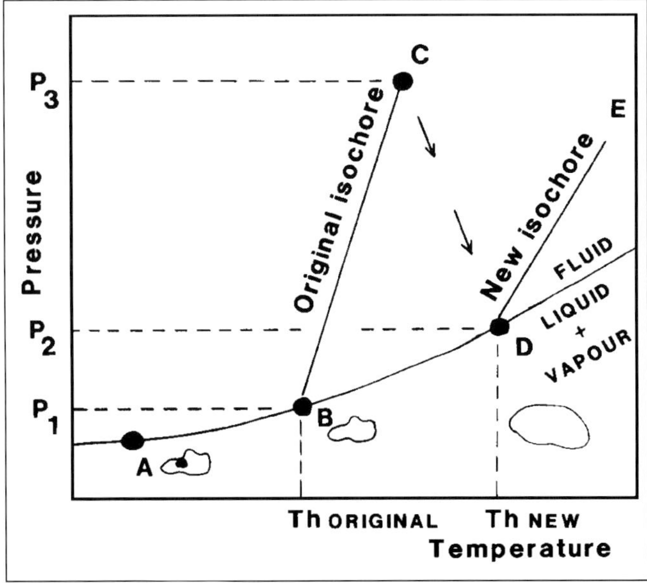

Figure 3. Pressure-Temperature phase relations of an aqueous fluid inclusion. When heated in the laboratory from surface conditions (A), the liquid-vapor curve is followed to the homogenization temperature $T_{h\ original}$, where the vapor bubble disappears (B). On further heating (as in further burial), the pressure-temperature relationship in the inclusion follows the isochore (constant fluid density) and the internal pressure increases P_1-P_3. At C, the inclusion stretches or fractures in order to release the internal overpressure down to P_2. If stretched by shape change, its volume increases, so the homogenization temperature rises to $T_{h\ new}$ (D). Continued burial and heating will follow a new isochore D-E corresponding to the new fluid density.

trapped at about the same temperature. Inclusion workers attempt to select primary unaltered FIs from these populations to obtain accurate temperature and salinity measurements. Inevitably these are biased towards the start of quartz growth. Raman probe analyses of oil field basin FIs show appreciable methane contents, meaning that a pressure correction is not necessary to obtain a hotter trapping temperature than the measured homogenization temperature (T_h) (Hanor 1980).

WHY RESETTING MIGHT OCCUR: FLUID INCLUSION SYSTEMATICS

When a fluid inclusion is trapped during mineral growth in the subsurface, it is assumed that one aqueous fluid phase is present, without any free vapor. Pressures and especially temperatures are lower at the surface than in the subsurface reservoir; the recovery of a reservoir rock sample results in the contraction of the aqueous fluid in an FI, thus forcing a small vapor bubble to nucleate within the aqueous fluid inclusion. Heating the inclusion in a laboratory mimics the burial temperature of original growth, and the vapor bubble merges again into one fluid. This is the homogenization temperature ($T_{h\ original}$ on Figure 3) and represents the minimum trapping temperature of that particular inclusion and the minimum growth temperature of that mineral. This interpretation relies on several crucial assumptions:

(1) The pressure correction—the difference between homogenization and trapping temperatures (C minus B on Figure 3)—is small so that the temperature of entrapment approximates to that measured by $T_{h\ original}$;

(2) The inclusion has remained a closed system, neither losing nor gaining fluid or vapor;

(3) The inclusion has maintained a constant volume; and

(4) The fluid was trapped as a single aqueous phase.

In sedimentary basins, the aqueous fluid inside inclusions is often analyzed to be methane-rich, so that assumption 1 is true and any pressure correction can be neglected (Hanor 1980, Burley et al., 1989). However, both conditions 2 and 3 potentially can be violated for FIs in minerals grown in a subsiding sedimentary basin, where pressures and especially temperatures outside the inclusions increase due to additional burial after their growth. As an inclusion is buried and becomes heated, the aqueous fluid trapped within it attempts to expand, within the fixed volume of the inclusion. The pressure within the inclusion consequently increases until at some point the strength of the host mineral is exceeded, and the inclusion fluid expands either by increasing the volume of the inclusion cavity, or by losing some of its contents. Experimental studies show that this can occur in two ways: plastic deformation and stretching (Bodnar & Bethke,

1984); or brittle deformation and fracture decrepitation (Roedder, 1984).

For decrepitation to occur, microfractures must develop from the inclusion towards the lower pressures at the edge of the grain, or within another inclusion, permitting the inclusion to lose fluids to its external environment. Fractures of the correct diameter may even selectively pump water out of inclusions (Bakker & Jansen 1990). Once the fractures heal, the growth temperature, pressure and chemical composition of that inclusion's contents will have changed. Loss of fluid from the inclusion lowers the internal density of the inclusion and must result in a new, higher, homogenization temperature. Decrepitated inclusions can potentially be recognized during petrographic observation, as they are often surrounded by a halo of small secondary inclusions (Sterner & Bodnar 1989). As an alternative response to the increased internal pressure, the inclusion may deform plastically and non-elastically (stretch), so that its volume increases. This again results in a decrease in the density of the fluid and an increase in homogenization temperature (Figure 3, $T_{h\ new}$). As there has been no loss of fluid, the inclusion contents retain their original chemical composition. This size change is impossible to detect petrographically: for an inclusion with a salinity of 5 wt% NaCl, a density change of only 3% is needed to produce a temperature increase from 100 to 150°C, equating to a spherical diameter change from 5 to 5.05 μm.

EXPERIMENTAL DATA

The problem for fluid inclusion studies in situations of increasing temperature, such as subsiding hydrocarbon basins, is to decide the limits to the ability of each mineral to resist inclusion decrepitation or stretching with increased temperature. It is well established that minerals such as fluorite and sphalerite are weak (Bodnar and Bethke 1984). Laboratory experiments have also been undertaken to measure the internal overpressure required to decrepitate fluid inclusions in quartz. These experiments are usually undertaken by short term heating (days or months) of artificially grown inclusions in laboratory quartz. Robinson et al. (1992) have reviewed experimental results and have confidence in FI data partly because they have been impressed by reports from Bodnar et al., (1989) and from Sterner and Bodnar (1989) that emphasize large decrepitation pressures rather than stretching as FI resetting mechanisms for quartz. We have examined these and other workers publications, and come to differing conclusions on processes. The experimental values obtained depend upon the size of the inclusion (Bodnar et al., 1989). To decrepitate, small inclusions (5 μm) require internal overpressures of >250 MPa, whereas large inclusions (30 μm) require overpressures of only 85-100 MPa. Neither of these values is likely to be achieved in a sedimentary basin, where Burrus (1987) has shown that internal overpressures resulting from 70°C (2 km) of subsidence will only reach 40 MPa. Thus, subsidence-related decrepitation of inclusions in authigenic quartz can be rejected.

By contrast, Gratier (1982) and Gratier and Jenatton (1984) have experimentally observed plastic and diffusive deformation and stretching of fluid inclusions in quartz. These experiments on inclusions in synthetic quartz, at internal overpressures as low as 5 MPa and temperatures sometimes increased from only 212 to 217°C, produced shape changes from irregular towards spherical or negative crystal inclusion shapes within 120 days. Inclusions which changed shape and did not fracture were observed to increase their T_h, inferred by Gratier and Jenatton (1984) to be due to the density decrease of the fluid inside the inclusion. These changes in FI shape were deduced to have a kinetic barrier, so that FIs in synthetic quartz produced measurable changes within days, whereas natural FIs in quartz would have needed 4 years to achieve similar changes. This time dependency is important in our arguments for slow geological resetting rates in natural systems.

Bodnar et al., (1989) found to their surprise that many inclusions underwent non-elastic stretching before undergoing brittle fracture. Stretching occurred at low overpressures and was often followed by decrepitation upon increased heating. In their experiment, this behavior was confined to the smaller inclusions (<6 μm) in their experiment, similar in size to those occurring naturally in diagenetic quartz. Internal overpressures in these inclusions were thought by Bodnar et al., (1989) to be only 40 MPa, and so are achievable during burial. About 20% of the inclusions increased their homogenization temperatures by up to 30°C during this 12 month experiment. Although the majority showed no temperature increase, these experiments still demonstrate that the process of resetting by stretching of inclusions in quartz is feasible. Sterner and Bodnar (1989) found experimentally that this resetting process is time dependent, once it has begun. Then individual inclusions come closer to reequilibration in the new P-T conditions with increasing time. In laboratory experiments, very long times are a difficult variable to reproduce; by looking back in time to the geological record, we can see the end effect of any resetting processes. This is what we attempt to do in the rest of this paper.

PATTERNS IN FLUID INCLUSION TEMPERATURE DISTRIBUTIONS

Roedder (1984) explains the different types of information available from FIs. Here we are only concerned

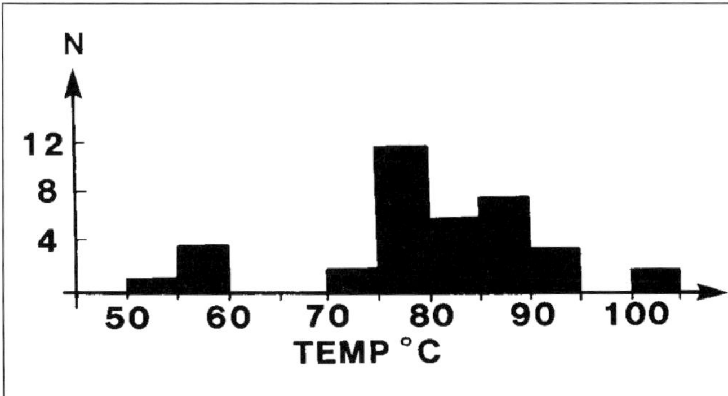

Figure 4. Idealized FI homogenization temperature distribution. There are isolated low temperature readings, one or more modes and a skewed high temperature tail. Such patterns need more than 40 readings at each depth to be visualised.

with homogenization temperatures which are commonly interpreted to yield the growth temperature of diagenetic quartz (Haszeldine et al., 1984a, Ehrenberg 1990, Glasmann 1989a, b, Burley et al., 1989). A cartoon histogram of FI homogenization data (c.f. Glasmann et al., 1989a, Haszeldine et al., 1992) is shown in Figure 4. This shows one or more modes, with isolated readings at lower temperatures and a skewed tail toward hotter temperatures. Empirically, we find that this pattern requires a large number of readings (greater than 40, preferably >100) to be measured at each individual depth. This pattern is blurred if data from different depths are merged, and the full pattern does not appear in smaller data sets. We propose that $T_{h\ min,\ mode\ and\ max}$ in these spectra are important.

EMPIRICAL GEOLOGICAL DATA

We have compiled all published FI data from North Sea quartz cements and combined these with our own measurements. When homogenization temperature is plotted against present-day burial depth of sandstones, a pattern of increasing temperature with present depth is seen (Figure 5). The minimum temperature of each distribution also increases with depth. When a 35°C/km temperature gradient, typical of today's North Sea is projected for comparison with present-day rock temperatures (Figure 5), we observe that some modes coincide very closely with the rock temperature today, whereas others are too cool. Also apparent is that temperature maxima in FI histograms rarely exceed today's rock temperature.

This simple co-relation of depth and FI temperature made us suspect that FI temperatures are partly a record of today's temperatures rather than of ancient temperatures. However, there is no simple graphical fit to these data (Figure 5); fits can be made ranging from a 35°C/km temperature gradient to an 18°C/km

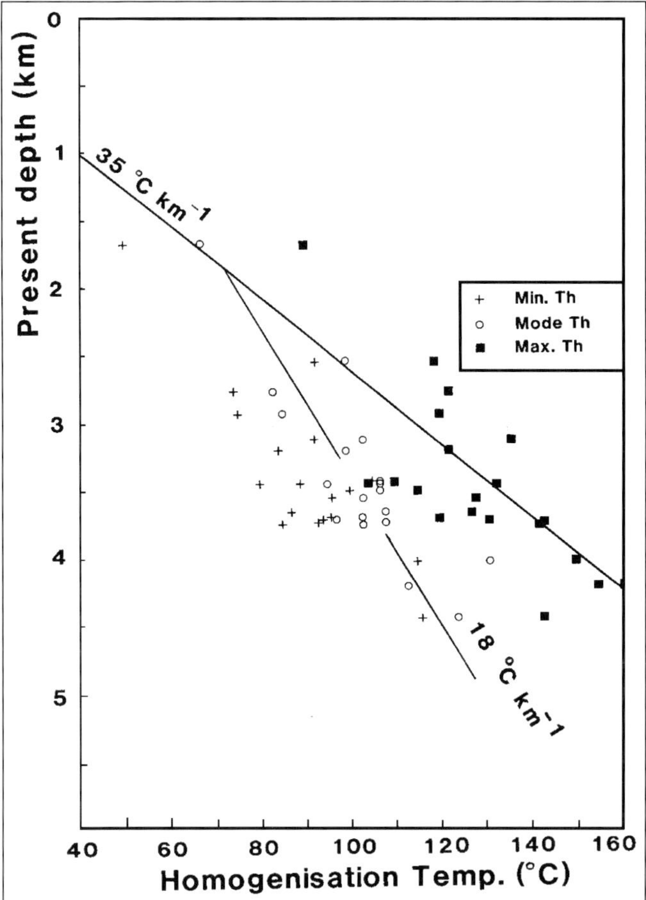

Figure 5. Compilation of published and unpublished FI homogenization temperatures from the North Sea and Haltenbanken (offshore northern Norway). The graph shows minimum, mode, and maximum temperatures plotted against present-day burial depth. All areas today have a temperature gradient of about 35° C/km (superimposed for comparison) and are at their maximum burial depth. A range of line fits can be proposed, with gradients ranging 18–35° C /km ; those of more than 35° C/km or less than 18° C/km can be excluded. Data from Haszeldine et al., (1992), Glasmann et al., (1989a,b), Hogg (1989), Jourdan et al., (1987), Konnerup-Madsen and Dypvik (1988), Malley et al., (1986), Walderhaug (1990); unpublished Glasgow University data from Macaulay (Fulmar), McLaughlin (South Brae), and Osborne (Emerald); see Table 1.

temperature gradient. The full interpretation is necessarily more complex (see below).

EVIDENCE THAT HOMOGENIZATION TEMPERATURES ARE RESET

It has been established experimentally (Bodnar and Bethke 1984) and geologically (Goldstein 1986, Barker and Goldstein 1990) that FIs in soft cleaved minerals such as carbonates are reset and show a good correlation with maximum paleotemperature of a sediment. Metamorphic quartz FIs are also reset during experimental or natural overheating (Sterner and Bodnar 1989). It is therefore logical to ask if FIs in diagenetic quartz can also be reset during burial and heating after their growth. Several lines of geological evidence suggest that this has indeed taken place.

Temperature Profiles

Quartz cement volume increases with present-day depth in Middle and Upper Jurassic sandstones from the North Sea (Figure 1) and so is here inferred to grow gradually through a depth range. Thus inclusions within the overgrowth would be expected to increase in homogenization temperature from the detrital grain rim towards the last formed edge of the diagenetic overgrowth. We have looked for this effect in inclusions from Emerald, NW Hutton, and South Brae fields, and have reexamined data from Alwyn South field (Hogg 1989). In none of these is any systematic temperature profile observed within overgrowths. Only in the Tartan field study (Burley et al., 1989) has any systematic increase in FI temperatures been shown throughout an overgrowth. Even there, no oscillations of temperature (from hot and cool fluids) are recorded. The Tartan case, however, may be complicated because the cooler FIs are very saline whereas the hotter ones are not. Osborne and Haszeldine (1993) point out that saline FIs are reset less than pure water, so that this alternative to hot fluid pulses is feasible in Tartan. We note also that the Tartan FI temperature paleogradient is 34°C /km , just as it is today, and the hottest FIs are within 10°C of present reservoir temperatures (Burley et al., 1989, figures 13, 15). These features suggest that Tartan FIs are reset.

Lowest FI Temperatures

The lowest homogenization temperatures provide minimum values for the commencement of overgrowths. Figure 5 shows that minimum temperatures today increase with present-day sandstone depths. One interpretation of this is that quartz growth in the same sandstone facies in different oil fields, and even within the same oil field, was initiated at different temperatures. If an equilibrium temperature gradient is assumed, this implies that some sandstones grew quartz cement at 2 km, whereas others had no quartz cement until 4 km. When we remember that the percentage of quartz cement today clearly increases with depth (Figure 1), and if this quartz cement commences at around 70°C, we consider the variable temperature and depth interpretation of FI evidence to be incorrect. A more satisfactory explanation is that quartz cement in these sandstones has been, and still is, commencing growth at around 70°C (65–90°C). This is shown by the intercepts on Figure 1. If however the original (coolest temperature) FIs are continually resetting with depth to increase their apparent temperature, then there can be no certain FI record of these lowest temperatures. Indeed, we see from Figure 5 that $T_{h\ min}$ does increase progressively with today's depth (not paleodepth). If FIs at grain-overgrowth boundaries did indeed grow at different temperatures in the past from a rising hot fluid, it is difficult to see how the present-day quartz-depth profile and the $T_{h\ min}$ profile could be preserved through varying subsidence. These FIs cannot be responding to a present-day hot rising fluid, for the deeper FI temperatures are cooler than rock temperature. This is discussed further, below.

Hot Paleotemperatures

Quartz FI temperature modes in these North Sea sandstones are always at or below present-day reservoir temperature and generally increase with present-day depth (Figure 5). Some FI maxima exceed present-day rock temperature. These do not reflect uplift, but are thought to result from FI stretching (see below). If quartz started growing when a sandstone was at a shallower depth, the FI temperatures from that time that we record today would have been too hot when related to a 35°C /km temperature gradient. One interpretation of this is to invoke a period of hot fluid circulation, which heated the sandstone and also transported the quartz cement (Haszeldine et al., 1984a,b, Jourdan et al., 1987, Burley et al., 1989). This leads to problems. First, a huge volume of fluid (at least 10^6 pore volumes) is needed to precipitate the volumes of quartz cement present, due to the very low solubility of quartz in oil-field brines. Such large volumes of fluid are not available during burial diagenesis (Haszeldine et al., 1984b; Aplin et al., this volume). Secondly, if quartz cement was transported by hot fluid, then the fluid should be cooling to precipitate quartz (Figure 6). A cooling fluid would fix a high apparent temperature gradient in the fluid inclusions. By contrast, if a rising fluid was keeping its heat and not cooling, its temperature would not change much vertically, and a low apparent temperature gradient would be fixed. The North Sea FI data apparently record a low to equilibrium 35-18°C/km gradient (Figure 5) implying a fluid rising but not cooling to precipitate quartz. If a rising fluid had cooled beneath the surface, then a high temperature gradient and high temperatures would exist in the shallow basin (Figure 6). Notice that these temperatures lie to the right of the line in Figure 6 whereas the data lie to the left. We have attempted to backstrip our samples to infer convincing episodes of hot fluids in the geological past, which are preserved unaltered in the FIs today—but without success, as the subsidence histories of the North Sea and the Norwegian margin are so different.

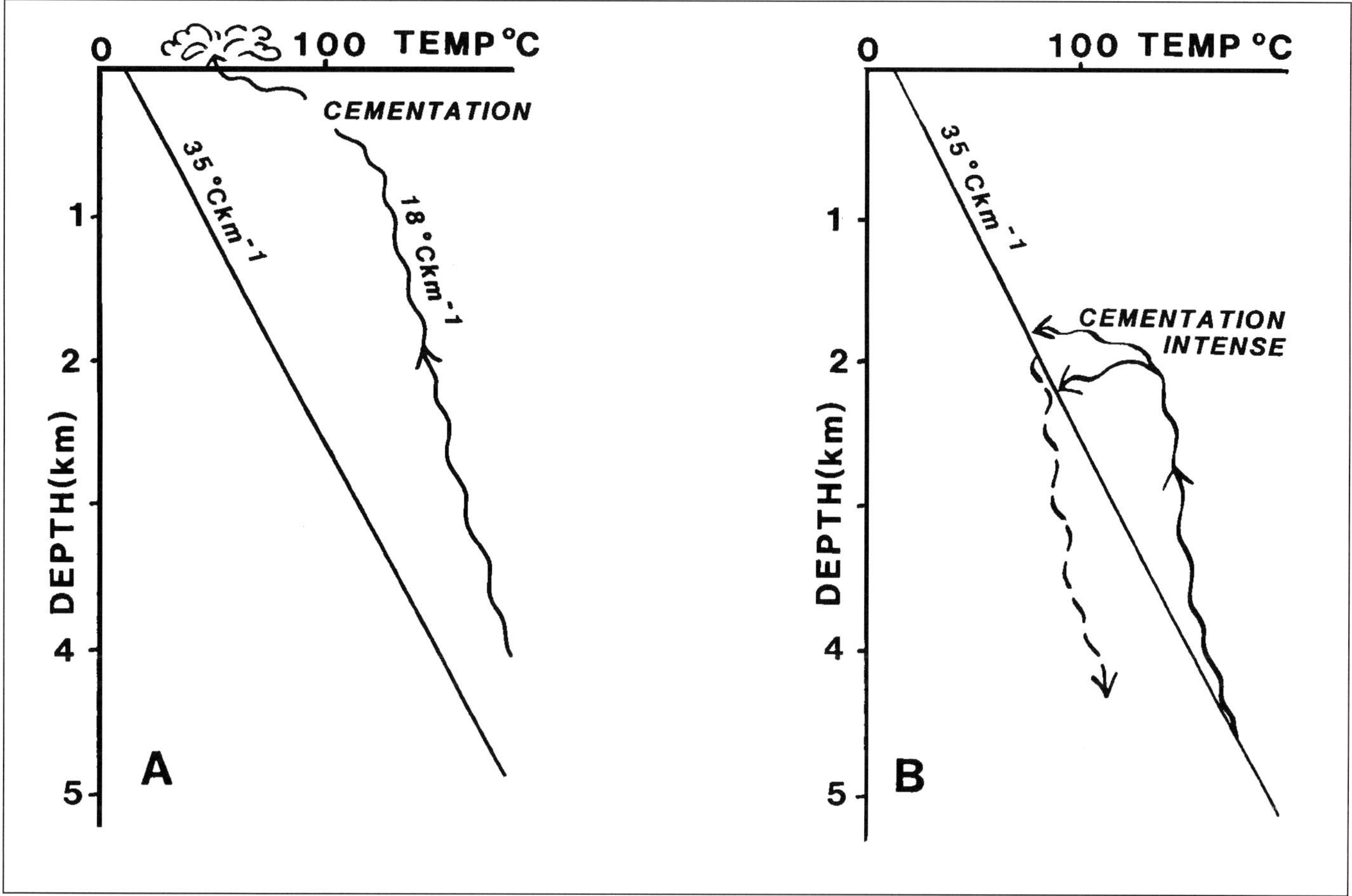

Figure 6. Cartoons illustrating possible temperature gradients for rising hot fluids; these do not match figure 5 and are rejected. A: Rising hot fluids cementing near the surface and discharging as hot springs. B: Rising hot fluids cooling and cementing in the subsurface. The rising fluid is shown with an approximate 18°C./km gradient to enable comparison with figure 5. Both A and B gradients lie to the right of the 35°C/km line indicating hydrothermal conditions, rather than to the left as in figure 5. In alternative B, most silica cementation must have occurred during the rapid cooling high gradient at around 2km. We find no FI evidence of such a high paleo-gradient within individual oil fields (eg. 34°C/km paleogradient between up and down-thrown fault blocks in Burley et al., 1989). A very fortuitous combination of post-cementation subsidence must also be proposed for both Haltenbanken and the North Sea to produce the present-day trends of gradually increasing quartz content with depth (Figure 1).

Shallow Veins

In the Fulmar field, there are subvertical veins 0.5–5 cm wide filled with carbonate and silica, ptygmatically compacted in silty sediment (Johnson et al., 1986). This texture clearly demonstrates that the veins formed at shallow burial depths (< 1 km), whilst the surrounding sediment was compacting. Macaulay (in press) has measured FI temperatures and quartz oxygen isotope ratios from these veins. The compaction of ptygmatic folds and $\delta^{18}O$ data are interpreted to indicate a growth temperature of 40° C from sea water. By contrast, the FIs in these quartz veins record a temperature mode of 119° C which compares well with present-day reservoir temperature (129° C). We infer that these FIs were formed around 40°C and have been reset during burial.

Shallow Overgrowths

Walderhaug (1990) studied quartz FIs from a Middle Jurassic sandstone in the Haltenbanken basin (offshore Norway; Ehrenberg 1990). He observed minor quartz overgrowths which pre-dated calcite cement, and major quartz overgrowths which post-dated calcite cement. Quartz overgrowths both inside and outside the calcitic sandstone show FI temperatures with indistinguishable modes of 94 and 100° C, and a maximum T_h of 118° C. The present reservoir temperature is also 118° C. Walderhaug (1990) suggests that quartz started growth above 90° C and the calcite cement post-dated this. We interpret that the FIs in these minor quartz overgrowths formed around 70° C throughout the sandstone (see Ehrenberg's compilation from this same area on Figure 1). One sandstone

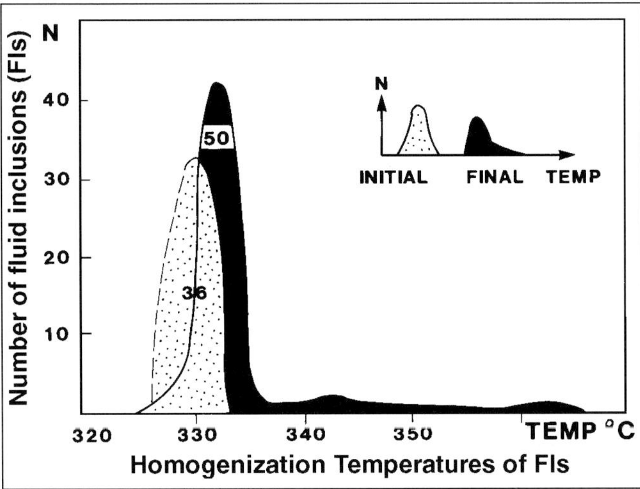

Figure 7. Histogram of homogenization temperatures of fluid inclusions in quartz before and after heating. Experimental data from Bodnar et al., 1989 showing that FI temperatures from 36 FIs before heating are cooler than 50 different FIs in the same sample after heating for one year at 40 MPa internal pressure. This shows that quartz FIs can reset slowly when heated above growth temperatures (c.f. Gratier and Jenatton 1984).

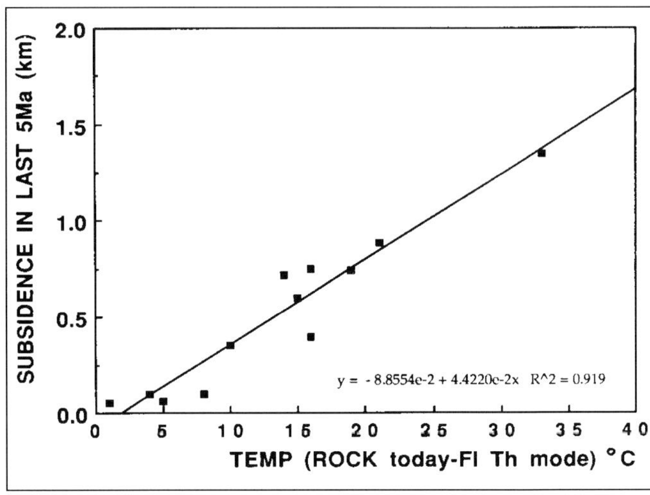

Figure 8. Plot of subsidence in last 5 m.y. against today's rock temperature minus FI temperature mode. The temperature difference between today's rock temperature and FI temperature is interpreted as a lag in resetting. This shows that a lag in FI resetting is directly proportional to rate of subsidence. Data from Table 1.

was occluded by calcite at relatively shallow depths. The other sandstone continued to cement with quartz until hydrocarbon charging. All inclusions subsided with the sandstone and their homogenization temperatures have been reset towards present-day reservoir temperatures.

An analogous effect is recorded in diagenetic calcites from the Middle and Upper Jurassic, Upper Viking Graben by Moge and Pagel (1990), where pre-quartz and post-quartz calcite cements are isotopically distinct but show identical ranges of homogenization temperatures. These have respective modes of 5° C above, and identical with, the intervening quartz (105° C).

Experimental Data

Bodnar et al. (1989) undertook experiments in which artificially grown FIs in quartz were heated from 330–370° C for one year to induce excess pressures of 40 MPa inside the inclusions. The temperature distribution of homogenization temperatures was recorded before and after these runs. A distinct increase occurred in the FI mode (Figure 7). In a different experiment, Sterner and Bodnar (1989) heated and then cooled FIs in quartz from metamorphic temperatures to induce an internal pressure difference of 100 MPa. In both cases FI temperature modes were changed. We suggest that these experiments indicate that FIs in quartz are less susceptible to resetting by overheating than those in other minerals, but that their homogenization temperature will slowly increase on heating. Inclusions may even re-form to trap cooler temperature samples of a second more saline pore fluid.

Correlation With Subsidence

We have noted that the simple plot of FI temperature against present-day depth (hence temperature) shows an overall trend of increasing FI temperature with depth (Figure 5), with a wide scatter. This should not be interpreted as a simple control of FI temperature by present-day temperature. We noted from experimental evidence that FI temperatures show a small increase after an overheating time of only one year. Therefore, we have investigated the effect of geological time on particular FI populations. We have used subsidence (geohistory) curves for the few oil fields where we also know the present-day temperature and have fluid inclusion data. These curves enable us to estimate total subsidence during the past 5 m.y. (this is a crude measure of subsidence rate). We have plotted this value against the *difference* between modal FI temperatures in quartz cement and present-day sandstone temperatures. We find a strong linear correlation (Figure 8), significant at the 99% confidence level. Oil fields that have remained within 50 m of their present depth during the past 5 m.y. show T_h that are within a few °C of today's temperature. By contrast, oil fields that have subsided more rapidly (greater than 100 m) within the last 5 m.y. show T_h populations that are too cool, and lag the present-day

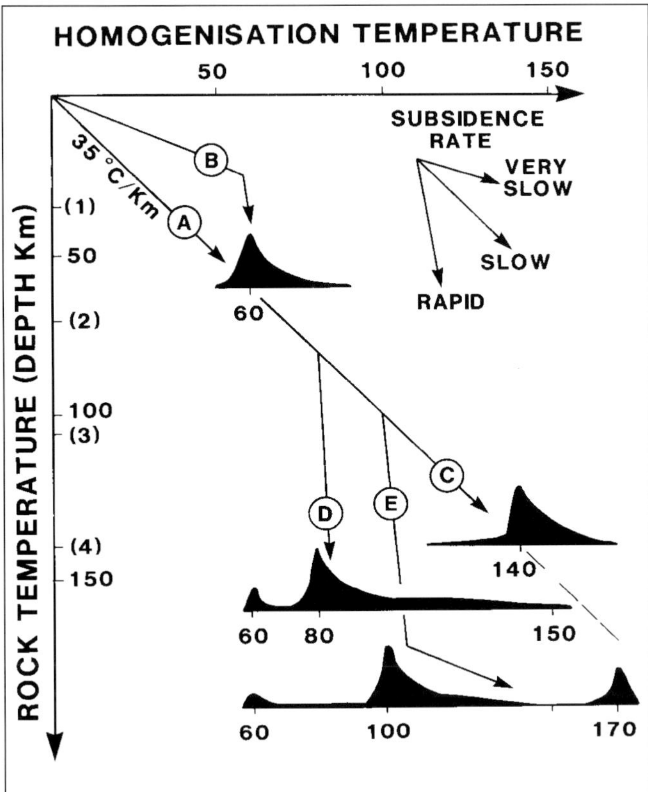

Figure 9. Schematic illustration of subsidence pathways responsible for different present-day FI homogenization temperature distributions. This assumes that quartz grows continually from 70° C, most FIs are trapped at at the grain-overgrowth boundary, and few FIs are trapped during burial; all FIs are reset. Compare actual sample data of Figure 5 with Figure 9, A and B slow and rapid arrival below 70° C threshold. C shows subsidence and resetting; D and E subsidence too fast for resetting.

temperature. The most extreme value, from the North Sea Central Graben was predicted to within 5° C by extrapolation from our slower subsiding dataset.

DISCUSSION

We believe that the monotonic increase in quartz cement volume with present-day depth is difficult to explain by an ancient mineralizing event because of the large volumes of fluid that must move through the rock and because the temperature gradient of less than 35°C /km recorded by fluid inclusion homogenization temperatures in the North Sea and Haltenbanken basins is evidence against the ascent of an ancient hot fluid cooling to precipitate quartz. The increase of quartz cement with depth makes it more probable that quartz grew (and is still growing) over a long time span, over a range of temperatures. Independent evidence for this process comes from a detailed isotopic study of the Magnus oil field, Northern North Sea (Macaulay et al., 1992), where stratified and distinct pore waters existed adjacent to each other for at least 35 m.y. This field contains the same diagenetic minerals as the others in our compilation, including 5–10% diagenetic quartz that must therefore have precipitated in a stationary pore fluid. If moving fluids did not transport diagenetic quartz, then diffusive supply of silica must be inferred. Mullis (1992) has shown numerically that diffusion is a geologically rapid process, and that the theoretical distance of silica transport is limited only by laboratory-derived solution residence times (rather than by geological information). We have presented several independent geological arguments that suggest that T_h values recorded today have been reset toward present-day temperatures. Finally, the relationship shown in Figure 7 suggests that an FI population needs increased temperature over a time span of at least 5 m.y. to complete re-equilibration. Osborne and Haszeldine (1993) discuss the resetting mechanism in detail, concluding that stretching and fluid inclusion shape change to produce a volume change of only 1–3% are quite sufficient to produce these temperature changes.

Diagenetic quartz overgrowths commence forming around a 70°C threshold temperature, perhaps depending upon the sediment's lithology and intergrain stress (overpressure). At such shallow depths, FI temperature distributions are not strongly controlled by slow or rapid subsidence (Figure 9A,B). If subsequent subsidence is slow, inclusions in a sandstone re-equilibrate and record maximum burial temperatures (Figure 9C). If the sandstone subsides rapidly, the homogenization temperatures lag behind reservoir temperatures because resetting needs time to complete (Figure 9D). Consequently these populations preserve temperature modes from depths where they were in equilibrium. If such a sandstone then subsides slowly at its maximum depth (Figure 9E), a high temperature skewed tail will gradually form by depleting the lower temperature FIs. This is the case in the Tartan field (Burley et al., 1989). The shape of an FI temperature distribution can therefore give information on pauses in subsidence and on the maximum temperature experienced, but only an approximate guide to the original growth temperature of the FI. The bulk fluid composition within the FI remains intact, although interpreted salinity values will need correction for density changes resulting from slight increases in FI volume. Relative salinities will be unaffected. We have not yet attempted extensive compilations of FI data in uplifted sandstones.

CONCLUSIONS

(1) The full temperature distribution of T_h measurements must be used when interpreting FI data. This entails at least 40 measurements for each sample depth, and no high temperature cutoff.

Table 1. Measured fluid inclusion temperatures, present-day temperature at sample depth, and subsidence rate during the last 5 m.y. Figures in column 3 represent the difference between present-day rock temperature and measured FI temperature mode. This selection is only for fields for which we have access to well locations, burial curves, temperatures, and accurately located single depth fluid inclusion temperature distributions. Thus much published data have been excluded. Temperatures and stratigraphy from Abbots (1991), Glasmann et al., 1989b (Heather field), Mark Osborne (unpublished data, 1992; Emerald field), Walderhaug (1990), Haszeldine et al. (1992; Thistle field), Hogg (1989; Alwyn South field), Glasmann et al. (1989a; Huldra field), O. McLaughlin (unpublished data, 1992; South Brae field), C. Macaulay (unpublished data, 1992; Fulmar), Burley et al. (1989; Tartan field), Ehrenberg (1990;Well I, Garn field), C. Macaulay (unpublished data, 1990; Magnus field), M. Osborne (unpublished data, 1992; UK Central North Sea, Quadrant 29 NE).

Field	Rock Temp. (°C)	FI Modal	Rock-FI	FI Depth (km)	5 m.y. Subsidence (km)
Heather	110	95	1	3.294	0.050
Emerald	60	67	5	1.650	0.060
Walderhaug	118	102	16	2.547	0.400
Thistle	104	83	21	3.400	0.881
Alwyn South	127	123	4	3.505	0.100
Huldra	150	131	15	4.000	0.600
South Brae	132	113	19	4.270	0.740
Fulmar	129	119	10	4.047	0.350
Tartan	109	101	8	3.653	0.100
Garn Well 1	143	127	16	3.702	0.750
Magnus	116	102	14	3.075	0.720
Quad 29 NE	183	150	33	5.480	1.350

(2) FI temperatures are slowly reset upwards to values in equilibrium with prevailing rock temperatures. This process takes at least 5 m.y., and may occur via stretching and shape changes.

(3) If subsidence is rapid, T_h values lag below rock temperature and preserve modal temperatures that record equilibrium at cooler temperatures.

(4) Re-equilibration at hotter temperatures gradually depletes the coolest first-formed FIs, which progressively stretch to produce a skewed tail to the population at each depth. Eventually the equilibrium FI mode is re-established within a few °C of rock temperature. This could find an application in estimating maximum burial temperatures of uplifted rocks, and in recording temperatures (hence times) of slow subsidence.

(5) There is an excellent correlation between the lag in T_h mode relative to present rock temperature, and subsidence during the past 5 m.y. This observation may have predictive value in reconstructing the most recent subsidence history, often the most important in terms of hydrocarbon generation.

(6) FI contents are intact, so that their relative salinity ranking is still feasible. However, corrections should be made if absolute salinity values are needed.

(7) There is no evidence of high paleotemperature gradients in FI populations from the North Sea and Haltenbanken Jurassic. This quartz did not grow by circulation of hot waters. We propose that quartz grew slowly in relation to burial depth, probably by some type of local diffusive supply.

ACKNOWLEDGMENTS

Mark Osborne was funded by NERC; the SURRC is supported by NERC and the Scottish Universities. Sovereign Oil provided Emerald core. Iain Samson (Windsor University, Canada) pointed us to the significance of Bodnar's work. Mark Wilkinson (Glasgow) commented on an early draft of this paper. Referees' comments by Jim Boles (Santa Barbara), Andrew Robinson, Norman Oxtoby, Andrew Hogg, Tim Primmer, and Craig Smalley (BP) improved our speculations.

REFERENCES CITED

ABBOTS, I.L. 1991, United Kingdom oil and gas fields, 25 years commemorative volume. Geological Society, London, 573 p.

BAKKER, R.J. and J.B.H. JANSEN, 1990, Preferential water leakage from fluid inclusions by means of mobile dislocations. Nature, v.345, p. 58-60.

BARKER, C.E. and R.H. GOLDSTEIN, 1990, Fluid inclusion technique for determining the maximum temperature in calcite and its comparison to the vitrinite reflection geothermometer. Geology, v.18, p.1003-1006.

BJØRLYKKE, K., T. NEDKVITNE, M. RAMM, and G.C. SAIGAL, 1992, Diagenetic processes in the Brent Group (Middle Jurassic) reservoirs of the North Sea—an overview. *In* A.C. MORTON, R.S. HASZELDINE, M.R. GILES, and S. BROWN, (Eds) Geology of the Brent Group. Geological Society

London, Bath UK. p.263-288.

BLANCHE, J.B. and J.H.McD. WHITAKER, 1978, Diagenesis of parts of the Brent Sand Formation. Journal Geological Society, London. v.135, p.73-82.

BODNAR, R.J. and P.M. BETHKE, 1984, Systematics of stretching of fluid inclusions. i). fluorite and sphalerite at one atmosphere confining pressure. Economic Geology, v.79, p.141-161.

BODNAR, R.J., P.R. BINNS, and D.L. HALL, 1989, Synthetic fluid inclusions vi). Quantitative evaluation of the decrepitation behaviour of fluid inclusions in quartz at one atmosphere confining pressure. Journal Metamorphic Geology, v.7, p.229-242.

BURLEY, S.D., J. MULLIS, and A. MATTER, 1989, Timing diagenesis in the Tartan Reservoir (UK North Sea); constraints from combined cathodoluminescence microscopy and fluid inclusion studies. Marine and Petroleum Geology, v.6, p.98-120.

BURRUS, R.C., 1987, Diagenetic paleotemperatures from aqueous fluid inclusions: re-equilibration of inclusions in carbonate cements by burial leaking. Mineralogical Magazine, v. 51, p. 477–481.

EHRENBERG, S.N., 1990, Relationship between diagenesis and reservoir quality in sandstones of the Garn Formation, Haltenbanken, Mid Norwegian Continental Shelf. American Association Petroleum Geologists Bull., v.74, p.1538-1558.

GILES, M.R., S. STEVENSON, S.V. MARTIN, S.J.C. CANNON, P.J. HAMILTON, J.D. MARSHALL, and G.M. SAMWAYS, 1992, The reservoir properties and diagenesis of the Brent Group; a regional perspective. *In* A.C. MORTON, R.S. HASZELDINE, M.R. GILES, and S. BROWN, (Eds) Geology of the Brent Group. Geological Society London, Bath UK., p.289-328.

GLASMANN, J.R., R.A. CLARK, S. LARTER, N.A. BRIEDIS, and P.D. LUNDEGARD, 1989a, Diagenesis and hydrocarbon accumulation, Brent Sandstone (Jurassic) Bergen high area, North Sea. American Association Petroleum Geologists Bull. v.73, p.1341-1360.

GLASMANN, J.R., P.D. LUNGERDARD, R.A. CLARK, B.K. PENNY, D. COLLINS, 1989b, Geochemical evidence for the history of diagenesis and fluid migration: Brent Sandstone, Heather Field, North Sea. Clay Minerals., v.24, p.255-284.

GOLDSTEIN, R.H. 1986, Re-equilibration of fluid inclusions in low temperature calcium carbonate cement. Geology. v.14, p.792-795.

GRATIER, J.P. 1982, Approche experimentale et naturelle de la deformation des roches par dissolution-cristallisation avec transfert de matiere. Bulletin de Mineralogie v.105, p. 291-300.

GRATIER, J.P. and L. JENATTON, 1984, Deformation by solution-deposition, and re-equilibration of fluid inclusions in crystals depending on temperature, internal pressure and stress. Journal of Structural Geology v.6, p.189-200.

HANOR, J.S., 1980, Dissolved methane in sedimentary brines: Potential effect on the PVT properties of fluid inclusions. Economic Geology, v.75, p.603-617.

HASZELDINE, R.S., I.M. SAMSON, and C. CORNFORD, 1984a, Dating diagenesis in a petroleum basin, a new fluid inclusion method. Nature, v.307, p.354-357.

HASZELDINE, R.S., I.M. SAMSON, and C. CORNFORD, 1984b, Quartz diagenesis and convective fluid movement: Beatrice oilfield, North Sea. Clay Minerals, v.19, p.391-402.

HASZELDINE, R.S., J.F. BRINT, A.E. FALLICK, P.J. HAMILTON, and S. BROWN, 1992, Open and restricted hydrologies in Brent Group diagenesis: North Sea. *In* A.C. MORTON, R.S. HASZELDINE, M.R. GILES, and S. BROWN, (Eds) Geology of the Brent Group. Geological Society London, Bath UK. p.401-420.

HOGG, A.J.C., 1989, Petrographic and isotopic constraints on the diagenesis and reservoir properties of the Brent Group sandstones, Alwyn South. Unpublished data PhD thesis Univ Aberdeen UK, 414p.

HOGG, A.J.C., E. SELLIER, and A.J. JOURDAN, 1992, Cathodoluminescence of quartz cements in Brent Group sandstones, Alwyn South. *In* A.C. MORTON, R.S. HASZELDINE, M.R. GILES, and S. BROWN, (Eds) Geology of the Brent Group. Geological Society London, Bath UK. p.421-440.

JOHNSON, H.D., T.A. MACKAY, and D.J. STEWART, 1986, The Fulmar oil field (Central North Sea): Geological aspects of its discovery, appraisal and development. Marine and Petroleum Geology, v.3, p.99-125.

JOURDAN, A., M. THOMAS, O. BREVART, P. ROBSON, F. SOMMER, and M. SULLIVAN, 1987 Diagenesis as the control of Brent Sandstone Reservoir properties in the Greater Alwyn area. *In* J. Brooks and K. Glennie, (eds.). Petroleum Geology of Europe. Graham and Trotman, London. p.951-961.

KONNERUP-MADSEN, J., and H. DYPVIK, 1988, Fluid inclusions and quartz cementation in Jurassic sandstones from Haltenbanken, offshore mid-Norway. Bulletin de Mineralogie, v.111, p.401-411.

MACAULAY, C.I., R.S. HASZELDINE, and A.E. FALLICK, 1992, Isotopic evidence for 35 Ma stratified pore waters; Magnus Field, North Sea. American Association Petroleum Geologists Bull., v. 76, p. 1625-1634.

MACAULAY, C.I., in press, Hot fluid inclusions from early silica veins in the Fulmar field, evidence for resetting. Marine and Petroleum Geology.

MALLEY, P., A. JOURDAN, and F. WEBER, 1986, Etude des inclusions fluides dans les nourrisages siliceux des gres reservoirs de Mer du Nord: Une nouvelle lecture possible de l'histoire diagenetique de la region d'Alwyn. Academie des Sciences Paris, v.302, p.653-658.

MOGE, M. and M. PAGEL, 1990, Petrography, fluid inclusion and stable isotopes of Jurassic carbonate-cemented sandstones (Viking Graben, North Sea). Geochemistry of the Earth's Surface, 2nd International Symposium. Aix, France. p.243-245.

MULLIS, A., 1992, A numerical model for porosity modification at a sandstone-mudstone boundary by quartz pressure solution and diffusive mass transfer. Sedimentology, v.39, p.233-239.

OSBORNE, M., and R.S. HASZELDINE, 1993, Fluid inclusions in diagenetic quartz record oilfield burial temperatures, not precipitation temperatures. Marine and Petroleum Geology, v. 10.

ROBINSON, A.G. and GLUYAS, J.G., 1992, Model calculations of porosity loss in sandstones as a result of compaction and quartz cementation: Marine and Petroleum Geology, v.9, p.319-323.

ROBINSON, A., S. GRANT, and N. OXTOBY, 1992, Evidence against natural deformation of fluid inclusions in diagenetic quartz. Marine and Petroleum Geology, v. 9, p. 568–573.

ROEDDER, E. 1984, Fluid inclusions: *In* P.H. Ribbe, (ed.). Reviews in Mineralogy., v.12, Mineralogical Society, America, 644p.

SCOTCHMAN, I.C., L.H. JOHNES, and R.S. MILLER, 1989 Clay diagenesis and oil migration in Brent Group sandstones of NW Hutton Field, UK North Sea. Clay Minerals, v.24, p.339-374.

STERNER, S.M. and R.J. BODNAR, 1989, Synthetic fluid inclusions vii. Re-equilibriation of fluid inclusions in quartz during laboratory-simulated metamorphic burial and uplift. Journal Metamorphic Geology, v.7, p.243-260.

WALDERHAUG, O., 1990, A fluid-inclusion study of quartz-cemented sandstones from off-shore Mid-Norway, possible evidence for continued quartz cementation during oil emplacement. Journal Sedimentary Petrology v.60, p.203-210.

PART 2

Diagenesis and Basin Hydrodynamics

Chapter 5

Meteoric Water and Sandstone Diagenesis in the Western Canada Sedimentary Basin

Fred J. Longstaffe
Department of Geology
University of Western Ontario
London, Ontario, Canada

ABSTRACT

A review of available stable isotope geochemistry for diagenetic minerals from several Mesozoic sandstones and conglomerates in the Western Canada sedimentary basin has permitted the general nature of pore water evolution to be traced from deposition to the present time. The importance of meteoric water throughout the evolution of these rocks has been demonstrated by this analysis. Meteoric water was present during early diagenesis of some sandstones simply because they were deposited in fluvial environments. In other cases, deposition in transitional fresh-water to marine settings resulted in a variable but nevertheless significant meteoric contribution to early pore waters. Diagenesis of the Lower Cretaceous Clearwater Formation oil sands, for example, has been substantially influenced by the prevalence of meteoric water early in their history. Meteoric water was also introduced early into sandstones and conglomerates deposited in marine environments. In many cases, this situation reflected post-depositional meteoric infiltration related to sea-level fluctuations, but the Inland Seaway itself may also have been brackish at some times.

The diagenetic mineral record also indicated that $\delta^{18}O$ values of pore waters rose during burial. This change resulted mostly from mixing between connate waters in the sandstones and more ^{18}O-enriched waters from underlying Paleozoic carbonates. Significant contribution of ^{18}O-rich shale pore waters to the sandstones (via the smectite to illite reaction) could not be demonstrated.

The compositions of late diagenetic minerals confirmed that ^{18}O-enrichment of sandstone pore waters was reversed by large-scale, gravity-driven introduction of meteoric water—a response to uplift of the Western Canada sedimentary basin in early Eocene (post-Laramide) time. The influx of low-^{18}O water dominated late diagenesis of most sandstones, resulting in formation waters largely controlled by mixing between the post-Laramide, meteoric flux and earlier, more evolved pore waters.

INTRODUCTION

The fact that water of meteoric origin comprises a significant fraction of formation waters in many sedimentary basins has been amply demonstrated, particularly through the use of oxygen and hydrogen isotopes (Clayton et al., 1966). How and when meteoric water or its evolved equivalents came to occupy the pore space of these sandstones, and whether it mixed with, or entirely replaced, earlier pore waters, are issues of some importance (Graf et al., 1965; O'Neil et al., 1986; Knauth and Beeunas, 1986; Knauth, 1988). The Western Canada sedimentary basin provides a classic example of the substantial role that meteoric water can play during sandstone diagenesis. The most commonly recognized involvement is gravity-driven flow that began after sedimentation ceased, and uplift related to the Laramide Orogeny produced substantial topography in the early Eocene (Hitchon, 1969a,b; Tóth, 1980; Schwartz and Longstaffe, 1988).

The Western Canada sedimentary basin (see Macqueen and Leckie, 1992, figure 1) contains a thick sequence of Paleozoic carbonates and evaporites overlain by sandstones and shales of mostly Cretaceous to Tertiary age. Along its western margin, the basin can be up to 5000 m thick, but its thickness (and elevation) steadily decrease toward the east-northeast until Precambrian basement rocks are exposed in northeastern Alberta and in northern Saskatchewan and Manitoba. Several researchers have used potentiometric data to demonstrate the existence of flow systems with recharge in major highlands and discharge in major lowlands on both regional and local scales (Hitchon, 1969a,b; Tóth, 1978). Expansion of these meteoric, gravity flow systems to deep within the basin has resulted in mixing with, and/or flushing, of pre-existing formation waters of varying origin.

The development of hydrodynamic conditions within the Western Canada sedimentary basin has further implications. Some believe that northeastward flow of deeper groundwater initiated by uplift was preferentially concentrated in highly permeable Paleozoic carbonates. This "regional aquifer" drained source rocks, and transported and discharged hydrocarbons that comprise, in degraded form, the bitumen and heavy oil accumulations of eastern and northeastern Alberta (Hitchon, 1969b, Garven, 1989). Equally important are the chemical reactions and mass transport related to moving groundwater in the basin; advection and dispersion undoubtedly exerted a powerful control on diagenesis during the post-Laramide influx of meteoric water (Schwartz and Longstaffe, 1988).

Flow systems involved in diagenesis, especially its earlier stages, commonly have been supplanted by later events. However, the paragenesis of diagenetic minerals, coupled with their oxygen, hydrogen, carbon and/or strontium isotope compositions, can retain signatures characteristic of previous fluids entertained by sediments at a given locality. By extending such observations both vertically and laterally, some understanding of pore water evolution and its relationship to basin development through time and space, can be realized.

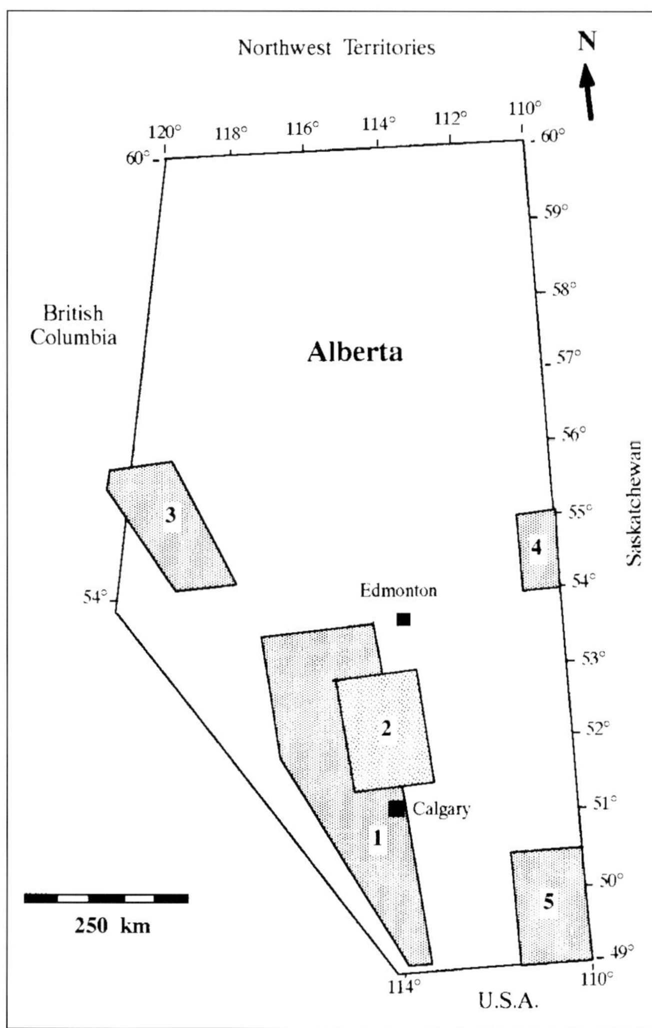

Figure 1. Location of study areas: 1–Belly River/Brazeau Group; 2–Viking Formation; 3–Alberta Deep Basin, including Marshybank and Cardium formations; 4–Clearwater Formation, Cold Lake area; 5–Milk River Formation.

This chapter reviews primarily stable isotope data available for diagenetic minerals from dominantly Mesozoic sandstones located in Alberta (Figure 1). How reliable are such data? Silicate minerals formed in equilibrium with diagenetic pore water obtain and retain oxygen isotope compositions characteristic of that fluid and crystallization temperature (O'Neil, 1987). Carbonates, and in particular calcite, are generally considered to be more susceptible to oxygen isotopic exchange (Clayton, 1959; Clayton et al., 1966) because of their greater tendency to recrystallization. Such behavior can be especially important if the carbonate formed early in diagenesis, prior to prograde (higher temperature) diagenesis (Longstaffe, 1983, 1987, 1989). Interpretation of hydrogen isotope data for clay minerals is more equivocal. Some data imply that hydrogen isotope exchange between clay minerals and sandstone pore waters can occur without a corresponding effect on the oxygen isotope composition of the clay (Wilson et al., 1987; Bird and Chivas, 1988; Longstaffe and Ayalon, 1990).

The following examples are offered as illustrations of three main points. First, meteoric water was abundant during early diagenesis of sandstones, even in units deposited in shallow marine environments. Second, meteoric water continued to play an important role during burial diagenesis. Third, meteoric water was an extremely important component of late diagenetic formation waters, introduced in the early Eocene as a consequence of the second (major) pulse of the Laramide orogeny.

EARLY (SHALLOW) DIAGENESIS

The presence of low-^{18}O pore waters during crystallization of early diagenetic siderite, calcite, chlorite, kaolinite, berthierine and/or smectitic clays has been recognized in conglomerates, sandstones and siltstones throughout Alberta. Localities include the gas-prone "Deep Basin" to the northwest, oil and gas-rich Cretaceous sandstones of central and southwestern Alberta, and much less deeply buried oil sands of northeastern and eastern Alberta (Figure 1) (Longstaffe, 1983, 1989; Longstaffe and Ayalon, 1987; Ayalon and Longstaffe, 1988; Machemer and Hutcheon, 1988; Tilley and Longstaffe, 1989; McKay et al., 1989; Bloch, 1990; Hart et al., 1992; Longstaffe et al., 1992a,b).

Some of these sandstones were deposited in continental settings, hence involvement of meteoric water is expected. Others were deposited in deltaic and coastal plain environments, where significant contribution of meteoric water early in diagenesis is also anticipated. But even sandstones deposited under marine conditions commonly contain minerals whose compositions indicate that meteoric waters were abundant early in diagenesis. In the examples that follow, we explore the regional influence and retained memory of connate meteoric waters in the diagenetic record, consider depositional versus early diagenetic but post-depositional origins for meteoric water, and finally illustrate the significant influence that meteoric water has had on the evolution of the oil sands.

Continental, Deltaic and Marine Sandstones: A Comparison

Longstaffe and Ayalon (1991) compared oxygen isotope compositions for diagenetic minerals from continental sandstones of the Upper Cretaceous, Belly River/Brazeau Group, underlying deltaic, nearshore and shoreline deposits of the basal Belly River Group sandstone (Longstaffe, 1986; Ayalon and Longstaffe, 1988), and dominantly marine sandstones of the Lower Cretaceous Viking Formation (Longstaffe and Ayalon, 1987), all from west-central and southwestern Alberta (Figure 1). Their main objectives were to deduce (1) whether connate water compositions had been preserved in the isotopic record provided by early diagenetic minerals, and (2) whether connate pore waters had influenced the isotopic compositions of later diagenetic minerals.

The continental sandstones of the Belly River/Brazeau Group were deposited in a fluvial setting typical of a low-lying coastal flood plain (Lerbekmo, 1963). Diagenesis was dominated by clay minerals, including Fe-rich chlorite ($\delta^{18}O$ = +3.3 to +9.3 ‰), illite-smectite (I/S, %I >> %S), chlorite-smectite (C/S) and illite ($\delta^{18}O$ = +9.5 to +10.7 ‰), kaolinite ($\delta^{18}O$ = +5.9 to +12.7 ‰), and illite-smectite (S/I, %S >> %I) and smectite ($\delta^{18}O$ = +11.0 to +14.0 ‰). Other phases included quartz overgrowths ($\delta^{18}O$ = +12.0 to +21.0 ‰) and calcite ($\delta^{18}O$ = +8.5 to +12.9 ‰; $\delta^{13}C$ = -14.3 to -2.0 ‰). Chlorite crystallized first, followed by quartz, I/S (≈40% smectitic layers, R1 ordering), calcite and kaolinite, and then S/I (≈80% smectitic layers, R0 ordering) and smectite. Highest temperatures (≈110°C) were reached in the southwestern portion of the study area, where R1-ordered I/S was detected.

Pore-water Evolution

Crystallization from meteoric water is strongly implied by the very low $\delta^{18}O$ values of early chlorite. If shallow diagenesis began at temperatures as low as 10±5°C, chlorite formation was possible from waters with $\delta^{18}O$ values as low as -16 ‰ (Stage I, Figure 2A). Shallow groundwater in the area, for comparison, has $\delta^{18}O$ values of -19 to -15 ‰ (Hitchon and Friedman, 1969; Schwartz et al., 1981). Chlorite formation most likely continued during burial, accompanied and followed by diagenetic quartz, as temperature and the $\delta^{18}O$ value of the pore water gradually increased (Stage II, Figure 2A).

The oxygen isotope composition of the pore water may have risen during burial diagenesis because of (1) external contribution of more ^{18}O-rich waters, or (2) sandstone-water interaction, such as alteration of rock fragments and feldspars, and isotopic exchange with ^{18}O-rich carbonate grains and cements (see "Burial Diagenesis"). The $\delta^{18}O$ results for diagenetic minerals formed during burial are entirely consistent with such a increase, but the magnitude of the ^{18}O-enrichment is difficult to estimate. Calculated pore water $\delta^{18}O$ values are as high as -3±1 ‰, assuming maximum temperatures of 110°C for crystallization of R1-ordered I/S (Figure 2B). In this system, such values for pore water would have been difficult to produce by rock-water interaction alone.

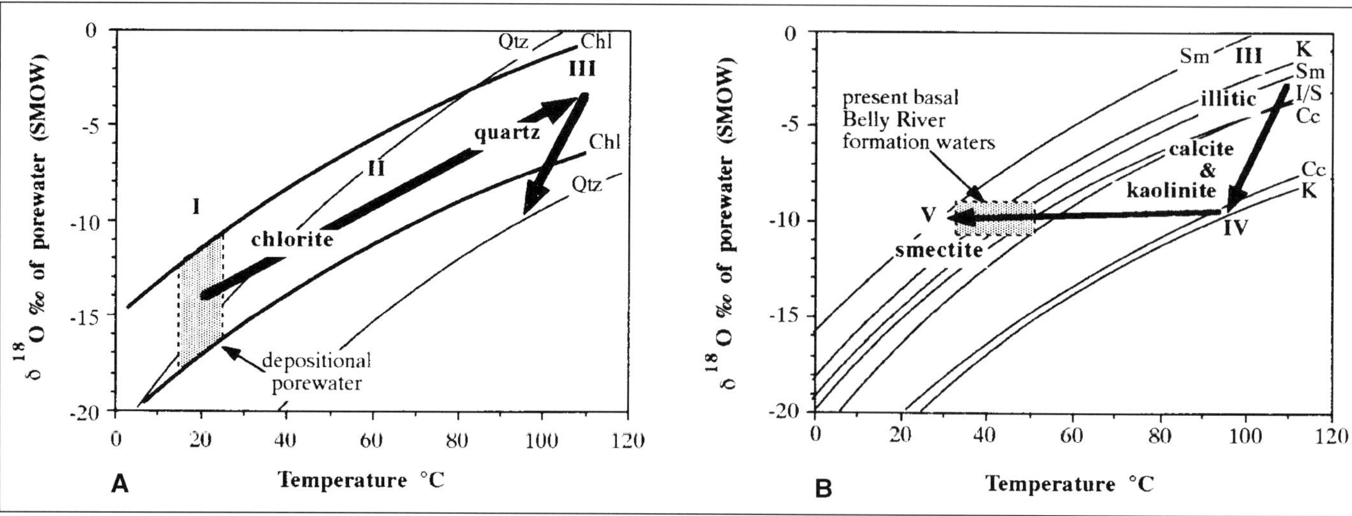

Figure 2(A). Interpretative porewater pathway (arrows) during crystallization of chlorite and quartz in continental sandstones of the Belly River/Brazeau Group. The pathway has been deduced using the paragenetic sequence and the range of oxygen isotope compositions obtained for each phase: Stage I, early (shallow) diagenesis; Stage II, burial diagenesis; Stage III, maximum burial. The curved lines (Chl) indicate solutions to the chlorite-water geothermometer of Savin and Lee (1988) for maximum and minimum chlorite $\delta^{18}O$ values. Similar curves are shown for quartz (Qtz), following Clayton et al. (1972), as modified by Friedman and O'Neil (1977). A temperature of 20±5°C was assumed for depositional waters (shaded area). After Longstaffe and Ayalon (1991). (B). Continuation of porewater pathway (arrows) from Figure 2A, tracing crystallization of late diagenetic I/S, kaolinite, calcite and smectitic clays from maximum burial (Stage III) through erosion (Stage IV) to present subsurface conditions (Stage V). A maximum burial temperature of 110°C has been assumed. Mineral-water geothermometers utilized were: illite-smectite (I/S, Sm), Yeh and Savin (1977), as corrected by Savin and Lee (1988); kaolinite (K), Land and Dutton (1978); calcite (Cc), O'Neil et al. (1969), as modified by Friedman and O'Neil (1977). Present formation water composition is indicated by the shaded area. After Longstaffe and Ayalon (1991).

A shift in pore water composition to lower $\delta^{18}O$ values, beginning at maximum burial (Stage III, Figure 2A) and accompanied by cooling as erosion of the uplifted basin proceeded (Stage IV, Figure 2B), is needed to explain the results for later diagenetic quartz, calcite and kaolinite. This behavior, and the crystallization of the paragenetically latest smectitic clays, has generally been attributed to regional, post-Laramide influx of meteoric water (see "Late Diagenesis").

Influence of Connate Water

The $\delta^{18}O$ values of equivalent diagenetic minerals from the continental Belly River/Brazeau Group sandstones, underlying deltaic, shoreline, and nearshore basal Belly River Group sandstones and marine sandstones of the Lower Cretaceous Viking Formation are compared in Figure 3. The average value for a given phase, whether formed early or late in diagenesis, is always highest in the Viking Formation and lowest in the Belly River/Brazeau Group. The ranges in $\delta^{18}O$ values for the same phases from continental and basal Belly River/Brazeau Group sandstones are very similar, but on average, slightly lower values typify continental deposits.

For minerals formed early in diagenesis, these differences are most simply explained by distinct compositions for depositional pore waters. Meteoric water would have been dominant in the continental sandstones whereas connate water in the underlying deltaic and shoreline sandstones would have originated from mixing of meteoric water and local sea water. The similarity in oxygen isotope compositions for early diagenetic chlorite throughout the Belly River/Brazeau Group sandstones suggests that fresh water rather than (Campanian) sea water was dominant in the basal sandstones, at least in facies where early chlorite was most abundant (distributary channels, etc.; Storey, 1982; Longstaffe, 1986).

In contrast, connate water in Viking Formation sandstones was probably dominated by (Albian) sea water—hence the ^{18}O enrichment of early diagenetic chlorite and quartz (by 4 to 6 ‰) relative to equivalent phases from the Belly River/Brazeau Group (Figure 3). An even larger difference might have been anticipated, except that (i) the Viking Formation sandstones may have experienced contribution of meteoric water at least locally during early diagenesis; and (ii) the Inland Seaway may have been brackish during Viking Formation deposition. Both possibilities have merit. In

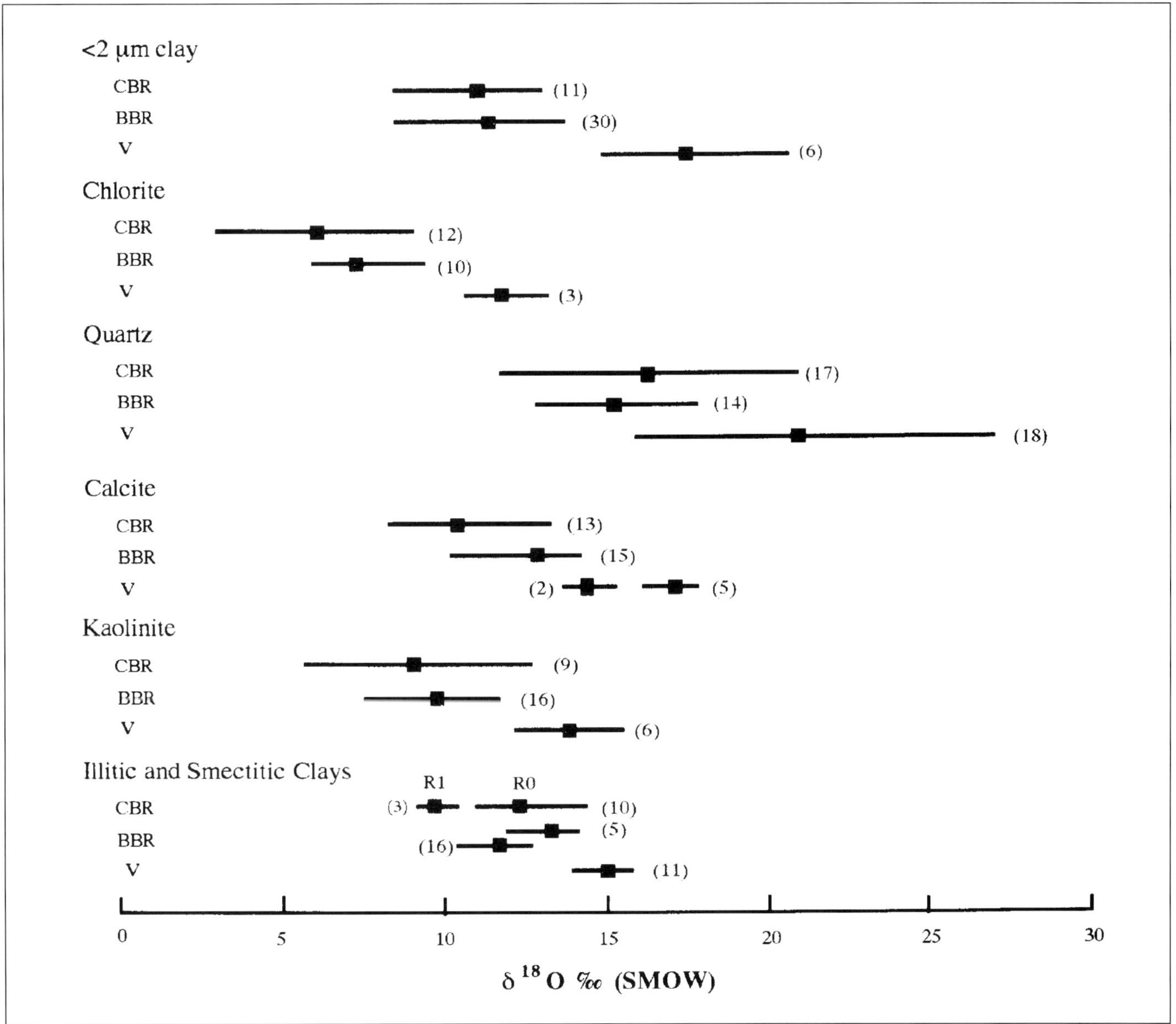

Figure 3. Range (bars) and mean (squares) $\delta^{18}O$ values for diagenetic minerals from continental sandstones of the Upper Cretaceous, Belly River/Brazeau Group (CBR), mixed meteoric-marine sandstones of the basal Belly River Group (BBR) and marine sandstones of the Lower Cretaceous, Viking Formation (V). The number of samples analyzed in each case is indicated in parentheses. After Longstaffe and Ayalon (1991).

the former case, Longstaffe and Ayalon (1987) commented on the occurrence of rooted intervals in some portions of the Viking Formation (see below). In the latter case, Kyser et al. (1993) have suggested that the Inland Seaway was decoupled from the open ocean at various times throughout the Cretaceous, and attained $\delta^{18}O$ values as low as -5 to -4 ‰ during the Greenhorn marine cycle in Saskatchewan. This timing would correspond well with the appearance of low ^{18}O early diagenetic minerals in marine Viking Formation sandstones, as well as Paddy Member tidal sediments from the Alberta Deep Basin (Longstaffe et al., 1992a).

Minerals formed later in diagenesis (illitic and smectitic clays, kaolinite, calcite) show the same pattern of oxygen isotope compositions. The highest values occur in the Viking Formation and the lowest values are found in continental sandstones of the Belly River/Brazeau Group, although the relative difference is smaller (2 to 3 ‰) than for early diagenetic minerals (Figure 3). The $\delta^{18}O$ values of present formation waters (Viking Formation sandstone: -9.5 to -5.6 ‰; basal Belly River Group sandstone: -10.4 to -9.2 ‰) confirm that the variation implied by the results for the diagenetic minerals is real. Our interpretation of this pattern is that some memory of differences in connate pore water composition has persisted throughout diagenesis, despite ^{18}O-enrichment during burial and then post-Laramide dilution by meteoric water.

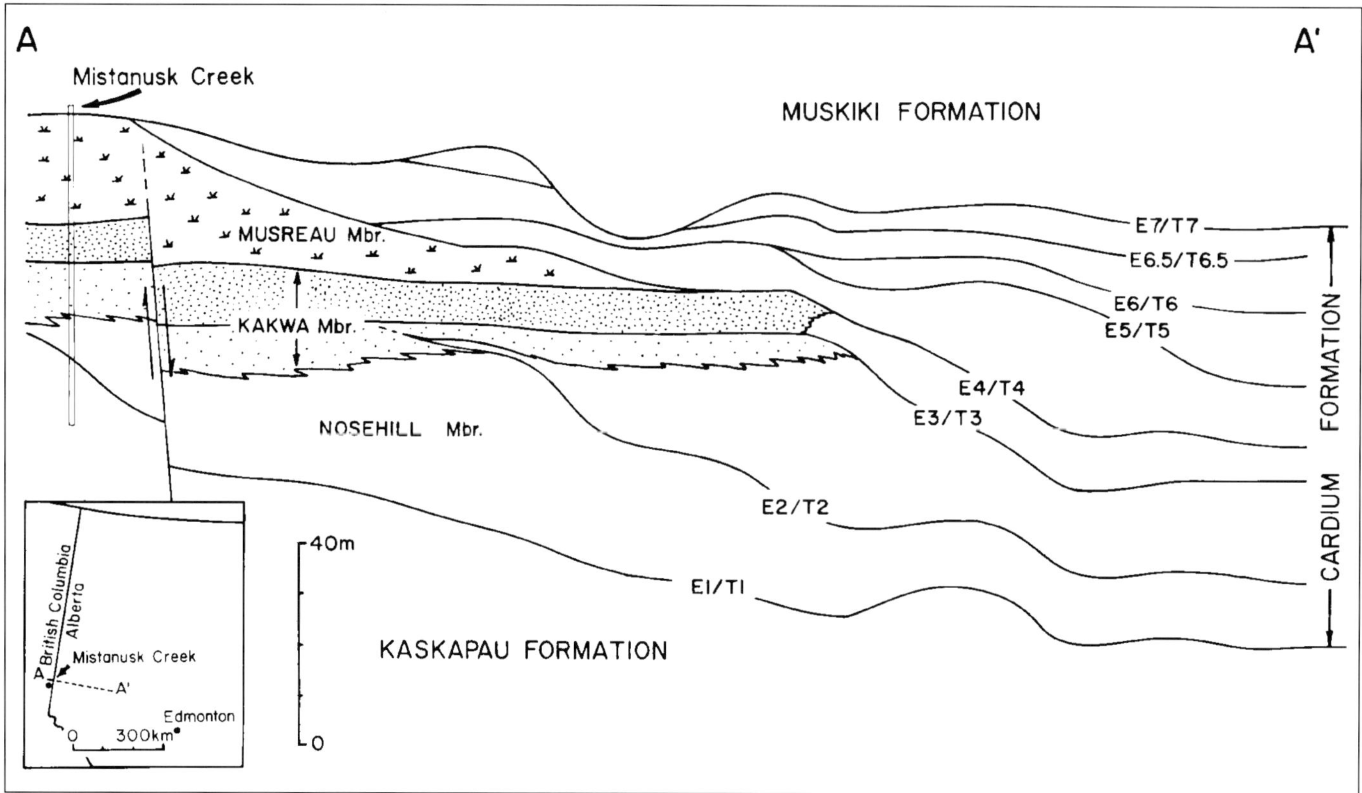

Figure 4. Summary stratigraphy of the Cardium Formation in the northern Rocky Mountain Foothills, as presented by Hart (1990), and modified from Plint et al. (1986, 1988). The numbered erosion surfaces separate the Cardium Formation into shoaling-upward successions. Shoreface sandstones of the Kakwa Member (stippled pattern) are overlain by the non-marine Musreau Member. From Hart et al. (1992).

Available chemical and isotopic data for current formation waters from these units support this observation. For example, Connolly et al. (1990a,b) concluded that the chemical and isotopic compositions of Viking Formation pore waters were best explained by mixing of evolved sea water and post-Laramide meteoric waters which had been modified by leaching of feldspar and clay minerals (see "Late Diagenesis"). Similar data for the basal Belly River Group pore waters contained no evidence for an earlier sea water component but such compositions do not necessarily indicate complete flushing of early pore waters. Connate waters in the basal Belly River Group sandstone were initially dominated by freshwater; sea water or its derivatives may never have comprised a significant fraction of the pore water at any stage of diagenesis (Longstaffe, 1986).

Domenico and Robbins (1985) showed that a steady-state mixing zone can persist between connate and introduced meteoric waters in deep, active flow systems, citing the Milk River Aquifer from southeastern Alberta as one example (Figure 1) (see also Longstaffe, 1984). The oxygen isotope data for sandstones from the Western Canada sedimentary basin in general support a conclusion that meteoric flushing of (evolved) connate waters, whether they be brines, sea water or freshwater, has not been complete (Longstaffe, 1984; Longstaffe et al., 1991).

Depositional Controls and Sea-Level Fluctuations

Depositional environment should exert an extremely important control on the composition of pore waters and associated early diagenetic phases. For example, in shallow marine clastic units, predictable changes should accompany the succession of transgressive and regressive facies that resulted from relative changes in sea level. Likewise, early diagenetic minerals in a facies assemblage caused by progradation into marine waters should show a systematic change in their oxygen isotope compositions as the influence of the freshwater phreatic lens waned seaward.

Hart et al. (1992) made a preliminary test of this idea, using early diagenetic siderite from an outcrop of Cardium Formation sandstone in the northern Rocky Mountain Foothills (Figure 4). They found that facies for which a greater freshwater influence is indicated from sedimentological data contain siderite with lower $\delta^{18}O$ values than more marine varieties (Figure 5). Similar observations have been reported for subsurface samples of Cardium Formation siderite from northwestern Alberta (Longstaffe et al., 1992a)

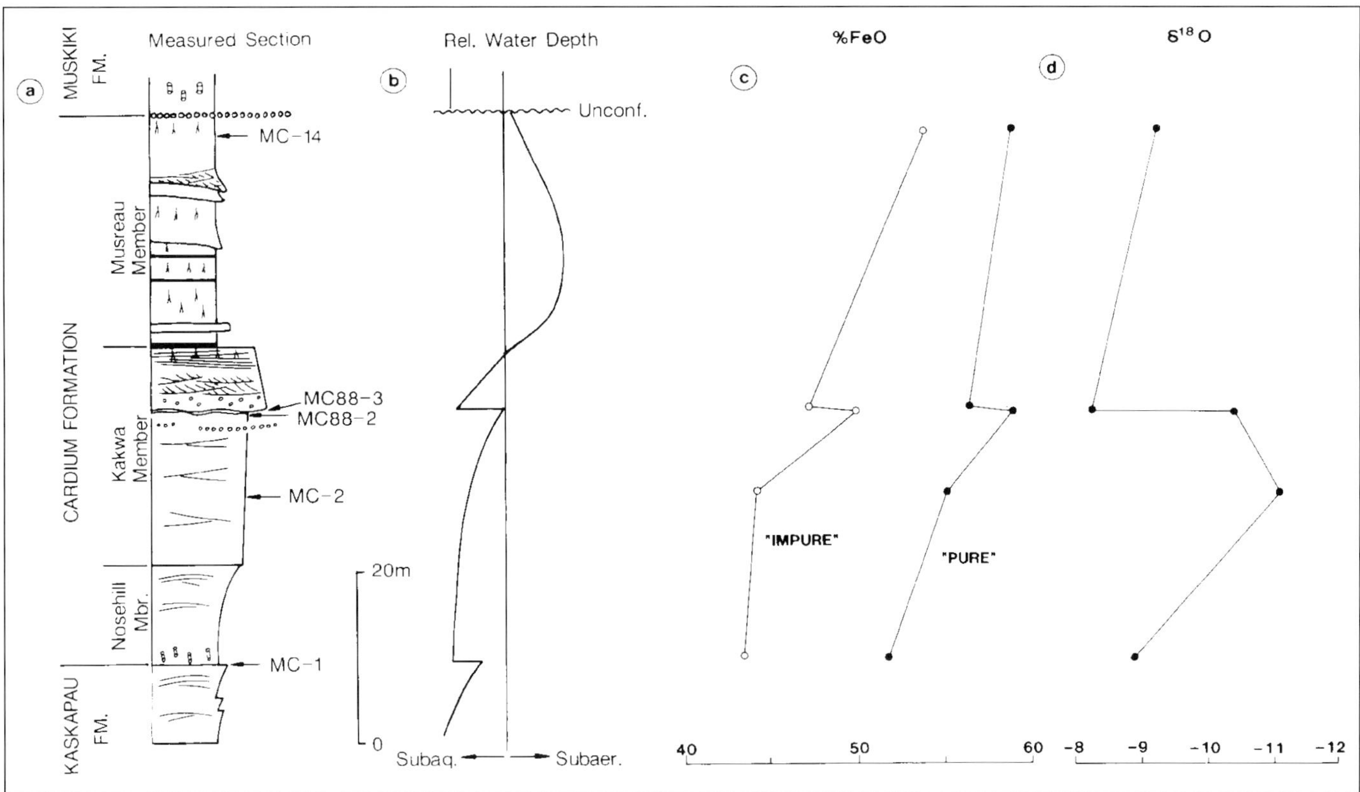

Figure 5(a). Stratigraphic section and sample locations for the Cardium Formation at Mistanusk Creek (see Figure 4 for location). (b). Relative water depth curve for preserved deposits, as deduced by Hart (1990) from stratigraphy and sedimentology. (c). Variations in molar percent FeO of siderite with stratigraphic depth. Compositions are given for the two early diagenetic varieties ("impure" and "pure") of siderite in each sample. (d). Variations in the $\delta^{18}O$ values (‰ PDB) of early diagenetic siderite with stratigraphic position. According to Hart et al. (1992), the decrease from sample MC-1 to MC-2 reflects an increasing freshwater influence as the shoreline prograded. The similarity in $\delta^{18}O$ values between samples MC-1 (near the base of the section) and MC88-3 (from the sandstone that draped the ravinement surface) is compatible with a return to deeper water conditions following transgression. The $\delta^{18}O$ value of the uppermost sample (MC-14) is consistent with a return to fresher water conditions, as would be anticipated for the non-marine Musreau Member, but the relatively small change indicates that brackish water conditions prevailed. From Hart et al. (1992).

(Figure 1). The generally Fe-rich nature of siderite from fresher water-dominated facies can be invoked to support the isotopic data. Freshwater siderite is commonly (but not always) characterized by a "purer" chemical composition (i.e., less Mg, Ca substitution for Fe) than more marine varieties (Mozley, 1989a,b). Hart et al.'s (1992) results are encouraging, although more detailed investigation is clearly required to properly evaluate this hypothesis.

Lowering of sea level, and associated development of emergent zones, may also have caused post-depositional, early infiltration of meteoric water into shallow marine sediments of the Western Canada sedimentary basin (Machemer and Hutcheon, 1988; McKay et al., 1989; Wickert et al., 1989; Longstaffe et al., 1992a). The cyclic nature of sea level fluctuation may have triggered many such invasions, resulting in a substantial flux of meteoric water during early diagenesis. Evidence for such behavior has been documented for two early phases of diagenetic siderite in marine shelf sandstones of the Upper Cretaceous Marshybank Formation (northwestern Alberta–northeastern British Columbia, Figure 1) (McKay, 1992; McKay, personal communication). At any particular location, the earliest siderite has considerable Ca-Mg substitution, $\delta^{18}O$ values (+28 ‰ SMOW) typical of precipitation from pore waters of marine origin, and $\delta^{13}C$ values (+5 ‰ PDB) suggestive of incipient microbial fermentation. The later (but still early) generation of siderite has much lower $\delta^{18}O$ (and $\delta^{13}C$) values that can only be explained by mixing between meteoric water and pre-existing connate waters of dominantly marine origin. A basinward waning of the post-depositional flux of meteoric water is indicated by data for siderite from several localities across the study area.

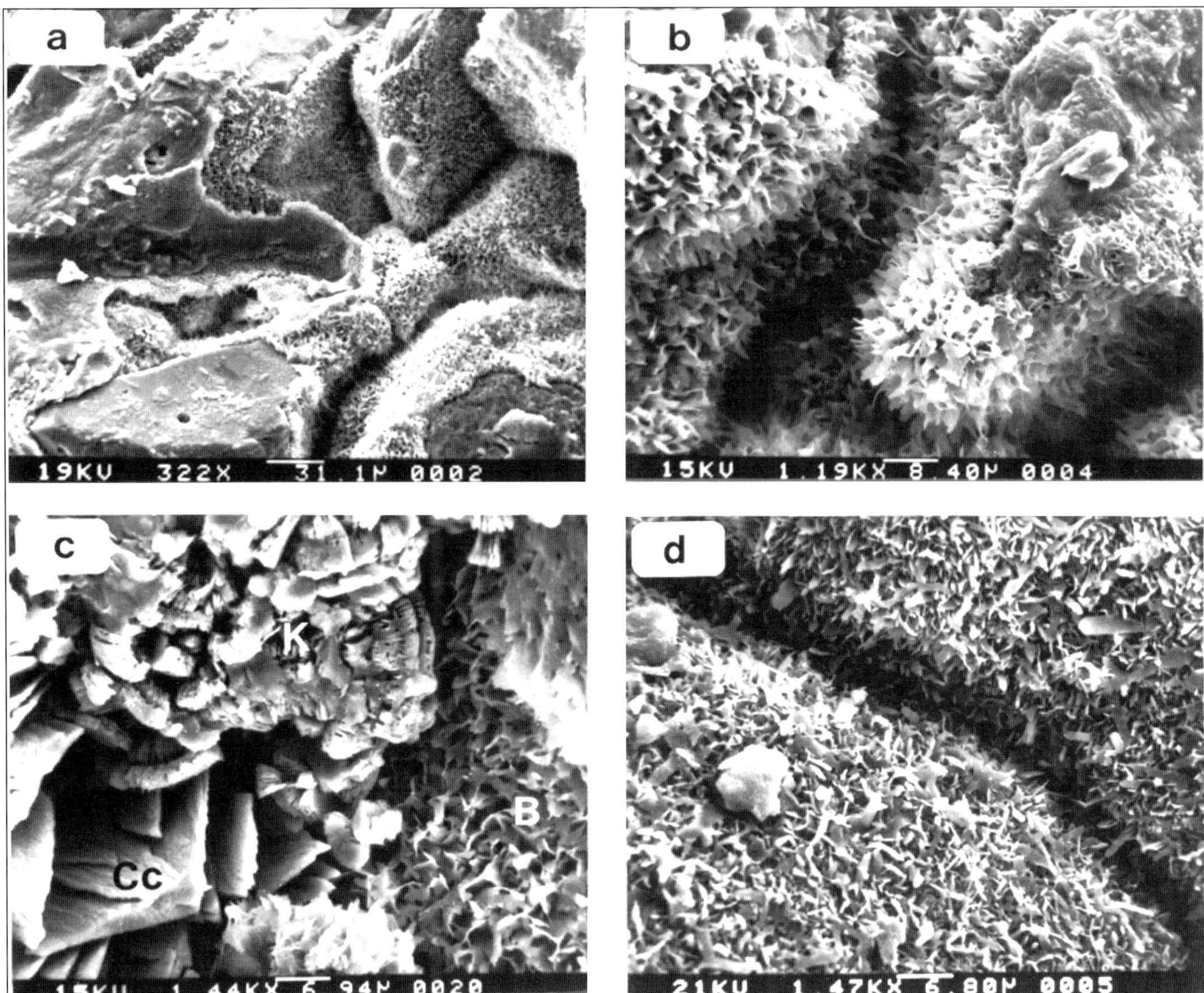

Figure 6(a). Grain-coating clays from bitumen-saturated sands of the Clearwater Formation; XRD results indicate that this sample is primarily a mixture of berthierine (75%) and smectitic clays (22%). (b). Magnified view of Figure 6a; primarily grain-coating berthierine is shown here. (c). Grain-coating clays (B = smectite/chlorite >> berthierine), diagenetic calcite (Cc) and rare diagenetic kaolinite (K) from bitumen-saturated sands of the Clearwater Formation. (d). Mixture of grain-coating smectitic clays and berthierine from bitumen-saturated sands of the Clearwater Formation; XRD data suggest that swelling clays comprise ≈85% of this assemblage.

Whether freshwater involved in crystallization of these low ^{18}O early diagenetic siderites (or equivalent phases) was of syndepositional or early, post-depositional origin remains difficult to know unequivocally. Hart et al. (1992) rejected a post-depositional influx because the exteriors of the compositional zoned, early diagenetic siderite are enriched in Ca and Mg relative to the cores, opposite to the pattern predicted if subsequent introduction of meteoric water had occurred (Mozley, 1989a,b). The universal applicability of such a test, however, remains uncertain since the chemical evolution of siderite may also depend on other parameters, such as development of pyrite even earlier in the paragenesis (McKay, 1992).

Clearwater Formation Oil Sands, Eastern Alberta

The Lower Cretaceous oil sands of the Clearwater Formation, Cold Lake area (Figure 1), provide a particularly significant example of the role that meteoric water has played during early diagenesis of many sands in the Western Canada sedimentary basin (Longstaffe, 1989; Racki, 1991; Longstaffe et al., 1992b). Moreover, because the Clearwater Formation deposits have not been as deeply buried as units further west, and because they were not so overwhelmingly affected by the post-Laramide diagenetic overprint (see "Late Diagenesis"), their early diagenetic character is more easily discerned.

The Clearwater sands contain an estimated 69 billion barrels of heavy oil or bitumen in place, and form part of Canada's enormous Mannville Group oil sand reserves. Because the API gravity of the hydrocarbons is so low, in situ thermal techniques such as steam stimulation are commonly used to mobilize bitumen for production. The permeability-altering, hot water-rock interactions that accompany this process are greatly influenced by the composition and distribution of diagenetic minerals, especially grain-coating berthierine and smectitic clays that dominate the portion of the Clearwater Formation described here.

The Clearwater Formation was deposited in a nearshore, deltaic complex, which prograded in a north-northeasterly direction (Harrison et al., 1981; Wightman et al., 1989). It varies in thickness from 35 to 65 m, thicker intervals resulting from stacking of successive deltaic lobes. Some of the highest quality oil sands occur in distributary channel and stream mouth-bar facies in the south-central portions of the complex. Further seaward, shales become more prominent, and interbedded sands are thin. The sands consist mainly of (feldspathic) litharenites, dominated by rock fragments (mostly volcanic, chert and shale). Thin calcite ± siderite-cemented intervals are ubiquitous. Present burial depths vary from ≈400 to 500 m, and post-Laramide erosion has removed no more than ≈1000 m of overlying strata (Hacquebard, 1977; Beaumont et al., 1985).

In bitumen-saturated sands, early diagenetic minerals include glauconite, di- and trioctahedral smectitic clays, berthierine, pyrite, siderite, and calcite (Figure 6). Alteration and partial dissolution of volcanic rock fragments and feldspars are abundant. Locally, concretionary carbonate cementation (mostly calcite) has partially replaced framework grains and filled pores, early in diagenesis. Glauconite/illite and/or smectitic clay contents commonly increase, and berthierine contents decrease, as calcite-cemented intervals are approached. Early framboidal pyrite postdated this stage of calcite cementation.

Berthierine is confined almost entirely to the south-central portions of the study area, commonly in stream-mouth bar and related facies, the same general area where Fe-poor, grain-coating calcite (with high $\delta^{13}C$ values) also appeared relatively early in diagenesis. Crystallization of euhedral pyrite, K-feldspar and silica overgrowths, siderite, zeolites, and in the northern portion of the deltaic complex, kaolinite (Figure 6c), also accompanied burial diagenesis, at the same time as continued dissolution of feldspars and volcanic rock fragments. Abercrombie (1989) suggested that the diagenetic quartz-K feldspar-zeolite assemblage represented maximum burial at about 80°C. Somewhat lower maximum burial temperatures (60-70°C) have been suggested from stratigraphic reconstructions (Longstaffe et al., 1992a).

The latest stages of diagenesis are commonly marked by Fe-rich calcite, and in the fringes of the deltaic complex, kaolinite. Kaolinite is much rarer within the thick, bitumen-saturated sands of the south-central portions of the delta; it formed only within calcite and/or siderite-rich zones of low bitumen-saturation or in sands close to the bitumen-water contact. Emplacement of hydrocarbons (subsequently altered to bitumen or heavy oil) greatly restricted further diagenetic mineralization.

Sands from the Clearwater Formation that remained water-saturated show the same general diagenetic pattern with a few important exceptions. Grain-coating clays like berthierine are rare, and early diagenetic, grain-coating calcite with high $\delta^{13}C$ values is almost never encountered.

Pore-water Evolution

A large range in $\delta^{13}C$ (-7.9 to +17.1 ‰) and to a lesser extent, $\delta^{18}O$ (+16.3 to +23.1 ‰) values was obtained for diagenetic calcite (Figure 7). Samples with "marine" $\delta^{13}C$ values (0±2 ‰) are most characteristic of early calcite in water-saturated and calcite-cemented sands from the edges of the deltaic complex, but also are present in south-central portions, mostly in upper parts of the bitumen-bearing interval. However, despite the "marine" carbon-isotope signature, pore waters were at least brackish; the $\delta^{18}O$ values of early cements (+20 to +23 ‰) are much lower than expected for low temperature crystallization from open marine, Cretaceous sea water.

Grain-coating diagenetic calcite, which is most abundant in lower portions of the bitumen- saturated, berthierine-bearing oil sands, is characterized by higher $\delta^{13}C$ and lower $\delta^{18}O$ values than the somewhat earlier calcite. Strongly positive $\delta^{13}C$ values in diagenetic carbonates are one by-product of microbial fermentation of organic matter (Irwin et al., 1977); such values have been previously reported for oil sands (Dimitrakopoulos and Muehlenbachs, 1987). That the $\delta^{18}O$ values of this calcite decreased as $\delta^{13}C$ values rose is consistent with increasing temperature during progressive burial and fermentation (Figure 7).

The low $\delta^{18}O$ values of the last generation of calcite indicate that a sizable fraction of meteoric water was also involved in its crystallization (Figure 7); meteoric water persisted throughout all stages of calcite cementation. However, this calcite commonly has $\delta^{13}C$ values < 0 ‰ (Figure 7). Low ^{13}C sources for carbonate ions may have included alteration of hydrocarbons during the aerobic microbial action and water-washing responsible for the current degraded conditions of the oil sands.

Oxygen isotope data were obtained for diagenetic smectitic clays ($\delta^{18}O$ = +18.0 to +18.8 ‰), berthierine (+6.6 to +12.5 ‰), K-feldspar (+15.9 to +17.0 ‰) and quartz (+21.0 to +23.2 ‰). These data (plus results for

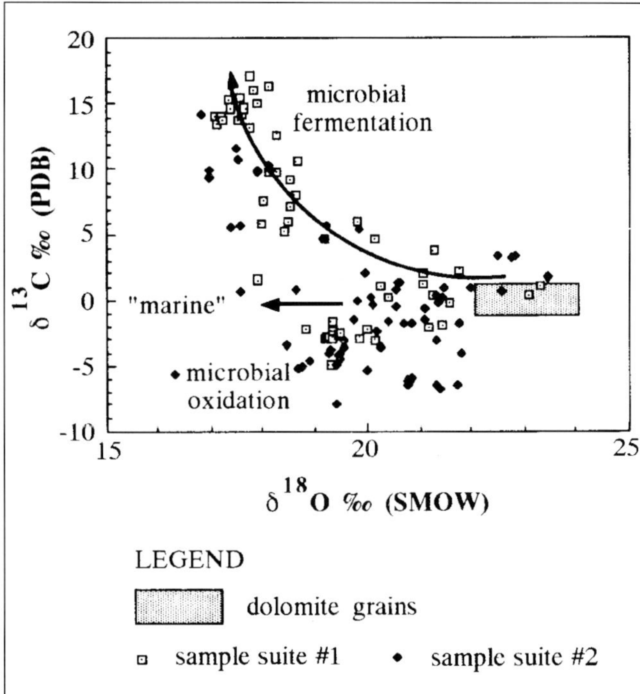

Figure 7. Carbon and oxygen isotope compositions of diagenetic calcite from Clearwater Formation sands and sandstones in the Cold Lake area. Samples from Suite #1 were restricted to oil-sands pilot sites; suite #2 contained samples collected throughout the deltaic complex. After Longstaffe et al. (1992b).

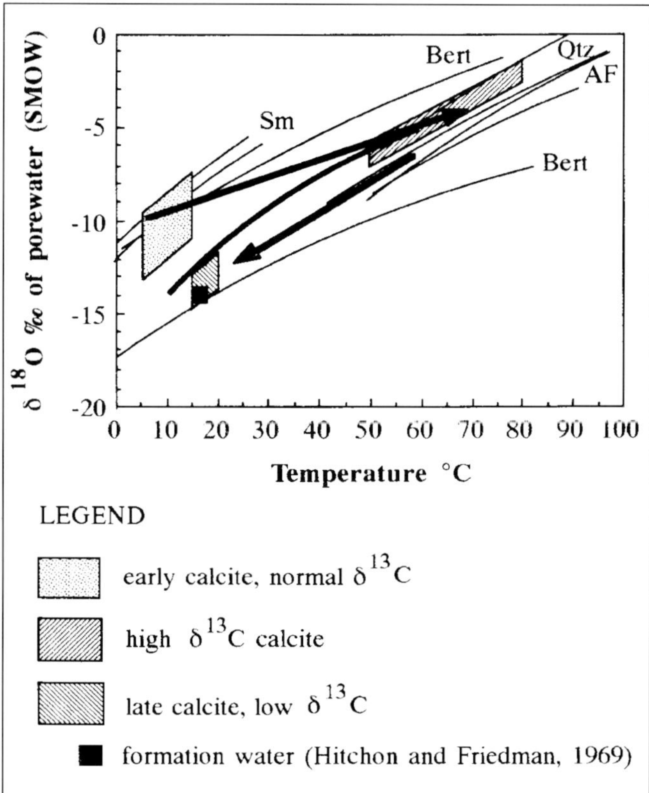

Figure 8. Idealized porewater evolution pathway for sands and sandstones of the Clearwater Formation in the Cold Lake area. Shaded areas indicate proposed conditions for crystallization of calcite. Sm = early smectitic clays; Bert = berthierine; Qtz = quartz; AF = K-feldspar. The following mineral-water fractionation curves were utilized: quartz-water and calcite-water, as for Figure 2; alkali feldspar-water, O'Neil and Taylor (1967); smectite-water, Savin and Lee (1988), adjusted for composition and interstratification, and berthierine-water, Longstaffe et al. (1992b). After Longstaffe et al. (1992b).

calcite) have been further interpreted using Figure 8. As for Figure 2, the idealized pore water evolution pathway (heavy arrows) was drawn to (i) intersect the mineral-water curves (thin curves) in a fashion compatible with the paragenetic sequence, (ii) not exceed burial temperatures of 70-80°C, and (iii) end at the current composition of Mannville Group formation water.

The main conclusion to be drawn from Figure 8 is that meteoric water was important throughout diagenesis of the Clearwater Formation sands. To initiate crystallization of grain-coating smectitic clays very early in diagenesis (10±5°C) required pore waters of about ≈-10 ‰ (mixed meteoric-marine waters containing a sizable fraction of freshwater). Similar waters were needed to form the normal $\delta^{13}C$ (0 ‰) calcite at the same stage of diagenesis. Even lower ^{18}O pore waters were involved in formation of some berthierine, assuming that crystallization occurred relatively very early in diagenesis. Continued berthierine crystallization as the thermal maximum was approached is consistent with the pore water pathway, as are the oxygen isotope compositions of the associated high ^{13}C calcite. While some shift towards more ^{18}O-rich pore water is required to explain the compositions of quartz and alkali feldspar formed at or near maximum burial temperatures, the isotopic signature remained strongly meteoric. Finally, the lower $\delta^{18}O$ values and temperature of present formation waters indicate that a considerable shift in pore water composition occurred following maximum burial. This change is most simply explained by increasing infiltration of low ^{18}O (local) meteoric water as erosion of overlying units progressed (see "Late Diagenesis"). The $\delta^{18}O$ values for the latest calcite cements are appropriate for crystallization from such waters (Figure 8).

Depositional environment largely controlled the distribution of brackish to meteoric pore waters and early diagenetic minerals in the Clearwater Formation sands. The earliest calcite concretions ($\delta^{13}C$ ≈ 0 ‰) have the highest $\delta^{18}O$ values (but still low relative to normal marine carbonate), and are most abundant in the fringes of the deltaic complex. These compositions can be explained by mixing between sea water and freshwater during very early diagenesis.

Likewise, the formation and distribution of early diagenetic smectitic clays and berthierine were probably related to syn- and early post-depositional conditions favored in distributary channel, stream-mouth bar and related facies within the thickest portion of the bitumen reservoir. A large fraction of meteoric pore water would have been trapped in these sands. As overlying sediments were deposited (a result of delta-lobe switching), reducing conditions should have been quickly established. The oxygen isotope results (Figure 8) may indicate that berthierine formed preferentially to smectitic clays in locations where the relative fraction of meteoric water in the connate pore fluids was highest. This observation is compatible with the pattern of increasing berthierine and decreasing smectitic clay content away from the "marine" ($\delta^{13}C \approx 0$ ‰) calcite concretions. In the meteoric water dominated, reducing environment, degradation of organic matter would have proceeded primarily by microbial fermentation, accounting for the appearance of low-Fe, high $\delta^{13}C$, grain-coating calcite in portions of this area where Fe-rich clays had not already coated framework grains. However, lower pore water $\delta^{18}O$ values associated with berthierine may also have resulted at least in part from mass balance effects of abundant, early diagenetic, ^{18}O-rich clay formation in a quasi-closed, sediment-water system.

The Origin of Berthierine

Reports of berthierine are relatively uncommon in the sandstone diagenetic record for several reasons. First, because of its 7Å character, it may have been overlooked. Second, berthierine can alter to chamosite beginning at temperatures as low as 70°C (Curtis et al., 1985; Jahren and Aagaard, 1989). Third, special circumstances are needed for primary crystallization of grain-coating berthierine in sands: (1) an Fe supply, (2) concentration of the Fe about sand grains, and (3) a reducing environment, or at least one in which the oxygen supply has been greatly restricted (Curtis and Spears, 1968). In the Clearwater Formation, dissolution of volcanic rock fragments would have provided abundant Fe. Grain-coating berthierine may simply be recrystallized "Fe-oozes" originally accreted about sand grains, a feature known from stream-mouth bar environments. But the general association between ^{13}C-rich calcite cements and berthierine deserves further consideration. For example, microbial processes may have been instrumental in the reduction of iron. Further, microbes whose activities contributed to precipitation of ^{13}C-rich calcite may have concentrated metals such as Fe on their membranes (Ferris et al., 1987), facilitating nucleation of berthierine.

BURIAL DIAGENESIS

Pore waters in sedimentary basins commonly evolve toward more ^{18}O-rich compositions during burial diagenesis (e.g., Figures 2, 8). In shale-dominated basins, including the Texas Gulf Coast and the Great Valley sequence of California, sandstone pore waters typically reach $\delta^{18}O$ values of +8 ‰. This result has been attributed to the conversion of smectite to illite in intercalated shales (Yeh and Savin, 1977; Suchecki and Land, 1983; Taylor, 1990). In the Western Canada sedimentary basin, the magnitude of pore water ^{18}O enrichment during prograde burial diagenesis has been difficult to measure directly, given the subsequent influx of meteoric water. Nevertheless, reasonable estimates have been possible using diagenetic minerals crystallized at or near maximum burial.

Longstaffe et al. (1992a) focused on the scale of pore water ^{18}O-enrichment in an investigation of the Alberta Deep Basin (Figure 1). Here, a stratigraphic succession of Permian to Upper Cretaceous sandstones was examined from the thick sandstone-shale sequence that overlies the Paleozoic carbonate rocks. The lowermost unit studied (the Permian Belloy sandstone) is situated within the uppermost portion of the carbonate sequence. All of the sandstones were diagenetically similar; early siderite or chlorite, kaolinite, quartz overgrowths, and then illitic clays formed with increasing burial, followed by crystallization of quartz druse, dickite/late kaolinite, calcite, and ankerite during post-Laramide uplift and erosion (Tilley, 1988). Integration of fluid-inclusion temperatures for diagenetic quartz and calcite (Tilley et al., 1989), available vitrinite-reflectance information, and the $\delta^{18}O$ values of diagenetic phases permitted the ^{18}O-enrichment of pore waters to be calculated with better resolution than in earlier studies.

How pore water $\delta^{18}O$ values at or near maximum diagenetic temperatures were deduced is illustrated in Figure 9 for the Lower Cretaceous Cadotte Member (Peace River Formation) sandstones and conglomerates. They were deposited as part of a prograding, high-energy shoreline (Leckie et al., 1990). Infiltration of meteoric to mixed meteoric-marine pore waters (−9±3 ‰) very early in diagenesis is strongly indicated by the very low $\delta^{18}O$ values of early siderite. Quartz druse crystallized much later, primarily in vugs created by partial dissolution of siderite concretions, and within horizontal fractures (Tilley, 1988). That vug-filling and fracture-filling quartz crystallized under virtually identical conditions is indicated by the similarity in their $\delta^{18}O$ values (+16.0 versus +16.4 ‰). Fluid inclusion results show that quartz druse crystallized at temperatures as high as 176°C, at or near the maximum temperatures experienced by these rocks. Under such conditions, pore water $\delta^{18}O$ values as high as

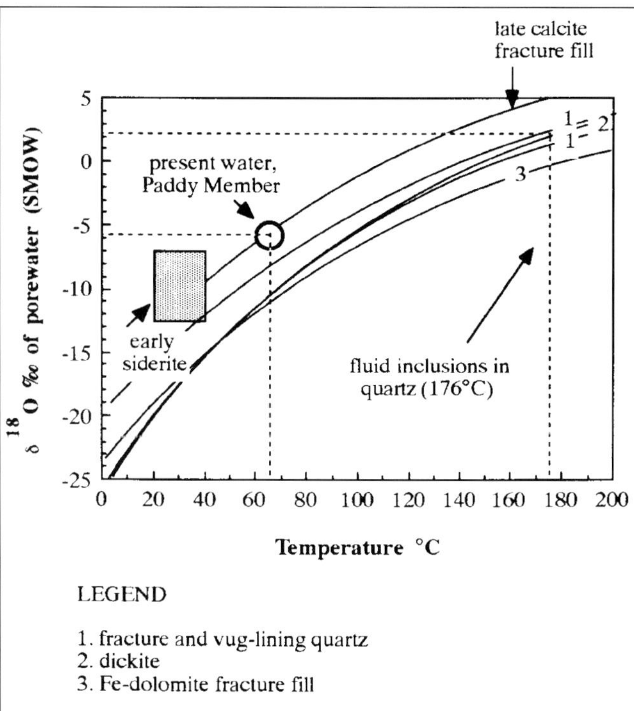

Figure 9. Oxygen isotope composition of porewater versus crystallization temperature for diagenetic minerals from the Cadotte Member. The mineral-water, oxygen-isotope fractionation equations used are listed in Figures 2 and 8, except for ankerite-water, which is from Dutton and Land (1985). The shaded area for early siderite is taken from Tilley and Longstaffe (1989). The average crystallization temperature for quartz druse, as determined from fluid inclusions (Tilley et al., 1989), is also indicated. The composition of formation water from the immediately overlying Paddy Member (Peace River Formation) sandstone is also shown (open circle). After Longstaffe et al. (1992a).

+2‰ are required (Figure 9). The $\delta^{18}O$ value of pore-filling dickite (+11.3 ‰) is also compatible with crystallization under such conditions, but lower temperature formation from lower ^{18}O pore waters cannot be ruled out in the absence of independent temperature data for dickite crystallization.

The Deep Basin investigation confirmed earlier observations that sandstone pore water $\delta^{18}O$ values in the Western Canada sedimentary basin were generally no higher than +2 to +3 ‰ at maximum burial. Depending on the absolute timing of diagenetic mineral formation, even lower values were possible. Only for sandstones adjacent to the Paleozoic sedimentary rocks could higher values (>+3 ‰) be inferred (Table 1).

Longstaffe et al. (1992a) commented on possible implications of this observation. First, the relationship between smectite-illite reactions in shales and pore water compositions in adjacent sandstones must be considered. Diagenetic illitic clays in typical Texas Gulf Coast shales ($\delta^{18}O \approx$ +16 to +19 ‰; Yeh and Savin, 1977) would be in isotopic equilibrium with pore waters of $\delta^{18}O \approx$ +5 to +10 ‰ at 130 to 170°C. Since pore waters in associated sandstones have such $\delta^{18}O$ values, expulsion of clay-mediated shale pore waters into the sandstones has been considered to be a significant process (Kharaka and Carothers, 1986; Taylor, 1990).

Shale is very abundant, and has been heated to equivalent temperatures, in many portions of the Mesozoic section in the Western Canada sedimentary basin, yet much lower maximum $\delta^{18}O$ values (<+3 ‰) for sandstone pore waters are apparently the rule. Several explanations for the difference are possible: (1) The high ^{18}O "signal" of pore waters derived from shales may have been diluted, since the connate waters in many sandstones contained a sizable fraction of meteoric water. (2) The smectite-illite reaction may have been relatively unimportant. Mesozoic shales in the Western Canada sedimentary basin contain abundant illitic clays, but excepting bentonitic horizons, an origin for the bulk of this material via the smectite to illite reaction is less than completely demonstrated. Conversion of smectite to illite in sufficient quantities to control shale pore water compositions may not have occurred. (3) The putative oxygen isotope signal resulting from the smectite-illite reaction need not have been the same as in other basins. Only a few oxygen isotope data are published for shales from the Western Canada sedimentary basin. Some values (+16 to +20 ‰) are similar to the Texas Gulf Coast and the Great Valley Sequence (Longstaffe et al., 1992a). But other samples of illite-smectite, for example, mudstones from the continental Belly River/Brazeau Group, have much lower $\delta^{18}O$ values. This clay, which crystallized at ≈80°C, has a $\delta^{18}O$ value of +12.6 ‰, implying a shale pore water composition of ≈-3 ‰ during the smectite-illite reaction (Longstaffe and Ayalon, 1991). This isotopic signal, if contributed to sandstone pore waters, would be impossible to distinguish from evolved meteoric water of other origins.

Besides the smectite-illite reaction, what other processes might have caused the change in pore water $\delta^{18}O$ values during burial of sandstones in the Western Canada sedimentary basin? Only carbonate cements, especially calcite, would have been sufficiently susceptible to oxygen isotope exchange to enrich sandstone pore waters in ^{18}O during prograde burial diagenesis. But the volume of calcite cement was too low to produce the necessary ^{18}O-enrichment, especially where pore water $\delta^{18}O$ values exceeded +3 ‰. Instead, the increase probably resulted mostly from mixing between evolving connate waters in the sandstones (in most cases initially of low ^{18}O character) and much more ^{18}O-rich waters introduced from underlying Paleozoic carbonates. The greatest effects should be apparent closest to the Paleozoic unconformity and generally diminish upwards in the section, except for specific localities where anomalously high fluid transmissivities facilitated more pervasive upward penetration of ^{18}O-rich brines.

This pattern was observed for the Alberta Deep Basin (Longstaffe et al., 1992a). Here, $\delta^{18}O$ values of ≈+9 ‰ were deduced for formation waters at maximum burial in the Permian Belloy Formation sandstone at the top of the Paleozoic section, decreasing to +7 ‰ in the Triassic Halfway Formation sandstone, near the base of the Mesozoic clastic sedimentary section, and then further declining to ≈ 0 ‰ for sandstones situated progressively higher in the section (Table 1). In more permeable portions of a few units (e.g., the Fahler Member conglomerates), ^{18}O-rich waters were able to penetrate higher into the Mesozoic section (Tilley and Longstaffe, 1989; Tilley et al., 1989).

The ^{18}O-rich character of waters originating from the Paleozoic units probably arose from at least two processes: evaporative brine formation and water/rock interaction. Trapping and evaporation of marine connate waters would be anticipated during formation of the Paleozoic interbedded carbonates, calcareous shales and evaporites. During initial stages of sea water evaporation, residual waters become enriched in ^{18}O and D, and in particular, $\delta^{18}O$ values ≥ +7 ‰ can be expected by 4X sea water concentration (Holser, 1979; Pierre et al., 1984). At higher concentrations, $\delta^{18}O$ and δD values generally decrease, but several different trajectories for the resulting "hooked" sea water evaporation curve have been proposed (Holser, 1979; Pierre et al., 1984; Knauth and Beeunas, 1986). Curves based on naturally occurring brines (evolved from normal sea water, $\delta^{18}O \approx 0$ ‰) suggest that concentrations well in excess of 45X are needed before pore water $\delta^{18}O$ values < 0 ‰ result. Connolly et al. (1990a) used chemical data to infer that such concentrations were atypical of brine endmembers in formation waters from central Alberta.

Post-depositional interaction between carbonates and pore waters could also generate the ^{18}O-rich formation waters that emanated from the Paleozoic section. Extensive dolomitization of the Paleozoic section in Alberta provides evidence for extensive water-rock interaction. Moreover, $\delta^{18}O$ values varying from +3 to +11 ‰ can be calculated for pore water equilibration with the average Paleozoic carbonate, depending on mineralogy, water-rock ratio and temperature (125 to 175°C assumed here).

LATE DIAGENESIS

Meteoric water comprises a significant fraction of current formation waters in sedimentary rocks of the Western Canada sedimentary basin (Clayton et al., 1966; Hitchon and Friedman, 1969). The consistently low ^{18}O character ($\delta^{18}O \approx$ +5 to +15 ‰) of late diagenetic clay, quartz and/or carbonate minerals from Cretaceous and Tertiary sandstones throughout Alberta testify to the pervasive impact that this meteoric water had on late diagenetic processes (e.g., Longstaffe, 1983, 1984, 1986). As mentioned in the Introduction, this influx of meteoric water is directly related to development of the Cordillera. The first phase of the Laramide orogeny caused thrusting and

Table 1. Calculated oxygen isotope composition of porewaters for Alberta sandstones.

Age	Unit	Connate Waters (‰ SMOW)	Maximum value acquired during burial (‰ SMOW)	Reference
Alberta Deep Basin, Northwestern Alberta-Northeastern British Columbia				
Late Cretaceous	Cardium Formation	–12 to –8	0±2	1
Early Cretaceous	Paddy Member	–4 ± 2	+2±1	1
	Cadotte Member	–9±3	+2±1	1
	Fahler Member	–10±3	+2±1 (anomaly to +8?)	2
	Bluesky Formation	–4±1	+1±2	1
(to Jurassic)	Cadomin/Nikanassin fms.	–8±3	+5±2	1,2
Triassic	Halfway Formation		+7±?	1
Permian	Belloy Formation		+9±?	1
West and West-Central Alberta				
Late Cretaceous	Brazeau (Belly River) Group	–14	–3±1	3
	basal Belly River Formation	–12	–1±1	4
Early Cretaceous	Viking Formation	–7 to –1	+2±1	5
Cold Lake Area, Northeastern Alberta				
Early Cretaceous	Clearwater Formation	–14 to –10	–3±1	6

1. Longstaffe et al. (1992a) 2. Tilley and Longstaffe (1989)
3. Longstaffe and Ayalon (1991) 4. Ayalon and Longstaffe (1988)
5. Longstaffe and Ayalon (1987) 6. Longstaffe et al. (1992b)

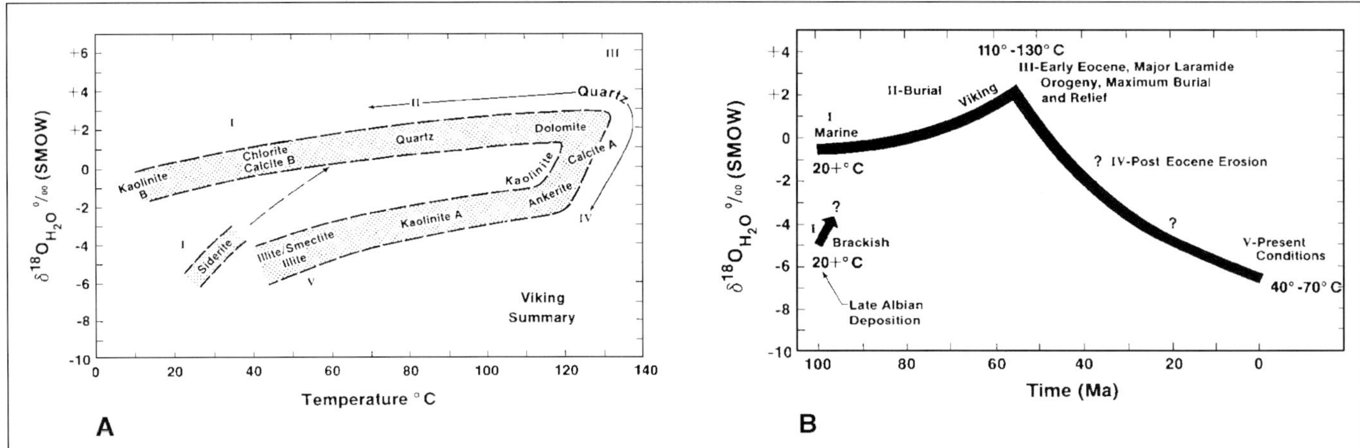

Figure 10(A). Idealized porewater evolution pathway ($\delta^{18}O$ of porewater versus temperature) for the Viking Formation sandstones and conglomerates, as deduced from the paragenesis and oxygen isotope compositions of diagenetic minerals. Stages I to V as in Figure 2. From Longstaffe and Ayalon (1987). (B). Idealized porewater evolution pathway ($\delta^{18}O$ of porewater versus time) for the Viking Formation sandstones and conglomerates, as deduced from the paragenesis and oxygen isotope compositions of diagenetic minerals. Stages I to V as in Figure 2. From Longstaffe and Ayalon (1987).

uplift along the eastern Cordillera in late Mesozoic time. To the east, continental deposition dominated throughout the Late Cretaceous and into the Early Tertiary as sediment was shed eastward into the downwarped sedimentary basin. Maximum burial of the Mesozoic sedimentary sequence probably occurred in the late Paleocene or early Eocene (Taylor et al., 1964; Hitchon, 1984). In the early Eocene, the major pulse of the Laramide Orogeny resulted in extensive deformation of the eastern Cordillera and significant uplift of the sedimentary basin in western Alberta. Uplift created a high potentiometric surface, causing large-scale, meteoric recharge of permeable sedimentary rocks in the basin (Hitchon, 1984). This process has continued, albeit gradually diminishing, as erosion of the accumulated Upper Cretaceous and Tertiary rocks proceeded (Taylor et al., 1964; Beaumont, 1981).

This major change in fluid regime has been dramatically recorded by late diagenetic minerals. As first shown in detail for Cretaceous sandstones of Viking Formation (Figures 10A,B), oxygen isotope compositions of most diagenetic phases formed at or following maximum burial can only be explained by a significant lowering of pore water $\delta^{18}O$ values beginning at, or shortly following, maximum burial. This scenario requires a pervasive influx of low-^{18}O meteoric water, and is compatible with geological development of the Western Canada sedimentary basin. Maximum burial coincided with maximum relief, which caused the gravity-driven influx of meteoric water. Hence, the "looped" pore water evolution pattern ($\delta^{18}O$ values gradually increasing during burial, and decreasing following maximum burial; e.g., Figures 2, 10A) developed that is so typical of Mesozoic sandstones from western and central Alberta. Cooling accompanied meteoric recharge and erosion, as sandstone pore waters evolved to current compositions.

Formation Water Evolution and Mixing

Current formation water compositions in sandstones of the Western Canada sedimentary basin are the cumulative result of mixing and water/rock interaction during early, burial and late diagenesis. The net effect, at least for oxygen and hydrogen isotopes, is the now familiar trend of formation water compositions for the "Alberta Basin", which verges from the meteoric water line towards higher $\delta^{18}O$ values for a given δD composition (Figure 11). More recent results for central Alberta follow the same general trend, intersecting the meteoric water line at $\delta^{18}O$ = -16.3 ‰ and δD = -121 ‰, virtually identical to the current weighted mean annual value for Edmonton rainfall (Connolly et al., 1990b).

Hitchon and Friedman (1969) invoked a particularly important role for mixing between evolved (connate) sea water and surface-derived meteoric water. Connolly et al. (1990a,b) developed this theme further, classifying formation waters from central Alberta (Figure 1) into three groups and two distinct fluid regimes on the basis of host lithology, chemistry and isotopic (Sr, O, D) composition.

The lowermost fluid regime consists of Groups I and II formation waters. Group I waters (av. $\delta D \approx$ -85 ‰; av. $\delta^{18}O \approx$ -5 ‰) are generally more saline, and are contained primarily in Devonian, Mississippian and lowermost Jurassic carbonates. Group II waters (av. $\delta D \approx$ -85 ‰; av. $\delta^{18}O \approx$ -7 ‰) are hosted by Middle Jurassic and Lower Cretaceous clastic units (the Ostracod, Glauconitic and Viking formations). The

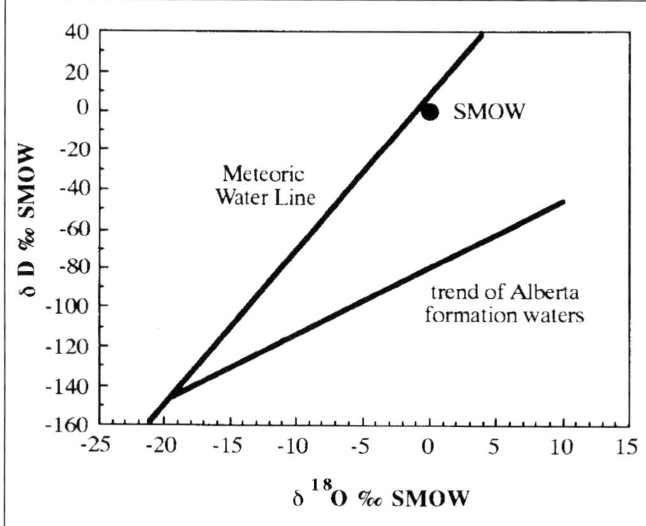

Figure 11. General relationship between the present-day Meteoric Water Line (after Craig, 1961) and formation waters in the Western Canada sedimentary basin, after Clayton et al. (1966) and Hitchon and Friedman (1969).

Groups I and II waters have $^{87}Sr/^{86}Sr$ ratios of 0.7076 to 0.7129, with formation waters from Devonian carbonates generally having the most radiogenic compositions. Most waters are more radiogenic than expected for sea water trapped during deposition, and also have higher $^{87}Sr/^{86}Sr$ ratios than coexisting carbonate cements. Of the units analyzed by Connolly et al. (1990b), only Precambrian basement rocks and Cambrian shales have $^{87}Sr/^{86}Sr$ ratios sufficiently high to account for the formation waters, although some Devonian shales probably also have appropriate Sr isotope compositions. These results suggest that fluid communication has occurred between Paleozoic carbonates (Group I) and Mesozoic clastic rocks (Group II). A mixing relationship between Group I and II formation waters is also indicated by their chemistry (e.g., Cl-Br, K-Br, Na-Br, Mg-Br, Ca-Br and $^{87}Sr/^{86}Sr$ versus 1/Sr). Connolly et al. (1990a,b) concluded that Group I and Group II waters are composed of a brine formed by evaporation of sea water beyond the point of halite saturation that was subsequently diluted as much as 50 to 80% by (post-Laramide) meteoric water.

The second, upper fluid regime contains Group III waters and is situated mostly in Upper Cretaceous sandstones (Belly River/Brazeau Group and Cardium Formation). Group III is characterized by dilute, meteoric waters (av. $\delta D \approx -110$ ‰; av. $\delta^{18}O \approx -12$ ‰) whose chemical compositions (Cl, Br, K, Na, Mg) are compositionally decoupled from the more saline, stratigraphically lower Groups I and II fluids. Further west, dilute waters with some Group III affinities were obtained at greater depths from the Jurassic Rock Creek Formation, but deeper penetration of meteoric water in more westerly portions of the basin is a predictable consequence of gravity-driven flow first initiated by deformation in the eastern Cordillera.

The $^{87}Sr/^{86}Sr$ ratios of Upper Cretaceous Group III formation waters (0.7058–0.7089) trend to much lower values than Groups I and II, a variation that is also reflected in late diagenetic carbonate and clay cements (Ayalon and Longstaffe, unpublished data; Connolly et al., 1990b). Lower $^{87}Sr/^{86}Sr$ ratios for the formation waters first appear in the stratigraphically high Lower Cretaceous sandstones, such as the Viking Formation. The lowest values are encountered in the uppermost Belly River/Brazeau Group (0.7058–0.7063). Such compositions lie well below those of Upper Cretaceous sea water, a result compatible with our earlier suggestion that sea water was not involved in diagenesis of these rocks. Connolly et al. (1990b) proposed that the low $^{87}Sr/^{86}Sr$ ratios reflect the emerging influence of a volcanic sediment provenance during Cretaceous time.

All three hydrochemical groups contain significant input of post-Laramide meteoric water. However, the chemically distinct nature of Group I and II versus Group III, and the mixing relationship between Groups I and II, have important implications. Garven (1989) proposed that the post-Laramide, regional gravity-driven flow system began to dissipate in Miocene-Pliocene time. By coupling hydrogen isotope compositions of the formation waters with paleoclimatic considerations, Connolly et al. (1990b) were able to deduce that only Group III waters have remained hydrodynamically connected to the post-Laramide recharge regime, and that hydrochemical isolation of Groups I and II probably began in the Pliocene. With the disappearance of the regional hydrodynamic drive, upward diffusional flow and density stratification then occurred within the lower hydrological regime.

Formation Water Evolution and Water/Rock Interaction

While mixing may have been the main control on current formation water compositions, chemical and isotopic effects arising from water/rock interaction during diagenesis cannot be ignored. Connolly et al. (1990a) showed that the chemistry of Group I waters had been modified by carbonate and clay mineral reactions and that Group II waters were affected by leaching of feldspars and clay minerals. Likewise, the $^{87}Sr/^{86}Sr$ ratios of many formation waters, which lay outside the range for sea water of any age, indicated extensive water/rock interaction in the form of dissolution of feldspar, micas, and clay minerals.

Hydrogen isotope exchange between diagenetic clays and formation waters may also have been a significant process, with little net effect on the water but

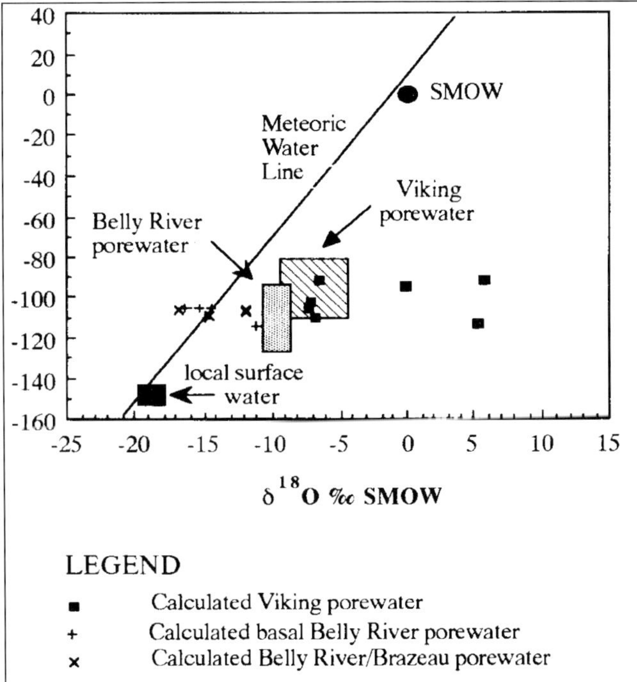

Figure 12. Comparison of measured oxygen and hydrogen isotope compositions for formation waters from the Viking Formation and Belly River/Brazeau Group with those calculated using the isotopic compositions of diagenetic kaolinite. The range of measured compositions for formation waters and surface water in the study areas are shown by the shaded boxes. Calculated formation water compositions (individual data points) were obtained by assuming that oxygen and hydrogen isotopic equilibrium was attained between kaolinite and formation water at present subsurface conditions. The hydrogen isotope, kaolinite-water equation of Lambert and Epstein (1980), and the oxygen isotope kaolinite-water equation of Land and Dutton (1978), have been used to perform the calculations. After Longstaffe and Ayalon (1990).

potentially a large influence on the clays (Longstaffe and Ayalon, 1990, Tilley and Longstaffe, 1989). This conclusion rests mainly on the remarkably small range of δD values determined for early and late diagenetic kaolinites in the Viking Formation (–132 to –112 ‰) and Belly River/Brazeau Group sandstones (–137 to –128 ‰). Kaolinite from both units is in, or close to, hydrogen isotopic equilibrium with current formation waters at present subsurface temperatures (Figure 12). By comparison, a proportionally much larger range in kaolinite $\delta^{18}O$ compositions was obtained for each unit (+13.8 to +26.9 ‰, Viking Formation; +5.9 to +12.7 ‰, Belly River/Brazeau Groups); a general state of oxygen-isotope equilibrium does not exist between kaolinite and the formation waters (Figure 12). Within the limits of available fractionation data, similar observations were made for some diagenetic illitic clays from the same units, although other samples of these clays are in both hydrogen and oxygen isotope equilibrium with present formation waters because of neoformation from these waters.

The general conclusion is that diagenetic clay minerals in sandstones from the Western Canada sedimentary basin may have exchanged hydrogen isotopes with current formation waters at temperatures as low as 40°C. Because this exchange is unaccompanied by significant changes in the original oxygen isotope composition of the clay, it occurred in the absence of dissolution-precipitation reactions. Accordingly, proton exchange reactions may control hydrogen isotope compositions of clay minerals in fluid-dominated (high water/rock ratio) sedimentary systems, and may have a substantial influence on the acidity of formation waters.

Did significant oxygen isotope exchange occur between formation waters and carbonate minerals in the pore system during the post-Laramide introduction of meteoric water? Connolly et al. (1990b) noted that for any particular locality in central Alberta, formation waters generally had oxygen isotope compositions either too low or too high to be in equilibrium with coexisting carbonates. Furthermore, Cl and Na versus $\delta^{18}O$ relationships showed fluid mixing to be the dominant control on the oxygen isotope composition of formation water, with only minor shifts attributable to exchange.

Given this scenario, current oxygen isotope equilibrium exists only for minerals crystallized directly from present formation waters. Results for last stage fracture-filling minerals in sandstones and conglomerates of the Alberta Deep Basin are consistent with this observation (Longstaffe et al., 1992a). The few isotopic data for formation waters from this vicinity show the same trend of increasing $\delta^{18}O$ values with stratigraphic depth as elsewhere in Alberta: Lower Cretaceous units, –10.8 to -5.7 ‰; Triassic units, –1.7 to +0.8 ‰; Paleozoic units, –1.2 to +7.2 ‰ (av. ≈ +3 ‰) (Hitchon and Friedman, 1969; Tilley and Longstaffe, 1989). An identical pattern can be predicted from the compositions of very late stage, fracture-filling carbonates (Table 2). These minerals are in oxygen isotope equilibrium with current formation waters (e.g., Figure 9 for fracture calcite from the Cadotte Member).

A more equivocal observation concerns some anomalously ^{18}O-rich, very late diagenetic calcites (+23.4 to +23.6 ‰) in conglomerates from the western portion of the Falher Member. During earlier burial diagenesis, these conglomerates acted as a preferred conduit for diagenetic fluids originating in underlying Paleozoic carbonates (Tilley and Longstaffe, 1989). The ^{18}O-rich, late diagenetic calcite may be the record of a still later episode of cross-formational flow. Formation waters in

Table 2. Porewater compositions and late fracture-filling carbonates, Alberta Deep Basin.

Stratigraphic Age	Unit	Phase	Calculated Pore Water (‰ SMOW)	Measured Pore Water (‰ SMOW)	Reference
Late Cretaceous	Cardium Formation	calcite	−13	−15.8 to −12.7	1,2
Early Cretaceous	Cadotte Member	calcite	−6	−5.7	3
	Fahler Member (siltstone)	calcite	−5		1
	Fahler member (conglomerate)	calcite**	+2 to +3(?)	+3.0 to +3.1*	1,2,3
(to Jurassic)	Nikanassin	ankerite	−1 to 0	+1±2	1
Triassic	Halfway Formation	calcite	−0.5 to +1.3	−1.7 to +0.8	1,2

*Underlying Devonian carbonates **Very late stage pore filler
1. Longstaffe et al. (1992a) 2. Hitchon and Friedman (1969) 3. Tilley and Longstaffe (1989)

Upper Devonian carbonates near this area have $\delta^{18}O$ values of +3.0 to +3.1 ‰ (Hitchon and Friedman, 1969), virtually identical to the compositions required for crystallization of the very late diagenetic calcite at current subsurface temperatures (Table 2).

We end our review with the general comment that integrated fluid inclusion, stable and radiogenic isotope, and chemical studies of fracture-filling minerals may hold the key to understanding how, when, and whether cross-formational flow takes place in sedimentary basins. Introduction of gravity-driven meteoric water into the system should be one of the simplest processes to characterize. But we also should be able to evaluate whether and when fluid movement occurred along orogenically induced faults, if vertical fracturing associated with overpressuring and then decompression of "basin pressure compartments" has been a significant process in vertical mass transport by aqueous media, and what role "seismic pumping" played in driving horizontal migration of formation fluids on both inter- and intra-basin scales.

ACKNOWLEDGMENTS

Funding from the Natural Sciences and Engineering Research Council of Canada (NSERC) and the Alberta Oil Sands Technology Research Authority (AOSTRA) is gratefully acknowledged, as is my considerable debt to many coinvestigators, especially Avner Ayalon, Barbara Tilley, Mira Racki, Cathy Connolly, Bruce Hart, Jennifer McKay and Guy Plint. I also thank Drs. Tony Fallick, Andrew Robinson and Craig Smalley for their helpful comments on an earlier version of this manuscript.

REFERENCES CITED

Abercrombie, H. J., 1989, Water-rock interaction during diagenesis and thermal recovery, Cold Lake, Alberta. Ph.D. thesis, University of Calgary, Canada.

Ayalon, A., and F.J. Longstaffe, 1988, Oxygen-isotope studies of diagenesis and pore water evolution in the Western Canada sedimentary basin: evidence from the Upper Cretaceous basal Belly River sandstone, Alberta. Journal of Sedimentary Petrology, v. 58, p. 489-505.

Beaumont, C., 1981, Foreland basins. Geophysical Journal of the Royal Astronomical Society, v. 65, p. 291-329.

Beaumont, C., R. Boutilier, A.S. MacKenzie, and J. Rullkotter, 1985, Isomerization and aromatization of hydrocarbons and the paleothermometry and burial history of the Alberta foreland basin. American Association of Petroleum Geologists Bulletin, v. 69, p. 546-566.

Bird, M.I., and Chivas, A.R., 1988, Stable-isotope evidence for low-temperature kaolinitic weathering and post-formational hydrogen-isotope exchange in Permian kaolinites. Chemical Geology (Isotope Geosciences Section), v. 72, p. 249-265.

Bloch, J., 1990, Stable isotopic composition of authigenic carbonates from the Albian Harmon Member (Peace River Formation): evidence of early diagenetic processes. Bulletin of Canadian Petroleum Geology, v. 38, p. 39-52.

Clayton, R.N., 1959, Oxygen isotope fractionation in the system calcium carbonate-water. Journal of Chemical Physics, v. 30, p. 1246-1250.

Clayton, R.N., J.R. O'Neil, and T.K. Mayeda, 1972, Oxygen isotope exchange between quartz and water. Journal of Geophysical Research, v. 77, p. 3057-3067.

Clayton, R.N., I. Friedman, D.L. Graf, T.K. Mayeda, W.F. Meents, and N.F. Shimp, 1966, The origin of saline formation waters, 1. Isotopic composition. Journal of Geophysical Research, v. 71, p. 3869-3882.

Connolly, C.A., L.M. Walter, H. Baadsgaard, and F.J. Longstaffe, 1990a, Origin and evolution of formation waters, Alberta basin, Western Canada sedimentary basin. I. Chemistry. Applied Geochemistry, v. 5, p. 375-395.

Connolly, C.A., L.M. Walter, H. Baadsgaard, and F.J. Longstaffe, 1990b, Origin and evolution of formation waters, Alberta basin, Western Canada sedimentary basin. II. Isotope systematics and water mixing. Applied Geochemistry, v. 5, p. 397-413.

Craig, H., 1961, Isotopic variations in meteoric waters. Science, v. 133, p. 1702-1703.

Curtis, C.D., and D.A. Spears, 1968, The formation of sedimentary iron minerals. Economic Geology v 24, p. 257-270.

Curtis, C.D., C.R. Hughes, J.A.Whiteman, and C.K. Whittle, 1985, Compositional variations within some sedimentary chlorites and some comments on their origin. Mineralogical Magazine, v. 49, p. 375-386.

Dimitrakopoulos, R.G., and K. Muehlenbachs, 1987, Biodegradation of petroleum as a source of ^{13}C-enriched carbon dioxide in the formation of carbonate cement. Chemical Geology (Isotope Geoscience Section), v. 65, p. 283-291.

Domenico, P.A., and G.A. Robbins, 1985, The displacement of connate water from aquifers. Bulletin of the Geological Society of America, v. 96, p. 328-335.

Dutton, S.P., and L.S. Land, 1985, Meteoric burial diagenesis of Pennsylvanian arkosic sandstones, southwestern Anadarko Basin, Texas. American Association of Petroleum Geologists Bulletin, v. 69, p. 22-38.

Ferris, F.G., W.S. Fyfe, and T.J. Beveridge, 1987, Bacteria as nucleation sites for authigenic minerals in a metal-contaminated lake sediment. Chemical Geology, v. 63, p. 225-232.

Friedman, I., and J.R. O'Neil, 1977, Compilation of stable isotope fractionation factors of geochemical interest, *in* M. Fleischer, M., ed., Data of Geochemistry, sixth edition: United States Geological Survey Professional Paper 440-KK, 12pp. + figures.

Garven, G., 1989, A hydrogeologic model for the formation of the giant oil sands deposits of the Western Canada sedimentary basin. American Journal of Science, v. 289, p. 105-166.

Graf, D.L., I. Friedman, and W.F. Meents, 1965, The origin of saline formation water, II. Isotopic fractionation by shale micropore systems. Illinois Geological Survey Circular 393, 32 pp.

Hacquebard, P. A., 1977, Rank of coal as an index of organic metamorphism for oil and gas in Alberta, *in* G. Deroo, T. Powell, B. Tissot, and R. McCrossan, eds., The origin and migration of petroleum in the Western Canadian sedimentary basin, Alberta: Geological Survey of Canada Bulletin 262, p. 11-22.

Harrison, D.B., R.P. Glaister, and H.W. Nelson, 1981, Reservoir description of the Clearwater oil sand, Cold Lake, Alberta Canada, *in* R.F. Meyer and C. T. Steele, eds., The Future of Heavy Crude Oils and Tar Sands: McGraw Hill, New York, p. 264-279.

Hart, B.S., 1990, The sedimentology and stratigraphy of the Upper Cretaceous Cardium Formation in northwestern Alberta and adjacent British Columbia. Ph.D. thesis, University of Western Ontario, London, Canada.

Hart, B.S., F.J. Longstaffe, and A.G. Plint, 1992, Evidence for relative sea level change from isotopic and elemental composition of siderite in the Cardium Formation, Rocky Mountain Foothills. Bulletin of Canadian Petroleum Geology, v. 40, p. 52-59.

Hitchon, B., 1969a, Fluid flow in the Western Canada sedimentary basin: 1. Effect of topography. Water Resources Research, v. 5, p. 186-195.

Hitchon, B., 1969b, Fluid flow in the Western Canada sedimentary basin; 2. Effect of geology. Water Resources Research, v. 5, p. 460-469.

Hitchon, B., 1984, Geothermal gradients, hydrodynamics, and hydrocarbon occurrences, Alberta, Canada. American Association of Petroleum Geologists Bulletin, v. 68, p. 713-743.

Hitchon, B., and I. Friedman, 1969, Geochemistry and origin of formation waters in the Western Canada sedimentary basin - I. Stable isotopes of hydrogen and oxygen. Geochimica et Cosmochimica Acta, v. 33, p. 1321-1349.

Holser, W.T., 1979, Trace elements and isotopes in evaporites, *in* R.G. Burns, ed., Marine Minerals: Reviews in Mineralogy, Mineralogical Society of America, p. 295-346.

Irwin, H., C. Curtis, and M.L. Coleman, 1977, Isotopic evidence for source of diagenetic carbonates formed during burial of organic-rich sediments. Nature, v. 269, p. 209-213.

Jahren, J.S., and P. Aagaard, 1989, Compositional variations in diagenetic chlorites and illites, and relationships with formation-water chemistry. Clay Minerals, v. 24, p. 157-170.

Kharaka, Y.K., and W.W. Carothers, 1986, Oxygen and hydrogen isotope geochemistry of deep basin brines, *in* P. Fritz, and J.Ch. Fontes, eds., Handbook of Environmental Isotope Geochemistry: Elsevier, Amsterdam, v. 2, p. 305-360.

Knauth, L.P., 1988, Origin and mixing history of brines, Palo Duro Basin, Texas, U.S.A. Applied Geochemistry, v. 3, p. 455-474.

Knauth, L.P., and M.A. Beeunas, 1986, Isotope geochemistry of fluid inclusions in Permian halite with implications for the isotopic history of ocean water and the origin of saline formation waters. Geochimica et Cosmochimica Acta, v. 50, p. 419-433.

Kyser, T.K., W.G.E. Caldwell, S.G. Whittaker, and A.J. Cadrin, 1993, Paleoenvironment and geochemistry of the northern portion of the Western Interior Seaway during Late Cretaceous time, *in* W.G.E. Caldwell and E.G. Kaufmann, eds., Evolution of the Western Interior Foreland Basin: Geological Association of Canada Special Paper (in press).

Lambert, S.J., and S. Epstein S., 1980, Stable isotope

investigations of an active geothermal system in Valles Caldera, Jemez Mountains, New Mexico. Journal of Volcanological and Geothermal Research, v. 8, p. 111-129.

Land, L.S., and S.P. Dutton, 1978, Cementation of a Pennsylvanian deltaic sandstone: isotopic data. Journal of Sedimentary Petrology, v. 48, p. 1167-1176.

Leckie, D.A., C. Singh, F. Goodarzi, and J.H. Wall, 1990, Organic-rich, radioactive marine shale: a case study of a shallow-water condensed section, Cretaceous Shaftesbury Formation, Alberta, Canada. Journal of Sedimentary Petrology, v. 60, p. 101-117.

Lerbekmo, J.F., 1963, Petrology of the Belly River Formation, southern Alberta foothills. Sedimentology, v. 2, p. 54-86.

Longstaffe, F.J., 1983, Diagenesis, IV. Stable isotope studies of diagenesis in clastic rocks. Geoscience Canada, v. 10, p. 44-58.

Longstaffe, F.J., 1984, The role of meteoric water in diagenesis of shallow sandstones: stable isotope studies of the Milk River aquifer and gas pool, *in* R.C. Surdam and D.A. MacDonald, eds., Clastic Diagenesis: American Association of Petroleum Geologists Memoir 37, p. 81-98.

Longstaffe, F.J., 1986, Oxygen isotope studies of diagenesis in the basal Belly River sandstone, Pembina I-Pool, Alberta. Journal of Sedimentary Petrology, v. 56, p. 78-88.

Longstaffe, F.J., 1987, Stable isotope studies of diagenetic processes, *in* T.K. Kyser, ed., Stable Isotope Geochemistry of Low Temperature Fluids: Mineralogical Association of Canada Short Course, v. 13, p. 187-257.

Longstaffe, F.J., 1989, Stable isotopes as tracers in clastic diagenesis, *in* I.E. Hutcheon, ed., Burial Diagenesis: Mineralogical Association of Canada Short Course, v. 15, p. 201-277.

Longstaffe, F.J., and A. Ayalon, 1987, Oxygen-isotope studies of clastic diagenesis in the Lower Cretaceous Viking Formation, Alberta: implications for the role of meteoric water, *in* J.D. Marshall, ed., The Diagenesis of Sedimentary Sequences: Geological Society Special Publication 36, p. 277-296.

Longstaffe, F.J., and A. Ayalon, 1990, Hydrogen-isotope geochemistry of diagenetic clay minerals from Cretaceous sandstones, Alberta, Canada: evidence for exchange. Applied Geochemistry, v. 5, p. 657-668.

Longstaffe, F.J. and A. Ayalon, 1991, Mineralogical and O-isotope studies of diagenesis and porewater evolution in continental sandstones, Cretaceous Belly River Group, Alberta, Canada. Applied Geochemistry, v. 6, p. 291-303.

Longstaffe, F.J., B.J. Tilley, A. Ayalon, and C.A. Connolly, 1992a, Controls on porewater evolution during sandstone diagenesis, Western Canada sedimentary basin: an oxygen isotope perspective, *in* D.W. Houseknecht and E. Pittman, eds., Origin, Diagenesis, and Petrophysics of Clay Minerals in Sandstones: SEPM Special Publication No. 47, p. 13-34.

Longstaffe, F.J., A. Ayalon, and M.A. Racki, 1992b, Stable isotope studies of diagenesis in berthierine-bearing oil sands, Clearwater Formation, Alberta, *in* Y.K. Kharaka and A.S. Maest, eds., Proceedings of the 7th International Symposium on Water-Rock Interaction, Park City, Utah, 1992; Rotterdam, Balkema Publishers, p. 955-958.

Machemer, S.D., and I. Hutcheon, 1988, Geochemistry of early carbonate cements in the Cardium Formation, central Alberta. Journal of Sedimentary Petrology, v. 58, p. 136-147.

McKay, J.L., 1992, Diagenesis of the Upper Cretaceous Marshybank Formation, northwestern Alberta - northeastern Britsh Columbia. M.Sc. thesis, University of Western Ontario.

McKay, J.L., F.J. Longstaffe, and A.G. Plint, 1989, Geochemical variations across marine-nonmarine transitions within the Bad Heart Formation, northwestern Alberta and northeastern British Columbia. Geological Association of Canada, Programme with Abstracts, v. 14, p. A14.

Macqueen, R. and D. Leckie, Introduction, *in* Foreland Basins and Fold Belts, edited by R.W. Macqueen and D.A. Leckie, AAPG Memoir 55, p. 1–8.

Mozley, P.S., 1989a, Relation between depositional environment and the elemental composition of early diagenetic siderite. Geology, v. 17, p. 704-706.

Mozley, P.S., 1989b, Complex compositional zoning in concretionary siderite: implications for geochemical studies. Journal of Sedimentary Petrology, v. 59, p. 815-818.

O'Neil, J.R., 1987, Preservation of H, C, and O isotopic ratios in the low temperature environment, *in* T.K. Kyser, ed., Stable Isotope Geochemistry of Low Temperature Fluids: Mineralogical Association of Canada Short Course, v. 13, p. 85-128.

O'Neil, J. R., and H.P. Taylor, Jr., 1967, The oxygen isotope and cation-exchange chemistry of feldspars. American Mineralogist, v. 52, p. 1414-1437.

O'Neil, J.R., R.N. Clayton, and T.K. Mayeda, 1969, Oxygen isotope fractionation in divalent metal carbonates. Journal of Chemical Physics, v. 51, p. 5547-5558.

O'Neil, J.R., C.M. Johnson, L.D. White, and E. Roedder, 1986, The origin of fluids in the salt beds of the Delaware Basin, New Mexico and Texas. Applied Geochemistry, v. 1, p. 265-271.

Pierre, C., L. Ortlieb, and A. Person, 1984, Supratidal evaporitic dolomite at Ojo de Liebre lagoon: mineralogical and isotopic arguments for primary crystallization. Journal of Sedimentary Petrology, v. 54, p. 1049-1061.

Plint, A.G., R.G. Walker, and K.M. Bergman, 1986, Cardium Formation 6. Stratigraphic framework of the Cardium Formation in the subsurface. Bulletin of Canadian Petroleum Geology, v. 34, p. 213-225.

Plint, A.G., R.G. Walker, and W.L. Duke, 1988, An outcrop to subsurface correlation of the Cardium Formation in Alberta, *in* D.P. James and D.A. Leckie, eds., Sequences, Stratigraphy, Sedimentology: Surface and Subsurface: Canadian Society of Petroleum Geologists, Memoir 15, p. 167-184.

Racki, M., 1991, Diagenesis of the Clearwater Formation, Cold Lake, Alberta. M.Sc. thesis, University of Western Ontario, London, Canada.

Savin, S.M., and M. Lee, 1988, Isotopic studies of phyllosilicates, *in* S.W. Bailey, ed., Hydrous Phyllosilicates (exclusive of micas): Reviews in Mineralogy, v. 19, p. 189-223.

Schwartz, F.W., and F.J. Longstaffe, 1988, Ground water and clastic diagenesis, *in* W. Back, J.S. Rosenshein, and P.R. Seaber, eds., Hydrogeology: Boulder, Colorado, Geological Society of America, The Geology of North America, v. O-2, p. 413-434.

Schwartz, F.W., K. Muehlenbachs, and D.W. Chorley, 1981, Flow-system controls of the chemical evolution of groundwater. Journal of Hydrology, v. 54, p. 225-243.

Storey, S.R., 1982, Optimum reservoir facies in an immature, shallow-lobate delta system: basal Belly River Formation, Keystone-Pembina area, *in* J.C. Hopkins, ed., Depositional Environments and Reservoir Facies in some Western Canadian Oil and Gas Fields: University of Calgary Core Conference, Calgary, Alberta, p. 3-13.

Suchecki, R.K., and L.S. Land, 1983, Isotopic geochemistry of burial-metamorphosed volcanogenic sediments, Great Valley sequence, northern California. Geochimica et Cosmochimica Acta , v. 47, p. 1487-1499.

Taylor, R.S., W.H. Mathews, and W.O. Kupsch, 1964, Tertiary, *in* R.G. McCrossan and R.P. Glaister, eds., Geological History of Western Canada: Alberta Society of Petroleum Geologists, p. 190-194.

Taylor, T.R., 1990, The influence of calcite dissolution on reservoir porosity in Miocene sandstones, Picaroon Field, offshore Texas Gulf Coast. Journal of Sedimentary Petrology, v. 60, p. 322-334.

Tilley, B.J., 1988, Diagenesis and porewater evolution in Cretaceous sedimentary rocks of the Alberta Deep Basin. Ph.D. thesis, University of Alberta, Edmonton, Alberta.

Tilley, B.J., and F.J. Longstaffe, 1989, Diagenesis and isotopic evolution of porewaters in the Alberta Deep Basin: the Falher Member and Cadomin Formation. Geochimica et Cosmochimica Acta, v. 53, p. 2529-2546.

Tilley, B.J., B.E. Nesbitt, and F.J. Longstaffe, F.J., 1989, Thermal history of Alberta Deep Basin: comparative study of fluid inclusion and vitrinite reflectance data. American Association of Petroleum Geologists Bulletin, v. 73, p. 1206-1222.

Tóth, J., 1978, Gravity-induced cross-formational flow of formation fluids, Red Earth region, Alberta, Canada: analysis, patterns, evolution. Water Resources Research, v. 14, p. 805-843.

Tóth, J., 1980, Cross-formational gravity-flow of groundwater: a mechanism of the transport and accumulation of petroleum (the generalized hydraulic theory of petroleum migration), *in* W.H. Roberts III, and R.J. Cordell, eds., Problems of petroleum migration: American Association of Petroleum Geologists Studies in Geology No.10, p. 121-167.

Wightman, D., B. Rottenfusser, J. Kramers, and R. Harrison, 1989. Geology of the Alberta oilsands deposits, *in* L. G. Hepler and C. Hsi, eds., AOSTRA Technical Handbook on Oil Sands, Bitumens and Heavy Oils: AOSTRA Technical Publication Series, v. 6, p. 1-9.

Wickert, L.M., F.J. Longstaffe, and S.G. Pemberton, 1989, A diagenetic investigation of sequence stratigraphy in the lower Cretaceous Clearwater Formation, Cold Lake oil sands, east-central Alberta. Geological Association of Canada, Program with Abstracts, v. 14, p. A86.

Wilson, M.R., T.K. Kyser, H.H. Mehnert, and J. Hoeve, 1987, Changes in the H-O-Ar isotope composition of clays during retrograde alteration. Geochimica et Cosmochimica Acta, v. 51, p. 869-878.

Yeh, H-W., and S.M. Savin, 1977, Mechanism of burial metamorphism of argillaceous sediments: 3. O-isotope evidence, Geological Society of America Bulletin, v. 88, p. 1321-1330.

Chapter 6

Diagenetic Pathways in Sedimentary Basins

Wendy J. Harrison and Regina N. Tempel
Department of Geology and Geological Engineering
Colorado School of Mines
Golden, Colorado, USA

ABSTRACT

Chemical reaction path models of mineral-fluid-gas equilibria can be constrained by data from hydrologic reconstructions of sedimentary basins to make predictions of diagenetic pathways. This approach allows estimates to be made of volumes of diagenetic products, their temporal order, and their consequent modifications of sediment porosity through geologic time. Diagenetic pathways for Wilcox Group and Frio Formation strata are presented as examples. A time-dependent diagenetic pathway for onshore Wilcox Group strata shows that episodic cementation is superimposed on a background of continuous chemical reactions and that significant changes in diagenetic style are related to changes in hydrologic regime during sediment burial. Calculations of the diagenetic consequences of fluid degassing during flow from the geopressured zone to the near-hydrostatic zone show that carbonates, clays, and quartz may precipitate alone or in combination, depending on the permitted physical conditions. These results can be used to estimate the attendant porosity loss in strata lying in the transition zone between geopressured and near-hydrostatically pressured pore fluids: maximum porosity decrease is between 3 and 4% while more likely porosity decreases are in the range 0.5 to 1%. Results from this study indicate that computer simulation is a critical step in the effective prediction of diagenetic pathways in sedimentary basins. The diagenetic makeup of a sedimentary rock is the result of combinations of processes and does not appear to be dominated by any one particular mechanism.

INTRODUCTION

Perhaps one of the most useful concepts brought to the discipline of sedimentary diagenesis in the last 10 years has been that of hydrologic regimes as proposed by Galloway (1984), who stated "a critical analysis of diagenetic processes within a realistic hydrologic context offers a major step in understanding and predicting diagenetic histories of sandstone reservoirs." Galloway's contribution was a simple statement of the obvious: "Such analysis necessitates a basic understanding of the hydrologic regimes and evolutionary pathways of sedimentary basins." Galloway (1984) identified three major hydrologic regimes in sedimentary basins—meteoric, compactional, and thermobaric—and identified the associated, commonly occurring diagenetic reactions, based on Gulf of Mexico basin petrographic data (Table 1). Water in each regime has a different origin and driving force and thus has characteristic hydraulic and chemical properties that cause variable diagenetic products when the water interacts with minerals in the rocks through which it flows.

Importantly, Galloway (1984) also noted that the hydrology of a basin evolves with the basin itself, such that composition and distribution of present day pore waters deduced from field measurements may not represent those waters once present in the pore system. The diagenetic alteration pattern observed in a rock is thus a summation of the effects of all the different pore fluids that have flowed through it during burial, in combination with attendant physical and geochemical conditions. A compilation of existing paleohydrologic analyses of sedimentary basins (Table 2) shows that two evolutionary pathways are particularly common, although different driving mechanisms may be involved. These two are replacement of early meteoric water by compactional/thermobaric waters and replacement of connate or compactional waters by meteoric waters, as presented within Galloway's framework (Figure 1). It is logical to expect that, if hydrologic variables are important in diagenesis, some common diagenetic patterns should arise if the effects of variables such as detrital composition and pressure-temperature paths can be unraveled. Predictable relationships must exist among basin evolution, paleohydrology, and certain aspects of chemical diagenesis in sedimentary rocks.

Given this conceptual framework, quantitative diagenetic predictions may be possible through combinations of field, experimental, and computational studies. Numerous papers during the 1980s have identified the paleohydrologic conditions leading to an observed set of diagenetic reactions; however, it is important to note that rarely is a single mineral characteristic of a particular hydrologic regime. Usually sets of minerals, as well as isotopic interpretations of water origin and temperature, are needed to provide convincing evidence for extant water chemistry (Milliken et al., 1981; Fisher and Land, 1986; Dutton and Land, 1988; Harrison, 1989). Advances in the recognition of paleohydrologic regimes from petrographic and geochemical studies of sedimentary diagenesis have been accompanied by advances in numerical simulation of hydrologic processes in sedimentary basins (Garven and Freeze, 1984a and b; Bethke, 1985; Garven 1989; Deming et al., 1990). These simulations have allowed prediction of paleohydrologic evolution in sedimentary basins (Bethke, 1986a and b; Bethke et al., 1988; Garven, 1989; Harrison and Summa, 1991). Accompanying advances in basin paleohydrology, calculations of complex chemical equilibria in aqueous systems are also possible following the development of geochemically sophisticated reaction path models (Wolery, 1979, 1983).

The combination of a powerful conceptual model with predictive abilities in paleohydrology and aqueous chemistry, and field evidence with which to calibrate the initial simulations, places the field of sedimentary diagenesis at an important turning point.

Table 1. Hydrologic Regimes and Diagenetic Features in Clastic Systems (modified from Galloway, 1984).

Regime	Hydrology	Diagenesis
Meteoric (< 3 km depth)	•infiltration of surface water •flow rates several meters/year-unlimited water supply •moves from recharge to discharge areas •low salinity	•clay fringes •detrital grain leaching •calcite, kaolinite, feldspar, zeolite, chalcedony cements
Compactional (surface to < 4 km)	•expulsion of trapped depositional water •reducing, evolved seawater* •moves upward & landward •flow rates few cm to mm.yr^{-1} •abnormal pressures possible •limited supply •wide range of salinities	•detrital grain & early cement •leaching •grain and cement replacements •chlorite, kaolinite, calcite, quartz feldspar cements •albitization of plagioclase
Thermobaric (>2.5–3 km)	•gas-rich dehydration waters •geopressured •flow rates < mm.yr^{-1} •limited supply •convective flow possible •wide range of salinities	•smectite to illite conversion •ferroan carbonate, kaolinite chlorite cements •complete albitization of all feldspars

*Evolved in the sense used here means that compactional water is related to seawater but compositional modifications have occurred as a result of a wide range of rock-fluid interactions, as well as water mixing processes.

Computational capability is critical because a multitude of variables affect chemical diagenesis in sediments, and these must be investigated systematically. In this chapter we present some of our initial calculations of quantitative diagenetic pathways by combining data from independent paleohydrologic and chemical simulations. In strict terms, coupling of the formal equations for fluid flow in porous media with chemical reactions between minerals, fluids, and gases is needed to make accurate predictions of porosity and permeability evolution and of attendant diagenetic changes. The development of combined chemical reaction-fluid flow models (e.g., Walsh et al., 1982; Chen et al., 1990) to undertake these kinds of simulations is in progress but few applications have thus far been described (Walsh et al., 1984; Meshri and Walker, 1990; Moore and Ortoleva, 1990). At this time, computer code is primarily developmental and not widely available. Simplifications in the chemical reaction and fluid flow components are required for computational efficiency, and the consequences of such simplifications for prediction of diagenetic reactions have not yet been evaluated.

Our approach is to use data obtained from paleohydrologic reconstructions of the Gulf of Mexico basin (Harrison and Summa, 1991) to constrain the configuration of reaction path models of mineral-water-gas interactions. Our calculations are calibrated

Table 2. Paleohydrologic evolution of major basin types.

Basin Example	Type	Primary water type	Primary Drive	Flow rate	Secondary water type	Secondary Drive	Flow rate	Paleohydrologic Evolution
Gulf of Mexico	Mature passive margin	Syn-depositional	Compaction	<1mm yr^{-1} to 10 cm yr^{-1}	Meteoric	Topography	10cm yr^{-1}	Meteoric water at elevated temperatures to >10m.yr^{-1} in Mesozoic; compactional geopressured since Tertiary restricting meteoric to shallow parts of basin (refs: 1, 2, 10).
Western Canadian basin	Mature interior foreland	Meteoric	Topography	<0.1 to >8m yr^{-1}	Syn-depositional	Compaction-compression	1-2 cm yr^{-1}	Syndepositional fluids moved by compaction in late Paleozoic. Tectonic compression in Laramide times mobilized compactional fluids that were mixed with and replaced by meteoric water driven by uplift on western margin (refs: 3-5,9,15).
Illinois	Cratonic	Syn-depositional	Compaction	0.2cm yr^{-1}	Meteoric	Margin tilting	5m yr^{-1}	Slow compaction in Paleozoic. Remobilization of syndepositional brines, flow towards north basin margin by Mesozoic tilting of south margin causing partial replacement and mixing of syndepositional and meteoric waters. Modern flow system reversed (refs: 6, 7, 10, 11).
Rhine graben	Early continental rift	Meteoric	Topography and thermal convection	0.1m yr^{-1}	Syn-depositional	Compaction	0.1mm yr^{-1}	Early stage compactional waters replaced by meteoric system and high thermal gradients initiating convectional circulation (refs: 8).
Appalachian and Arkoma basins	Interior foreland (active compressional phase)	Syn-depositional	Tectonic compression	-2cm yr^{-1}	Meteoric	Orogenic elevation	0m yr^{-1}	May have an early compactional phase prior to compression. Hydrology typical of slowly compacting basins. Meteoric regime emplaced by orogenic uplift replaces compactional system (refs: 9, 12, 13, 14).

References:
1. Harrison and Summa, 1991
2. Bethke et al., 1988
3. Garven and Freeze, 1984a,b
4. Garven , 1989
5. Toth, 1978
6. Bethke, 1985
7. Bethke et al., 1991
8. Person and Garven, 1988
9. Ge and Garven, 1988
10. Bethke, 1986b
11. Walther et al., 1990
12. Deming et al., 1990
13. Bethke and Marshak, 1990
14. Oliver, 1986
15. Bethke, 1989

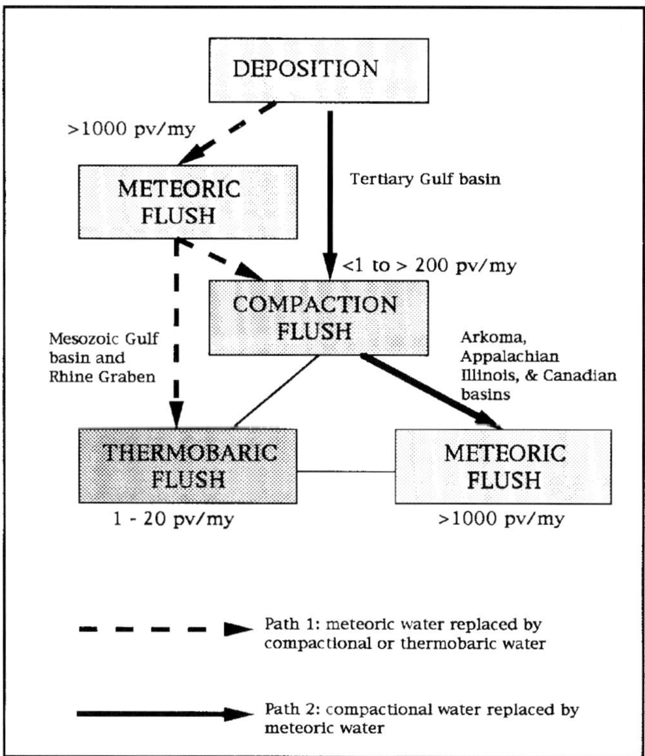

Figure 1. Galloway's generalized paths of hydrologic evolution for a sandstone (Galloway, 1984). Compilation of paleohydrologic studies (Table 2) reveals that two pathways are particularly common: compactional water replaced by meteoric water as shown by the solid bold line (Appalachians, Canadian Basin, Illinois Basin) and meteoric water replaced by thermobaric or compactional water as shown by the dashed bold lines (parts of Gulf of Mexico and Rhine graben). Other possible paths are shown by lighter-weight lines. Number of pore volumes per million years is calculated from specific discharge and/or flow rate data when given by the authors listed in Table 2. These values are averages and thus smaller as well as larger fluid amounts are possible. Each pore volume is 1000 cm³ (see text for further definition of pore volume).

using petrographic data from this basin, one of the best-documented sedimentary basins in the world in terms of the variables needed for our predictions. The approach presented in this chapter is simple, provides insight into complexities likely to be encountered in future advanced simulation, and also provides close approximations to observed diagenetic histories. The methodology we present can be applied to any sedimentary sequence of interest. We demonstrate our approach to combining independent calculations of hydrologic and chemical systems with two sets of calculations. The first presents a diagenetic pathway for the onshore Wilcox Group sandstone facies from time of deposition to present day. The second calculation involves the movement of pore fluid across a pressure gradient at the transition zone between abnormal and near-normal pressure in the offshore Frio shales.

GEOLOGY OF THE GULF OF MEXICO BASIN

The Gulf of Mexico basin has been characterized by continuous subsidence since its formation by rifting apart of the North and South American continents, possibly during Triassic times (Pindell, 1985). Evaporites, carbonates, and fluvial to marine clastics filled the large, shallow basin during Jurassic and Cretaceous times. From the early Tertiary and continuing to the present day, enormous quantities of terrigenous clastic sediment have been supplied to the basin from the continental interior and have been deposited as great prograding deltaic, coastal plain, and continental slope sequences (Hardin, 1962, Rainwater, 1967). Stratigraphically continuous sediment thicknesses may be as much as 17 km (Worzel and Watkins, 1973). The basin has been dominated by gravity tectonics including the massive mobilization of the Jurassic salts and development of contemporaneous growth faults which complicate the depositional sequences. Geothermal gradients are low to moderate, varying from 20°C km^{-1} offshore to 40°C km^{-1} onshore. Elevated heat flows were possible in the Jurassic post-rifting period (Smith et al., 1981). The Tertiary clastic sediments as well as the Jurassic carbonates and eolian sandstones (Jurassic Smackover and Norphlet Formations, for example) are the main hydrocarbon source and producing intervals (Kennicutt et al., 1992) .

The lithologic makeup of the fluvial, deltaic, and marine Tertiary sequences is simple: quartzose, arkosic, and lithic sandstones are interbedded with smectitic and illitic shales. Occasional coals are found in the fluvial facies. The Oligocene Frio and Catahoula formations are unique in that they have a significant volcaniclastic component. Hundreds of publications describe the geology, petrology, and diagenesis of the Tertiary sediments in the Gulf of Mexico basin (see, for example, Loucks et al., 1984; Land et al., 1987; Harrison and Summa, 1991; and references therein). Volumetrically important diagenetic reactions include grain leaching and replacement, authigenic kaolinite, quartz, and calcite, albitization of feldspar and late ankerite cements. Other phases, less abundant but common, include authigenic K-feldspar, albite, and chlorite. Analcime and clinoptilolite are found in the Catahoula and Frio Formations. The classic smectite to illite conversion dominates shale diagenesis at depths between 3 and 3.5 km. Secondary porosity estimates vary, depending on the vogue, but may be around 10% (Lundegard et al., 1985).

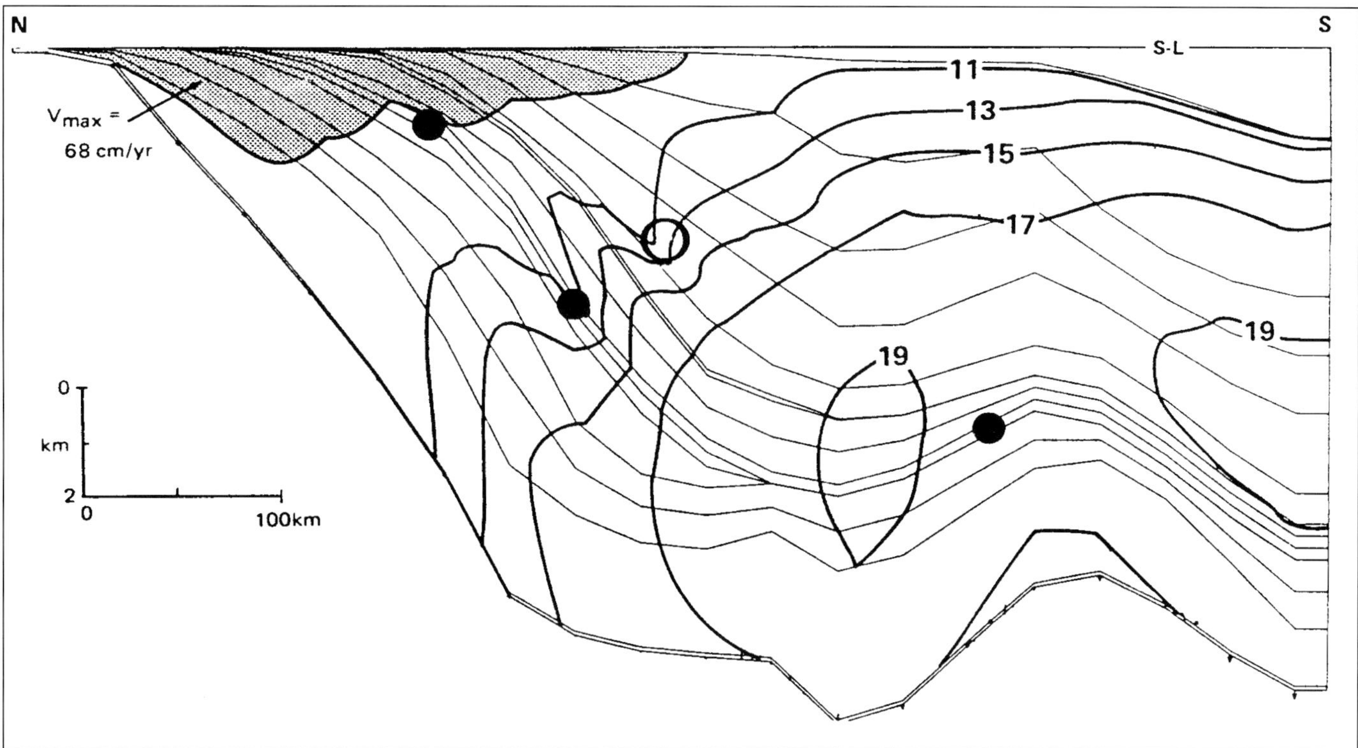

Figure 2. Simulated cross section through northern Gulf of Mexico Basin (Harrison and Summa, 1991) showing locations for which diagenetic pathways will be calculated. Solid circles are Wilcox Group locations. The open circle shows the Frio Formation transition zone from abnormal to normal fluid pressures. In this chapter, only two diagenetic pathways are presented: for the shallowest Wilcox location and for the Frio Formation transition zone region in the circle. The contours, which are fluid pressure gradient in MPa. km^{-1}, show the portion of the basin having fluid pore pressures significantly in excess of hydrostatic. (Conversion of pressure gradients in units of MPa.km^{-1} to psi.ft^{-1} is as follows: 10 MPa.km^{-1} = 0.47 psi.ft^{-1}, 16 MPa.km^{-1} = 0.7 psi.ft^{-1} and 23 MPa.km^{-1} = 1.0 psi.ft^{-1}.) Maximum flow velocity of 68 cm.yr^{-1} is shown in the meteoric regime.

PALEOHYDROLOGY OF THE GULF OF MEXICO BASIN

Present day pore waters vary from meteoric with a few hundred mg.kg^{-1} total dissolved solids (TDS) to saline brines derived from the Jurassic evaporites with TDS >250,000 mg.kg^{-1} (Morton and Land, 1987). The Gulf of Mexico basin is a classic example of a geopressured basin in which geopressure has been generated by compaction disequilibrium (Dickinson, 1953; Magara, 1971; Bethke, 1986a). Geopressured pore fluids are typically found below 2-4 km depth, separated from the overlying near-hydrostatically pressured sediments by a transition zone of variable thickness (Wallace et al., 1977). In the hydrostatic portion of the basin, meteoric waters are underlain by, and partially mixed with, waters expelled from the compacting shales and waters derived from salt dome dissolution (Wesselman, 1985; Morton and Land, 1987). The present day distribution of shallow near-hydrostatic and deeper geopressured sediments is not typical of Mesozoic and early Tertiary times, however, and paleohydrologic reconstructions show that the north central basin has only become strongly geopressured since the Miocene (Bethke et al., 1988; Harrison and Summa, 1991).

Three stages characterize the hydrologic evolution of the Gulf of Mexico basin (Harrison and Summa, 1991). The first stage, from Jurassic to early Tertiary times, was characterized by circulating meteoric waters, heated by elevated post-rifting geothermal gradients and mixed at depth with saline waters expelled from compacting Jurassic salts. During much of Tertiary time the second stage, involving development of geopressures and restricted meteoric water infiltration, developed as the clastic sediment supply increased. The final stage of widespread geopressure, fluid pressure gradients approaching lithostatic, and minimal meteoric water infiltration has been active since late Miocene times. Regional variations to this history have occurred within the basin depending upon the time and location of depocenter activity. There has been a clockwise migration of the major Tertiary clastic deltas from South Texas in Paleocene times to west of the present Mississippi delta in Pleistocene times (Hardin, 1962). Thus the onset of geopressuring was probably during the Eocene times in south Texas, Oligocene to Miocene times in the cen-

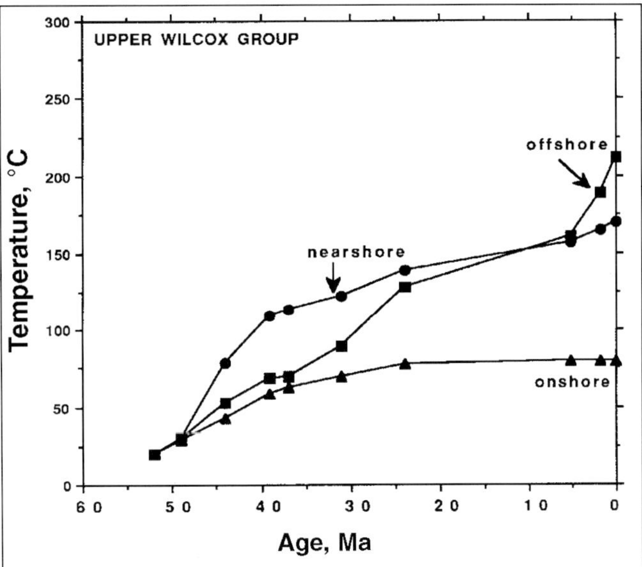

Figure 3. Temperatures for the three locations in the Wilcox Group shown in Figure 2 as a function of time since deposition at 52 Ma.

tral Texas region, and Pliocene-Pleistocene times in north Texas, Alabama, and Louisiana (Harrison and Summa, 1991).

Computational reconstructions of the hydrology of the Eocene Wilcox Group and Oligocene Frio Formation by Harrison and Summa (1991) are used in this study. Figure 2 shows locations (nodal blocks in the simulations) for which medium and fluid properties have been predicted through time. Many nodal blocks have similar flow and medium properties, and thus to be efficient in our chemical calculations we have chosen hydrologically distinctive locations at which to calculate diagenetic pathways: an onshore facies (predominantly sandstone) initially flushed by meteoric water prior to replacement by compactional water expelled from the offshore shales, a nearshore facies (sandstone and shale) which has been buried to moderate depths entirely in the compactional hydrologic regime, and an offshore shale which has been strongly geopressured for most of its burial history.

Figures 3 through 6 show the key physical variables in each location since deposition of the upper Wilcox strata 52 Ma. Inspection of these figures reveals differences in all the flow and medium properties between each location and differences among the diagenetic alteration patterns are to be expected. Of particular interest is Figure 6 which shows that the number of pore volumes of fluid varies by three orders-of-magnitude among the three facies and that 70% of the cumulative discharge has passed through the rocks in the first 15 m.y. following deposition.

COMPUTATIONAL METHODOLOGY

Conceptual Model

The methodology involved in integrating the hydrologic and reaction path calculations is straightforward. For a selected nodal block in the paleohydrologic simulations, fluid and medium properties are calculated for any user-specified time interval. To reduce the number of computational cycles we must make, we average all properties over a selected time interval that typically spans several millions of years and represents a major stratigraphic unit. The first step is thus to tabulate the porosities, temperatures, pressures, specific discharges, and linear velocities over the selected time step (Columns 1-6, Table 3).

The second calculation involves converting specific discharge rates to pore volumes based on a single pore volume of 1000 cm^3 (column 7, Table 3). This is necessary because the chemical calculations made using EQ3/6 (Wolery, 1979, 1983) equilibrate minerals, gases, and 1000 g of aqueous fluid; thus chemical calculations will be for reaction occurring in a single

Table 3. Medium and Flow Properties—Onshore Wilcox Group.

Interval (m.y.)	Porosity (%)	Temp. (°C)	Pressure (MPa)	Discharge ($cm^3.cm^{-2}$)	Velocity ($cm.yr^{-1}$)	Pore Volumes per m.y.
-52 to -49	52	24.2	3.35	$3.33x10^6$	2.35	1110
-49 to -44	43.5	35.5	5.3	$5.38x10^6$	2.6	1076
-44 to -39	35.5	50.5	8.55	$2.98x10^6$	0.492	596
-39 to -37	30	60.6	11.3	$0.15x10^6$	0.0011	75
-37 to -31	28	66	12.8	$0.19x10^6$	0.229	32
-31 to -24	26	73.1	14.7	$1.02x10^6$	0.918	146
-24 to -5.2	24.5	78.1	15.75	$2.19x10^6$	0.016	116
-5.2 to -1.8	23.5	79.3	15.6	$0.56x10^6$	0.71	165
-1.8 to 0.0	23	79.4	15.6	$0.23x10^6$	0.41	128

Columns show average data for a time interval. For example, a total discharge of $3.33x10^6$ cm^3 cm^{-2} applies to the 52 to 49 Ma time interval.

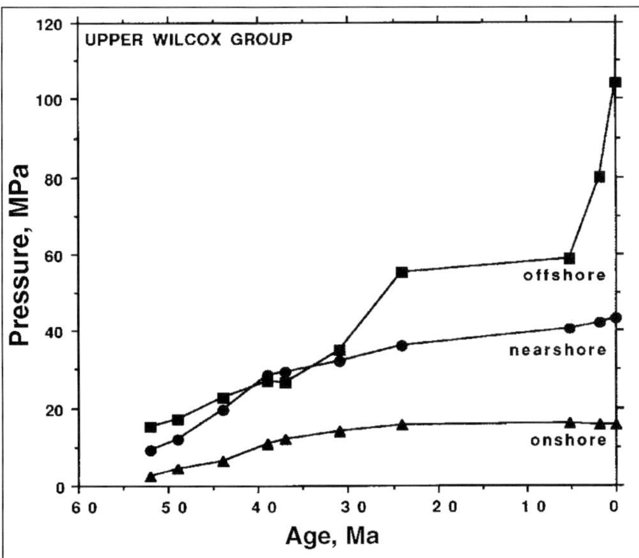

Figure 4. Pressures for the three locations in the Wilcox Group shown in Figure 2 as a function of time since deposition at 52 Ma.

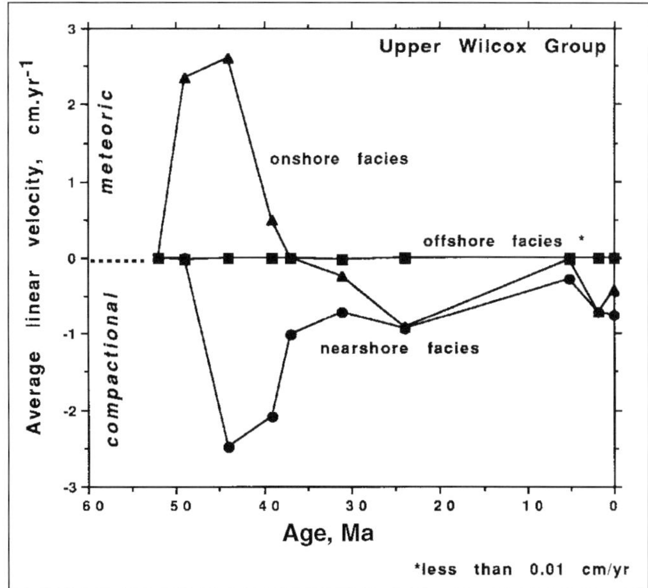

Figure 5. Average linear velocities for fluid flow at the three locations in the Wilcox Group shown in Figure 2 as a function of time since deposition at 52 Ma. Positive velocities indicate basinward, meteoric flow whereas negative velocities indicate landward, compactional flow. Fluid in the onshore facies sandstones is initially meteoric in origin and is replaced by landward flowing compactional fluids about 12 million years after deposition. Flow rates are so slow in the offshore shales (< 10^{-4} $cm.yr^{-1}$) that they plot at 0 $cm.yr^{-1}$ in this figure.

pore volume (assuming 1000 g is equivalent to 1000 cm^3). Here we use pore volume in the reservoir engineering context to mean the volume of pore space in an arbitrarily chosen volume of porous rock.

The third step involves the reaction path calculations. The calculations are closed system, equilibrium, isothermal, and isobaric at the average temperature and pressure for the time interval. Pressure is used to constrain the partial pressures of gases in equilibrium with the water. Reaction proceeds by the incremental addition of reactant minerals to the water until the entire rock volume has equilibrated with 1000 g of water. Clearly, this situation does not occur in nature, based on textural and mineralogical relationships among detrital and authigenic phases. The amount of mineral reaction (i.e. the extent of reaction as monitored by the reaction progress variable, Z_i) that is permitted with 1000 g of water is determined empirically as being the largest volume possible before the product mineral assemblage no longer matches mineral assemblages commonly reported by petrographers in the Gulf of Mexico basin. Our initial approach was to use rate constants (Lasaga, 1984) to determine the amount of dissolution that should occur in the time it takes to replace a single pore volume calculated from the linear velocities in Table 3, however, we found that because the flow rates are very slow we could essentially react the entire rock volume with the pore fluid. The small amount of rock reaction that appears reasonable thus translates to a difference between the total surface area of the rock and the reactive surface area, the latter being as much as 1000 to 10,000 times smaller. Differences of these magnitudes are also reported from experimental and other modeling studies (Bruton, 1986; Helgeson et al., 1988; White and Peterson, 1990).

The fourth step is to take the volumes of dissolved reactant and precipitated product minerals in a single pore volume and multiply these volumes by the number of pore volumes available in a selected time step (Column 7, Table 3). This calculation allows us to obtain a new, modified, rock composition and porosity which will be used starting again as input to step three with a higher temperature, pressure, and new number of pore volumes. If any phase is completely dissolved by the amount of fluid available during a given time step, a new reaction path calculation is made, thus the number of reaction path calculations exceeds the number of time increments shown in Table 3. Typically, between 15 and 20 EQ3/6 calculations are made to obtain a complete diagenetic history for a given location.

Uncertainties and Assumptions

A number of uncertainties and assumptions underlie this approach. In addition to the accuracy of the paleohydrologic reconstructions (see Harrison and Summa, 1991 for a discussion of the paleohydrologic sensitivity analysis) and the thermodynamic and kinetic data in the reaction path calculations, these

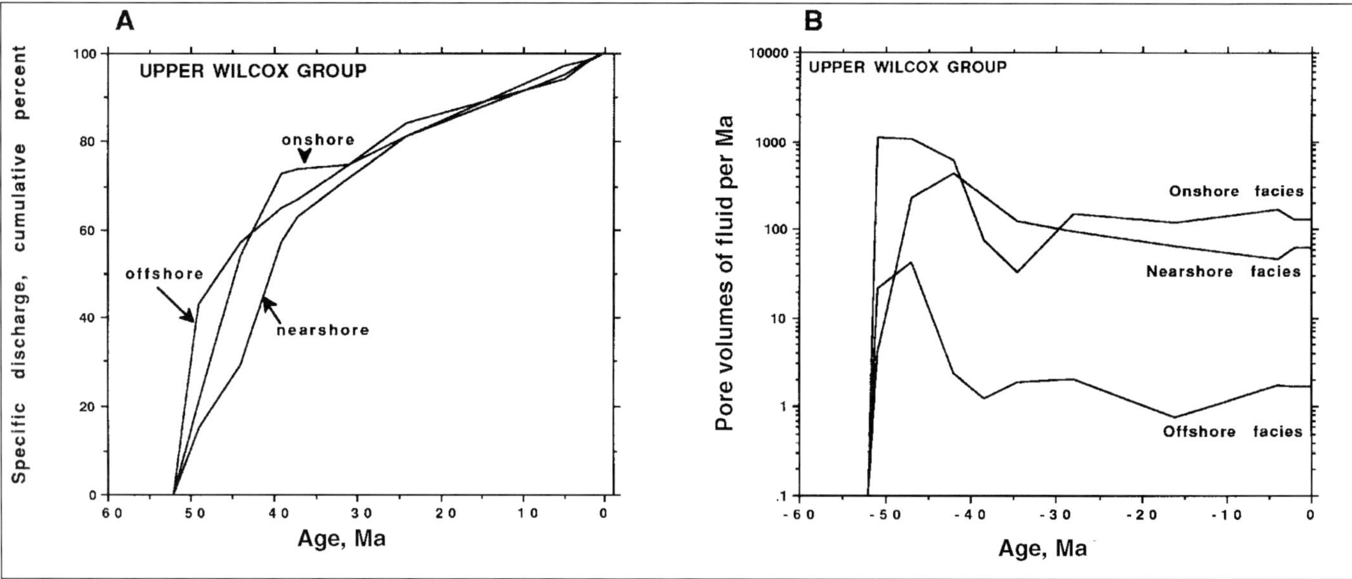

Figure 6. (A) Cumulative percent specific discharge ($cm^3.cm^{-2}.yr^{-1}$), and (B) Pore volumes of fluid for the three locations in the Wilcox Group shown in Figure 2 as a function of time since deposition at 52 Ma. Although there are differences of orders-of-magnitude in discharge among the three locations, 70% of all the available fluid at each location flows through the rocks in the first 15 m.y. after deposition.

uncertainties include:

(1) the extent of rock reaction allowed,

(2) the assumption of isothermal/isobaric conditions for any time step,

(3) each pore volume of water is "new", not evolved, and only the dissolved gas composition changes between time steps,

Table 4A. Input Water Compositions—Onshore Wilcox Group.

Meteoric*	mg kg^{-1}	Compactional**	mg kg^{-1}
Ca^{2+}	0.70	Ca^{2+}	1490
Mg^{2+}	0.20	Mg^{2+}	151
Na^+	2.20	Na^+	29800
K^+	0.60	K^+	230
Cl^-	0.34	Cl^-	46300
SO_4^{2-}	3.6	SO_4^{2-}	18
SiO_2	tr	SiO_2	84
Fe^{2+}	tr	Fe^{2+}	tr
Al^{3+}	tr	Al^{3+}	tr
pCO_2	0.0032 MPa	pCO_2	0.05 MPa
pH	5.5	pH	6.5
T°C	24-50	T°C	50-79

Table 4B. Hypothetical Sandstone Composition—Onshore Wilcox Group.

Quartz: 37.5%
Plagioclase (An50 Ab50): 18.75%
K-feldspar: 18.75%
Porosity: 25%

*Meszaros (1981) **Morton and Land (1987)

(4) kinetics of precipitation and dissolution are not important on the time scale of the simulation,

(5) no pressure solution is accounted for,

(6) there is no feedback between mineral reaction and porosity modification in the chemical model and permeability and fluid flow in the hydrologic model, and,

(7) porosity change due to compaction is only accommodated in the flow calculations which yield the pore volumes.

All of the above, with the exception of 5 , 6, and 7, can be investigated with sensitivity analyses, and can be calibrated by comparing predicted diagenetic pathways with the observed ones. This is an advantage of the approach, not a limitation. Assumption (3) results in calculation of the most optimistic scenario, i.e., the most amount of chemical diagenesis that can occur. Correct manipulation of chemical reaction plus pressure solution in the presence of moving fluid requires a more sophisticated combined flow-reaction simulation (Chen et al., 1990; Dewers and Ortoleva, 1990).

Calculations

We demonstrate our approach to combining independent calculations of hydrologic and chemical systems with two sets of calculations. The first presents a diagenetic pathway for the onshore Wilcox Group sandstone facies from time of deposition to present day. This calculation demonstrates the time-dependent diagenetic changes that occur for one selected set of physical and chemical conditions. The second calculation involves the movement of pore fluid across a pressure gradient at the transition zone between

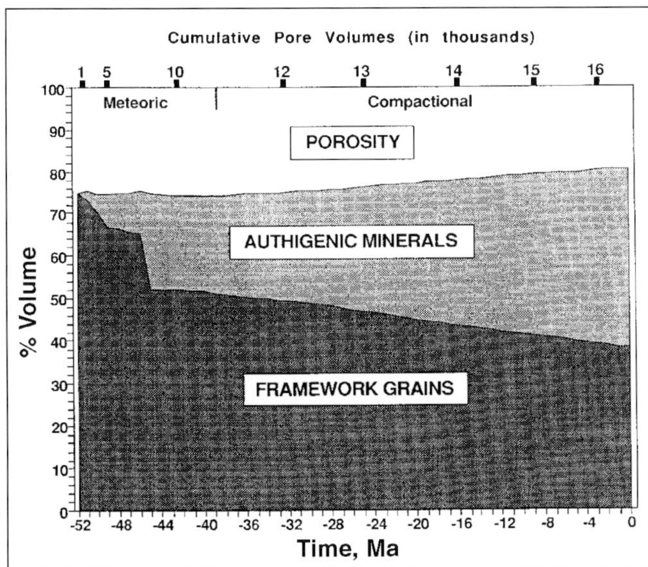

Figure 7. Calculated diagenetic pathway for onshore facies Wilcox Group sandstone. Overall there is a 5% porosity loss due to all non-compactional diagenetic changes. The relationship between age and the volume of fluid that has passed through the rock is not linear and is a function of the flow rate and specific discharge variation shown in Table 3.

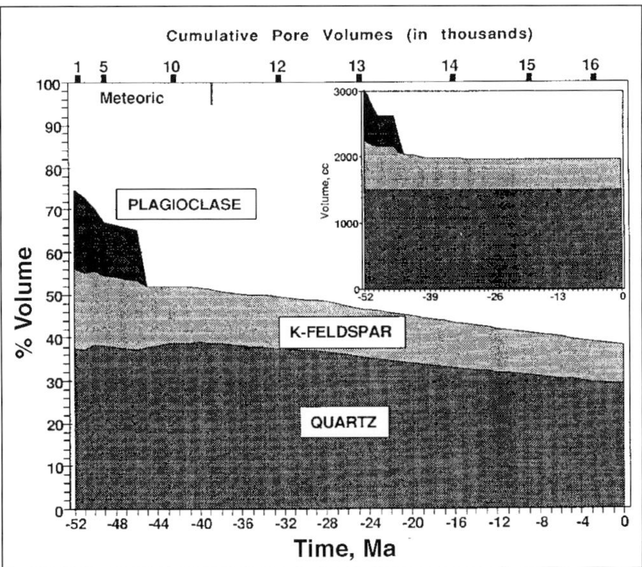

Figure 8. Calculated changes in detrital mineralogy during the diagenetic pathway for onshore facies Wilcox Group sandstone. Plagioclase dissolves completely in the first few million years after burial. K-feldspar dissolves more rapidly in the meteoric regime but continues to dissolve throughout most of the pathway. Quartz remains stable throughout. The insert shows the absolute volumes from which the volume percentages are derived. Decreases in the volume percent of quartz result from the increasing total volume of the system as higher molar volume authigenic phases replace the detrital minerals.

abnormal and near-normal pressure in the offshore Frio shales. This calculation demonstrates the use of sensitivity analyses to place limits on the chemical reactions that can occur within a small interval of the total diagenetic pathway.

RESULTS

Diagenetic Pathways in the Wilcox Group

Input Data for Onshore Wilcox Group: The diagenetic pathway is constrained by data listed in Table 3. The starting sandstone composition was assumed to be an arkose (with no iron-bearing detrital phases for simplicity in preliminary calculations) having an initial intergranular volume of 25% (Table 4). The early stages of diagenesis occurred while the sandstone was being invaded by meteoric water, whereas the later stages occurred as shale-derived compactional water flushed the sandstone (Table 4; Figure 5). The total pressure was used to constrain partial pressures of the gases CO_2 and CH_4 and in this particular pathway the CH_4:CO_2 ratio was fixed at 95:5 (molar) based on typical measurements of CO_2 at the shallow depths of burial experienced by this particular sandstone (Franks and Forester, 1984).

Diagenetic Pathways for Shallow, Onshore Wilcox Group

Figure 7 shows how the composition of the model sandstone changes from time of deposition to the present day. Porosity decreases from 25% to 20% as a result of authigenic mineral formation. Figure 7 also shows the dramatic replacement of the framework grains by authigenic phases. At deposition framework grains make up 75% of the rock but are reduced to 40% of the rock at the present time. The percentage of authigenic phases increases, correspondingly, from 0% to 40% during the burial history. Plagioclase feldspar is the least stable component in this pathway and is completely dissolved within the initial 7 million years of burial (Figure 8). The early authigenic minerals derive most of their components from plagioclase dissolution. Potassium feldspar persists over the length of the pathway although more dissolution occurs in the meteoric system than in the compactional system. In the final stages of compaction, K-feldspar is essentially stable. The volume of detrital quartz remains constant (Figure 8): decreases in the volume percent of quartz result from the increasing total volume of the system as higher molar volume authigenic phases replace the detrital minerals. We select volume percent as a graphical variable because this parameter is routinely determined during petrographic analysis.

Figures 9 and 10 show more detail of the diagenetic pathway for the onshore Wilcox facies. Quartz

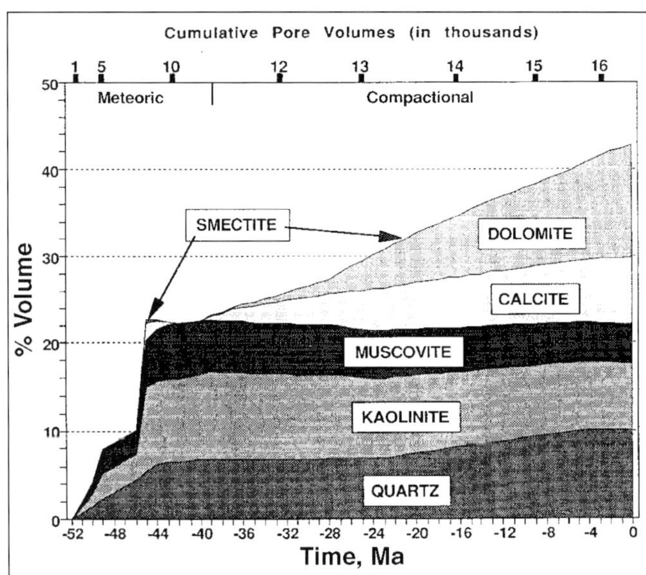

Figure 9. Authigenic phases produced during diagenesis of onshore, Wilcox Group sandstone. Muscovite (illite-see text) and kaolinite are produced continuously in the meteoric regime as a result of feldspar breakdown. Early calcite and smectite are products of plagioclase dissolution. Two episodes of carbonate cementation are evident: 1-2 % calcite cement is meteoric in origin, whereas as much as 20% is produced later, after the invasion of compactional waters. A similar trend is seen in development of quartz cement. Limitations to the supporting thermodynamic data base probably result in the prediction of dolomite as opposed to the more commonly reported ankerite.

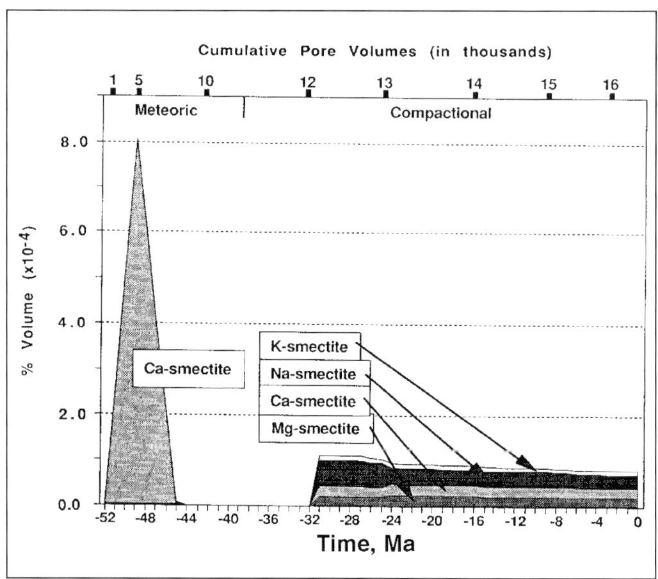

Figure 10. Composition of diagenetic smectite produced during diagenesis of onshore, Wilcox Group sandstone. The predicted smectite is a solid solution whose components are 33% beidellite and 67% montmorillonite in the meteoric regime and 50% montmorillonite (e.g. $Ca_{.165}Mg_{.33}Al_{1.67}Si_4O_{10}(OH)_2$), 30% beidellite (e.g. $Ca_{.165}Al_{2.33}Si_{3.67}O_{10}(OH)_2$ and 20% nontronite (e.g. $Ca_{.165}Fe_2^{+++}Al_{.33}Si_{3.67}O_{10}(OH)_2$) in the compactional regime.

cement, kaolinite and muscovite (a thermodynamic proxy for illite: unreliability in basic thermodynamic quantities for illite as well as lack of chemical data on exact authigenic illite compositions in Gulf of Mexico sediments justifies this substitution of muscovite for illite) form as a result of feldspar dissolution. The rate of kaolinite formation is greatest during the first 7 million years of burial and decreases during the last 5 million years of burial. Two episodes of carbonate cementation are observed. Initially, a few percent calcite cement are produced as a result of plagioclase dissolution with the carbonate ions being provided by CO_2 dissolved in the meteoric water (43-47 Ma). During the transition from meteoric to compactional systems, this calcite is dissolved. In later stages of burial, after the replacement of the meteoric water by compactional waters is complete, a second, volumetrically significant carbonate cementation event occurs involving initially calcite (beginning at 39 Ma), then calcite and dolomite (beginning at 37 Ma). Quartz cement in this simulation has two origins: several percent of quartz precipitate as a result of feldspar dissolution; and in the compactional regime quartz cement continues to form when feldspars no longer dissolve, this quartz cement originating from silica dissolved in the pore waters (Table 4A). Our calculations show that there is adequate water movement in the compactional hydrologic system to cause significant quartz cementation, however, at this time we have not evaluated the source of the dissolved components in the Morton and Land (1987) waters. The final authigenic assemblage in this onshore Wilcox sandstone is thus 13% dolomite (ferroan carbonate/ankerite?), 7% calcite, 4% illite, 6% kaolinite, and 10% quartz. A small amount of smectite is formed as the plagioclase dissolves (Figure 10). When precipitated in the meteoric regime it is a calcic composition whereas that produced in the compactional regime is predominantly a Na-smectite, in agreement with observations by Hanor and McManus (1985) in shallow aquifers in Louisiana.

Variations in water chemistry are shown in Figure 11. These variations reflect the composition of the water in equilibrium with the rock at each time step, i.e., the evolved composition at the end of a single pore volume calculation. The partial pressure of CO_2 is fixed in any time step by the total pressure constraint discussed previously. The abrupt change from meteoric to compactional compositions is an assumption we have made: more gradual mixing and replacement could be tested in subsequent calculations.

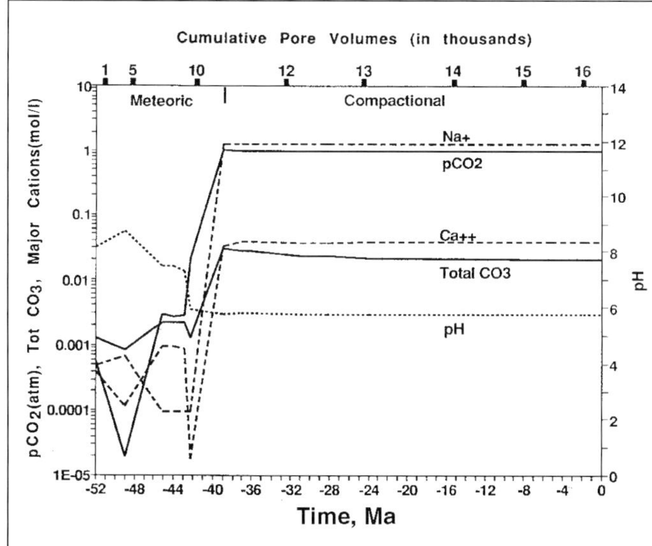

Figure 11. Composition of pore waters during diagenesis of onshore, Wilcox Group sandstone. This figure shows calculated water compositions at the end of each time increment after reaction with the sandstone. The partial pressure of CO_2 was fixed during each reaction step and only varied between reaction steps.

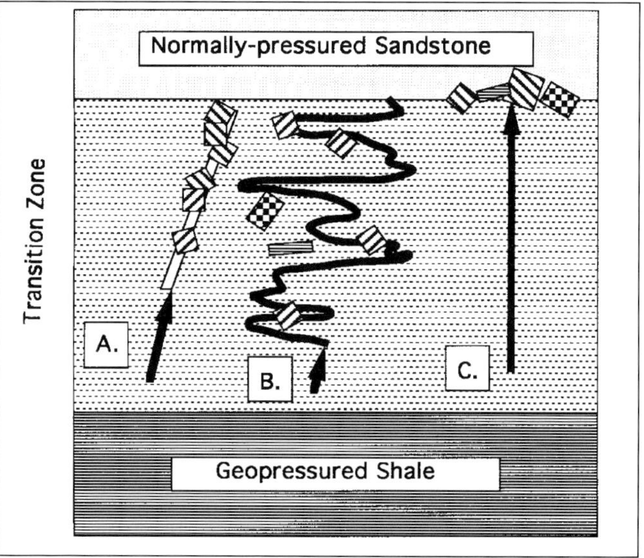

Figure 12. Schematic representation of several possible scenarios at the transition zone between geopressured and near-hydrostatically pressured sediments. A. Rapid degassing as pore fluid moves through fractures in shale may or may not be accompanied by cooling: diagenetic minerals precipitate in fluid conduit. B. Fluid takes a tortuous path through shale, reacting with minerals in shale and cooling along flow path: diagenetic minerals are dispersed within the shale. C. No rock reaction occurs on fluid flow path through shale: all reaction is confined to the sandstone/shale interface.

Sensitivity Analyses

Sensitivity analyses show that initial rock composition and paleohydrology—the latter also defining fluid type, temperature, and pressure—influence the chemistries and volumes of authigenic mineral phases in a sandstone. Work is in progress to compare diagenesis in the onshore, nearshore, and offshore facies rocks and in three Wilcox sandstone compositions—a quartzose subarkose, a lithic arkose, and a feldspathic litharenite—and some of these results are summarized below. Continued sensitivity analyses will consider the role of mineral surface area and the effects of pCO_2, pCH_4, and organic acids on diagenesis.

An overall increase in the feldspar content of a sandstone results in an increase in the amount and diversity of the authigenic clay assemblage. Greater amounts of kaolinite correspond with increasing detrital potassium feldspar content. Likewise, the amount of Ca-smectite increases with increasing amounts of detrital plagioclase. Chlorites are observed to form from compactional fluids where the feldspar content of the framework grains has been increased.

Changes in paleohydrology, with accompanying differences in fluid type, temperatures, and pCO_2, influence the amounts and compositions of the clays. In the meteoric regime, Ca-smectites dominate, whereas in the compactional regime, Mg-smectites are the primary clay. In the nearshore facies, Na-smectites precipitate early in the burial history and are replaced by Ca-smectites, which are in turn replaced by Mg-smectites late in the burial history. Compositional difference in the nearshore facies result in greater amounts of Ca-smectite and Mg-smectite precipitating as the amount of plagioclase feldspar increases. In the quartz-rich sandstone, a late stage kaolinite is precipitated. The amount of quartz cement varies with the amount of detrital feldspar, the amount of silica in the pore waters, and the temperature history during burial. Amounts of quartz cement of up to 12% have been calculated for the nearshore diagenetic pathway: despite reduced rates of fluid flow, elevated temperatures promote silica cementation. Sensitivity analyses on the behavior of iron in our predicted diagenetic pathways shows a complex distribution of iron among smectites, chlorites, illites, oxides, and sulfides.

Comparison With Petrographic Descriptions

Petrographic descriptions of paragenetic sequences in the Wilcox Group provide the calibration for our model diagenetic pathway. Our simplified pathway is in good agreement with observations of reactions. We are limited by lack of petrographic data for shallow Wilcox Group sandstones as most of the classic diagenetic studies (Boles, 1978; Loucks et al., 1984; Fisher and Land, 1986) involve deeper rocks (generally 2 to >4 km burial depth) which thus do not have compa-

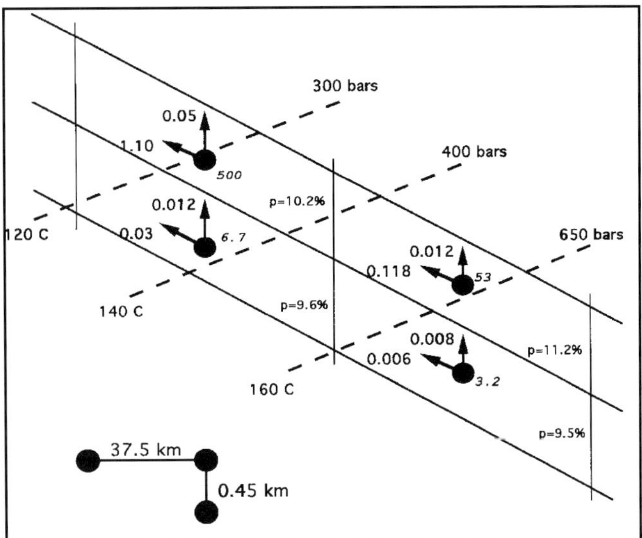

Figure 13. Flow and medium properties in four nodal blocks defining the transition zone in Frio Formation shales (location shown in Figure 2). Conversion of pressure in bars to MPa is 10 bars = 1 MPa. Large solid circles are centers of nodal blocks and are the points for which fluid properties are calculated. Bold arrows denote cross-stratal (vertical) and strata-parallel (approximately horizontal) velocities in cm•yr^{-1}; p denotes porosity, in percent; italicized numbers are the cumulative number of pore volumes of fluid for a 4 m.y. time step flowing through each nodal block. More details on the construction of the paleohydrologic model are given by Harrison and Summa (1991).

rable burial histories. Interpretation of the earliest, shallow diagenetic reactions is not always unequivocal, although Stonecipher and May (1990) has shown that careful petrographic analysis can reveal syndepositional modifications. Maximum temperature of burial is only 80°C in our simulated pathway and depths greater than 1.5 km are not reached. Reactions such as albitization and smectite conversion to illite are not predicted in our pathway. Our calculations do not allow the distinction between various textural modes-of-occurrence and thus we cannot specify whether the authigenic clays are replacive or pore-filling. We predict a maximum of 10% quartz cement, which is reasonable given the temperatures and depths of this pathway, the arkosic composition of the starting composition, and the fact that the silica is derived from both feldspar dissolution as well as the compactional waters. Significantly higher amounts of quartz cement are not commonly reported at depths shallower than 1.5 km (McBride, 1989). Our predicted smectite compositions are comparable to those analyzed by Hanor and McManus (1985) in shallow diagenesis in the Louisiana Cretaceous aquifers: Ca-smectite is found in the shallowest alteration assemblages and becomes more sodic downdip from the recharge area. All the carbonate cement in this diagenetic pathway comes from plagioclase dissolution in conjunction with dissolved calcium, magnesium and carbonate ions: there is no detrital carbonate in the starting rock composition. Dolomite is rarely reported in the Gulf of Mexico Tertiary formations, however, ferroan calcite and ankerite cements are common in the early Tertiary rocks in particular (Boles, 1978; Fisher and Land, 1986; Land et al., 1987). We assume that the dolomite is being precipitated in our models because at this time we have not included more than trace amounts of iron in the chemical system and because we do not have the appropriate thermodynamic data for ankerite or ferroan calcites. Both current limitations are subjects of ongoing studies.

The dissolution of plagioclase and K-feldspars and the precipitation of calcite, illite, and kaolinite, either replacing the original grains or occurring as discrete pore-filling phases, are commonly reported diagenetic consequences of meteoric water infiltration into feldspathic sandstones. Our simulated diagenetic pathway shows that for conditions of relatively slow fluid flow (a few cm.yr^{-1}) in this particular hydrologic regime, a sandstone can increase its porosity from 25% to 46% by detrital grain dissolution in as little as 7 million years if authigenic phase precipitation is inhibited or if dissolved components are removed from the local system. Porosities of 46% are, of course, unlikely in the subsurface. A more realistic interpretation of this result should include further compaction of the rock concurrently with grain dissolution. However, in the event that authigenic minerals do precipitate in situ, an overall reduction of porosity by about 2% could occur in the same time interval.

Diagenetic Consequences of Transition Zone Degassing

As pore fluids flow from areas of geopressure to areas where pore pressures are near-hydrostatic, dissolved gases may be exsolved. The loss of such gases causes changes in the pore water chemistry that may result in mineral precipitation and dissolution along the flow path (Bruton and Helgeson, 1983). This degassing fluid may maintain its temperature, and/or cool, and/or react with the rock (sandstone/shale/limestone) through which it flows, resulting in the conceptual scenarios shown in Figure 12. In this conceptualization we place the transition interval within a shale and propose a normally-pressured sandstone although other lithologic combinations might be possible. Minerals may be precipitated along fractures, may be dispersed in the shale, or may be precipitated at a sandstone-shale contact. Fluid expulsion from shale has been implicated in sandstone cementation (e.g., Boles and Franks, 1978) and more recently in hydrologic sealing of shales (Hunt,

1990). By using reaction path calculations we can place boundary conditions on the diagenetic consequences of the scenarios shown in Figure 12. These calculations involve combinations of polybaric, polythermal, isothermal, and fluid-rock interaction reaction paths.

Input Data For Transition Zone Calculations

Hydrologic conditions at the transition zone between geopressured and near-hydrostatic regions of the Gulf of Mexico basin are shown in Figure 13. From the data in this figure showing the four nodal blocks that make up the transition interval in the lower part of the Oligocene Frio Formation (Figure 2), we can derive the temperature and pressure gradients at the present day needed to configure reaction path models. A pressure decline of 69 to 35 MPa (690 to 350 bars) coupled with cooling from 160 to 120°C are the maximum changes predicted. Flow rates vary from as little as 0.006 cm.yr^{-1} in the geopressured zone to 1.10 cm yr^{-1} in the near-hydrostatic zone. Data in Figure 13 are for the present day; the hydrologic simulations indicate that these conditions represent the maximum pressures, temperatures, and flow rates in this geologic unit and that such conditions have been in existence for no longer than about 4 m.y. (Harrison and Summa, 1991). Prior to 4 million years ago (Ma) this location was at near-hydrostatic conditions. It is reasonable to take present day flow rates and calculate the number of pore volumes of fluid moving across the transition zone, if we assume that such discharges can have been in existence for no more than 4 m.y. For the range of flow rates in Figure 13 and shale porosities of 9–11%, we obtain between 2 and 500 pore volumes of fluid (1000 cm^3 per pore volume) that may have flowed across the transition zone in 4 m.y., with amounts of 10–100 pore volumes being most likely.

The composition of the water assumed to be passing across the transition zone and the mineralogy of the shale through which the fluid passes are shown in Table 5. The key reaction path variable is gas pressure. We calculated the mineralogical consequences of gas loss for a fluid initially in equilibrium with CO_2 and CH_4 at a total pressure of 69 MPa (690 bars) which re-equilibrated at a total gas pressure of 35 MPa (350 bars) after flowing across the transition zone. The fraction of CO_2 in the gas was between 0.5 and 5 mole % (Franks and Forester, 1984). All calculations were polybaric with respect to the gas partial pressure only and no pressure effects on mineral solubilities or intra-aqueous reactions were considered. The reaction path is thus one of sliding gas fugacities. We also included isothermal, polythermal, and fluid-rock interactions in some calculations. As previously, we assume that the rates of mineral precipitation and dissolution are fast relative to the time it takes to fill a pore volume with fluid. Based upon calculations using the data of Lasaga (1984) and flow rates in Figure 13, such an assumption appears valid.

Diagenetic Consequences of Degassing at the Transition Zone

Figure 14 shows the minerals precipitated and changes in fluid chemistry over the pressure interval

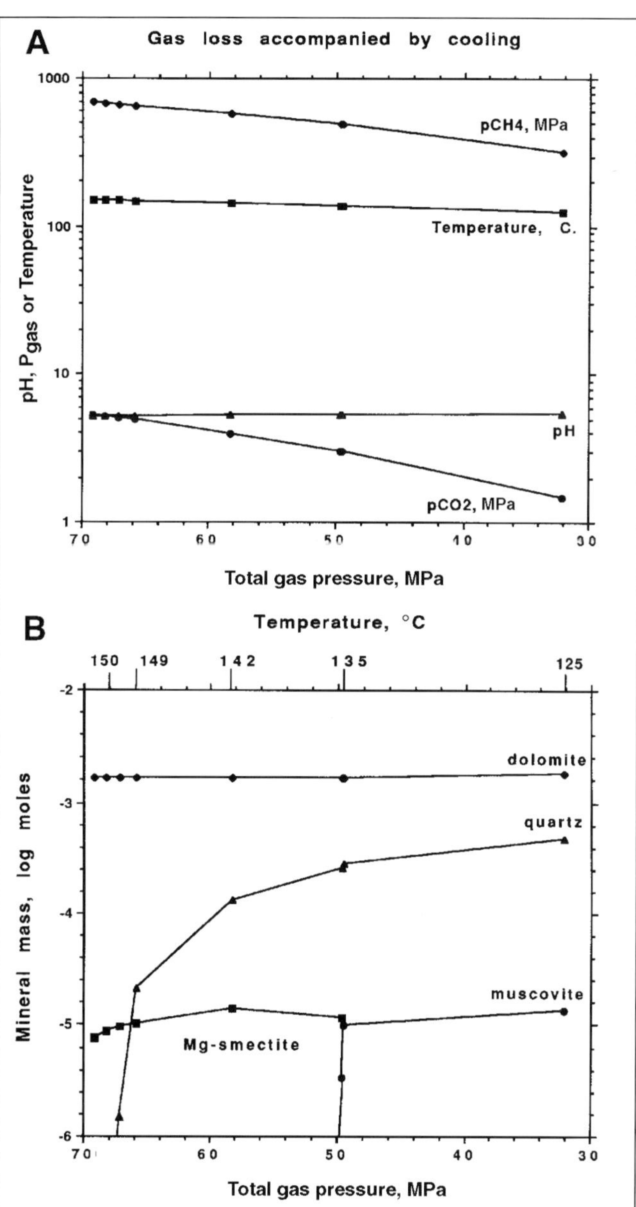

Figure 14. Reaction path calculations for a single pore volume of fluid flowing across pressure and temperature gradients equivalent to those at the transition zone. Conversion of pressure in bars to MPa is 10 bars = 1 MPa. No shale-fluid interactions are allowed. A. Fluid chemistry. B. Minerals precipitated. Most of the dolomite precipitated results from supersaturation of the hypothetical fluid as it is expelled from the source shale (see text). The smectite composition is a montmorillonite ($Mg_{.165}Mg_{.33}Al_{1.67}Si_4O_{10}(OH)_2$) .

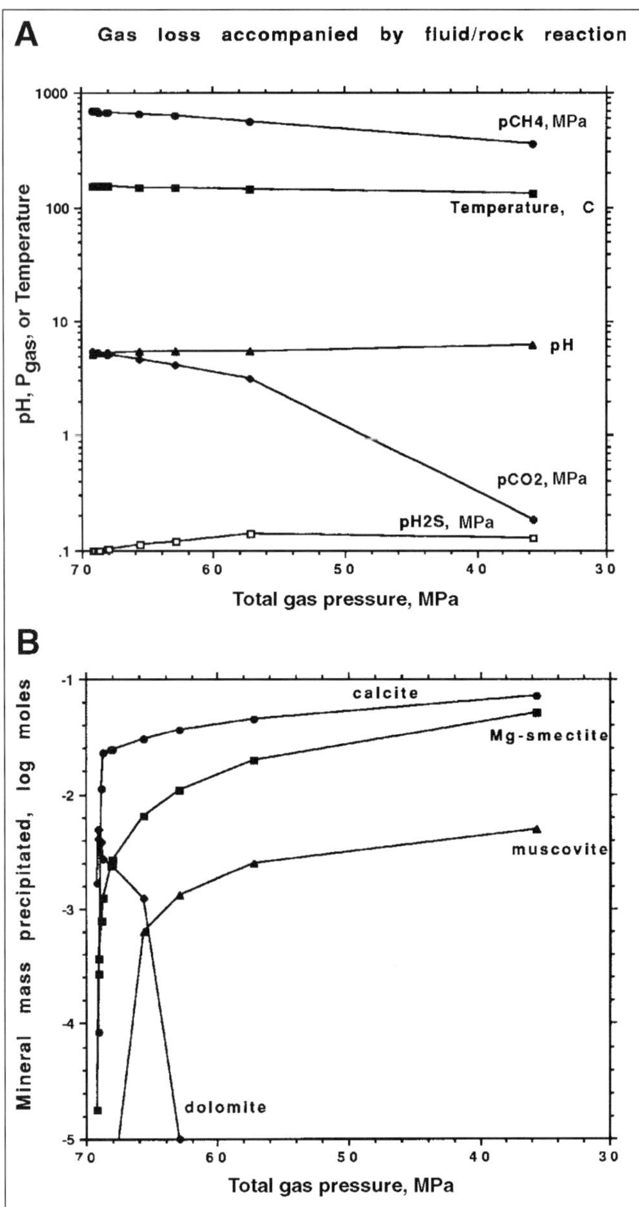

Figure 15. Reaction path calculations for a single pore volume of fluid flowing across pressure and temperature gradients equivalent to those at the transition zone and where shale-fluid interactions are allowed. Conversion of pressure in bars to MPa is 10 bars = 1 MPa. A. Fluid chemistry. B. Minerals precipitated. The smectite composition is a montmorillonite ($Mg_{.165}Mg_{.33}Al_{1.67}Si_4O_{10}(OH)_2$) .

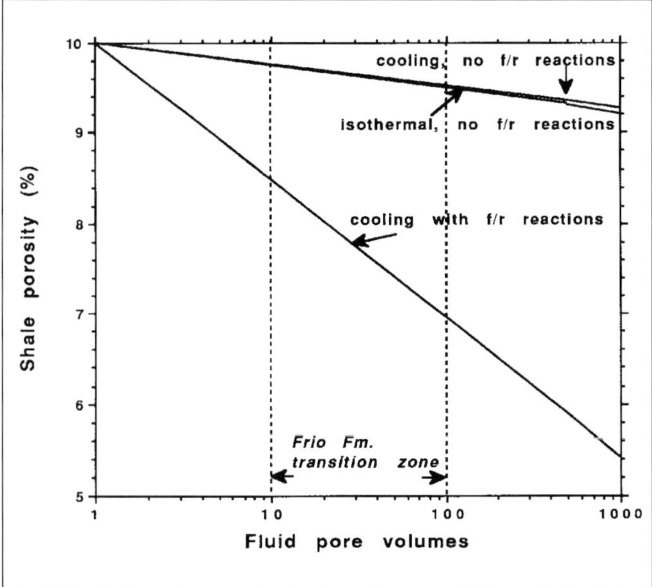

Figure 16. Porosity loss attributed to mineral precipitation as fluids are degassed at the transition zone between normal and geopressured regions. Up to 3.5% porosity decline can be caused by such a mechanism under optimal conditions; porosity losses of less than 1% are more probable based on calculations of fluid volumes shown in Figure 13. The principal minerals precipitated are carbonates and clays.

of the selected part of the lower Frio Formation. The calculations are for a single pore volume of water and include both decreases in pressure and temperature, but no interaction with the rock is permitted. As is to be expected, the loss of gas results in a small increase in pH and in the precipitation of the carbonate minerals, in this instance dolomite, although calcite and ankerite could precipitate from a slightly different starting water composition. The precipitation of significantly smaller volumes of quartz and clays results from the cooling of the water alone and not the pH change. Slightly larger amounts of carbonate precipitate from the water if degassing is isothermal, a result of the retrograde solubility of carbonate minerals, but under these conditions no silicates are precipitated (not shown). Similar calculations, allowing the fluid to interact with the shale through which it passes, result in higher volumes of precipitated minerals as the fluid causes the alteration of some of the detrital feldspars to new phases with higher molar volumes (Figure 15).

With potentially as many as 500 pore volumes of fluid passing across the transition zone in the last 4 m.y., we calculated the total volume of precipitated minerals and the resultant loss in shale porosity assuming a compacted value of 10% for this burial depth (Figure 13). Our results are shown in Figure 16. Unless the fluid reacts with the shale during degassing, the impact on porosity is minimal, < 1.0 % porosity reduction. If fluid-rock interaction does occur there is a greater porosity loss of up to 4 %. In the calculations shown in Figures 14 and 15, the fluid composition has been derived through a pre-equilibration step with a hypothetical shale. In this step the fluid is allowed to supersaturate with respect to dolomite because of uncertainties in the precipitation kinetics of this mineral. Thus, a portion of the mineral precipitate along the depressurization path actually occurs at the start of the path as a result of the initial supersaturation. If this scenario is unrealistic then the amount of precipitation errs on the high side and the impact of this process on porosity has been overestimated.

Calculations of this nature indicate that carbonate and silicate minerals may precipitate as fluids flow from geopressured to near-hydrostatically pressured regions (see also Bruton and Helgeson, 1983). However, cementation by degassing fluids is not likely to be a very effective mechanism of porosity loss, although we are unable to quantify the attendant permeability reduction. The reason is not that the conceptual model of sandstone or shale cementation is poor, rather, it is that hydrologically reasonable conditions do not permit water volumes to be high enough such that the mechanism could be effective. From quantification of the hypothesis we find that pore fluid degassing can only be a partial cause of sandstone/shale permeability loss and cannot be a dominant process.

DISCUSSION AND CONCLUSIONS

Paleohydrologic analysis of this sedimentary basin provides an overall perspective on controls on clastic sediment diagenesis at a broad scale. Similarities between sedimentary strata with respect to their hydrologic evolution will result in certain similarities in sediment diagenesis, although these may not be evident if there are significant variations among detrital mineralogy and thermal history. Paleohydrologic analysis also points to processes that must be significant during sediment diagenesis. Inspection of Figure 6 suggests that most open-system diagenesis, where mass transfer occurs by advection, must occur early in the burial history when flow rates are relatively rapid. Open system diagenesis is followed by a prolonged period when mineral-water equilibration occurs in a more closed chemical system and where chemical diffusion may be an important process. Even under these conditions, transport of material in solution still continues and can be effective in producing diagenetic alteration. In the Wilcox Group example we present here, 4% of the quartz cement is attributed to feldspar dissolution during relatively early open-system diagenesis, while the remaining 6% is the result of precipitation from slower-moving, but higher-temperature, compactional fluids. The behavior of both carbonates and clays shows an additional feature of our predicted diagenetic pathways: episodic cementation related to significant changes in hydrologic conditions is superimposed on a background of continuous reactions (Figures 9 and 10). In a sedimentary basin that has tectonic reactivation of the hydrologic system, additional periods of open-system diagenesis may occur later in the burial history although we do not show these simulations in this chapter.

The combination of chemical calculations constrained by physical data from hydrologic reconstructions and petrographic observations makes it possible to quantify the diagenetic products that may be produced along specific pressure and temperature paths during sediment burial. Furthermore, sensitivity analysis can be used to place limits on cement volumes, the variables that control cement types, and the resultant porosity changes that may occur. Although our calculations are incomplete, it is clear that this approach provides new insight into relative efficiencies of the processes controlling chemical diagenesis. Inspection of Figures 7 through 10, for example, shows that cementation and porosity loss are not continuous processes during burial: diagenesis appears to be episodic in nature, with major cementation events being related to significant changes in hydrologic evolution which, in turn, are related to broader aspects of basin evolution. In the calculations presented in this paper we deal only with diagenetic pathways in the shallow Wilcox: presumably, we will see additional episodic cementation/dissolution events as we incorporate the effects of hydrocarbon generation deeper in the sedimentary sequence.

Table 5A. Input Water Composition—Frio Formation Transition Zone.*

Parameters	mg kg^{-1}
Ca^{2+}	1490
Mg^{2+}	51
Na^{+}	29800
K^{+}	230
Cl^{-}	46300
SO_4^{2-}	18
SiO_2	84
Fe^{2+}	tr
Al^{3+}	tr
pCO_2	0.5 MPa (5 bars)
pCH_4	68.5 MPa (685 bars)
pH	6.0
T °C	160-125

Table 5B. Hypothetical Shale Composition—Frio Formation.

Quartz: 40.5%	Kaolinite: 3.6%
Plagioclase (An50:Ab50): 16.2%	Calcite and dolomite: 0.9%
K-feldspar: 21.6%	Pyrite: 1.8%
Illite: 3.6%	Mg-smectite: 1.8%
Porosity: 10%	

*Modified from Morton and Land (1987).

Figure 16 shows that it is unlikely that degassing at the transition zone (as represented by the Gulf of Mexico Frio Formation) can be an effective mechanism for creating seals (c.f. Hunt, 1990) although certainly such fluid could, however, contribute a few percent of mostly carbonate cement, perhaps distributed in microfractures in shale or in patches at sand-shale contacts. We find further evidence to support the suggestion of Harrison and Thyne (1991) that the diagenetic makeup of a particular sediment is most likely the result of many processes that have acted during burial and that to assign a single process as being dominant places an obligation to demonstrate that such a process can be quantitatively effective.

ACKNOWLEDGMENTS

This work has been supported by grants to the first author from Texaco Exploration and Production Technology Division, Mobil Research and Development Company, and the National Science Foundation NSF-8916440. The authors appreciate thoughtful reviews by Carol Bruton, Edward Warren, and Andrew Robinson.

REFERENCES CITED

Bethke, C.M., 1985, A numerical model of compaction-driven groundwater flow and heat transfer and its application to the paleohydrology of intracratonic sedimentary basins: Journal of Geophysical Research, v. 90, p. 6817-6828.

Bethke, C.M., 1986a, Inverse hydrologic analysis of the distribution and origin of Gulf Coast- type geopressured zones: Journal of Geophysical Research, v. 91, p. 6535-6545.

Bethke, C.M., 1986b, Hydrologic constraints on the genesis of the Upper Mississippi Valley mineral district from Illinois basin brines: Economic Geology, v. 81, p. 233-249.

Bethke, C.M., 1989, Modelling sub-surface flow in sedimentary basins: Geologische Rundschau, v. 78, p. 124-154.

Bethke, C.M. and S. Marshak, 1990, Brine migrations across North America—the plate tectonics of groundwater: Annual Reviews of Earth and Planetary Sciences 1990, v.18, p. 287-351.

Bethke, C.M., W.J. Harrison, C. Upson, and S.P. Altaner, 1988, Supercomputer analysis of sedimentary basin: Science, v. 239, p. 261-267.

Bethke, C.M., J.D. Reed, and D.F. Oltz, 1991, Long range petroleum migration in the Illinois basin, American Association of Petroleum Geologists Bulletin, v. 75, p. 925-945.

Boles, J.R., 1978, Active Ankerite cementation in the sub-surface Eocene of Southwest Texas; Contrib. Mineralogy and Petrology, v. 59, p. 281–292.

Boles, J.R. and S.G. Franks, 1978, Clay diagenesis in Wilcox sandstones of Southwest Texas: implications of smectite diagenesis on sandstone cementation: Journal of Sedimentary Petrology, v. 49, p. 55-70.

Bruton, C.J., 1986, Predicting mineral dissolution and precipitation during burial: Synthetic diagenetic sequences: Proceedings of the Workshop on Geochemical Modeling, LLNL Publ. CONF-8609134, p. 111-120.

Bruton, C.J. and H.C. Helgeson, 1983, Calculation of the chemical and thermodynamic consequences of differences between fluid and geostatic pressure in hydrothermal systems: American Journal of Science, v. 283-A, p. 540-588.

Chen, W., A. Ghaith, A. Park, and P.J. Ortoleva, 1990, Diagenesis through coupled processes: modeling approach, self-organization, and implications for exploration, *in* I. Meshri, ed., Prediction of reservoir quality through chemical modeling: American Association of Petroleum Geologists Memoir 49, p. 103-131.

Deming, D., J.A. Nunn, and D.G. Evans, 1990, Thermal effects of compaction-driven groundwater flow from overthrust belts: Journal of Geophysical Research, v. 95, p. 6669-6683.

Dewers, T. and P.J. Ortoleva, 1990, Interaction of reaction, mass transport, and rock deformation during diagenesis: mathematical modeling of intergranular pressure solution, stylolites, and differential compaction/cementation, *in*, I. Meshri, ed., Prediction of reservoir quality through chemical modeling: American Association of Petroleum Geologists Memoir 49, p. 147-161.

Dickinson, G., 1953, Geologic aspects of abnormal reservoir pressures in Gulf Coast Louisiana: American Association of Petroleum Geologists Bulletin, v. 37, p. 410-432.

Dutton, S.P. and L.S. Land, 1988, Cementation and burial history of a low permeability quartz arenite, Lower Cretaceous Travis Peak Formation, East Texas: Geological Society of America Bulletin, v. 100, p. 1271-1282.

Fisher, R.S. and L.S. Land, 1986, Diagenetic history of Eocene Wilcox sandstones, south central Texas: Geochimica et Cosmochimica Acta, v. 50, p. 551-563.

Franks, S.G. and R.W. Forester, 1984, Relationships among secondary porosity, pore-fluid chemistry and carbon dioxide, Texas Gulf Coas, in, D.A. McDonald and R.C. Surdam, eds., Clastic Diagenesis: American Association Petroleum Geologists Memoir 37, p. 63-81.

Galloway, W.E., 1984, Hydrogeologic regimes of sandstone diagenesis, in D.A. McDonald and R.C. Surdam, eds., Clastic Diagenesis: American Association Petroleum Geologists Memoir 37, p. 3-13.

Garven, G., 1989, A hydrogeologic model for the formation of the giant oil sands deposits of the western Canadian sedimentary basin: American Journal of Science, v. 289, p. 105-166.

Garven, G. and R.A. Freeze, 1984a, Theoretical analysis of the role of groundwater flow in the genesis of stratabound ore deposits 1. Mathematical and numerical model: American Journal of Science, v. 284, p. 1085-1124.

Garven, G. and R.A. Freeze, 1984b, Theoretical analysis of the role of groundwater flow in the genesis of stratabound ore deposits 2. Quantitative results: American Journal of Science, v. 284, p. 1125-1174.

Ge, S. and G. Garven, 1988, Tectonically induced transient groundwater flow in foreland basins, in R.A. Price , ed., Origin and Evolution of Sedimentary Basins and their Energy and Mineral Resources: American Geophysical Union Geophysical Monograph Series 48, p. 145-157.

Hanor, J.S. and K.M. McManus, 1985, Sediment alteration and clay mineral diagenesis in a regional groundwater flow system, Mississippi Gulf Coastal Plain: Gulf Coast Association of Geological Societies, Transactions, v. 38, p. 495-502.

Hardin, G.C., 1962, Notes on the Cenozoic sedimentation in the Gulf Coast geosyncline USA, in, E.H. Rainwater and R.P. Zingula , eds., Geology of the Gulf Coast and Central Texas and guidebook of excursions: Houston Geological Society for the 1962 Annual Meeting of the Geological Society of America and Associated Societies p. 1-15.

Harrison, W.J., 1989, Modeling fluid/rock interactions in sedimentary basins, in, T.A. Cross, ed., Quantitative Dynamic Stratigraphy: New York, Elsevier, p. 195-231.

Harrison, W.J. and G.D.Thyne, 1991, Predictions of diagenetic reactions in the presence of organic acids: Geochimica et Cosmochimica Acta, v. 56, p. 565-586.

Harrison ,W.J. and L.L. Summa, 1991, Paleohydrology of the Gulf of Mexico Basin: American Journal of Science, v. 291, p. 109-176.

Helgeson, H.C., W.M. Murphy, and P. Aargaard, 1988, Thermodynamic and kinetic constraints on reaction rates among minerals and aqueous solutions II Rate constants, effective surface area and the hydrolysis of feldspar: Geochimica et Cosmochimica Acta v. 48, p. 2405-2433.

Hunt, J.M., 1990, Generation and migration of petroleum from abnormally pressured fluid compartments: American Association of Petroleum Geologists Bulletin, v. 74, p. 1-12.

Kennicutt, M.C., II, T.J. McDonald, P.A. Comet, G.J. Denoux, and J.M. Brooks, The origins of petroleum in the northern Gulf of Mexico: Geochimica et Cosmochimica Acta, v. 56, p. 1259- 1280.

Land, L.S., K.L. Milliken, and E.F. McBride, 1987, Diagenetic evolution of Cenozoic sandstones, Gulf of Mexico basin. Sedimentary Geology, v. 50, p. 195-225.

Lasaga, A.C., 1984, Chemical kinetics of water-rock interactions: Journal of Geophysical Research, v. 89, B6 p. 4009-4025

Loucks, R.G., M.M. Dodge and W.E. Galloway, 1984, Regional controls on diagenesis and reservoir quality in Lower Tertiary sandstones along the Texas Gulf Coast, in, D.A. McDonald and R.C. Surdam, eds., Clastic Diagenesis: American Association of Petroleum Geologists Memoir 37, p. 15-47.

Lundegard, P.D., L.S. Land, and W.E. Galloway, 1985, Problem of secondary porosity: Frio Formation (Oligocene) south Texas: Geology, v. 12, p. 399-492.

Magara K., 1971, Permeability considerations in generation of abnormal pressures: Society of Petroleum Engineers Journal, v. 11, p. 236-242.

McBride, E.F., 1989, Quartz cement in sandstones, a review: Earth-Science Reviews v. 26, p. 69- 112.

Meshri, I. D. and J. M. Walker, 1990, A study of rock-water interaction and simulation of diagenesis in the Upper Almond Sandstones of the Red Desert and Washakie basins, Wyoming, in, I. D. Meshri and P.J. Ortoleva, eds., Prediction of reservoir quality through chemical modeling: American Association of Petroleum Geologists Memoir 49, p. 55-70.

Meszaros, 1981, Atmospheric Chemistry: Elsevier Scientific Publishing Co., New York, 210p.

Milliken, K.L., L.S. Land, and R.G. Loucks, 1981, History of burial diagenesis determined from isotopic geochemistry, Frio Formation, Brazoria County, Texas: American Association of Petroleum Geologists Bulletin, v. 65, p. 1397-1413.

Moore, C.H. and P.J. Ortoleva, 1990, Effects of fluid and rock compositions on diagenesis: a modeling investigation, in, I. Meshri, ed., Prediction of reservoir quality through chemical modeling: American Association of Petroleum Geologists Memoir 49, p. 131-147.

Morton, R.A. and L.S. Land, 1987, Regional variations in formation water chemistry, Frio Formation (Oligocene) Texas Gulf Coast: Association Association of Petroleum Geologists Bulletin, v. 71, p. 191-206.

Oliver, J., 1986, Fluids expelled tectonically from orogenic belts: their role in hydrocarbon migration and other geologic phenomena: Geology v. 14, p. 99-102.

Person, M. and G. Garven, 1988, Hydrologic constraints on the thermal evolution of the Rhine Graben, in, A.E. Beck, G. Garven, and L. Stegna, eds., Hydrogeologic regimes and their subsurface thermal effects: American Geophysical Union Geophysical Monograph 47, p. 35-58.

Pindell, J.L., 1985, Alleghenian reconstruction and subsequent evolution of the Gulf of Mexico, Bahamas and Proto-Caribbean: Tectonics, v. 4, p. 1-39.

Rainwater, E.H., 1967, Regional stratigraphy of the Gulf Coast Miocene: Gulf Coast Association of Geological Societies, Transactions, v. 1, p. 81-124.

Smith, D.L., W.T. Dees, and D.W. Harrelson, 1981, Geothermal conditions and their implications for basement tectonics in the Gulf Coast Miocene: Gulf Coast Association of Geological Societies, Transactions, v. 31, p. 181-190.

Stonecipher, S.A. and J.A. May, 1990, Facies controls on early diagenesis: Wilcox Group, Texas Gulf Coast, in, I. D. Meshri and P.J. Ortoleva, eds., Prediction of reservoir quality through chemical modeling: American Association of Petroleum Geologists Memoir 49, p. 25-44.

Toth, J., 1978, Gravity-induced cross-formational flowof formation fluids, Red Earth Region, Alberta, Canada: Analysis, patterns and evolution: Water Resources Research, v. 14, p. 805-843.

Wallace, R.H. Jr., R.E.Taylor, and J.B. Wesselman, 1977, Use of hydrogeological mapping techniques in identifying potential geopressured-geothermal reservoirs in the lower Rio Grande embayment Texas: Third Geopressured-Geothermal Energy Conference, v.1, p. GI-1 - GI 88.

Walsh, M.P., L.W. Lake, and R.S. Schechter, 1982, A description of chemical precipiation mechanisms and their role in formation damage during stimulation by hydrofluoric acid: Journal of Petroleum Technology, September 1982, p. 2097-2112.

Walsh, M.P., S.L. Bryant, R.S. Schechter, and L.W. Lake, 1984, Precipitation and dissolution of solids attending flow through porous media: American Insitute of Chemical Engineers Journal, v. 30, p. 317-328.

Walther, L.M., A.M. Stueber, and T.J. Huston, 1990, Br-Cl-Na systematics in Illinois basin fluids: constraints on fluid origin and evolution: Geology, v. 18, p. 315-318.

Wesselman, J.B., 1985, Structures, temperatures, pressures and salinities of Cenozoic aquifers of South Texas—map: US Geological Survey Hydrologic Investigations Atlas H654.

White, A.F. and M.L. Peterson, 1990, Role of reactive-surface area characterization in geochemical knietic models, Chapter 35 in "Chemical modeling of aqueous systems II" D.C. Melchior and R.L. Bassett, eds., American Chemical Society Symposium Series 416: American Chemical Society, Washington D.C., p. 461-477.

Wolery, T.J., 1979, Calculation of chemical equilibria between aqueous solutions and minerals—The EQ3/6 software package: Lawrence Livermore National Laboratory Report UCRL-52658.

Wolery, T.J., 1983, EQ3NR: A computer program for geochemical aqueous speciation-solubility calculations - User's guide and documentation: Lawrence Livermore National Laboratory Report, UCRL-53414.

Worzel, J.L. and J.S. Watkins, 1973, Evolution of the northern Gulf Coast deduced from geophysical data: Gulf Coast Association of Geological Societies, Transactions, v. 23, p. 84-91.

Chapter 7

Stress-Induced Fluid Flow in Rifted Basins

R.T. van Balen and S.A.P.L. Cloetingh
Tectonics/Structural Geology Group
Vrije Universiteit
Amsterdam, The Netherlands

ABSTRACT

We present a model for sudden changes of fluid expulsion rates and overpressures in rifted sedimentary basins. The numerical model couples the stretching mechanism for rifted basin formation with fluid conservation and flow equations for compacting sediments.

Forces operating at plate boundaries exert stresses in the lithosphere, affecting the shape of basins and sedimentation rates. We demonstrate that these stresses affect the hydrodynamics of sedimentary basins, with possible implications for the diagenesis of sediments, faulting, and localization of economic resources. Increases in the level of compressive intraplate stresses induce flank uplift and increased subsidence in the basin center, thereby causing a contemporary basin-wide increase in compaction driven flow in the center and deeper penetration of meteoric water at the flanks of the basin. An increase in tensile intraplate stresses causes subsidence at the flanks of the basin and differential uplift in the basin center, which leads to a contemporary basin-wide decrease in compaction driven flow and reduced penetration of meteoric water.

INTRODUCTION

Fluids in sedimentary basins are mobile on geological time scales and create and localize hydrocarbons and ore deposits. Two major sources exist for fluids in sedimentary basins: connate fluids (pore water) and meteoric water. The connate fluids are driven out during burial of the sediments and are generally overpressured because of a continuous increase in loading by the sedimentation process and restricted flow possibilities. Meteoric water penetrates the basin from the flanks where sediments are exposed. Maximum flow velocities in the meteoric domain can be of the order of a meter per year, while in the compaction domain maximum velocities are only of the order of centimeters per year (Bethke et al., 1988). The boundary between meteoric and compaction domains is controlled by the permeability distribution, pressure differences, and buoyancy (Bjoerlykke et al., 1989; De Vries, 1989), as the density of fresh water is lower than for connate fluid. This is important as influx of fresh water in sediments which previously contained connate fluid has significant implications for their diagenesis (Bjoerlykke et al., 1989).

The sediments in the basin can be subdivided into permeable (sandstones, carbonates) and almost impermeable (shales, evaporites) strata. The interlayering of the strata controls the location of overpressures and the directions of flow. Thick shale deposits are in general overpressured due to the combination of low permeability and high porosity at deposition. The overpressures can approach the total vertical stress or overburden pressure exerted by the overlying column of rock and water, causing the sediments to have porosities much higher than expected (Smith, 1971). The highest overpressures are found in basins with high sedimentation rates and in basins which contain shales and salt in, for example, the Niger Delta, the Gulf of Mexico and the North Sea basins, and can be as high as 100 MPa. (Bethke et al., 1988; Bethke, 1989; Ungerer et al., 1987; Burrus and Audebert, 1990). Underpressuring also may occur in sediments with low permeabilities like shales because of erosional unloading (Neuzil and Pollock, 1983; Bethke, 1989; Kreitler, 1989), causing slight decompaction of sediments and degassing of the pore fluid.

Sedimentation rate and permeability distribution are controlling factors for the generation of overpressures (Bethke, 1985). The extent and velocities of flow in the meteoric water domain depend on the topography and the amount of sediment exposure. Intraplate stresses in the lithosphere change the basin shape (Cloetingh, 1986), thereby changing the sedimentation pattern of the basin, the topography, and the amount of exposure at the flanks of the basin. As a result, variations in the level of intraplate stresses also perturb the hydrodynamics of sedimentary basins. Although considered to be important during the basin formation stage, intraplate stresses, until recently, generally have been ignored in models for the postrift evolution of sedimentary basins. In this chapter, we investigate the effect of intraplate stresses on the compactional hydrodynamics of sedimentary basins, using transient simulations of basin development and fluid flow. Subsequently we discuss the implications for diagenesis and the localization of economic resources.

INTRAPLATE STRESSES AND RIFTED BASIN EVOLUTION

The stretching model (McKenzie, 1978) provides a convenient mathematical formulation to describe the process of rifted basin formation. Subsidence during rifting is caused by isostatic adjustment of the thinned crust. A result of the thinning process is the elevation of temperatures in the lithosphere. The postrift tectonic subsidence of rifted sedimentary basins can be largely explained by subsequent cooling and associated thermal contraction, sedimentary loading, and flexural bending due to concentrated vertical loads. Thickening of the lithosphere caused by cooling results in a widening of the basin during the postrift evolution (Watts, 1982), which combined with an increasing sedimentary load, produces a continuous onlap for the stratigraphy at the flank of the basin. The stretching model predicts an exponential decay of the postrift tectonic subsidence. Short-term deviations of this pattern can be explained in terms of relative sea level changes. Such deviations are associated with a sequence of onlaps and offlaps in the stratigraphy of the basin. Glacio-eustatics (Pittman and Golovchenko, 1983) and intraplate stresses can both explain second and third order changes in relative sea level (Cloetingh, 1991). These intraplate stresses are induced by plate tectonic forces (slab pull and ridge push) and can reach magnitudes up to several hundreds MPa (Cloetingh and Wortel, 1985; Stephenson and Lambeck, 1985).

Figure 1 demonstrates the effect of intraplate stresses lithosphere during the rifting phase. During the post-rift evolution, increases in the level of compressive intraplate stresses result in an uplift of the flank and deepening of the center of the basin, thus producing an offlap at the flank and increased sedimentation at the center. Increases in tensile intraplate stresses produce increased subsidence at the flank and uplift of the center of the basin, resulting in an increased onlap at the flank and decreased sedimentation at the basin center (Cloetingh et al., 1985). The actual magnitudes of the stress- induced uplifts and subsidence depend on the ratio of stress and lithosphere strength (Cloetingh, 1991), and vary between several tens of meters for subtle changes in stress levels (order of magnitude a few tens of MPa) to magnitudes of several hundred meters to a few km when the stress levels approaches the lithospheric strength.

Expected changes in the hydrodynamics of a sedimentary basin due to intraplate stresses are shown schematically in Figure 2. Deepening of the basin center caused by compressive stresses induces higher sedimentation rates and, therefore, results in an increase in compaction driven flow and overpressures. The resulting flank uplift induces an increase in the penetration of meteoric water. Uplift in the basin center and subsidence at the basin flank caused by tensile intraplate stresses results in a contemporary decrease in compaction induced flow and overpressures, and a decrease in the extent of the meteoric water domain. The sign and magnitude of intraplate stresses change on a time scale of a few million years (Philip, 1987). Therefore, the hydrodynamics of basins may change on the same time scale. Basin-wide thermal, diagenetic and ore-forming events might be explained by this mechanism, as these fluids transport heat and minerals.

NUMERICAL SIMULATION OF STRESS INDUCED FLUID FLOW IN RIFTED BASINS

Modeling Of Basin Subsidence

We have adopted the depth-dependent stretching model proposed by Royden and Keen (1980) to calculate the basin subsidence due to rifting and cooling of the lithosphere. A listing of the stretching values used is given in Table 1. Different values for crustal (δ) and sub-crustal (β) stretching factors are a general outcome of rifted basin studies, and can be explained in terms of heating in the sub-crustal lithosphere (Kooi, 1991).

The flexural deformation of the basin, which includes deformation due to intraplate stresses, is determined by assuming that the lithosphere acts as an elastic beam. The deflections of a beam with lateral varying flexural rigidity D on top of a fluid asthenos-

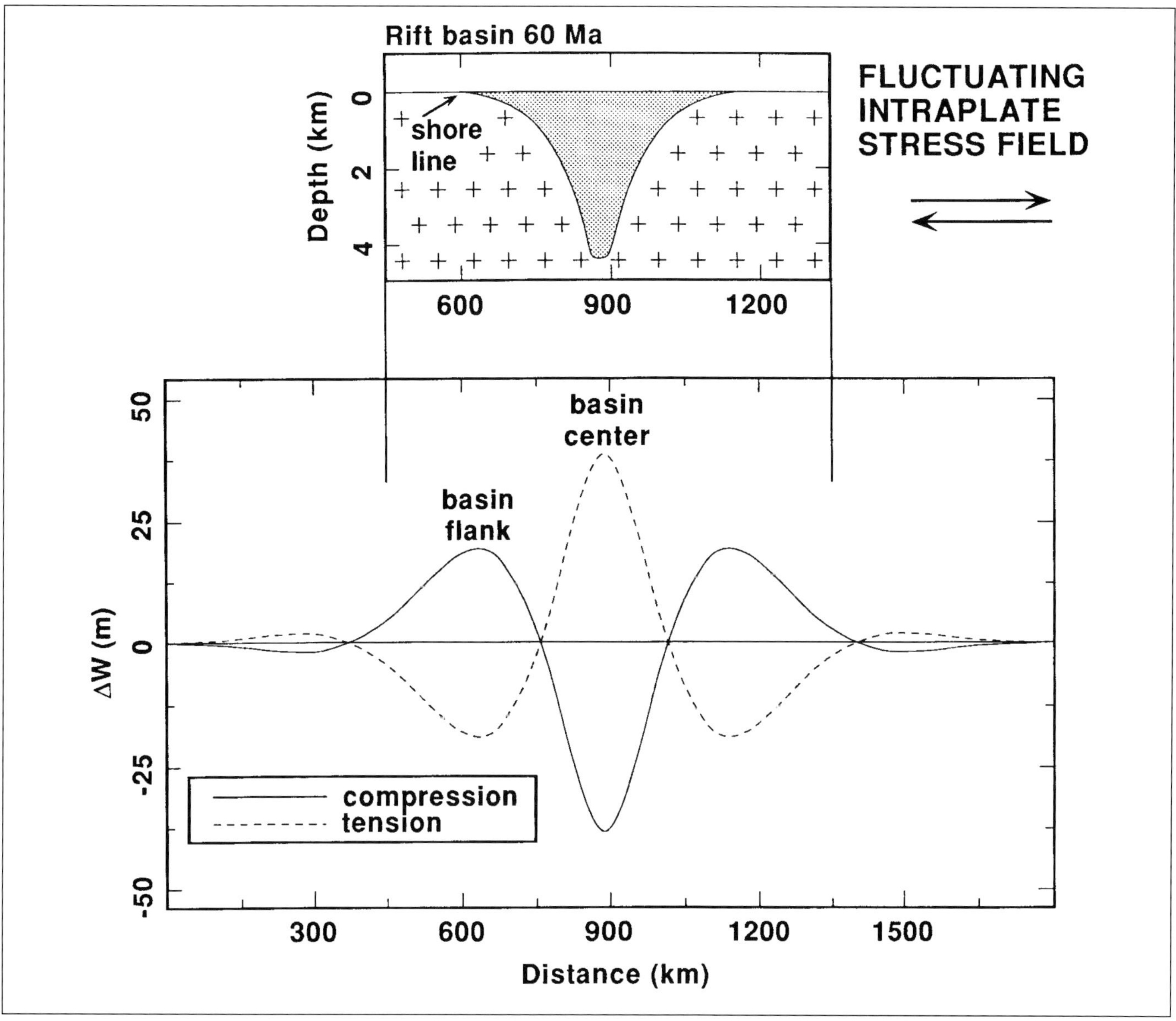

Figure 1. Flexural deflections at a rifted sedimentary basin caused by changes in the level of intraplate stress. Uplift is positive, subsidence negative. Above: a 60-m.y.-old basin initiated by rifting. Loading of sediments induces flexural isostasy. Below: vertical deflections caused by 100 MPa compression (solid line) and 100 MPa tension (dashed line) adopting an elastic rheology for the lithosphere. Compression results in an uplift of the basin flank and subsidence of the basin center. Tension induces subsidence at the basin flank and uplift at the basin center. After Cloetingh et al. (1985).

phere with density ρ_a, bending in response to a horizontal force F and vertical load q is given by (Cloetingh, 1988):

$$\frac{\delta^2}{\delta x^2}\left(D(x)\frac{\delta^2 w}{\delta x^2}\right)+F\frac{\delta^2 w}{\delta x^2}+\left(\rho_a-\rho_{fill}\right)gw(x)=q(x) \quad (1)$$

with ρ_{fill} denoting the density of the basin fill, and g the gravitational acceleration. The horizontal force F is equal to the product of the intraplate stress and effective elastic thickness T_e of the lithosphere. The flexural rigidity D depends on the effective elastic thickness of the lithosphere and the elastic material properties E_c (Young's modulus) and v (Poisson's ratio) :

$$D=\frac{E_c T_e^3}{12\left(1-v^2\right)}$$

Equation (1) is solved with an implicit finite-difference method (see also Bodine, 1981). The effective elastic thickness of the lithosphere is taken to correspond to the depth to the 450°C isotherm of the cooling lithosphere.

Compaction-Driven Flow

Consolidation may be defined as the result of all processes causing the progressive transformation of an argillaceous sediment from a soft mud to a mudstone or shale (Skempton, 1970). These processes include inter-particle bonding, desiccation, cementation, and in particular, the squeezing out of pore water under the increasing weight of overburden. In fact, the process of mechanical compaction dominates to such a degree for clays that consolidation and compaction are almost synonymous (Skempton, 1970). The resulting bulk volume reduction is due to reorientation, cleavage and fracturing of brittle grains, and pseudo-plastic deformation of ductile grains (Bjoerlykke et al., 1989). In this study we therefore only consider mechanical compaction.

Although porosity decreases with burial depth, undercompacted shales are frequently encountered in

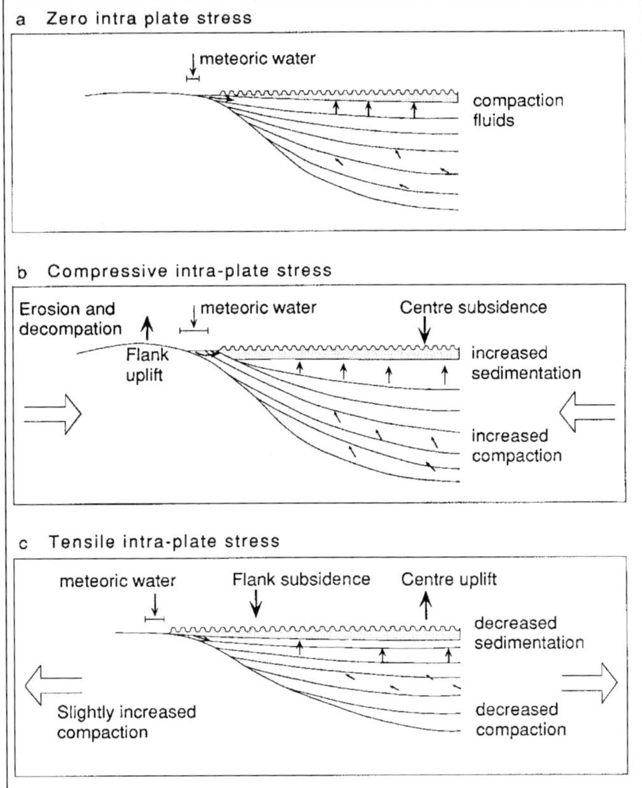

Figure 2. Schematic representation of the effect of intraplate stresses on the hydrodynamics of a rifted sedimentary basin. (a) Hydrodynamic situation in the absence of intraplate stresses. Meteoric water enters the basin from the flank and compaction driven fluid flow directions are mainly vertical. (b) The effect of a compressive intraplate stress field. Sediments are eroded at the basin flank and redeposited basinwards, creating an offlap in the stratigraphy. Meteoric water influx increases, while contemporary compaction driven flow velocities increase. (c) The result of a tensile intraplate stress regime. Additional subsidence at the basin flank results in an increased onlap in the stratigraphy. Compaction driven flow slows down and reduced penetration of meteoric water occurs.

sedimentary basins. Undercompaction is a result of high intergranular fluid overpressure, which hampers the compaction process (Smith, 1971). This shows that porosity also depends on the fluid pressure, which can not be accounted for by assuming that these sediments are a special kind of deposit, with their own particular depth-porosity dependence (Thorne, 1985). Therefore, an effective pressure-porosity relationship (Smith, 1971; Shi and Wang, 1986; Ungerer et al., 1990) is more appropriate than depth-porosity functions frequently employed in backstripping procedures.

In our modeling, we use the equation given by Shi and Wang (1986):

$$\phi_z = \phi_0 e^{-bP_{eff_z}}$$

$$P_{eff_z} = \int_0^z (1-\phi_z)\rho_s g dz + \int_0^z (\phi_z)\rho_{fl} g dz - P_{fl}$$

(Terzaghi law)

with b a sediment-dependent empirical parameter, ϕ the porosity, ρ_s the density of sediments and ρ_{fl} the density of the fluid (water).

The permeability k of the sediments is dependent on porosity:

$$k_{hor} = 10^{-a+b\phi}, k_{vert} = ck_{hor}$$

with a, b and c sediment-dependent empirical parameters and k_{hor} and k_{vert} denoting the permeability along and orthogonal to the strata. The empirical parameters for the porosity and permeability equations are taken from Shi and Wang (1986) and Bethke (1985), respectively.

The equation for compaction-driven flow for slightly compressible fluids is (Bethke, 1985; see also the Appendix) :

$$\phi\beta \frac{\delta(\Phi + \rho_{fl} V_{zm})}{\delta t} = \nabla^T \left\{ \frac{k}{\mu} [\nabla(\delta\Phi)] \right\} - \frac{1}{(1-\phi)} \frac{\delta\phi}{\delta t}$$

Table 1. Extension factors used in calculations.

x (km)	δ	β
150	1.00	1.00
160	1.05	1.25
170	1.10	1.47
180	1.13	1.60
190	1.14	1.60
200	1.20	1.63
210	1.30	1.63
220	1.40	1.70
230	1.60	1.70
350	1.60	1.70

For the numerical modeling of fluid flow we use the implicit Galerkin finite-element technique with a Lagrangian reference frame.

RESULTS

In the simulations we have restricted ourselves to the compaction driven flow problem. The total simulated time span of basin development is 35 m.y., including an initial rifting phase of 20 m.y. during which subsidence is caused by stretching of the crust. The sediment input is listed in Table 2. The adopted values for all other parameters are given in Table 3. At every time step the basin is filled to sea level. Figure 3 shows the results of the simulations in terms of the synthetic stratigraphy, the fluid velocity field, and the overpressures. The insets show a detailed view of the stratigraphy at the flank of the basin.

The predicted basin stratigraphy and fluid overpressures and flow velocities in the absence of intraplate stresses are shown in Figure 3a. The maximum overpressure, which is located at the lower right corner of the profile, is 1.8×10^4 Pa. The maximum flow velocity at the top of the profile is 1.1×10^{-9} m/s (3.5 cm/yr), while at the bottom velocities are of the order of 6.7×10^{-11} m/s (0.2 cm/yr). The inset shows a continuous onlap for the stratigraphy at the flank of the basin. Figure 3b depicts the effect of an increase in the level of compressive intraplate stress, increasing linearly from zero at 30 m.y. to 300 MPa compression at 35 m.y. The maximum overpressure in this case is 2.4×10^4 Pa. The maximum flow velocities at the top and bottom are respectively 1.4×10^{-9} m/s and 8.5×10^{-11} m/s. The offlap produced at the basin flank is shown in the inset. The effect of a tensile intraplate stress, increasing linearly form zero at 30 m.y. to 300 MPa tension at 35 m.y., is shown in Figure 3c. The maximum overpressure equals 1.4×10^4 Pa. Maximum flow velocities at the top and bottom of the cross-section are 7.8×10^{-10} m/s and 5.0×10^{-11} m/s. The increased onlap produced by tensional stresses at the basin flank, is depicted in the inset.

Comparison of the numerical modeling results demonstrates that, compared with the zero intraplate stress case, a compressive intraplate stress results in an increase in fluid overpressures and flow velocities, while tensile intraplate stresses induce the opposite effect. As shown by the simulations displayed in Fig. 3, the stress-induced perturbations in fluid pressures and flow velocities can be up to the order of 30%.

Because intraplate stress events have a finite duration, stresses are ultimately relaxed, inducing differential vertical motions and patterns of sedimentation opposite to those induced by the initial increase of the stress level. Therefore, stress-induced fluid overpressures and flows take the shape of a pulse, of which the first part occurs simultaneously with the increasing intraplate stress and the second part is synchronous with the stress drop. This feature has interesting implications for the hydrodynamics of the basin. In the case of relaxation of compressive intraplate stress, meteoric water might get trapped at the basin flank, while overpressures and flows at the basin center are

Table 2. Sediment input.

time (m.y.)	sediment type
0.0–10.0	sands
10.0–12.5	shales
12.5–15.0	sands
15.0–17.5	silts
17.5–20.0	sands
20.0–25.0	silts
25.0–27.5	sands
27.5–30.0	silts
30.0–32.5	sands
32.5–35.0	silts

Table 3. Numerical values adopted in calculations.

Material	Parameter	Value
Lithosphere	ρ_a	3.3 g/cm^3
	ρ_c	2.9 g/cm^3
	th_l	125 km
	th_c	35 km
	T_a	1350 °C
	κ	$7.5\times10^{-7} m^2/s$
	α	3.2×10^{-5} °C^{-1}
	E_c	$7.0\times10^{10} Pa$
	ν	0.25
fluid	ρ_{fl}	1.024 g/cm^3
	β	$4.3\times10^{-10} Pa^{-1}$
	μ	$5.0\times10^{-4} Pa\ s$
shales	ϕ	$0.6e^{-5.0\times10^{-8} P_{eff}}$
	$\log k_{hor}$	$-19+8\phi$
	k_{hor}/k_{vert}	10.0
	ρ_s	2.72 g/cm^3
silts	ϕ	$0.5e^{-3.0\times10^{-8} P_{eff}}$
	$\log k_{hor}$	$-16+6\phi$
	k_{hor}/k_{vert}	5.0
	ρ_s	2.68 g/cm^3
sands	ϕ	$0.5e^{-2.5\times10^{-8} P_{eff}}$
	$\log k_{hor}$	$-13+2\phi$
	k_{hor}/k_{vert}	2.5
	ρ_s	2.65 g/cm^3

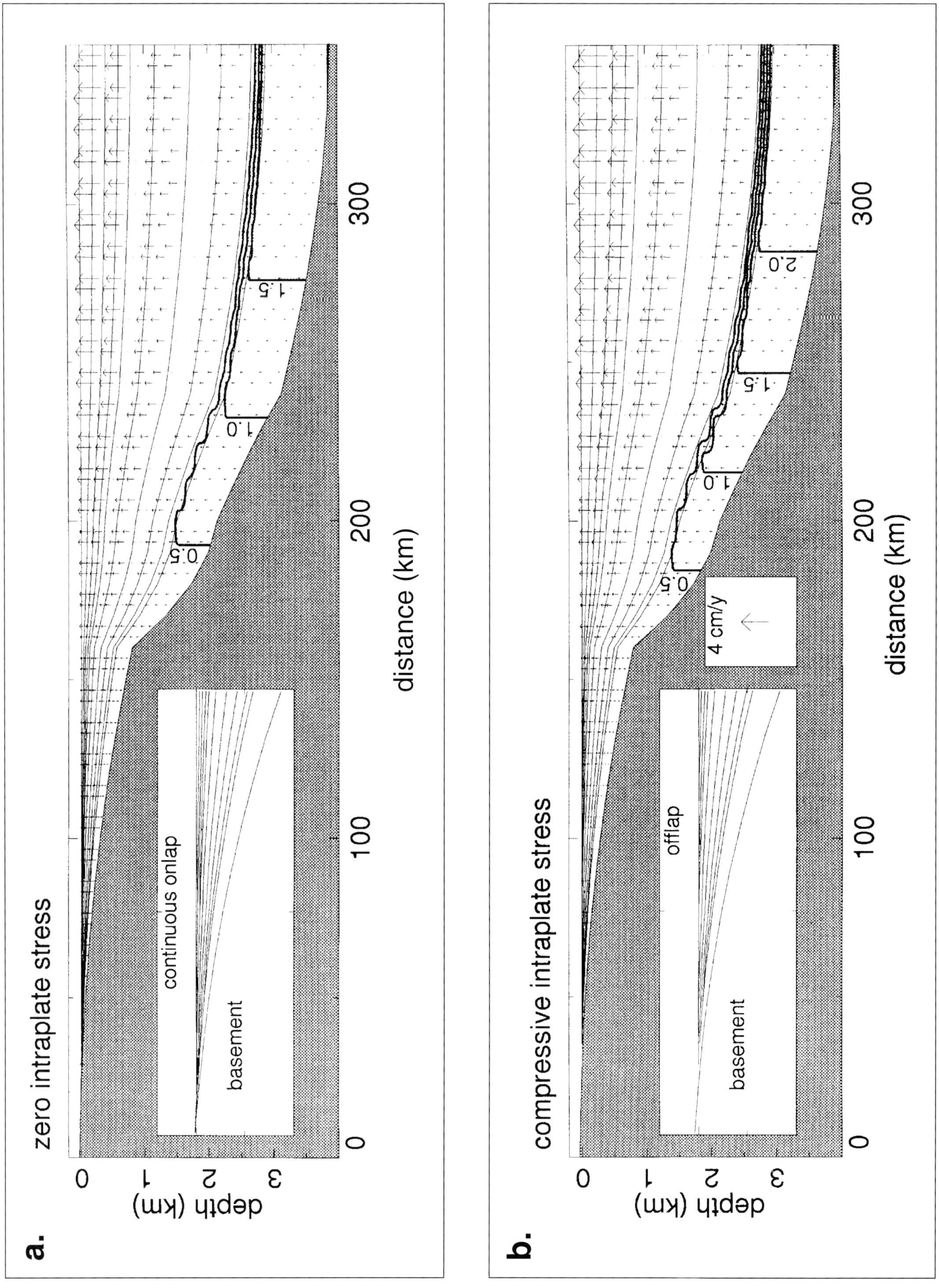
a.
zero intraplate stress
depth (km)
0
1
2
3
distance (km)
0
100
200
300
continuous onlap
basement
0.5
1.0
1.5
b.
compressive intraplate stress
depth (km)
0
1
2
3
distance (km)
0
100
200
300
offlap
basement
4 cm/y
0.5
1.0
1.5
2.0

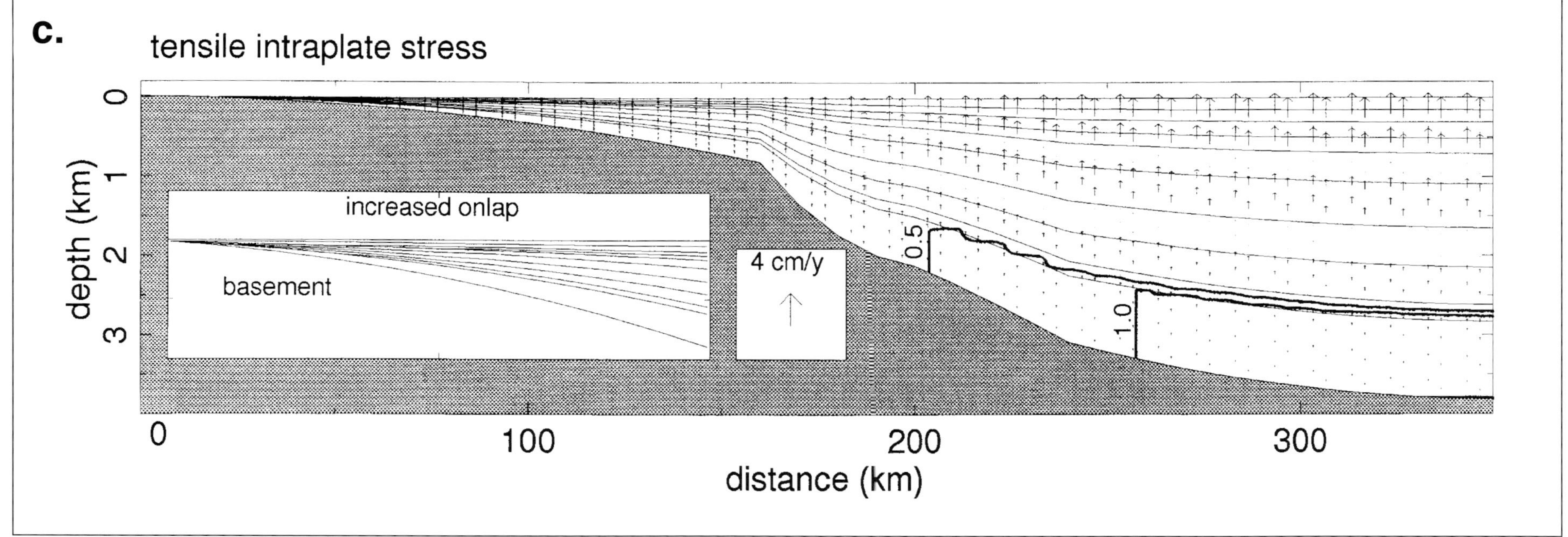

Figure 3. Modeling results: predicted stratigraphy (thin lines) compaction driven flow velocities (arrows) and overpressures (solid lines) for 35 m.y. old rifted basin. The ages of the stratigraphic lines coincide with the timing of changes in the sediment input as listed in Table 2. Contour interval for overpressures is 5.0×10^3 Pa. The velocity field has been scaled the same way as the cross section (vertical exaggeration is ±25 times). The insets show a detailed view of the stratigraphy at the flank of the basin at a location between 20 and 120 km on the horizontal axis in the main picture. (a) Compaction driven flow velocities and overpressures in the absence of intraplate stresses. The maximum overpressure is 1.8×10^4 Pa. The maximum flow velocities in top and bottom of profile are respectively 1.1×10^{-9} m/s and 6.7×10^{-11} m/s. The inset shows the continuous onlap developed at the flank of the basin. (b) The result of a compressive intraplate stress, increasing linearly from zero at 30 m.y. to 300 MPa compression at 35 m.y. The maximum overpressure is 2.4×10^4 Pa. The maximum flow velocities are 1.4×10^{-9} m/s and 8.5×10^{-11} m/s at the top and bottom of the profile. The inset depicts the stratigraphic offlap which results from the flank uplift. (c) The effect of a tensile intraplate stress, increasing linearly from zero at 30 m.y. to 300 MPa tension at 35 m.y. The maximum overpressure is 1.4×10^4 Pa. The maximum flow velocities at the top and bottom of the cross section are 7.8×10^{-10}m/s and 5.0×10^{-11} m/s. The inset shows the continuous onlap produced at the flank of the basin.

less than normal. Relaxation of tensile intraplate stresses leads to an increase of overpressures and flows, while the extent of the meteoric water domain increases.

The actual effect of intraplate stresses on the hydrodynamics of a sedimentary basin depends on the fill, the amount of stretching, and the age of the basin. The synthetic young rifted basin of Figure 3 does not contain an evaporitic layer or a thick shaly sequence and, therefore, the calculated overpressures are low from a mechanical point of view. Furthermore, vertical deflections caused by intraplate stresses are larger when the basin is older because of a larger sedimentary load (Cloetingh, 1988).

Our model does not take into account that intraplate stresses also occur in the sediments, which cause a horizontal component of compaction. This will lead to a further increase in the enhancement of compaction driven flows and overpressures by intraplate stresses.

As shown by Bethke et al. (1988), eustatic drops in sea level lead to an increased influx of meteoric water and a slight decrease in overpressure. Our modeling results show that the effect of intraplate stresses on the hydrodynamics of a basin differs from the predictions of models involving eustatic sea level changes.

IMPLICATIONS

A number of implications of stress-induced fluid flow in sedimentary basins are summarized in Figure 4. These include the faulting and fracturing of strata and the alteration of the heat flow pattern, the diagenesis of sediments, and the localization of economic resources in a rifted sedimentary basin.

Meteoric water flow is closely linked to the paleogeography and tectonic uplifts around the basin. Meteoric water is sometimes encountered far away from the shoreline position (Bjoerlykke et al., 1989). The meteoric water flow is particularly important in relation to dissolution of feldspar and carbonate in sandstone, causing secondary porosity (Bjoerlykke et al., 1989; Harrison, 1989). Meteoric water also often contains dissolved carbonates. Along the boundary between meteoric and connate domains dolomite may be formed (Harrison, 1989), which can become very thick because of stress-induced relative sea level variations.

Jurassic reservoir rocks in the North Sea Basin show evidence of flushing by meteoric waters, which penetrated shortly after deposition and uplift (Bjoerlykke et al., 1989). Our model gives a possible explanation for the relationship between tectonics and influx of meteoric waters. Recent work (e.g. Kooi and Cloetingh, 1989; Cloetingh et al., 1990; Ziegler, 1990) has shown that the late Neogene dramatic increase in subsidence in the North Sea Basin can be explained by a compressive intraplate stress event, which is contemporary with the late orogenic phase in the Alpine domain. It is to be expected that this feature was accompanied by a diagenetic and thermal event. Jessop (1989) explained the heat flow pattern in the British part of the North Sea Basin by a landward directed fluid flow from the center of the basin. This could be the result of increased compaction rates caused by an increase in the level of compressive intraplate stress. The high overpressures in the Viking and Central grabens (Buhrig, 1989; Chiarelli and Duffaud, 1980; Ungerer et al., 1987) could be a very recent phenomena.

Mississippi Valley type (M.V.T.) strata-bound ore deposits are formed along the margins of large basins under conditions of high temperatures at relatively shallow depths of about 1 km. Both gravity and compaction driven flow may have caused their formation (Garven and Freeze, 1984; Bethke and Marshak, 1990). The formation of M.V.T. deposits can be explained by cumulative episodic pulses of enhanced compaction driven fluid flow (Cathles and Smith, 1983; Cathles, this volume) with time intervals of about 1 m.y. and increase in fluid velocities by a factor of several thousands. Although the exact origin of these pulses is uncertain, episodic dewatering of geopressured zones when fluid pressures approach lithostatic conditions seems to be a plausible explanation. These conditions can be reached during periods of rapid sedimentation (Ravenhurst and Zentilli, 1987), occurring during a compressive intraplate stress event. Stresses accumulating in the flexed lithosphere are eventually relaxed by faulting. During buildup of stresses, overpressures increase because of increased sedimentation rates. Upon fault-

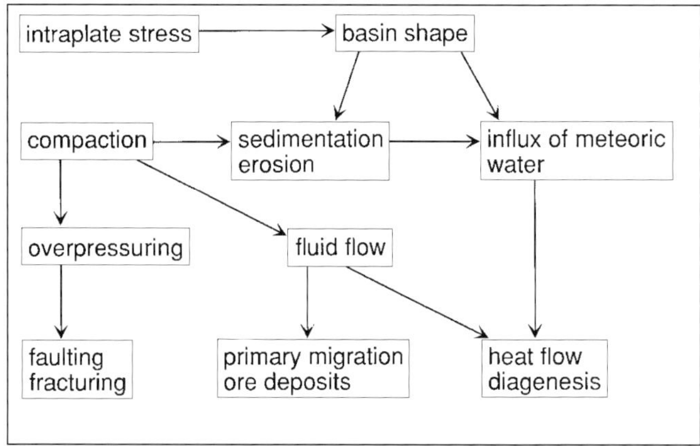

Figure 4. Implications of stress-induced fluid flow events. Increased overpressures during a compressive intraplate stress event enhance the development of fractures and faults. The induced change in fluid flow velocities and size of the meteoric water domain during an intraplate stress event affects the primary migration of hydrocarbons, the transportation of ore forming minerals, the diagenesis of sediments, and the heat flow in the sedimentary basin.

ing, permeable pathways are created for the fluids to escape to the surface. Bethke and Marshak (1990) have suggested that M.V.T. deposits have been formed by moving brines from surrounding orogenies with the topography as the most important "driving force" for the migration. As pointed out by Deming and Nunn (1991), this mechanism may be effective only when the flow is channelized in the discharge zone. Topography-driven recharge could operate in conjunction with the compaction driven flow caused by stacking of thrust sheets on to the foreland (Oliver, 1986; Ge and Garven, 1989). Also in this tectonic setting intraplate stresses exerted by the orogenies into the foreland basins could lead to additional subsidence and increased sedimentation of an episodic nature.

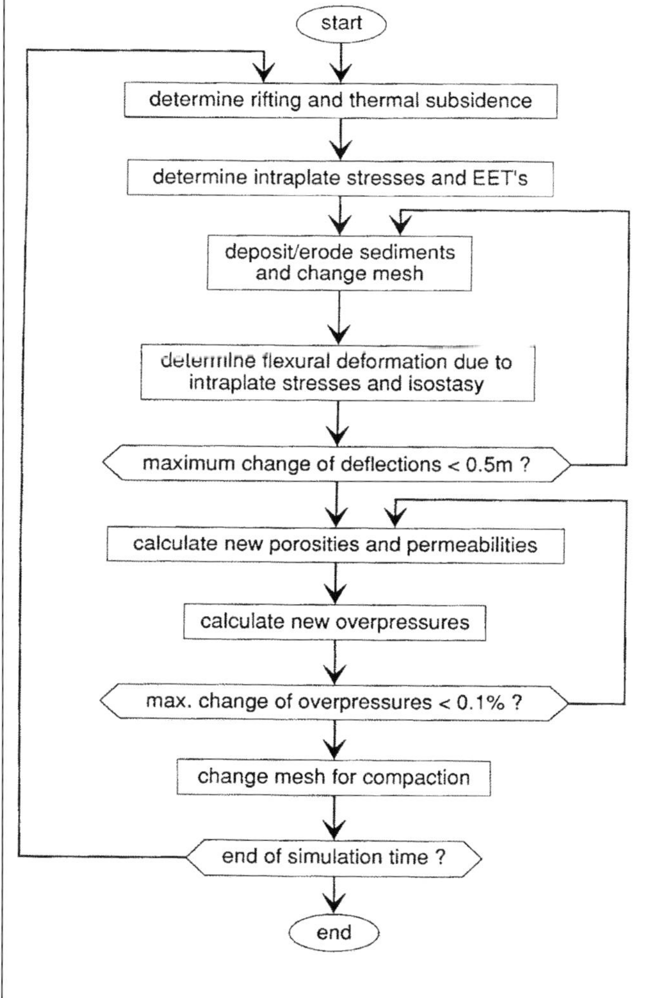

Figure 5. Flow chart denoting the structure of the program. The program consists of two main parts, of which the first module determines the basin development while the second module calculates the hydrodynamics of the basin. The inner loop in the basin evolution part determines the flexural isostasy. The inner loop in the hydrodynamics part is used to solve the nonlinearity resulting from the mutual dependence of fluid pressures and sediment porosity.

Diagenesis, generation, and primary migration of hydrocarbons depend partly on the temperature distribution and compaction driven flow. Previous studies have shown that, with the exception of sudden fluid expulsions, compaction driven flow during (post-rift) thermal subsidence of an extensional basin is too slow to cause significant perturbations of the temperature field (Cathles and Smith, 1983; Bethke, 1985; Deming et al., 1990). As faulting is enhanced by fluid pressures (Hubbert and Rubey, 1959), a sudden increase in fluid overpressures will trigger (growth-) faulting in sedimentary basins. Our modeling has shown that as a result of short-term changes in the level of intraplate stresses during the post-rift evolution of extensional basins compaction driven flow and the build up of overpressures can occur in pulses. Erosional unloading caused by uplift of the basin margin due to an increase in the level of compressive intraplate stresses could also explain degassing of the pore fluid in sediments with a low permeability (Neuzil and Pollock, 1983; Bethke, 1989; Kreitler, 1989).

CONCLUSIONS

Intraplate stresses modulate the shape and subsidence patterns of rifted sedimentary basins, affecting sedimentation patterns and the extent of the meteoric water domain. As a result, intraplate stresses alter the hydrodynamics of sedimentary basins on a time scale of a few million years. Compressive intraplate stresses could promote the development of fluid overpressures, triggering faulting and fracturing of strata. The increased overpressures also can lead to an increase of the compaction driven fluid flow velocities, increased temperatures and heat flow in the basin center, and enhancement of primary migration, diagenesis, and transportation of metallic-rich brines. The contemporary increase in influx of meteoric water can result in a decrease in temperatures and heat flow and may affect the diagenesis of sediments at the flank of the basin. The model predicts that during a tensile intraplate stress event flow velocities in the compaction domain will be decreased simultaneously with reduced penetration of meteoric water. These conditions could lead to a slight elevation of temperatures and heat flow in the depocenter and depression of both quantities at the flank of the basin.

ACKNOWLEDGMENTS

Partial support for this research was provided by The Netherlands Ministry of Economic Affairs. Discussions with Prof. Dr. J.J. De Vries and Dr. H. Kooi are appreciated. We benefitted from constructive reviews by Dr. J. Burrus, Dr. A. Robinson and Dr. N. Wilson.

REFERENCES CITED

Bethke, C.M., 1985, A numerical model of compaction-driven groundwater flow and heat transfer and its application to the paleohydrology of intracratonic sedimentary basins: Journal of Geophysical Research, v. 90, p. 6817-6828.

Bethke, C.M., 1989, Modeling subsurface flow in sedimentary basins: Geologische Rundschau, v. 78, p. 129-154.

Bethke, C.M., W.J. Harrison, C. Upson, and S.P. Altaner, 1988, Supercomputer analysis of sedimentary basins: Science, v. 239, p. 261-267.

Bethke, C.M., and S. Marshak, 1990, Brine migrations across North America - the plate tectonics of groundwater: Annual Review of Earth and Planetary Science, v. 18, p. 287-315.

Bjoerlykke, K., M. Ramm, and G.C. Saigal, 1989, Sandstone diagenesis and porosity modification during basin evolution: Geologische Rundschau, v. 78, p. 243-268.

Bodine, J.H., 1981, The thermo-mechanical properties of the oceanic lithosphere: Ph.D. Thesis Columbia University, 332 pp.

Buhrig, C., 1989, Geopressured Jurassic reservoirs in the Viking Graben: modelling and geological significance: Marine and Petroleum Geology, v. 6, p. 31-48.

Burrus, J. and F. Audebert, 1990, Thermal and compaction processes in a young rifted basin containing evaporites: Gulf of Lions, France: American Association of Petroleum Geologists Bulletin, v. 74, p. 1420-1440.

Cathles, L.M. and A.T. Smith, 1983, Thermal constraints on the formation of Mississippi Valley-Type lead-zinc deposits and their implications for episodic basin dewatering and deposit genesis: Economic Geology, v. 78, p. 983-1002.

Chiarelli, A., and F. Duffaud, 1980, Pressure origin and distribution in Jurassic of Viking Basin (United Kingdom-Norway): American Association of Petroleum Geologists Bulletin, v. 64, p. 1245-1266.

Cloetingh, S., 1986, Intraplate stresses: a new tectonic mechanism for fluctuations of relative sea level: Geology, v. 14, p. 617-620.

Cloetingh, S., 1988, Intraplate stresses: a new element in basin analyses, in K. Kleinspehn and C. Paola, ed., New perspectives in basin analyses: New York, Springer, p. 305-330.

Cloetingh, S., 1991, Tectonics and sea level changes: a controversy ?, in D. Mueller, H. Weissel and J. McKenzie, ed., Controversies in modern geology: a survey of recent developments in sedimentation and tectonics, London, Academic Press, p. 249-277.

Cloetingh, S., F.M. Gradstein, H. Kooi, A.C. Grant, and M. Kaminski, 1990, Plate reorganizations: a cause of rapid late Neogene subsidence and sedimentation around the North Atlantic ? : Journal of the Geological Society of London, v. 147, p. 495-506.

Cloetingh, S., H. McQueen and K. Lambeck, 1985, On a tectonic mechanism for regional sea level variations: Earth and Planetary Science Letters, v. 75, p. 157-166.

Cloetingh, S., and R. Wortel, 1985, Regional stress field of the Indian plate: Geophysical Research Letters, v. 12, p. 77-80.

Deming, D., J.A. Nunn, and D.G. Evans, 1990, Thermal effects of compaction-driven groundwater flow from overthrust belts:

Deming, D., and J. A. Nunn, 1991, Numerical simulations of brine migrations by topographically driven recharge: Journal of Geophysical Research, v. 96, p. 2485-2499.

De Vries, J.J. , 1989, Groundwater hydraulics, Third edition, Free University, Amsterdam, 67 pp.

Garven, G. and R.A. Freeze, 1984, Theoretical analysis of the role of groundwater flow in the genesis of stratabound ore deposits: American Journal of Science, v. 284, p. 1085-1174.

Ge, S. and G. Garven, 1989, Tectonically induced transient groundwater flow in foreland basin, in R.A. Price, ed., Origin and evolution of sedimentary basins and their energy and mineral resources, American Geophysical Union, Geophysical Monograph, v. 48, p. 145-157.

Harrison, W.J., 1989, Modeling fluid/rock interactions in sedimentary basins, in T.A. Cross, ed., Quantitative Dynamic Stratigraphy, Prentice Hall, p. 195-231.

Hubbert, M.K. and W.W. Rubey, 1959, Role of fluid pressure in mechanics of overthrust faulting, I: Mechanics of fluid-filled porous solids and its application to overthrust faulting: Bulletin of the Geological Society of America, p. 115-166.

Hughes, T.J.R., 1987, The finite element method. Prentice Hall International Inc, New Jersey.

Huyakorn, P.S. and G.F. Pinder, 1983, Computational methods in subsurface flow, Academic Press Inc, New York.

Jessop, A.M., 1989, Hydrological distortian of heat flow in sedimentary basins: Tectonophysics, v. 164, p. 211-218.

Kooi, H., 1991, Tectonic modelling of extensional basins, the role of lithospheric flexure, intraplate stress and relative sea-level change, Ph.D. Thesis, Vrije Universiteit, Amsterdam, 183 pp.

Kooi, H. and S. Cloetingh, 1989, Intraplate stresses and the tectono-stratigraphic evolution of the central North Sea: American Association of Petroleum Geologists Memoir, v. 48, p. 541-558.

Kreitler, C.W., 1989, Hydrology of sedimentary basins: Journal of Hydrology, p. 29-53.

McKenzie, D.P., 1978, Some remarks on the development of sedimentary basins: Earth and Planetary Science Letters, v. 40, p. 25-31.

Neuzil, C.E. and W. Pollock, 1983, Erosional unloading

and fluid pressures in hydraulically "tight" rocks: Journal of Geology, v. 91, p. 179-193.

Oliver, J., 1986, Fluids expelled tectonically from orogenic belts: Their role in hydrocarbon migration and other geologic phenomena: Geology, p. 99-102.

Philip, H., 1987, Plio-quaternary evolution of the stress field in Mediterranean zones of subduction and collision, Annales Geophysicae, v. 5, p. 301-320.

Pittman, W.C., and X. Golovchenko, 1983, The effect of sea level change on the shelf edge and slope of passive margins: The Society of Economic Paleontologists and Mineralogists Special Publication 33, p. 41-58.

Polivka, R.M. and E.L. Wilson, 1976, Finite element analyses of non-linear heat transfer problems: Rep. no. UC SESM 76-2, Department of civil engineering, University of California, Berkely, California.

Ravenhurst, C.E., and M. Zentilli, 1987, Hot brine evolution in Fundy/Magdalen Basin, in C. Beaumont and A.J. Tankard, ed., Sedimentary basins and basin-forming mechanisms, Canadian Society of Petroleum Geologists Memoir 12, p. 335-349.

Royden, L. and C.E. Keen, 1980, Rifting processes and thermal evolution of the continental margin of eastern Canada determined from subsidence curves: Earth and Planetary Science Letters, v. 51, p. 343-361.

Shi, Y. and C. Wang, 1986, Pore pressure generation in sedimentary basins: overloading versus aquathermal: Journal of Geophysical Research, v. 91, p. 2153-2162.

Skempton, A.W., 1970, The consolidation of clays by gravitational compaction. Quarterly Journal of the Geological Society of London, v. 125, p. 373-411.

Smith, J.E., 1971, The dynamics of shale compaction and evolution of pore-fluid pressures: Mathematical Geology, v. 33, p. 239-263.

Stephenson, R, and K. Lambeck, 1985, Isostatic response of the lithosphere with in-plane stress: application to central Australia: Journal of Geophysical Research, v. 90, p. 8581-8588.

Thorne, J.A., 1985, Studies in stratology: the physics of stratigraphy, Ph.D. Thesis, Columbia University, 523 pp.

Ungerer, P., J. Burrus, B. Doligez, P.Y. Chenet, and F. Bessis, 1990, Basin evaluation by integrated two-dimensional modeling of heat transfer, fluid flow, hydrocarbon generation, and migration. American Association of Petroleum Geologists Bulletin, v. 74, p. 309-335.

Ungerer, P., B. Doligez, P.Y. Chenet, J. Burrus, F. Bessis, E. Lafargue, G. Giroir, O. Heum and S. Eggen, 1987, A 2-D model of basin scale petroleum migration by two-phase fluid flow. Application to some case studies, in B. Doligez, ed., Migration of hydrocarbons in sedimentary basins, Paris, Edition Technip.

Watts, A.B., 1982, Tectonic susbsidence, flexure and global changes of sealevel: Nature, v. 297, p. 469-474.

Ziegler, P.A., 1990, Geological atlas of Western and Central Europe, second and revised edition, Shell Internationale Petroleum Maatschappij / Geological Society of London, 240 pp.

APPENDIX

Hooke's law give the relationship between elastic stress and strain (for notation see Table 4):

$$\delta\sigma = E\delta\varepsilon$$

by definition:

$$\varepsilon = \frac{\delta V}{V};\ \delta\sigma = -\delta P_{fl} \Rightarrow E = \frac{-\delta P_{fl}}{\delta V / V} \quad (1)$$

$$\delta(\rho V) = 0 \Rightarrow \delta\rho_{fl} = -\rho_{fl}\frac{\delta V}{V} \Leftrightarrow \delta\rho_{fl} = \rho_{fl}\frac{\delta V}{V - \delta P_{fl}}\delta P_{fl} \quad (2)$$

(1) combined with (2) gives the equation of state for a slightly compressible fluid (Bethke, 1985):

$$\delta\rho_{fl} = \rho_{fl}\frac{1}{E}\delta P_{fl} \Leftrightarrow \delta\rho_{fl} = \rho_{fl}\beta\delta P_{fl} \quad (3)$$

$$\delta m = \delta(\rho_{fl}V) = \rho_{fl}\delta V + V\delta\rho_{fl} \quad (4)$$

$$V = \phi V_b \quad (5)$$

$$\delta V = \delta(\phi V_b) = V_b\delta\phi + \phi\delta V_b \quad (6)$$

assuming incompressible grains:

$$\delta V = \delta V_b$$

combined with (6) gives:

$$\delta V = \frac{V_b}{1-\phi}\delta\phi \quad (7)$$

the change in mass due to flow:

$$\delta m = V_b\nabla^T(\rho_{fl}q_{fl})\delta t = V_b\delta t\left[\rho_{fl}\nabla^T q_{fl} + q_{fl}\nabla^T(\rho_{fl})\right] \quad (8)$$

of which the last term can be neglected (De Vries, 1989).

Darcy's law states that:

$$q_{fl} = -\frac{k}{\mu}\nabla P_{fl} \quad (9)$$

Substituting (9) into (8); (5), (6), and (7) into (4); and equating (8) and (4) gives (conservation of mass):

$$V_b \rho_{fl} \nabla^T \left(\frac{k}{\mu} \nabla P_{fl} \right) = \frac{\rho_{fl} V_b}{1-\phi} \delta\phi + \phi V_b \rho_{fl} \beta \delta P_{fl} \Leftrightarrow$$

$$\phi \beta \delta P_{fl} = \nabla^T \left(\frac{k}{\mu} \nabla P_{fl} \right) \delta t - \frac{1}{1-\phi} \delta\phi$$

Taking the derivatives with respect to time and taking into account that this equation is to be solved in a Lagrangian reference frame that deforms due to compaction, and assuming flows are only driven by overpressure, this equation resolves into:

$$\phi \beta \frac{\delta(\Phi + \rho_{fl} g V_{zm})}{\delta t} = \nabla^T \left(\frac{k}{\mu} \nabla \Phi \right) - \frac{1}{1-\phi} \frac{\delta\phi}{\delta t}$$

With $P_{fl} = \rho_{fl} g z + \Phi$, therefore $V_{zm} = \delta z$.

For the solution of the fluid flow equation, we have adopted the implicit Galerkin finite-element technique with a Lagrangian reference frame. The nonlinearity that results from the dependence of porosity and permeability on fluid pressure, with fluid pressure depending on the change in porosity, is solved with a Picard iteration. Triangular elements with three pressure nodes and linear shape functions are used, which is adequate for transient fluid flow (Huyakorn and Pinder, 1983). Boundary conditions are zero overpressure for nodes at the basement. The conductivity matrix is kept symmetric, implying that inside an element the permeabilities k_{hor} and k_{vert} are constant. The mass matrix is approximated by a lumped diagonal matrix, which causes a small loss of accuracy (Polivka and Wilson, 1976). The resulting matrix equations are solved using Cholesky factorization (Hughes, 1987).

The finite-element mesh coincides with the basin stratigraphy. Because of continuous sedimentation, the height of the elements in the top layer grows until either the type of sediment changes or the elements become too high. In both cases, a new layer is added to the mesh. Each element can represent one type of sediment. A typical computer simulation, as shown in Figure 3, consists of a mesh with 2500 elements and takes 6 hours of computation time on a Sun-sparc workstation or 2.5 hours on an IBM 3090 mainframe. A flow chart of the developed program is shown in Figure 5. The program consists of two main parts, of which the first part determines the basin evolution while the second part calculates the hydrodynamics of the basin. Both are executed every time step. The basin evolution module contains one inner loop that determines the isostatic adjustment of the basin due to the changed sedimentary load. As the kind of material filling in the depression created by isostasy is a priori known, the problem can only be solved by an iteration. The inner loop in the hydrodynamics module of the program solves the nonlinearity mentioned above. The initial time step of 10,000 years is automatically reduced when the solution becomes unstable.

Table 4. Notation.

Symbol	Meaning
D	flexural rigidity ($N\ m$)
E	Young's modulus of water (Pa)
E_c	Young's modulus of lithospheric material (Pa)
F	horizontal force (N)
P_{fl}	total fluid pressure (Pa)
P_{LITH}	lithostatic or overburden pressure (Pa)
P_{eff}	effective pressure (Pa)
T_e	effective elastic thickness (m)
T_a	temperature at lithosphere-asthenosphere boundary (°C)
V	fluid volume (m^3)
V_b	bulk volume (m^3)
V_{zm}	displacement due to compaction, relative to a fixed point (m)
a,b,c	parameters for empirical laws relating porosity to effective pressure and permeability to porosity.
g	acceleration of gravity (m/s^2)
m	fluid mass in an elemental volume (kg)
k	permeability (m^2)
k_{hor}, k_{vert}	horizontal and vertical permeability (m^2)
q	vertical load due to sediments and sea water (Pa)
q_{fl}	Darcian fluid velocity (m/s)
t	time (s)
th_l, th_c	thickness of lithosphere, thickness of crust
v	Poisson's ratio
w	vertical deflection (m)
Φ	fluid overpressure, pressure above hydrostatic (Pa)
α	thermal expansion (°C^{-1})
β	fluid compressibility (Pa^{-1})
κ	thermal diffusivity (m^2/s)
ε	elastic strain
ϕ, ϕ_0	porosity, porosity at sedimentation
ρ_a	density of asthenosphere (kg/m^3)
ρ_c	density of crust (kg/m^3)
ρ_{fill}	density of basin fill (kg/m^3)
ρ_s	density of grains (kg/m^3)
ρ_{fl}	fluid density (kg/m^3)
μ	fluid viscosity ($Pa\ s$)
σ	elastic stress (Pa)
∇^T, ∇	divergence, gradient

Chapter 8

A Discussion of Flow Mechanisms Responsible for Alteration and Mineralization in the Cambrian Aquifers of the Ouachita-Arkoma Basin-Ozark System

L. M. Cathles
Cornell University
Ithaca, New York, USA

ABSTRACT

Cathodoluminescent microstratigraphy in epigenetic dolomite cements correlates over a north-south distance of >275 km across the Ozark Mountains. Trace elements in dolomite show coherent regional variations. Ubiquitous coeval Pb-Zn mineralization contains fluid inclusions with homogenization temperatures >100°C. All suggest the flow of brines (mainly) north from the Arkoma Basin through Cambrian sandstone and carbonate aquifers. The observation that brines still fill Ordovician and Cambrian strata ringing the Ozark Plateau constrains the cumulative flow that has occurred. If the Mid-Continent sediment cover was thick and insulating, brine flow driven by topographic differences in hydrologic head could have been slow enough to avoid salt flushing and still accommodate the fluid inclusion homogenization data. If the cover was thermally conductive and thin, as seems most geologically reasonable, flow at the rates required to explain the homogenization temperatures would have quickly flushed salt from the aquifers in contradiction to present observations. Topographically driven hydrologic flow across the Arkoma basin could not have continued uninterrupted for protracted periods. Given the present high permeability of the Pb-Zn deposits, there is no obvious way to limit or pulse cross-basin hydrologic flow. The simplest explanation is that brines were expelled by compaction or gas displacement. High temperature, low salinity fluid inclusions in the Ozark Cambrian aquifers probably represent the incursion of meteoric water into outcrop areas warmed by pulses of brine outflow. Channeling of fluid flow and the relation of alteration and fluid inclusions to flow channels need to be further investigated theoretically and in the field.

INTRODUCTION

The flow of brine north from the Arkoma Basin has produced remarkably extensive and coherent alteration in Upper Cambrian sediments across Arkansas, Missouri, eastern Kansas, and eastern Oklahoma (Figure 1). The alteration includes: (1) A correlatable cathodoluminescent microstratigraphy in hydrothermal epigenetic dolomite cements over a N-S distance of ~275 km (Figure 1B; Voss and Hagni, 1985; Rowan, 1986; Farr, 1989), (2) almost ubiquitous traces of Pb-Zn mineralization coeval with deposition of the dolomite cement (Voss et al., 1989; Coveny et al., 1987), and four or more major Mississippi Valley-type (MVT) Pb-Zn mining districts, (3) fluid inclusion homogenization temperatures in four mining districts that range from 60 to 180 °C (Leach and Rowan, 1986; Rowan and Leach, 1989), (4) a northward increase in K/Cl in fluid inclusions with constant Na/Cl ratios (Viets and Leach, 1990), and (5) a regular northward decrease in Fe, Mn and $^{87}Sr/^{86}Sr$, and an increase in the Sr contents of epigenetic dolomite within 1 m of the Lamotte Sandstone contact (Gregg and Shelton, 1989a, b). Carbon and oxygen isotopic ratios in the epigenetic dolomites vary in a fashion consistent with thermal excursion(s) from 50° up to 130° and then back to 50°C (Gregg and Shelton, 1989a). The spread in fluid inclusion homogenization temperatures, banding and dissolution in the dolomite cement stratigraphy, episodes of metal deposition with dissolution in between (Sverjensky, 1981; Hagni, 1976), and the wide range of lead isotopic ratios in single galena crystals (Hagni, 1976; Hart et al., 1981) indicate variable physical or chemical conditions. Structural control of brine flow from the Arkoma Basin by the Reelfoot Rift (Farr, 1989) and brine flow south from the Illinois Basin (Gregg and Shelton, 1989b) are suggested by the pattern of trace elements in the dolomite cements. The Ozark-Arkoma system is similar in many respects to other basin margin systems, especially for example the Marathon-Permian Basin-Llano uplift, the Western Canada Basin, and the Appalachian basin, all of which host MVT lead-zinc deposits.

Two flow mechanisms are generally considered possible explanations for the MVT lead-zinc deposits and their associated epigenetic alteration: Bethke (1986); Bethke et al. (1988); and Garven (1984, 1985) proposed that MVT deposits are produced when basin brines are moved out of a basin by gravity flow from highland areas such as the Ouachita mountains. This hypothesis is favored by many for the Ozark Mountains (e.g., Leach and Rowan, 1986) because the mineralization and alteration found there require the throughput of large fluid volumes. Given enough time, gravity flow can propel almost unlimited quantities of fluid. The other flow mechanism that has been suggested for MVT deposits and their associated alteration is the expulsion of brines by compaction or processes related to sediment heating such as gas generation and water displacement from accumulating and deforming sediment piles (Sharp, 1978; Cathles and Smith, 1983; Oliver, 1986; Rich, 1927). The volume of brine that can be expelled by these mechanisms is limited but adequate to produce the observed mineralization and alteration. The purpose of this chapter is to examine these competing mechanisms of fluid flow with reference to the alteration and mineralization in the Ouachita-Arkoma Basin-Ozark system.

THE OUACHITA-ARKOMA BASIN-OZARK SYSTEM

The flow system considered in this paper is shown in Figure 1. What is today the Ouachita orogenic belt was a rifted Atlantic-type margin in early and middle Paleozoic time (Houseknecht,1986; Arbentz; 1989). The 0–150-m-thick, quartzose Lamotte sandstone was laid down unconformably on Precambrian basement and covered by shallow marine carbonates and shales from Cambrian to Early Pennsylvanian time. The maximum thickness of these strata is 1.5 km. Overthrusting from the south in Pennsylvanian time flexed the margin down to form a classic foreland basin. Sedimentation increased abruptly at ~310 Ma (Arbenz, 1989, figure 7). Five and a half kilometers of Atokan sediments were deposited in just 5 m.y. Portions of these and earlier marine sediments were thickened tectonically as they were caught up in the Ouachita thrusting and folding. The deformation and sedimentation ceased shortly after Atokan time.

The Cambrian and Ordovician stratigraphy of the area is shown in Figure 2. From a diagenesis–fluid-flow perspective, perhaps the most remarkable aspect of the transect shown in Figure 1 is the continuity of the microstratigraphy displayed by epigenetic dolomite in the Upper Cambrian to basal Atokan units. The Lamotte Sandstone appears to have been the main conduit for the flow of mineralizing brines, although brines also moved through the overlying Cambrian dolomites and underlying granitic basement. Hydrothermal dolomite cements in these strata have a characteristic cathodoluminescent banding. As indicated in Figure 1B, hydrothermal dolomite banding and cement dissolution events can be correlated over the ~70-km-long Viburnum Trend (Voss and Hagni, 1985), between the Viburnum Trend and the Northern Arkansas Pb-Zn district (Rowan, 1986), and for 70 km or so north and west of the Viburnum Trend (Farr, 1989; Voss et al., 1989; Gregg, 1985), a distance of at least 275 km along this section.

The nature of the cathodoluminescent banding varies systematically. The number and intensity of bands increases abruptly near the Bonneterre-Lamotte transition and then decrease upward with much

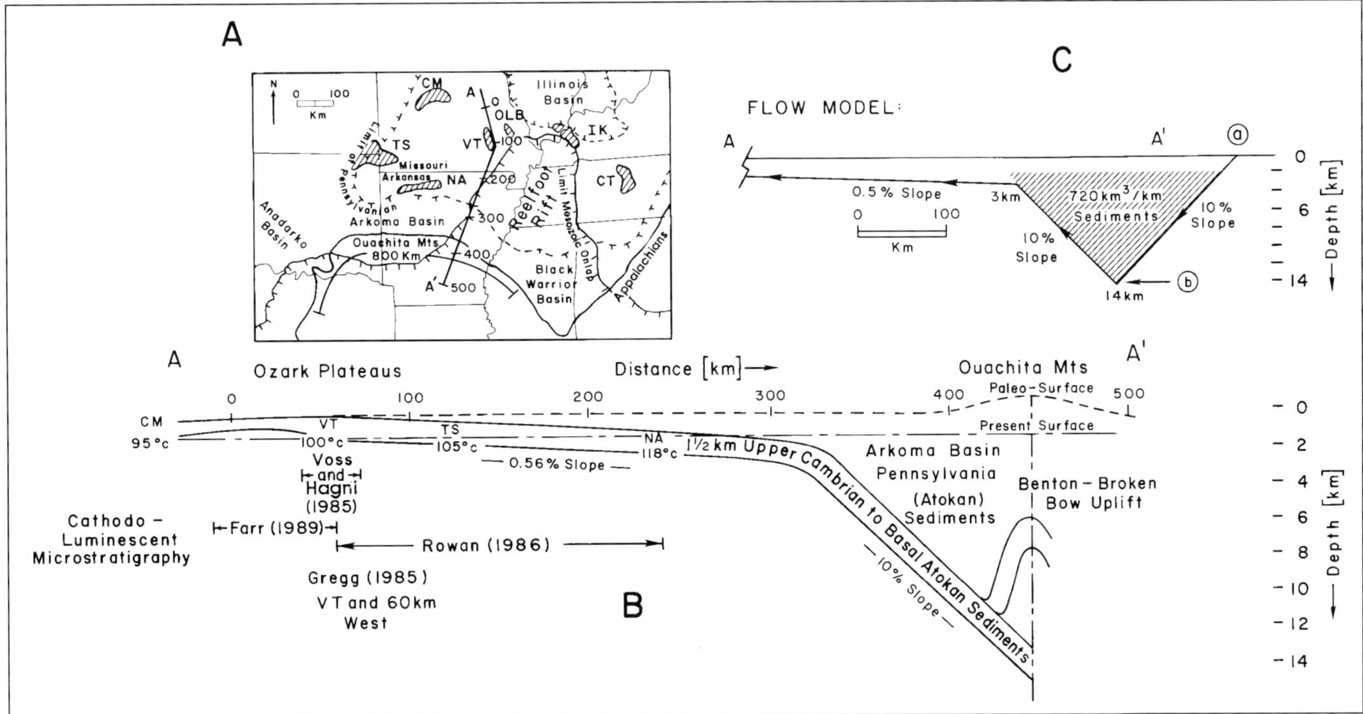

Figure 1. Mississippi Valley-type Pb-Zn (MVT) districts (A) and the continuity of the cathodoluminescent microstratigraphy in hydrothermal dolomite cements in the Mid-Continent (B) are shown in relation to the subsurface geology (B). The base map (A) is after Houseknecht (1986, Figure 1). The cross section (B) is constructed from sections in Thacker and Anderson (1977), Houseknecht (1986), and Arbenz (1989). MVT districts are after Viets and Leach (1990, figure 1) and are labeled: CM= Central Missouri, VT=Viburnum Trend, OLB=Old Lead Belt, NA=Northern Arkansas, IK=Illinois-Kentucky, CT=Central Tennessee, TS=Tri-State. The flow paths used in the calculations presented in Figure 5 are labeled in (C).

fewer and lower intensity bands in shale units (Farr, 1989). These relationships are expected because: (1) iron above a certain concentration inhibits luminescence, (2) Fe and Mn (the main fluorescent agent) are consumed by dolomitization, and (3) shale inhibits fluid throughput and therefore alteration (Farr, 1989). In the middle and lower Bonneterre the number of bands and the percentage of dull bands increase from the St. Francois Mountains (which lie between the Viburnum Trend (VT) and Old Lead Belt (OLB) districts in Figure 1A) to both the south and east. This is consistent with relatively Fe- and Mn-rich brines reaching the St. Francois Mountain area both by flowing directly north out of the Arkoma and by flowing from the Arkoma northeast along the Reelfoot Rift and then east along the northwest striking Broomfield Lineament that connects the Reelfoot with the St. Francois Mountains area. Deposition of distinct sulfide minerals occurs district wide between particular cathodoluminescent zones (Voss et al., 1989).

An excellent and comprehensive review of all aspects of the chemical and isotopic alteration of the area that supports and augments the inferences drawn from the dolomite cement microstratigraphy has been given by Gregg and Shelton (1989a). Trace elements in the hydrothermal dolomite within 1 m of the Bonneterre-Lamotte transition show an increase in Fe and Mn and a decrease in Sr from the St. Francois Mountains to the south (Gregg and Shelton, 1989b). The pattern is absent 3 to 6 m above the contact, however, indicating less flow in that part of the Bonneterre. Similar Fe, Mn, and Sr trends from the St. Francois Mountains to the northeast suggests that brines arrived in the same cathodoluminescent time-bracket from the north. Flow from the Illinois Basin as well as from the Arkoma basin seems to have contributed to alteration and mineralization (Gregg and Shelton, 1989a,b). The decrease in Fe and Mn upward in the Bonneterre from the Lamotte sandstone, as well as Mn, Fe, Zn, Na, and K enrichment in shales higher in the stratigraphy, suggest that mineralizing fluids upwelled from the Lamotte and Bonneterre in the Viburnum trend area (Gregg and Shelton, 1989a). Detailed study of the concentration of these elements in the most fractured parts of shales above the Buick mine, where upward leakage was perhaps greatest, show that enrichment of Na, Mn, Fe and Zn was followed by depletion to background levels (Panno et al., 1988). Depletion did not occur in the less fractured parts of the shale. The leaching of the fractured parts of the shales may be related to the dissolution of sulfides and hydrothermal dolomite in the ore zones;

both could be caused by meteoric inflow.

As shown in Figure 3, fluid-inclusion homogenization temperatures from different mining districts in the region range from 60° to 180°C. Fluid inclusions in baroque dolomite in the upper Bonneterre formation in the Reelfoot Rift suggest two much hotter episodes of heating, the first with mean homogenization temperatures of 235°C and the second with mean temperatures of 170°C (Tobin, 1991). Fluid inclusions in the core of the Ouachita Mountains have even higher homogenization temperatures up to 300°C (Shelton et al., 1986). A general decrease in temperature across the ore district is suggested by these data, especially if flow was directed from the Reelfoot rift, and the Ouachita homogenization temperatures are indicative of the temperatures of fluids entering the Cambrian aquifers at the base of this sediment pile.

There is a large scatter in the homogenization data from the mining districts in Missouri and Arkansas, however, and the scatter is far bigger than the difference in modal temperature between districts. Leach and Rowan (1986) attribute this to inclusion necking, but it could also reflect real temperature variations associated with pulsing flow. The reality of the modal trend of 0.09°C/km they deduce from the mining district data in Figure 3 could certainly be questioned

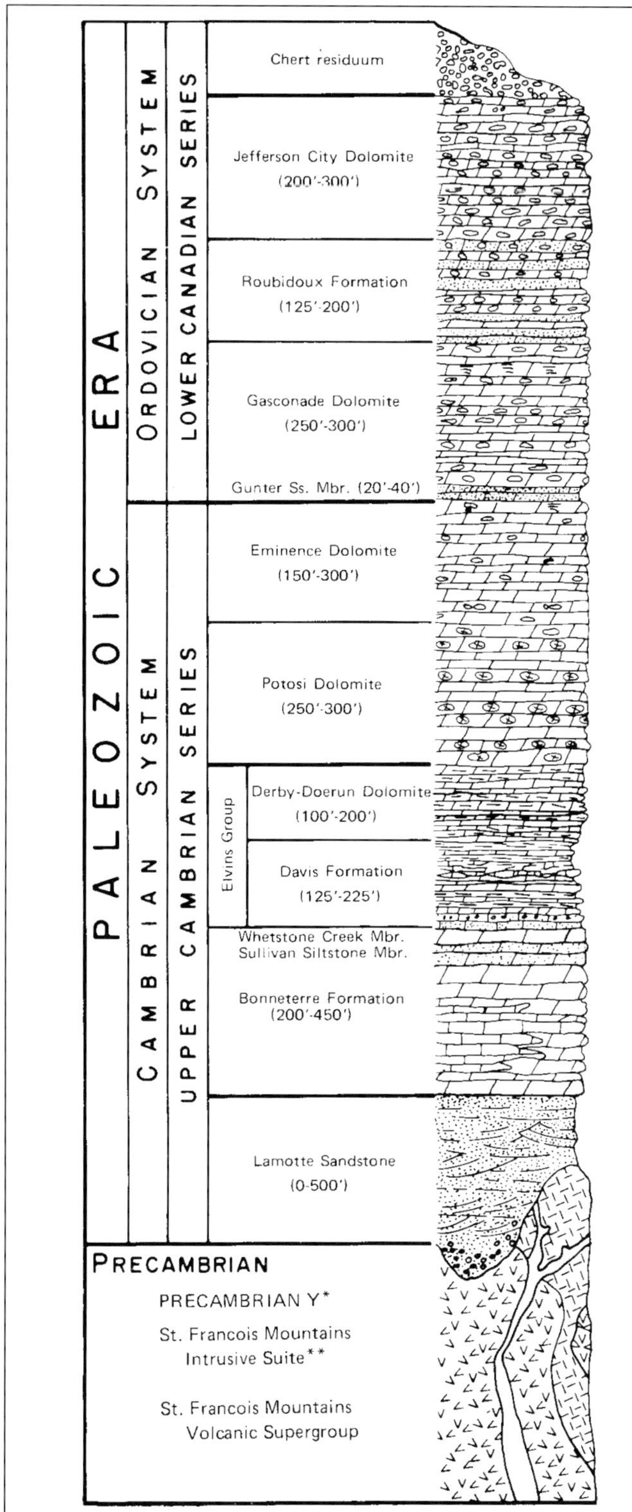

Figure 2. Paleozoic stratigraphy in the St. Francois Mountains area from Kisvarsanyi (1976, p. 2).

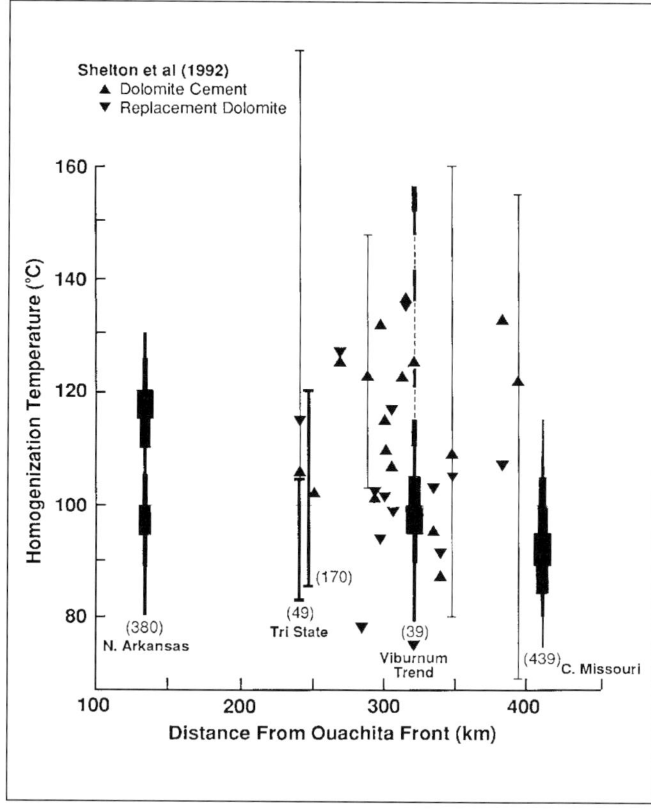

Figure 3. Homogenization temperatures of fluid inclusions in sphalerite from some of the mining districts shown in Figure 1, and in epigenetic dolomite from the Bonneterre (Cambrian) aquifers in southeastern Missouri. Mining district data are from Leach and Rowan (1986) and are indicated by heavy lines or boxes. The width of box indicates number of inclusions homogenizing within the temperature range indicated by the length of the box. The dolomite inclusion data cover the area indicated by the dashed box in Figure 4 and are from Shelton et al. (1992).

(e.g., Gregg and Shelton, 1989a; Shelton et al., 1992), and for this reason the line connecting the modal data drawn in Leach and Rowan's original figure has been omitted. There is a general consensus that the sites of ore deposition were never buried by more than 1 to 1.5 km of sediment. Mineralized areas seem therefore to have been at temperatures considerably above those that would be predicted by their burial depths at the time of mineralization. This is the most significant point shown by Figure 3. Almost everyone attributes this thermal elevation to fluid flow in the Lamotte and other Cambrian aquifers.

The present-day flow system in the Ozark area is an interesting one. Water recharges and flows radially out from the Ozark Plateau through Cambrian to Mississippian strata, labeled upper Cambrian to basal Atokan sediments in Figure 1B and known hydrologically as the Ozark Plateaus Aquifer System (Jorgensen et al., 1988). At the same time, flow sweeps east through the deep Western Interior Plains Aquifer System from catchments in the Front Range of the Rocky Mountains, picks up salt in Kansas, and upwells where it meets the outward flowing waters of the Ozark Plateaus Aquifer System (Jorgensen et al., 1988; Banner et al., 1989). One of the brine springs that express this upwelling is "old Boonslick" where Daniel Boone "settled and manufactured salt" (Figure 4) (see Shepard, 1907, p 78). Where strata carrying the eastward-moving waters approach the surface in eastern Kansas and Oklahoma, the surface temperature gradient is increased to ~40°C km^{-1} (Stavnes and Steeples, 1982; Luza et al., 1984).

It is important to recognize that basin brines today occupy the deeper parts of the Cambrian aquifers and almost completely ring the Ozark Plateau. This is shown clearly in recent maps by Jorgensen et al. (1986) and older maps published by Dott and Ginter (1930). Data from these sources are summarized in Figure 4. As stated by Dott and Ginter, "The most striking feature of the map [Figure 4] is the seeming relationship between the concentration of [Cambrian and] Ordovician waters [e.g., waters contained in Cambrian and Ordovician strata] and outcrops of Ordovician rocks." Equivalent fresh water head maps make it clear that the reason for this relationship is that fresh waters recharging outcrop areas such as the Ozark Plateau have, over time, swept out the brine (Jorgensen et al., 1986). Present-day flow directions inferred from the contours in equivalent fresh water head are shown by arrows in Figure 4. Meteoric recharge has not swept brines from the same aquifers deeper in the Illinois, Arkoma, and Oklahoma basins. In fact the flow from the Ozark Plateau meets the flow from the Rocky Mountains in the areas of present brine upwelling (Jorgensen et al., 1986).

The geology and geochemistry reviewed above indicate that starting in Permian time, basin brines moved, perhaps in pulses, from the Arkoma basin (and other basins) through the Lamotte Sandstone and other Cambrian strata to Mississippian aquifers and produced regional, coherent chemical and isotopic alteration and extensive lead-zinc mineralization. Rich deposits of lead and zinc accumulated where the aquifers cropped out (e.g., the St. Francois Mountain area), were cut by permeable units (e.g., the reefs of the Viburnum Trend), or where they were intensely fractured. Meteoric water enters these same outcrop areas today, and brine has been displaced by fresh water. This may also have occurred in the past, in between pulses of brine outflow. At the same time there is today a regional cross-basin hydrologic flow system carrying waters from the Rocky Mountains to Kansas and Missouri, and this flow is expelling brines in salt springs such as old Boonslick. The present-day Mid-Continent thus provides a good backdrop for discussing the nature of the brine flow that occurred there in the past and produced extensive and coherent alteration and mineralization.

THERMAL CONSTRAINTS ON FLUID FLOW

Both cross-basin hydrologic flow and basin dewatering must produce flows sufficient to explain the fluid inclusion homogenization temperatures in Figure 3. We start with an analysis of this thermal constraint following the approach developed by Cathles and Smith (1983). We ask the simple question: How fast must waters flow out of the Arkoma Basin to heat the Cambrian aquifers in the Ozarks to temperatures of 80° to 120°C? The critical issue is the heating of the aquifer system. If the aquifer has the required temperatures, fluids will be able to move locally upward in fractures or permeable cross-cutting strata and deposit minerals with little loss in temperature, recording the aquifer temperatures as they do so. If the aquifer system is not heated sufficiently, local fluid upwelling at any rate will not be able to produce fluid inclusions with the required homogenization temperatures. To put the problem in a present-day context, temperatures in the Arbuckle or Rice aquifers in Kansas would need to be 80° to 120°C at ~1 km depth. They are ~40°C (25° to 30°C above ambient) at this depth today. Weak flow through these aquifers elevates the thermal gradient near where the aquifers crop out to ~40°C km^{-1}, but these gradients do not extend to depth. Much greater flow rates would be required to bring waters of 80° to 120°C within a kilometer of the surface.

How much faster the flow must have been to accommodate the temperatures and depths thought to have existed during Pb-Zn mineral deposition in Missouri and Arkansas can be determined by calculating the temperature profile in a basal aquifer or

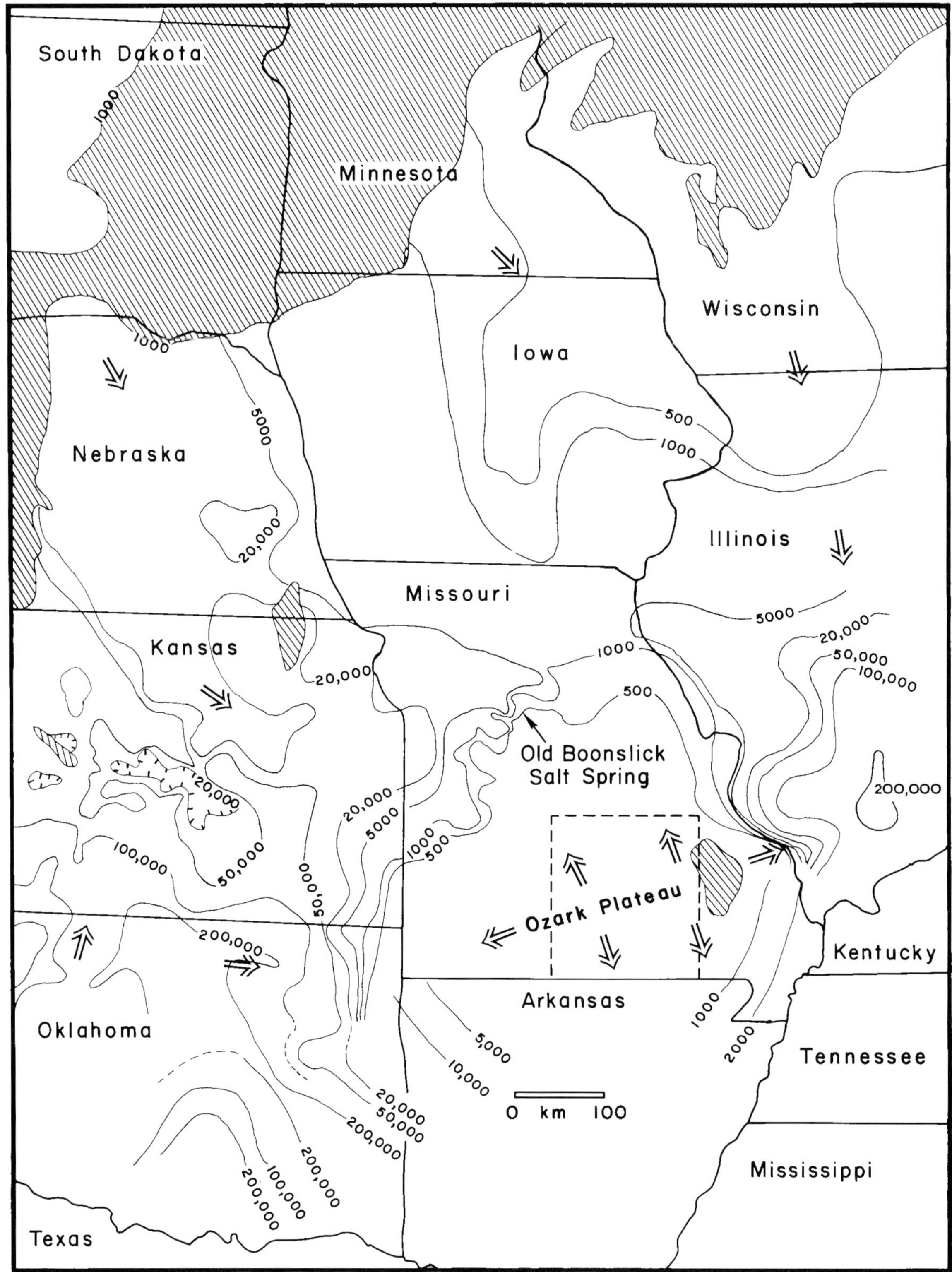

Figure 4. Dissolved solids concentration (ppm) in Cambrian and Ordovician sediments in the Mid-Continent area from Jorgensen et al. (1986) and, in southern Oklahoma, Arkansas, and southeastern Missouri, from Dott and Ginter (1930). Present directions of fluid flow, inferred from contours of fresh water equivalent head given by Jorgensen et al. (1986), are shown by arrows. Arrow length has no significance. Present day flow originates from areas where Cambrian and Ordovician sediments outcrop (hachured areas) or nearly outcrop. This flow has flushed brines from the Cambrian and Ordovician strata as shown in the figure and discussed in the text. The dashed box indicates the area where detailed studies of drill core fluid inclusions and alteration have been carried out by Gregg, Shelton, and others.

aquifer system by daisy-chaining analytical solutions for steady state temperatures along aquifer segments of uniform dip (Cathles, 1987). Analytical solutions for each linear (uniform dip) segment of the aquifer are joined together by requiring continuity of flow and temperature. The temperature at only one location along the aquifer need be specified. The temperature profile then depends on the thermal conductivity, K, of the sediments above the aquifer, the regional heat flow, j_o , and the total flow per unit strike length through the aquifer or aquifer system, Q.

Both j_o and K are parameters of critical importance, especially in the flat portion of the flow path across the Ozark plateaus. The Cambrian to Mississippian cover in this area consists today of a basal sandstone overlain by carbonates with some shales. In the St. Francois Mountains area, this stratigraphy is ~80% carbonate, ~12% sandstone, and ~8% shale (Figure 2). The thermal conductivity of a lithologic sequence can be estimated using the porosity-dependent rock type thermal conductivities and porosity-depth relations in Royden and Keen (1980). I calculate using Royden and Keen's relations that the effective thermal conductivity of a 2.5-km-thick column of sediment consisting of 80% carbonate, 12% sandstone, and 8% shale is ~5.6 x 10^{-3} cal/cm-s-°C (5.6 Thermal Conductivity Units, or TCU). Without compaction, the thermal conductivity would be ~5 TCU. The column would have to be 53% shale to have a thermal conductivity of ~4 TCU. If the sediments now eroded from the Ozark plateaus were dominantly shale, a value as low as ~4 TCU might be approached. It is more likely, however, that these shallow water sediments would have been sands or carbonates, in which case the effective thermal conductivity of a 2.5 km section would have been between ~5.6 TCU (all carbonates) and ~6.5 TCU (50% carbonates, 50% sand). A lower bound for the thermal conductivity of the sediments covering the Ozark plateaus at the time of brine movement is therefore probably ~4 TCU and the most reasonable estimate ~5.6 TCU or greater. Heat flow in Missouri today is ~1.3 x 10^{-6} cal/cm^2-s or 1.3 Heat Flow Units (HFU; Kron and Stix, 1982). There is no reason for it to have been significantly different in Permian time.

Figure 5 shows the thermal profile along the "Cambrian aquifer loop" produced by cross-basin flow of various magnitudes. In Figure 5A, K=5.6 TCU, j_o= 1.3 HFU, and Q[cm^3/cm-s] is specified in terms of the dimensionless parameter α, where α = K/Q c tan β, and tan β is the dip of the steep Ouachita-Arkoma aquifer segments (=0.1) and c is the heat capacity of water (= 1 cal/cm^3-°C). Temperatures in the Cambrian aquifers (Lamotte Sandstone, Eminence Dolomite, etc.,) are not elevated significantly at any flow rates for these parameter values. The narrow Ouachita-Arkoma Basin complex is cooled by the high flow rates (low values of α) before aquifers in the Ozarks can be significantly warmed.

Figure 5B shows that if K=4 TCU, cross-basin flow can warm the Ozark Cambrian aquifers to the levels indicated by the modal fluid inclusion homogenization data in Figure 3 in the south end of the transect near the Northern Arkansas District. However, even very large uniform flow rates cannot produce the temperature increases required by the modal (let alone maximum) fluid inclusion homogenization data in the north near the Central Missouri District. Uniform cross basin hydrologic flow thus cannot match the modal fluid inclusion temperatures over the entire Ozark transect for the depth of burial selected, even in the unlikely event that the effective thermal conductivity of the cover was as low as 4 TCU. The Ouachita-Arkoma belt is too narrow to sufficiently heat waters flowing uniformly across it at the rates that would be required to warm the margin to the levels observed.

It is important to note that if the thermal conductivity of the Ozark cover could be lowered just slightly below 4 TCU, or if the cover thickened, the requisite temperatures could be attained with very slow or no fluid flow. For example, if the effective thermal conductivity of the Ozark cover were 3 TCU and the regional heat flow 1.5 HFU, the unperturbed thermal gradient would have been 50°C/km and temperatures adequate to explain the Ozark fluid inclusion homogenization temperature data could be obtained under 1.5 to 2 km of cover with no fluid flow at all. A similar result could be obtained if the cover were substantially thicker. In either case, fluid movements could have been slow and the required Pb, Zn, Mn, etc. brought in over a protracted period of time.

If the Ozark cover were not thick and insulating, cross-basin flow could still produce the required temperatures in the Ozark aquifers if the flow were nonuniform in time or space. If the flow is sufficiently focused once it leaves the deep warm sections of the flow loop under the Ouachita Mountains and Arkoma Basin, the temperatures required by the fluid inclusion data can be attained at the Central Missouri Pb-Zn district and other distal Ozark locations. The Ouachita-Arkoma flow rates that raise the temperatures of the 3-km-deep slope break the most (to ~180°C above surface ambient) are characterized by α=~-2 (see Figure 5A). If the flow is then focused by a factor of 40, daisy-chain calculations indicate that distal locations are warmed to the extent required by the modal fluid inclusion data.

The requisite distal temperatures could also be attained by transient (pulsed) flow. Across-basin flow will initially displace deep, warm brines from the Arkoma just as will compactive or other types of basin dewatering. The initial flow of warm brine could heat the Ozark aquifers to temperatures

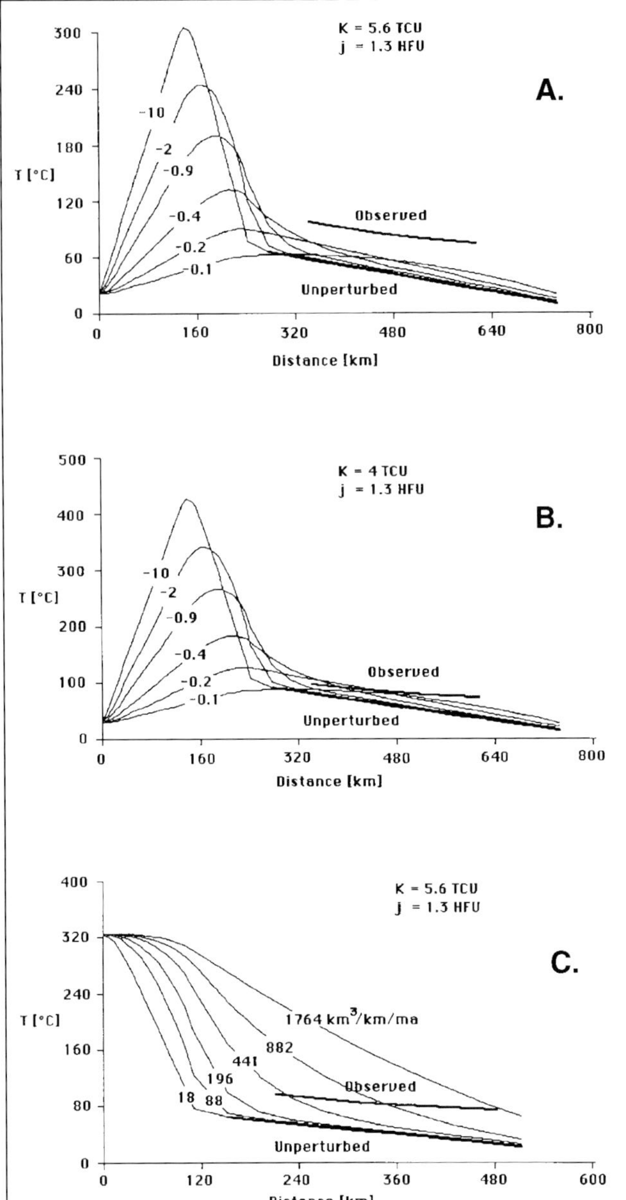

Figure 5. (A) Temperature profile in degrees Centigrade above ambient along flow path "a" (which starts at the surface in the Ouachita highlands, moves to 14 km depth, then to 3 km depth and then to discharge in Ozark plateaus as shown in Figure 1C) calculated for various values of the dimensionless flow rate parameter α. Smaller values of α indicate larger flow rates. Regional heat flow j = 1.3 x 10^{-6} cal/cm-sec or 1.3 HFU; the thermal conductivity of the cover, K=5.6 x 10^{-3} cal/cm-sec-°C or 5.6 TCU. On this and the other figures the normal unperturbed temperatures along the Cambrian aquifers are indicated by the lower heavy line labeled "unperturbed." The modal homogenization temperatures in Figure 3 with an ambient temperature of 20°C subtracted are indicated by the upper heavy line labeled "observed." (B) Temperature profiles for various values of α along loop a in Figure 1 for K= 4 TCU and j_o=1.3 HFU. (C) Temperature profile (in degrees Centigrade above ambient) along flowpath "b" (which starts at 14 km depth and then follows flowpath "a" as shown in Figure 1C) calculated for various values of α under the same conditions as Figure 5A. Fluids are assumed to enter the Cambrian aquifers at 14 km depth at equilibrium temperatures for that depth (325°C above ambient). The flow rates, Q, shown in this figure correspond to the values of α given in Figures 5A and 5B. For example α= -0.1 corresponds to Q = 1744 km^3/km/m.y. The formula relating α and Q is given in the text.

approaching those in the deepest parts of the Ouachita-Arkoma basin. With continued flow, those deep locations will be cooled by cold meteoric inflow and temperatures in the Ozark aquifers will drop. Temperatures in the Ozark aquifers will thus rise and fall in a transient fashion. The maximum transient temperatures that could be attained at various flow rates can be estimated if brine is introduced to the deep extension of the Ozark Cambrian aquifers at temperatures normal at those depths under no flow, thermal equilibrium conditions (path b in Figure 1B). Figure 5C shows that such flow can elevate temperatures in the Cambrian aquifers to the highest end of the modal ranges indicated in Figure 3. The required temperatures in the south can be produced by flow rates of 441 km^3/km/m.y. (α=-0.4). In the north, flow rates of 1764 km^3/km/m.y. (α=-0.1) are required. Temperatures in the south and north are thus compatible with a flow in the Cambrian aquifers that concentrates by a factor of 4 as it proceeds north. If the aquifers in question have a cumulative thickness of 100 m, the Darcy flow rate would have increased northward from 4.4 to 17.6 m yr^{-1}.

There is general agreement regarding the flow rate required to heat a basin margin. For example the flow rates required to significantly warm the outflow margin of the Illinois Basin (slope ~0.75%) in Bethke's (1986, Figure 11b) analysis of the Upper Mississippi Valley Pb-Zn district in Wisconsin was ~540 km^3/km/m.y. The long, flat, deep flow segment across the Illinois Basin avoids the heating problems caused by the narrowness of the Ouachita-Arkoma system and allows sufficiently warm fluids to enter the margin aquifers at the rates required. Once warm enough fluids are obtained, however, the flow rates needed to warm the margin aquifers are similar to those we estimate. The fundamental parameter controlling margin heating is Q. My estimates of Q are minimum estimates. To the extent that flow is less concentrated in basal units, conductive thermal losses would be greater (because the heat has less distance to diffuse to the surface), and Q must be greater for the same basin margin temperature perturbation.

SALINITY CONSTRAINTS ON CUMULATIVE FLUID FLOW

Requisite flow rates of ~441 $km^3/km/m.y.$ and the present-day pore water salinity distribution place severe constraints on how possible flow mechanisms could have operated in the Ouachita-Arkoma-Ozark area. In cross-basin hydrologic flow, fresh meteoric water will rapidly displace brines and dissolve any salt contacted. For example, if the Arkoma-Ouachita sediments contained 10 volume-% evaporites and the flow contacted these evaporites, cross-basin flow at 441 $km^3/km/m.y.$ would dissolve and carry away all the salt in 1.25 m.y. In this calculation we assume 10% of the 720 km^3/km Ouachita-Arkoma section is salt, that salt has a density of 2 g cm^{-3}, and that the efflux has a salt-saturated salinity of 26 wt%. The salt is dissolved when the salt transported out of the basin in time t, 0.26 Q t, equals the amount of salt in the basin, 720 (0.1) (2 g cm^{-3}). If the flow rate were 88 $km^3/km/m.y.$ (α=-2), the same amount of salt would be flushed in 6.25 m.y.

The dilemma such fast flushing of salinity poses for cross-basin hydrologic flow is as follows. Erosion is a slow process. Once the topography to "drive" cross-basin hydrologic flow is established, it will take tens of millions of years to remove. The permeability of the MVT deposits we have been discussing remains for the most part very high today. Six and a half tons of water were pumped for every ton of ore mined in the Old Lead Belt in 1943 (Weigel, 1943). Two years of pumping were required to achieve a 700 ft drawdown prior to mining in the Viburnum Trend (Bullock, 1973). Thus, although there may have been permeability reduction where metals were deposited in disseminated fashion in matrix pores (~90% of the ore is of this type; personal communication from Jay Gregg, 1992), the fracture and breccia pathways that channeled fluids to these areas remains high. The flow paths have clearly not been plugged by alteration or mineral precipitation. Also it is clear from the position of the deposits that brine discharged upward through them to the surface at the time of mineralization. Consequently it is difficult to see how brines could remain in the basins surrounding the Ouachitas if cross basin hydrologic flow at the rates required to account for the modal homogenization temperatures in MVT deposits in the area were ever established. The lack of a full (or even partial) salinity flush from the deep parts of the basins surrounding the Ozark plateaus that are thought to have produced the MVT mineralization in the region (the Illinois and Arkoma basins) appears to pose a fundamental difficulty for the cross-basin hydrologic flow model in the Ouachita-Arkoma-Ozark area and elsewhere (Cathles, 1987). Either a way must be found to turn off cross-basin hydrologic flow shortly after it is established by tectonic events, or a way must be found to explain why the basin brines have not been flushed by such flow.

Compactive dewatering does not have a salinity flush problem because fresh water does not enter the basin and only a portion of the pore waters are compactively expelled. On the other hand, as discussed by Cathles and Smith (1983), transient compactive expulsion is constrained by two kinds of heat balance considerations. If the accumulation of the 720 km^3/km of sediments shown in Figure 1C occurred in 10 m.y. (probably a generously rapid rate), for example, and caused 10% compaction, fluids would be expelled at ~7 $km^3/km/m.y.$ This is ~1.6% of the rate required to warm the margins (441 $km^3/km/m.y.$) by the requisite amount. Dewatering of the Arkoma must be focused ~62 fold on the margins or temporally episodic if the dewatering model is to warm the Ozark Plateaus Aquifers as observed. Furthermore each dewatering pulse must be of sufficient volume to warm the escape aquifers. If the Cambrian Ozark aquifers cumulatively had an average thickness of 100 m over the 330 km from their bend in Figure 1B to the Central Missouri MVT District, a volume of water ~33 km^3/km would be required for each pulse. The 720 km^3/km sediment prism in Figure 1b could thus contribute about 2 pulses of fluid. If there were progressive channeling of brine flow paths to the north, which is reasonable to expect and indicated by the pattern of geochemical alteration (channeling along Reelfoot rift), more pulses could be delivered. For example, if flow were focused by a factor of four by channeling, 8 pulses of warm brine could be delivered to areas as far north as Central Missouri.

FLOW REQUIREMENTS OF MINERALIZATION AND ALTERATION

The limited quantities of brine that might be compactively expelled from the Ouachita-Arkoma system are adequate to produce the mineralization and alteration observed in Arkansas and Missouri. If we take the axis of the Broken Bow uplift as the southern boundary of the sediment volume that might contribute fluids to the Ozark Plateau (with the logic that any sediments south of this divide would expel their fluids south, not north), then the Ouachita-Arkoma sediment volume below 2 km depth is closely approximated by a triangular wedge 12 km deep and 120 km wide (Figure 1C). The volume of such a wedge is 720 km^3 per kilometer strike length. The pertinent strike length of the Arkoma-Black Warrior-Ouachita system that contributed fluids to northern Arkansas and Missouri is ~800 km (see Figure 1) so the total volume contributing fluids to the Ozark area is ~ 566,000 km^3.

The Viburnum Trend deposits contain ~30 x 10^6 tonnes lead (Wiegel, 1965). The Old Lead Belt on the eastern side of the St. Francois Mountains produced about 9 x 10^6 tonnes of lead (Snyder and Gerdemann, 1968). The Tri-State District produced about 14 x 10^6 tonnes of zinc (80%) and lead (20%) (Brockie et al., 1968). The other Central Missouri and Northern Arkansas mining districts contained smaller amounts of metal. All told, the mineralization in Arkansas, Missouri, and immediately adjacent areas totals perhaps ~80 x 10^6 tonnes of combined Pb and Zn.

Basin brines can easily contain ~200 ppm dissolved Pb or Zn (Carpenter et al., 1974). If just 10 ppm lead plus zinc precipitated from basin brines propelled across the Ozarks, 8000 km^3 of brine would be required to precipitate the known resources of the area. This volume of brine represents a brine-filled porosity of less than 1.5% of the 566,000 km^3 of sediments defined above. The sediment volumes identified would have had little difficulty supplying the brine volumes required for mineralization. Quantitative discussion of how basin brines behave as ore fluids has been offered by Sverjensky (1984).

The brines contained in the identified sediment volumes are also sufficient to produce the alteration observed in the Cambrian aquifers of the Ozarks. Gregg (1985) pointed out the extensive nature of the epigenetic dolomite cements at the base of the Bonneterre. He estimated that ~100 km^3 of Bonneterre limestone was dolomitized over a >16,600 km^2 area near the St. Francois Mountains and that 35,000 km^3 of brine would be required by this dolomitization. Even this seemingly very large brine volume represents the brine in pores representing just 6% of the 566,000 km^3 sediment volume. Clearly either cross-basin topographically driven flow or compaction could relatively easily supply the volumes of brine required for alteration and mineralization without exhausting the supply of brine in the Ouachita-Arkoma system.

DISCUSSION, SUMMARY, AND CONCLUSIONS

The calculations presented above are selected to make several specific points. Many different combinations of flow path, focusing, overburden thickness, thermal conductivity, and heat flow could have been selected. The calculations are simple enough using the daisy-chain method that an interested reader can easily explore these possibilities and, I hope, will be encouraged to do so by this discussion.

The calculations indicate the importance of parameters perhaps not emphasized enough in the basin modeling literature. The thermal conductivity of margin cover is critical to the thermal impact of flow. An insulating cover in the Ozarks sufficiently thick to create temperatures similar to those indicated by fluid inclusions in the Cambrian aquifers without fluid flow does not appear to be geologically reasonable, but since the cover has been removed by erosion it is impossible to be completely certain. If the cover insulation was inadequate, the shallow dip of the margin requires rapid flow rates of hundreds of km^3/km/m.y. to elevate aquifer temperatures to the levels indicated by the homogenization data.

In this context, a critical observation is the present distribution of brine in the Cambrian to Mississippian strata in basins surrounding the Ozark plateaus. Across-basin topography-driven flow at 100s of km^3/km/m.y. would flush the brines from the Cambrian and Mississippian aquifers in only a few million years. The fact that brine has not been flushed from these aquifers, but has been flushed by oppositely directed meteoric recharge entering outcrop areas in the Ozark plateaus and is being flushed by cross-basin flow from the Rocky Mountains, requires some explanation. The minimum explanation is that the cross-basin topography-driven flow, if it existed in the Ouachita-Ozark system, was pulsed or episodic.

Several geological arguments suggest that cross-basin hydrologic flow was not operative during mineralization. The very rapid Atokan sedimentation and tectonic thickening of the Permian section during the Ouachita orogeny almost certainly overpressured fluids in the Ouachitas and related basins such as the Arkoma. Overpressuring would have prevented any cross-basin flow during active tectonism and sedimentation. The maximum vitrinite reflectance ($R_o \geq$ 3.0 %) and the highest metamorphic grade occur in the core of the Ouachitas (see Arbentz, 1989). This is not compatible with the cooling by meteoric inflow that would be predicted if cross basin flow were to occur at rates sufficient to warm the Ozark plateaus. Fluid inclusions in Ba-rich adularia dikes and quartz veins in the core of the Ouachita Mountains suggest the outflow of metamorphic fluids at temperatures of up to 300°C (Shelton et al., 1986).

On balance the geological and geochemical evidence seems most compatible with the episodic dewatering hypothesis. Seismic pumping cannot expel sufficient volumes of brine to warm the Cambrian aquifers. Tectonic pulses are of too long duration to be thermally effective. There is no easy way to produce pulsating cross-basin topography-driven flow, since the flow paths have not been sealed by alteration or mineral deposition at their discharge ends. Compaction could easily have expelled a volume of brine sufficient to account for the observed alteration and mineralization.

One argument used against episodic brine expulsion is that the Mid-Continent deposits do not appear to have been temperature anomalies at the time of ore deposition. The calculations presented in Figure 5 also clearly show that the deposits need not have been local temperature anomalies. Whether or not this is so will depend on how localized the escape from the Cambrian aquifers was and how high (near the paleo-surface) the deposits were in the stratigraphic column at that time. If the deposits were near the Cambrian aquifer source under ~1 km of cover, local thermal anomalies would be lost in the scatter of the homogenization data, and the most obvious signal would be the elevation of temperatures with perhaps some hint of a slight up-dip decrease in the mode of the homogenization data.

Compactive dewatering might have been assisted by a gas drive. The lack of water in the Arkoma (a basin almost literally filled with gas) raises the intriguing possibility that gas generation may have expelled brine, a possibility suggested long ago in general and specifically for the Ouachitas by Rich (1927). Rates of gas-driven brine expulsion are competitive with compaction for moderate kerogen grades and normal sedimentation rates. If the gas generation or gas-driven fluid expulsion occurred in pulses, as particularly organic rich strata were heated or as seals ruptured, for example, the rate of expulsion could have been substantially faster.

Regardless of their cause, pulses of fluid flow are suggested by a variety of observations and have geochemical advantages when it comes to alteration. For example, Sverjensky (1981) documents at least eight pulses of mineralization interspersed with periods of ore dissolution in the Buick mine of the Viburnum trend. Hagni (1976) notes eight intervals of chalcopyrite and pyrite deposition, six intervals of sphalerite deposition, and five intervals of galena and quartz deposition in the Tri-State district of Kansas, Oklahoma, and Missouri. Hagni (1976) comments (p. 487): "The deposition of metallic sulfides took place over a long period of time. This condition is attested to by the many repetitions in the paragenetic sequence... and is suggested by the wide range of Pb isotopes recorded from the center to the edge of a single crystal... The sulfides were even subject to periods of corrosion between periods of deposition...". The corrosion could be produced by meteoric recharge of the Ozark Cambrian aquifers (as is happening today) between pulses of brine expulsion. The low Fe and Mn content of sparry dolomites in the back reef area of the Viburnum Trend suggest this recharge (Buelter and Guillemette, 1988). The pressure-temperature incompatible mineralogy (Barton, 1981) and stable isotopic evolution (reviewed above) also indicate pulses of fluid flow with minerals deposited at different temperatures along perhaps many prograde-retrograde thermal excursions. Pulses of fluid flow can achieve a cumulatively greater alteration because alteration occurs mainly when fluids flow through temperature gradients and the thermal front of an expulsive pulse provides a steeper gradient than the 0.09°C/km average paleogradient across the Ozarks. Pulses of flow could also mature hydrocarbon gases that would react with sulfate to provide the reduced sulfur for base metal sulfide deposition (Anderson, 1991).

Some aspects of the fluid inclusion data just briefly mentioned in the literature may provide critical evidence. Gregg and Shelton (1989a) point out that early fluid-inclusion studies by Roedder and later studies by Bauer found a high-temperature, low-salinity population of fluid inclusions. Fluid inclusions at any given homogenization temperature in fact range from low to high (brine) salinities. The lower salinity inclusions occur in the dolomite cements from late diagenesis through ore deposition time (Shelton et al., 1992).

There are at least three possible explanations for the low salinity inclusions. They could record pulses of metamorphic fluids from the core of the Ouachita Mountains. Fluid inclusions with homogenization temperatures up to 300°C and salinities of 3 wt% NaCl equivalent are found in Ba-rich adularia in organic-rich seams in Cambrian shales on the shores of Lake Ouachita in Arkansas (Shelton et al., 1986). Alternatively they could reflect pulses of cross-basin hydrologic flow. In either of these two cases, pulses of low salinity fluid would flush brine out of flow channels and trap low salinity fluid inclusions. Brines would re-enter the flow channels from less permeable zones after the low salinity and heat pulses had passed. Finally the lower salinity inclusions could be produced by the flow of meteoric water into discharge areas warmed by a pulse of brine outflow. This in fact seems to me the most likely explanation. The 25,000 km study area in southeastern Missouri that Gregg, Shelton, and others have most intensively analyzed, and which now provides the best documentation of the lower salinity inclusions, lies entirely within those parts of the Ozark Plateau where the recharge of meteoric water has displaced brine (dashed box in Figure 4). Thus most if not all of the samples containing lower salinity inclusions could have been affected by meteoric recharge between pulses of brine expulsion. The isotopic composition of the low salinity fluid inclusions and their abundance pattern should distinguish these possibilities.

Observations, modeling, and the possible explanations for the lower salinity fluid inclusions indicate the importance of fluid focusing. Geologic observations suggest some focusing of flow in the Reelfoot rift. Channeling could reconcile steady cross-basin topography-driven flow and the modal fluid inclusion data, but the flow rates would have to be optimum and the lack of salinity flushing would still require flow to be episodic. Channeling in the deeper parts of the aquifers directly affects the combinations of rate of fluid expulsion and overburden thickness that are required to satisfy measured homogenization temperatures in the dolomite in the Ozark plateaus. Definition of the pathways of fluid movement in the discharge areas is necessary for the refined interpretation of fluid inclusion salinity and alteration data.

The above discussion underscores the interrelated regional nature of basin alteration. Low salinity fluid inclusions in Ba-rich adularia in the metamorphic core of the Ouachita Mountains may be related to low salinity, warmer fluid inclusion in the Ozarks, for example. Perhaps we should look past local variations and place alteration in a *far* larger geographic context than has seemed reasonable in the past. The regional fluid flow models discussed here and those that are now appearing in the literature are helpful in encouraging us to do this.

In summary, simple analytical "daisy chain" models show that if the Ozark plateaus' cover was thermally conductive and thin as seems most geologically reasonable, continuous flow at the rates required to explain fluid inclusion homogenization temperatures would have quickly flushed salt from the aquifers in contradiction to present observations. Flow of mineralizing and dolomitizing brines across the plateaus must have been episodic, regardless of the flow mechanism. The most plausible way to produce this kind of flow is episodic compactive expulsion. Meteoric recharge into discharge sites between expulsive pulses is indicated by geochemical zoning in the back reef areas and could explain the low salinity fluid inclusions and the dissolution of sulfides and dolomite.

The analysis depends critically on the following physical factors: (1) the thermal conductivity of Ozark cover, (2) the present day brine distribution in the deeper portions of the Cambrian aquifers surrounding the Ozarks, (3) the focusing of channeling of flow in three dimensions, and (4) the rate of meteoric recharge between pulses of brine discharge. Better definition of the thermal conductivity, pattern of alteration, pattern of pore water salinity, the paragenesis of salinity variations in fluid inclusions, the isotopic composition of the low salinity inclusion fluids, and the regional and local relationships between diverse kinds of alteration and diagenesis in the Mid-Continent could distinguish episodic topographically driven flow from episodic compactive expulsion followed by meteoric recharge. The economic value of base metal cements in the Ouachita-Arkoma Basin-Ozark system has led to unusually good definition of the alteration pattern in this area. Broader and better definition of the pattern could provide particularly clear insights into how fluids move in sedimentary basins, an issue of fundamental concern to both the hydrocarbon and minerals industries.

ACKNOWLEDGMENTS

I would like to thank the Gas Research Institute and the 11 oil company sponsors of the Global Basins Research Network for support of research related to this publication. I thank Jack Oliver for drawing my attention to Rich's paper on gas expulsion, Dave Houseknecht for pointing out the dry, gas filled character of much of the Arkoma Basin today, and Bill Holser, Kevin Shelton, an anonymous reviewer, and especially Jay Gregg for very helpful reviews and suggestions. I thank Kevin and Jay for providing a prepublication copy of their 1992 fluid inclusion study.

REFERENCES CITED

Anderson, G. M., 1991, Organic maturation and or precipitation in southeast Missouri, Economic Geology, v. 86, p 909-926.

Arbenz, J. K., 1989, The Ouachita system, in Bally, A. W., and A. R. Palmer, eds., The Geology of North America- An Overview: Boulder Colorado, Geological Society of America, The Geology of North America, v. A, p 371-396.

Banner, J. L., G. J. Wasserburg, P. F. Dobson, A. B. Carpenter, and C. H. Moore, 1989, Isotopic and trace element constraints on the origin and evolution of saline ground waters from central Missouri, Geochimica et Cosmochimica Acta, v. 53, p 383-398.

Barton, P. B., 1981, Physical-chemical conditions of ore deposition, in D. T. Rikard and R. E. Wickam, eds., Chemistry and Geochemistry of Solutions at High Temperatures and Pressures, Pergamon, New York, p. 509-526.

Bethke, C. M., 1986, Hydrologic constraints on the genesis of Upper Mississippi Valley mineral district from Illinois Basin brines, Economic Geology, v. 81, p. 233-249.

Bethke, C. M., W. J. Harrison, C. Upson, and S. P. Altaner, 1988, Supercomputer analysis of sedimentary basins, Science, v. 239, p. 261-267.

Brockie, D. C., E. H. Hare Jr., and P. R. Dingess, 1968, The geology and ore deposits of the Tri-State District of Missouri, Kansas, and Oklahoma, *in* J. D. Ridge, ed., Ore Deposits of the United States, The Graton-Sales Volume, Volume I: The American Institute of Mining, Metallurgy and Petroleum Engineers, Inc., New York, p. 400-430.

Buelter, D. P., and R. N. Guillemette, 1988, Geochemistry of epigenetic dolomite associated with lead-zinc mineralization of the Viburnum Trend, Southeast Missouri, *in* Shukla, V., and P. A. Baker, eds., Sedimentology and Geochemistry of Dolostones: SEPM Special Publication 43, p. 85-93.

Bullock, R. L., 1973, Mine-plant design philosophy evolves from St. Joe's New Lead Belt operations, Mining Congress Journal, v. 59, p. 20-29.

Carpenter, A. B., M. L. Trout, and E. E. Pickett, 1974, Preliminary report on the origin and chemical evolution of lead- and zinc-rich oil field brines in Central Mississippi, Economic Geology, v. 69, p. 1191-1206.

Cathles, L. M., and A. T. Smith, 1983, Thermal constraints on the formation of Mississippi Valley-type lead-zinc deposits and their implications for episodic basin dewatering and deposit genesis, Economic Geology, v. 78, p. 983-1002.

Cathles, L. M., 1987, A simple analytical method for calculating temperature perturbations in a basin caused by the flow of water through thin, shallow-dipping aquifers, Applied Geochemistry, v. 2, p. 649-655.

Coveny, R. M., E. D. Goebel, and V. M. Ragan, 1987, Pressures and temperatures for aqueous fluid inclusions in sphalerite for Mid-Continent country rocks, Economic Geology, v. 82, p. 740-751.

Dott, R. H., and R. L. Ginter, 1930, Iso-con map for Ordovician waters, American Association of Petroleum Geologists Bulletin, v. 14, p. 1215-1218.

Farr, M. R., 1989, Compositional zoning characteristics of late dolomite cement in the Cambrian Bonneterre formation, Missouri: implications for parent fluid migration pathways, Carbonates and Evaporites, v. 4, p. 177-194.

Garven, G., 1984, Theoretical analysis of the role of groundwater flow in the genesis of stratabound ore deposits: 2. quantitative results, American Journal of Science, v., 284, p. 1125-1174.

Garven, G., 1985, The role of regional fluid flow in the genesis of the Pine Point Deposit, Western Canada Sedimentary Basin, Economic Geology, v. 80, p. 307-324.

Gregg, J. M., 1985, Regional epigenetic dolomitization in the Bonneterre Dolomite (Cambrian), southeastern Missouri, Geology, v. 13, p. 503-506.

Gregg, J. M. and K. L. Shelton, 1989a, Geochemical and petrographic evidence for fluid sources and pathways during mineralization and lead-zinc mineralization in southeast Missouri: a review, Carbonates and Evaporites, v. 4, p. 153-175.

Gregg, J. M. and K. L. Shelton, 1989b, Minor- and trace-element distributions in the Bonneterre Dolomite (Cambrian), southeast Missouri: Evidence for possible multiple-basin fluid sources and pathways during lead-zinc mineralization, Geological Society of America Bulletin, v. 101, p. 221-230.

Hagni, R. D., 1976, Tri-state ore deposits: the character of their host rocks and their genesis, in K. H. Wolf, ed., Handbook of Strata-bound and strataform Ore Deposits, Elsevier, New York, v. 6, p. 457-494.

Hart, S. R., N. Shimizu, and D. A. Sverjensky, 1981, Lead isotope zoning in galena: an ion mocroprobe study of a galena crystal from the Buick Mine, southeast Missouri, Economic Geology, v. 76, p. 1873-1878.

Houseknecht, D. W., 1986, Evolution from passive margin to foreland basin: the Atoka Formation of the Arkoma Basin, south central U.S.A., Special Publications of the International Association of Sedimentologists, v. 8, p 327-345.

Jorgensen, D. G., J. O. Helgesen, R. B. Leonard, and D. C. Signor, 1986, Equivalent freshwater head and dissolved-solids concentration of water in rocks of Cambrian, Ordovician, and Mississippian age in the northern Mid-Continent, U.S.A., Miscellaneous Field Study Map MF-1835B, U. S. Geological Survey.

Jorgensen, D.G., J. Downey, A.R. Dutton and R. W. Maclay, 1988, Region 16, Central and nonglaciated plains, in Back, W., J.S. Rosenshein, and P. R. Seaber, eds., Hydrogeology: The Geology of North America, Boulder Colorado, Geological Society of America, v. O-2, p. 141-156.

Kisvarsanyi, E. B., ed., 1976, Studies in Precambrian Geology, Missouri Department of Natural Resources, Report of Investigation Number 61, 190 p.

Kron, A. and J. Stix, 1982, Geothermal gradient map of the United States, National Geophysical Data Center, National Oceanic and Atmospheric Administration, Data Mapping Group Code D62, Boulder, Colorado.

Leach, D. L., and E. L. Rowan, 1986, Genetic links between the Ouachita foldbelt tectonism and the Mississippi Valley-type lead-zinc deposits of the Ozarks, Geology, v. 14, p. 931-935.

Luza, K. V., W. F. Harrison, G. A. Laguros, M. L. Prater, and P. K Cheung, 1984, Geothermal resources and temperature gradients of Oklahoma, Oklahoma Geological Survey, Norman, Oklahoma, Map GM 27.

Oliver, J., 1986, Fluids expelled tectonically from orogenic belts: their role in hydrocarbon migration and other geologic phenomena, Geology, v. 14, p. 99-102.

Panno, S. V., G. Harbottle, and D. V. Sayre, 1988, Distribution of selected elements in the shale of the Davis Formation, Buick Mine area, Viburnum Trend, southeast Missouri, v. 83, p. 140-152.

Rich, J. L., 1927, Generation of oil by geologic distillation during mountain building, American Association of Petroleum Geologists Bulletin, v. 11, p.1139-1149.

Rowan, E. L., 1986, Cathodoluminescent zonation in hydrothermal dolomite cements: relationship to Mississippi Valley-type Pb-Zn mineralization in southern Missouri and northern Arkansas, in Hagni, R. D., ed., Proceedings, Process Mineralogy IV: Including Applications to Precious Metals and Cathodoluminescence, American Institute of Mining Engineers, New Orleans, p. 69-87.

Rowan, E. L., and D. L. Leach, 1989, Constraints from fluid inclusions on sulfide precipitation mechanisms and ore fluid migration in the Viburnum Trend District, Missouri, Economic Geology, v. 84, p. 1948-1965.

Royden, L., and C. E. Keen, 1980, Rifting process and thermal evolution of the continental margin of eastern Canada determined from subsidence curves, Earth and Planetary Science Letters, v. 51, p. 343-361.

Sharp, J. M., 1978, Energy and momentum transport model of the Ouachita basin and its possible impact on formation of economic mineral deposits: Economic Geology, v. 72, p. 474-479.

Shelton, K. L., R. M. Bauer, and J. M. Gregg, 1992, Fluid inclusion studies of regionally extensive dolomites, Bonneterre Dolomite (Cambrian), southeast Missouri: Evidence of multiple fluids during dolomitization and lead-zinc mineralization, Geological Society of America Bulletin, v. 104, p. 675-683.

Shelton, K. L., J. M. Reader, L. M. Ross, G. W. Viele, and D. L. Seidemann, 1986, Ba-rich adularia from the Ouachita Mountains, Arkansas: Implications of a postcollisional hydrothermal system, American Mineralogist, v. 17, p. 916-923.

Shepard, E. M., 1907, Underground waters of Missouri, U. S. Geological Survey Water Supply Paper, no. 195, 224 p.

Snyder, F.G., and P.E. Gerdemann, 1968, Geology of the Southeast Missouri Lead District, in J. D. Ridge, ed., Ore Deposits of the United States, The Graton-Sales Volume, Volume I: The American Institute of Mining, Metallurgy and Petroleum Engineers, Inc., New York, p. 326-358.

Stavnes, S., and E. W. Steeples, 1982, Geothermal resources of Kansas, Kansas Geological Survey, The university of Kansas, Lawrence, Kansas.

Sverjensky, D. A., 1981, The origin of a Mississippi Valley-type deposit in the Viburnum Trend, southeast Missouri, Economic Geology, v. 76, p. 1848-1872.

Sverjensky, D. A., 1984, Oil field brines as ore-forming solutions, Economic Geology, v. 79, p. 23-37.

Thacker, J. L., and K. H. Anderson, 1977, The geologic setting of the southeast Missouri lead district-regional geologic history, structure and stratigraphy, Economic Geology, v. 72, p. 339-348.

Tobin, R. C., 1991, Diagenesis, thermal maturation and burial history of the Upper Cambrian Bonneterre dolomite, southeastern Missouri; an interpretation of thermal history from petrographic and fluid inclusion evidence, Organic Geochemistry, v. 17, p 143-151.

Viets, J. G., and D. L. Leach, 1990, Genetic implications of regional and temporal trends in ore fluid geochemistry of Mississippi Valley-type deposits in the Ozark region, Economic Geology, v. 85, p.842-861.

Voss, R. L., R. D. Hagni, and J. M. Gregg, 1989, Sequential deposition of zoned dolomite and its relationship to sulfide mineral paragenetic sequence in the Viburnum Trend, southeast Missouri, Carbonates and Evaporites, v. 4, p. 195-209.

Voss, R. L., and R. D. Hagni, 1985, The application of cathodoluminescence microscopy to the study of sparry dolomite from the Viburnum Trend, southeast Missouri, in Hausen, D. M., and O. C. Kopp, eds., Proceedings, Paul F. Kerr Memorial Symposium: New York, American Institute of Mining Engineers, Processes in Mineralogy, v. 5, p. 51-68.

Weigel, W. W., 1943, Mine drainage, Southeast Missouri Lead District, Transactions of the American Institute of Mining Engineers, v. 153, p. 74-81.

Weigel, W.W., 1965, The inside story of Missouri's exploration boom-Part 1, Engineering and Mining Journal, v. 166, p.77-88.

Chapter 9

Circulation Of Saline Ground Water In Carbonate Platforms— A Review and Case Study From the Bahamas

Fiona F. Whitaker and Peter L. Smart
Department of Geography
University of Bristol
Bristol, U.K.

ABSTRACT

In open marine carbonate platform systems there is a potential for dolomitization by circulation of saline ground waters because of the high magnesium content of sea water and the long residence time of the carbonate sediments in the shallow subsurface. Saline ground water circulation is driven by hydraulic head, which may be generated either by differences in sea-surface elevation caused by tides, winds or ocean currents, or by differences in water density related to salinity and/or temperature. Buoyant circulation occurs where saline ground waters are entrained by the fresh ground water circulation, while the increase in salinity of bank surface waters due to evaporation induces reflux. Finally, thermal contrasts between cold ocean waters and ground waters warmed by the geothermal heat flux gives rise to convective circulation. The scale of any platform, the steepness of its margins, and relative position of sea level are major controls on the type of circulation that may occur. In addition, climate, the distribution of permeability in the carbonate sediments, and geographic position in relation to other land masses and the atmospheric and oceanic circulations are significant.

Active circulation of saline ground water occurs beneath North Andros Island on the Great Bahama Bank in the form of major outflows ($0.3–3.9 \times 10^5$ $m^3 day^{-1}$) from karstic blue holes along the margin of the Tongue of the Ocean. The elevated salinity of these waters indicates that they derive by reflux from the Great Bahama Bank, and flow eastward under North Andros Island. Ground water temperatures are, however, low, suggesting that cold normal salinity sea waters from depth within the adjacent oceans are also involved in the circulation. It is concluded that this circulation results from the combined effects of differences in temperature, salinity, and elevation potential. Numerical modeling of saline ground water circulation is needed in order to understand the sensitivity of the circulation to the differing controlling parameters, and may enable prediction of the three-dimensional distribution of resultant diagenetic alteration.

INTRODUCTION

The diagenesis of carbonate rocks depends upon the mineralogy of the sedimentary carbonates and on the geochemistry of the associated pore waters, which together control the thermodynamic potential for diagenesis (James and Choquette, 1984). However, in the absence of fluid circulation to supply the reactants and remove the products of the diagenetic reactions, this potential is rapidly exhausted, limiting the extent of diagenetic change. The importance of fluid circulation has been widely recognized in meteoric environments (Hanshaw and Back, 1980; Longman, 1980; Budd, 1988; Craig, 1988). In contrast, much less is known about circulation of saline ground waters, despite studies suggesting its importance in precipitation of marine calcite cements (Aissaoui et al., 1986; James and Ginsburg, 1979) and particularly dolomitization (Fanning et al., 1981; Saller, 1984; Simms, 1984). In this chapter, we review the mechanisms that are capable of driving large-scale circulation of saline ground waters through carbonate platforms and identify important controls on their operation, illustrating this using a well-constrained modern case study from the Bahamas. Emphasis is placed on the potential importance of saline ground water circulation as an explanation of large-scale, near-surface replacement dolomitization.

Modern dolomites have been reported forming in many different environments, including the deep sea sub-floor (Kelts and Mackenzie, 1982; Baker and Burns, 1985), supratidal flats (Shinn et al., 1965), sabkhas (Illing et al., 1965) and ephemeral hypersaline lakes (Von der Borch and Lock, 1979). These modern dolomites have provided new insights into the geochemical processes controlling dolomitization, however they are not analogous to the thick sequences of replacement dolomites that dominate the fossil record (Land, 1985; Given and Wilkinson, 1987). Some massive dolomites clearly result from diagenesis under elevated temperatures and pressures associated with circulation of hydrothermal fluids at depth (Mattes and Mountjoy, 1980; Zenger, 1983), while others are associated with circulation of basinal evaporitic brines (Kendall, 1989; but see Land 1985). In many cases however, dolomitization clearly pre-dates burial and is not associated with evidence of evaporitic conditions (Enos 1988, Dawans and Swart, 1988). The environments of formation of these dolomites remain contentious (Machel and Mountjoy, 1986). Following Runnels (1969) and Hanshaw et al. (1971), formation in the fresh-salt water mixing zone was widely accepted (Badiozamani, 1973; Choquette and Steinen, 1980), but increasingly this paradigm has been questioned because of the relatively restricted vertical extent and apparent absence of dolomites in modern mixing zones (Machel and Mountjoy, 1986; Hardie, 1987). More recently attention therefore has been focussed on environments where circulation of near-normal sea water can occur.

Despite the considerable thermodynamic drive to precipitate dolomite from sea waters (Berner, 1971), laboratory experiments have consistently failed to synthesize significant volumes of dolomite from normal sea water under near-surface conditions (Lippmann, 1973). Current approaches to this problem have concentrated on trying to identify modifications to sea water which permit the apparent kinetic barriers to dolomitization to be overcome. Such fluids include those with elevated Mg/Ca ratios, those depleted in sulphate, and those enriched with organic matter (reviewed by Machel and Mountjoy, 1986). However, other workers, including Land (1985) and Hardie (1987), have suggested that dolomite may actually precipitate from unmodified sea water at a rate that, while insufficient to produce measurable quantities of dolomite on the time-scale of laboratory experiments, may be sufficient to generate extensive dolomitization over geological time-scales. This suggestion accords with the fact that carbonate rocks spend the majority of their pre-burial history bathed in ground waters of near sea water composition (Matthews and Froelich, 1987). Although sea water is a rich source of magnesium, an estimated 650 pore volumes are required to dolomitize a single volume of carbonate sediment with a porosity of 40% (Land, 1985). Furthermore, if only 10% of the available magnesium is used (Simms and Hardie, 1983), this estimate increases to 6500 pore volumes. Even over an extended period, rates of diffusion are many orders of magnitude too slow to satisfy these mass-balance requirements. Therefore the prime requirement for extensive dolomitization is the existence of a transport mechanism capable of circulating large volumes of sea water through carbonate buildups.

In this chapter we review the processes that are capable of driving large-scale circulation of saline ground water through carbonate build-ups, consider how these processes may be identified in modern and ancient systems, and identify important controls on their operation. We then examine a well-constrained case study of modern-day saline ground water circulation through the Great Bahama Bank, which is shown to be considerably more complex than simple end-member systems. Finally, we consider the potential of ground water numerical modeling of flow systems for prediction of the distribution and extent of dolomitization.

PROCESSES OF SALINE CIRCULATION IN CARBONATE BUILDUPS

Sea water can circulate through carbonate platforms in response to the hydraulic head generated by differences in the distribution of both sea-surface elevation and fluid density. Four saline ground water circulation systems can be distinguished based on their

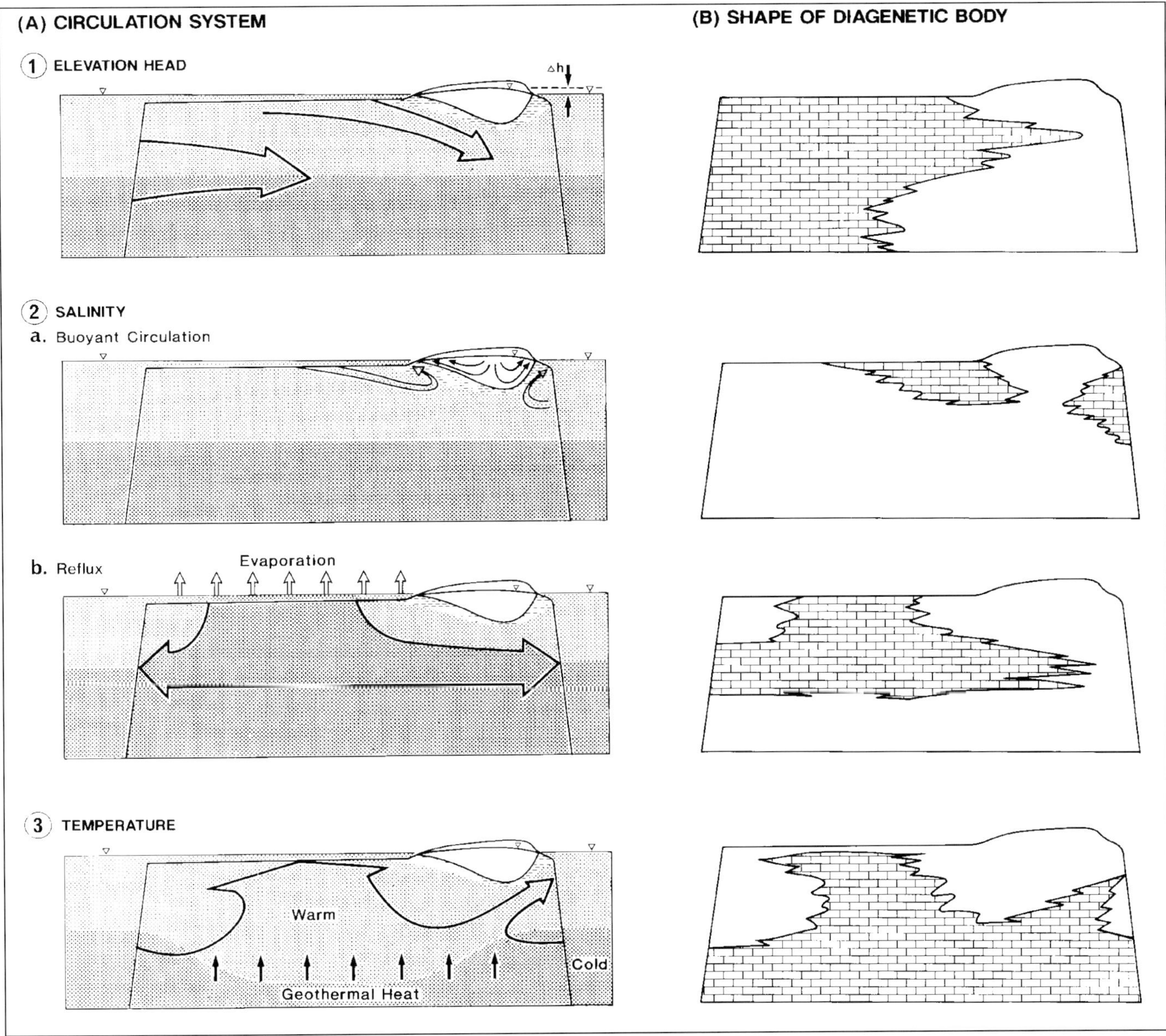

Figure 1. (A) Schematic diagram illustrating the processes that may drive circulation of saline ground waters in carbonate platforms. No pattern = fresh water; broken lines = mixing zone; light stipple = lower density saline water; heavy stipple = higher density saline water. (B) Hypothetical diagenetic bodies that might result from the saline ground water circulations of Figure 1A, assuming that no geochemical evolution is required to drive diagenesis, and that reactants are exhausted along the flow path.

driving mechanisms (Figure 1). Elevation potential can be generated on three different time scales: by tidal conditions (diurnal), by wind and wave set-up (days to months or longer), and by ocean currents (thousands/millions of years). Density-driven flows may result from contrasts in salinity due to both dilution of saline water by fresh water (buoyant circulation) and to evaporative concentration causing increased density (reflux), and also from differences in water temperature that cause convection. In the following sections fundamental controls on the operation of these four types of circulation system are described, with examples from modern and fossil platforms.

Elevation Potential

A difference in sea surface elevation from open ocean to shallow banks, across an emergent island or the whole of a carbonate platform, is capable of driving saline ground water flow (Figure 1A). Such gradients may be generated by ocean tides as bottom friction retards and dampens the tidal response of sea water on shallow banks compared to the open ocean, causing pumping of sea water through the upper part of the platform. Tidal pumping is most significant at the margins of rimmed carbonate platforms (Matthews, 1974), and where circulation is restricted on one side of an island or reef. This process causes movement of sea water through reefs (Buddemeir and Oberdorfer, 1986), where it may cause precipitation of marine calcite cements (e.g. Aissaoui et al., 1986), and also through supratidal sediments, where it results in dolomitization (Carballo et al., 1987). However, tidally-driven flow is unlikely to generate large-scale saline circulation, and the effects may be limited to surficial sediment veneers on the shallow banks and to platform margins.

Wind- and storm-controlled differences in sea-surface elevation between leeward and windward shores of reefs and islands and across isolated platforms may be more significant in driving large-scale circulation. For instance, Atkinson et al. (1981) measured a wave-generated set-up of 30-50 cm across the windward reef at Enewetak Atoll, while similar magnitude variations in sea-level on the coastal shelf of western Australia are generated by the northerly progression of weather systems (Wolanski, 1981). In low-relief coastal environments wind-driven set-up can cause extensive flooding, with sea water extending for several kilometres across the flat sabkhas of the Arabian Gulf (Patterson and Kinsman, 1982). Differences in water surface elevation can also be generated across islands, such as North Andros, Bahamas, where winter flooding caused by onshore winds generates high water-levels on the west coast, which are maintained over large areas of the tidal flats by ponding behind levees (Hardie and Ginsburg, 1977). Such trans-island gradients have been cited to account for the windward/leeward contrast in the distribution of Pleistocene dolomites in Barbados (Humphrey, 1988).

Finally, regional oceanic circulation may result in local differences in sea-surface elevation. In zones of coastal upwelling such as the Coral Sea along the shelf margin of the Australian Great Barrier Reef (Marshall, 1986), differences in sea-surface elevation are generated across the reef margin. A similar process has been invoked by James and Ginsburg (1979) to explain pervasive marine cementation on the oceanic flank of the Belize barrier reef. Variations in sea surface elevation may also be generated across straits with large currents such as the Gulf Stream in the Straits of Florida (Maul, 1986). Although ground water flows driven by such ocean circulations are much less frequently recognised than those resulting from wind-driven forcing, they may be equally or more significant since they are likely to operate over large areas and be maintained for long periods.

Elevation potential is generated by extrinsic factors which control astronomical tides, and the oceanic and atmospheric circulations. However, the effect of these is dependent on the morphology of the platform, particularly the presence of topographic irregularities such as rimmed margins and tidal flat levees which can cause ponding and channelling of waters. This function is highly dependent on the relative position of sea level, which controls the extent of platform exposure. The occurrence of these circulation systems in the fossil record may be difficult to identify. However, flow is concentrated within the platform flanks (e.g. Playford, 1984) and, in the case of trans-island or trans-bank flow, generates an asymmetric diagenetic body facing the principle wind or current direction (Figure 1.1). This asymmetry may also be reflected in the sedimentology and facies distribution of the host carbonates.

Buoyant Circulation

Where the upper part of the platform is emergent, meteoric recharge leads to the development of a fresh-water lens overlying saline ground waters. Discharge from the lens in response to recharge entrains underlying saline ground waters. Dispersive mixing between the two water bodies results from fluctuations in the position of the interface in response to oceanic tides, barometric pressure, and variations in the rate of fresh water recharge (Bear and Todd, 1960). The brackish water is hydrodynamically unstable and rises along the base of the fresh water lens towards the margins of the island where it discharges via brackish springs (Kohout, 1960; Alfirevic, 1966). The discharge of saline water through the mixing zone requires a compensatory flow of sea water into the platform at depth (Cooper, 1959; Figure 1.2a). While the role of the mixing zone as a pump for circulating mixed waters is generally recognised (Kohout and Kline, 1977; Sanford and Konikow, 1989), the implications for flow in the saline zone are rarely considered (Simms and Hardie, 1983). However, Vahrenkamp et al. (1991) have suggested that dolomitization of the Little Bahama Bank occurred during prolonged periods of subaerial exposure, when saline water was drawn into the platform beneath an extensive fresh water lens.

Two factors control the potential for buoyant circulation of saline ground waters: the relation between the morphology of the platform and sea-level, and the climatic regime. In isolated platforms, buoyant circulation is only effective on a large-scale during low sea stands, since subaerial exposure is required to develop an extensive and thick fresh water lens. Under steady-state conditions, discharge from the lens is proportional to effective recharge, which will increase with total rainfall and is inversely related to evapotranspiration. Thus, circulation would be expected to be more active beneath the better developed fresh water lenses of the northern Bahamian islands than those of the southern Bahamas and Turks and Caicos, where the size of the fresh water lens is limited by low effective rainfall (Cant and Weech, 1986). Where relief is limited, inland lagoons may develop and evaporative loss of fresh water will segment and greatly reduce the lens size, as occurs on the island of San Salvador (Davis and John-

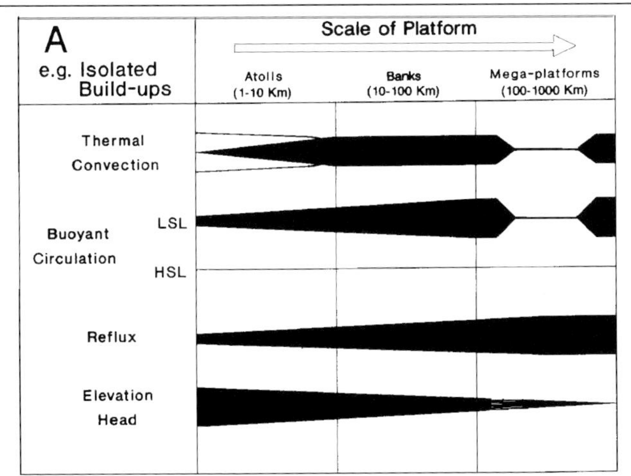

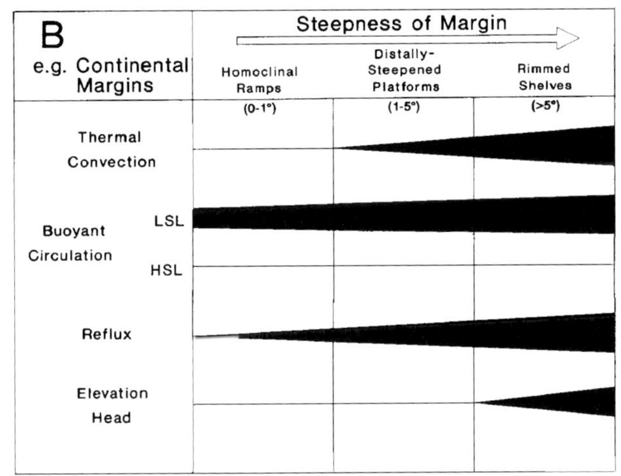

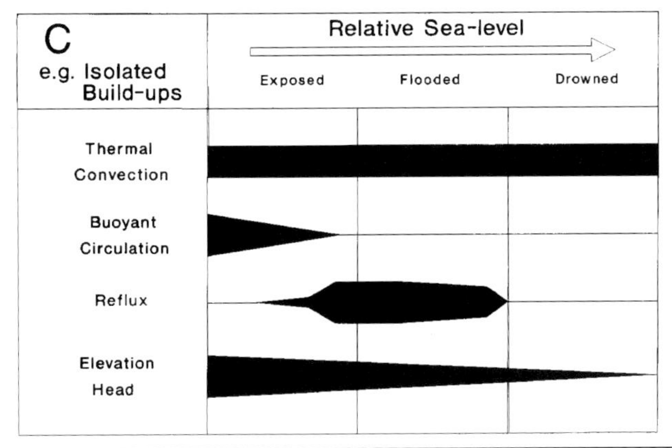

Figure 2. Schematic model of the probability of saline groundwater flow in response to thermal convection, buoyant circulation, reflux and elevation potential, in (A) platforms of varying scale (exemplified for isolated platforms), (B) platforms with margins of varying steepness (exemplified for shelves and ramps developed on continental margins), and (C) with different relative sea levels. Note that thermal convection may be substantially increased if the platform comprises an atoll built on a volcanic basement (Figure 2A). Also, as illustrated for mega-platforms, processes operating within the marginal part of platforms may differ from those in the central areas, particularly when the margins are steep. Note that at this stage any attempt to quantify the relative magnitude or volumetric extent of these circulation systems is premature (see text); the breadth of the lines in Figure 2 thus only indicates the relative probability of operation, and not the comparative groundwater flux.

son, 1988). In contrast to isolated platforms, in ramps the zone of circulation is able to migrate with changing sea level, affecting a large volume of sediment. In continental margin settings, buoyant circulation will frequently be the dominant mechanism driving saline ground water circulation because the discharge of fresh water is large due to the extensive inland catchment, as reported for the northeast coast of Yucatan by Hanshaw and Back (1980).

In the fossil record, buoyant circulation may be identified by the associated paleo-exposure surface and distinctive cements, which represent the former position(s) of the fresh water phreatic and mixing zones (Horbury and Adams, 1989; Walkden and Williams, 1991). As buoyant circulation is more active around the margins of an island, the resulting diagenetic body within the saline zone should be toroidal in shape (Figure 1.2a), with greater vertical extent and occurring at shallower depth nearer to the paleo-shoreline (Bouvier et al., 1990). However, in continental situations where the mixing zone circulation involves large volumes of water (Kohout and Kline, 1977; Enos, 1988), this body may extend a considerable distance seaward of the shoreline. In addition, saline ground water movement driven by buoyant circulation occurs at a greater depth beneath lower permeability carbonates which are able to host a thicker fresh water lens (Vacher, 1978).

Reflux

Evaporation of shallow waters increases their salinity and generates potential for density-driven downward movement (reflux), displacing less dense saline waters at depth. This will continue until the refluxing waters reach a depth where their density is equivalent to that of waters in the adjacent ocean basins, when lateral flow will occur (Figure 1.2b).

This flow system was originally proposed for very restricted environments such as the hypersaline lakes of Bonaire (Deffeyes et al., 1965, although subsequent work by Lucia, 1968, has cast doubt on their interpretation). Hypersaline conditions have been widely reported in environments with very restricted sea water circulation such as behind the dune barriers of the MacLeod salt basin western Australia (Logan, 1987). While in some cases these hypersaline waters have been traced in the sub-surface (e.g. Amdurer and Land, 1982), in others the salinity signal is apparently lost (e.g. Lucia, 1968), probably by dilution with relatively large volumes of normal sea waters circulating within higher permeability horizons (Simms, 1984). However, evaporative pumping driving flow in the opposite direction has also been reported from hypersaline environments (Hsu and Schneider, 1973; Rougerie, 1985). The direction and velocity of net ground water flow is therefore dependent upon the relative density and water surface elevation differences between evaporating lagoon waters and adjacent water bodies. However, Simms (1984) has demonstrated theoretically and using scaled sand-box models that only very small increases in salinity (to 38–42‰) are required to cause large-scale reflux of waters through carbonate platforms. Thus, this type of reflux can potentially affect large volumes of host carbonates and may be of more widespread significance than generally is recognised.

The magnitude and distribution of reflux depends primarily on the distribution of fluid density and permeability within the platform (Neild, 1968). The prime requirement for accumulation of elevated salinity waters is restriction of flow. This may result from physical barriers to surficial water movement such as barrier reefs (Adams and Rhodes, 1960) and beach dune complexes (Logan, 1987), or simply from long residence times of shallow water bodies, as on the present-day Great Bahama Bank (Broecker and Takahashi, 1966). Thus the morphology of the platform top and the relative position of sea level are of critical importance. Extensive areas of restrictive circulation are more likely to be developed on flat-topped banks or atolls with a central lagoon than on ramps, where they tend to be limited to coastal margins, but the former are also more likely to be abandoned by changes in sea -level. Hypersalinity is generally associated with aridity and high rates of evaporation, but waters of slightly elevated salinity can accumulate even in relatively humid environments such as the present-day Bahamas because of strong seasonality in the rainfall.

In ancient buildups, the presence of early diagenetic evaporite minerals (or later pseudomorphs) may be indicative of hypersaline reflux, with dolomitization favored both by high salinities and increased Mg/Ca ratio due to the precipitation of gypsum (Folk and Land, 1975). The geochemistry of the secondary carbonates formed may reflect that of the original fluids; for example dolomites would have lowered Sr/Ca ratios (Holser, 1979), but frequently chemical and isotope composition are reset by later recrystallization (Land, 1980). Sedimentological features of the paleosurface of the platform may thus be more useful, as for instance for the ramp system described by Wood and Wolfe (1969). Reflux of waters of only slightly elevated salinity is more difficult to detect as the chemical and isotopic shift from normal sea water is small. Nonetheless, in the Eocene dolomites from Enewetak, fluid inclusion analysis suggests formation from reflux waters twice as saline as sea water (Goldstein et al., 1991), rather than from thermal convection of normal sea water as originally proposed by Saller (1984). In many cases the shape of the diagenetic body may be more diagnostic. Flow occurs in cylindrical plumes extending downwards from the surface of the build-up (Figure 1.2b), with horizontal flow at depth where stratiform bodies may result. Over the longer-term, shifting of the position of the plumes may produce massive bodies (Simms, 1984).

Thermal Convection

Convective circulation of sea waters through carbonate platforms occurs in response to the horizontal density gradient between the surrounding cold, dense ocean waters, and ground waters that are warmed by the geothermal heat flux. Warming of ground waters reduces density and results in upward convective flow (Figure 1.3), with colder ocean waters being drawn in at depth. When the platform is submerged these waters will discharge from the upper surface, particularly above the area of maximum heating (Wilson et al., 1990). However, on platform emergence, upward movement will be constrained by the presence of a fresh water lens, and saline waters will flow laterally to discharge on the margins of the platform.

Thermally driven circulation can be recognised from ground water temperature profiles, as described from the Florida Peninsula by Kohout (1960). These typically show a decrease in temperature with depth (in opposition to the geothermal gradient) and relative warming from the periphery to the centre of the platform (Kohout, 1967; Kohout et al., 1977). In Florida thermally-driven circulation appears to operate to a depth of 1000 m, and is aided by interaction with buoyant circulation beneath the well-developed fresh water lens. Saline thermal ground waters discharging from submarine springs off the west coast of Florida are depleted in magnesium (Fanning et al., 1981), indicating active precipitation of dolomite. Negative geothermal gradients have also been observed by Swartz (1958) within Pacific atolls including Enewetak and Bikini, and thermally driven circulation systems have been invoked by a number of authors to account for the dolomitization of late Tertiary and Eocene atolls in the Pacific (Saller, 1984; Samaden et al., 1985; Aharon et al., 1987). These ground waters may also be geochemically modified where platforms are developed on a volcanic core. For instance in Mururoa Atoll silica, nitrates and phosphates derived from weathering of the andesitic core are carried upward by thermal convection (Rougerie, 1985).

The magnitude of thermally driven circulation is controlled by the lateral temperature gradient between ocean and platform waters, and thus would

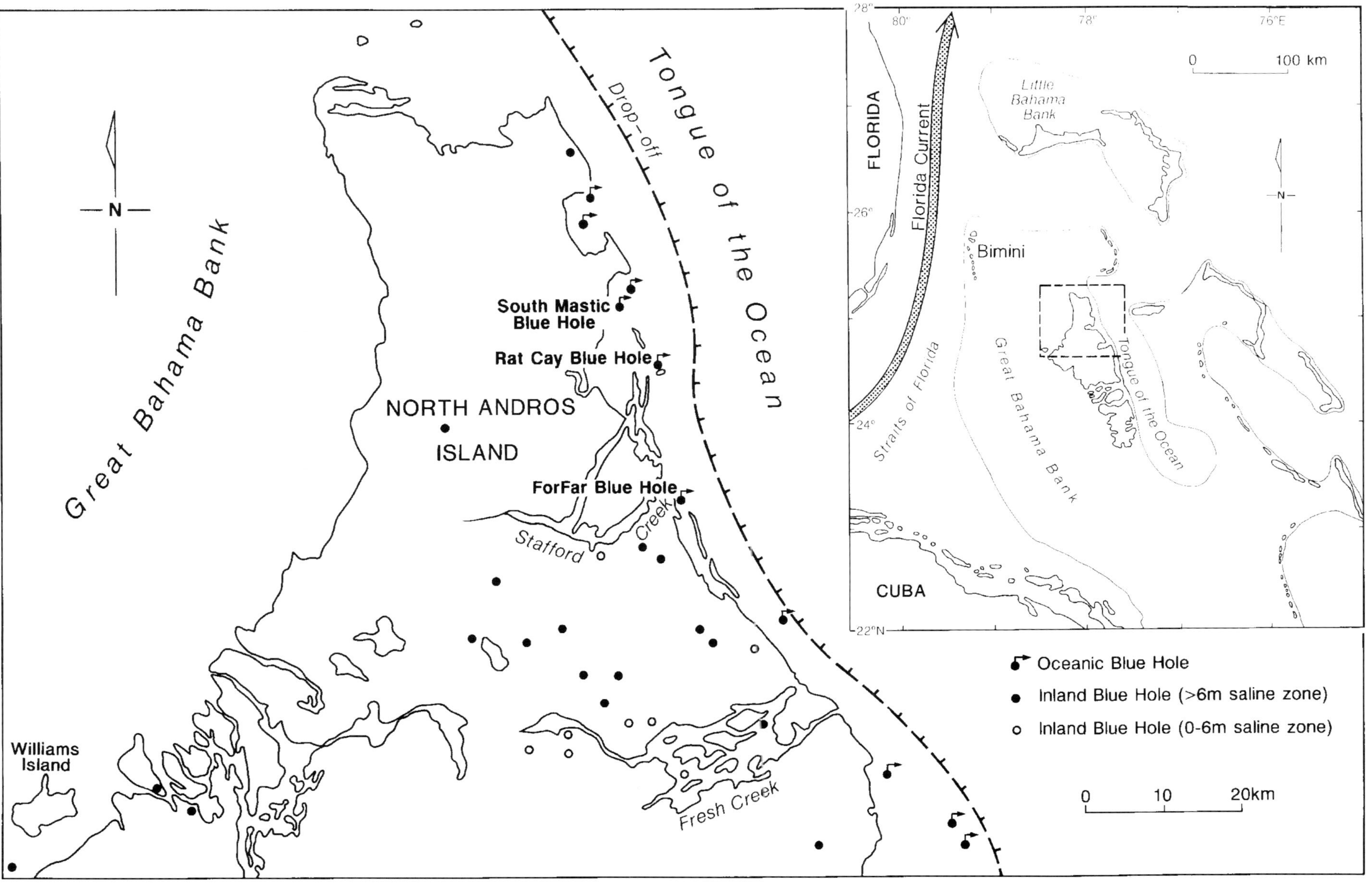

Figure 3. Map of North Andros study area showing discharging oceanic blue holes and inland blue holes, which fully penetrate the saline zone.

be expected to increase with the vertical height of the platform above the sea floor, and be greater in steep-flanked, rimmed shelves (Simms, 1984). Flow is enhanced by the presence of a thermal anomaly, such as a volcanic pedestal that drives active circulation beneath Niue Atoll in the Pacific (Aharon et al., 1987), and circulation will become more vigorous during periods of rejuvenation of volcanic activity (Rodgers et al., 1982). The degree of horizontal penetration of thermally driven flow is dependent upon the relative vertical and horizontal permeability of the carbonates (Wilson et al., 1990), and may be confined to the flanks of the platform where vertical permeability is well-developed, for instance by bank-marginal fracturing, or lateral changes in facies.

In ancient buildups, the distribution of paleo-temperatures derived from fluid inclusions and stable isotopes in cements and replacement dolomites may provide evidence of thermal convection, while trace element distribution may be indicative of hydrothermal involvement (Aharon et al., 1987, but see Haszeldine and Osborne, this volume, for problems involving the interpretation of fluid inclusions in buried sediments.) In addition the shape of the dolomite body formed by thermal convection may be distinctive, with 'rootless' plumes extending upwards from the base of the sequence, as in the Jurassic Vajont limestone in northern Italy (Zempolich and Hardie, 1991). On a larger scale, thermal convection generates a characteristic 'mushroom' shape. Thus in the Triassic Latemar platform (Wilson et al., 1990) the thermal drive for circulation within the platform can be clearly demonstrated from the geochemistry and spatial distribution of the resultant dolomites.

SUMMARY OF FACTORS CONTROLLING THE POTENTIAL FOR SALINE GROUND WATER CIRCULATION

The potential for saline ground water circulation driven by differences in density (thermal convection, buoyant circulation and reflux) and elevation potential (in response to oceanic and atmospheric circulation) varies with the scale and the steepness of the margins of the platform, and the relative position of sea level (Figure 2).

Scale of the Platform

The effect of scale is illustrated in Figure 2A with reference to isolated platforms. These comprise a continuum ranging from atolls (1–10 km across), such as Enewetak and Mururoa, which frequently have volcanic basements, through platforms such as the Bahamas (10–100 km across), to the mega-platforms (100–10,000 km across) that are known in the geological record and that hosted extensive epeiric seas (for example the lower Ordovician carbonates of Newfoundland, Pratt and James, 1986). With the exception of flow driven by elevation potential which is likely to be greatest in smaller platforms, saline ground water circulation will tend to be more active in larger platforms. These are likely to have greater temperature contrasts between ground waters and adjacent ocean waters, and larger surface areas, important both for fresh water lens development during low sea stands, and for evaporative concentration of waters generating reflux during highstands. However, in mega-platforms, thermal convection and buoyant circulation will tend to be restricted to areas around the platform margins unless the ratio between horizontal and vertical permeability is very large (Phillips, 1991). At all scales the occurrence of buoyant circulation is critically dependent on exposure, which can only occur during periods of low sea level. Thus Figure 2A presents two alternative models of circulation at high and low sea stands. Note also that while the small scale of atolls may restrict thermal convection, the presence of a volcanic core associated with a sea-floor hot-spot will greatly increase the geothermal heat flux and may drive very active circulation (Aharon et al., 1987).

Steepness of Platform Margins

A second important control on the circulation of saline ground water is the steepness of the slopes surrounding the platform. This is illustrated in Figure 2B for intermediate-scale build-ups (10^1–10^2 km across) developed on continental margins. These range from rimmed shelves with steep margins, such as developed on the coast of southern Florida and eastern Australia, to distally-steepened platforms, as in northeast Yucatan, and homoclinal ramps (slope angle $<1°$) such as the Trucial coast of the Arabian Gulf. Saline ground water circulation will tend to be more active in steep-margined shelves where the horizontal temperature gradient will be greatest. Flow driven by buoyant circulation will be particularly active in all continental margin settings due to the large inland catchment area. The steep and often porous margins of rimmed shelves allow saline water to readily enter the platform at depth in response to buoyant circulation in the mixing zone, while in homoclinal ramps the circulation will be limited by the lower permeability of the deeper water sediments, the sedimentary structure which gives preferential flow along the low angle bedded units, and the longer flow path from the ocean margin to the mixing zone. Waters capable of reflux may develop in restricted shelf lagoons behind shelf rims, and to a lesser extent on low angle ramps where circulation in areas of shallow water is sluggish. Local differences in elevation may also be established across these rims, but large-scale circulation of saline water is commonly severely restricted by the presence of considerable fresh water elevation potential onshore.

Relative Sea Level

A third important control on saline ground water circulation is the relative position of sea-level (Figure 2C), which depends on the balance between subsidence rates and the rate of carbonate production, and also on eustatic sea-level changes which may be both large magnitude and relatively rapid when dominated by glacio-eustacy. With the exception of thermal convection, circulation systems will be most active within platforms which are either exposed or flooded to a shallow depth. Buoyant circulation will cease when emergent islands become too small to support a fresh water lens (Budd and Vacher, 1991), but other flow systems will continue to operate despite flooding, and reflux is likely to be particularly active if sea water circulation is restricted on the shallow flooded banks. With continued submergence the platform is drowned and thermal convection, which is largely independent of sea-level, becomes dominant. Indeed convective circulation may continue as the platform is buried, although flow rates in buried carbonates will tend to be slower and involve much smaller volumes of fluid (Giles, 1987). Where marginal slopes are low angle, as in homoclinal ramps, variations in relative sea level will simply result in the migration of zones affected by buoyant circulation and reflux up and down the ramp, generating a complex diagenetic sequence superimposed on the pattern of shifting facies belts.

Platform Geography

In addition to these three major controls on saline ground water flow, small-scale surficial features of a platform will influence saline ground water flow, particularly that driven by differences in elevation potential or reflux, because bank-top sea water circulation is very sensitive to the presence of barriers such as reefs and sand bodies. The geographic location of the platform in relation to atmospheric and oceanic circulations, and to other land masses will have an impact not simply on the potential for flow in response to water surface elevation gradients, but also on climate which influences the salinity of shallow sea waters and the extent of fresh water lens development. Finally, it is important to remember that with large-scale (glacio-eustatic) sea-level changes, the climate of an individual platform may vary significantly between high and low sea stand conditions, particularly for enclosed basins or extensive platforms of high relief.

Permeability Distribution

While the potential for saline ground water flow is generated by differences in elevation potential and density, the rate and distribution of the resultant ground water flow is dependent on the magnitude and distribution of permeability. In carbonates, permeability is not simply dependent upon the grain-size distribution of particular facies (Enos and Sawatsky, 1981) because this may be strongly modified during early diagenesis (James and Choquette, 1984). In general high secondary porosity generated by meteoric diagenesis gives high rates of flow, and sea-level history is thus of considerable importance. At a larger scale, the lateral and vertical distribution of different facies is critical in controlling the degree of connectivity between individual transmissive units. Thus the presence of a laterally continuous horizon of low permeability may severely restrict vertical flow. For example, in the present MacLeod Basin, Australia, the development of a seal on the lagoon floor caused by gypsum precipitation restricts reflux of evaporitic brines (Logan, 1987). Of particular importance is the development of karstic porosity, which increases the transmissivity of the carbonates by orders of magnitude (Atkinson and Smith, 1976), allowing circulation of substantial volumes of ground water to be driven by relatively small differences in elevation or density (Drogue, 1989). Such porosity is often concentrated at or below laterally extensive exposure surfaces, or past positions of the fresh water lens (Craig, 1988) or mixing zone (Smart and Whitaker, 1988; Bouvier et al., 1990), which may then act as important routes for integration of regional-scale ground water flow. A similar role is played by tectonic openings such as the extensional fractures that often develop at the edges of steep rimmed platforms (Playford, 1984; Smart et al., 1988). In general, the facies distribution and architecture of rimmed shelves favor circulation of saline ground waters, with flow particularly active at the margins, while those of low-angle ramp systems lead to flow concentration in particular transmissive units with reduced saline circulation and a greater meteoric influence.

SALINE GROUND WATER CIRCULATION—A CASE STUDY FROM THE BAHAMAS

Introduction

The aim of this case study was to determine whether present-day circulation of saline ground water occurs beneath the Great Bahama Bank and to identify which of the various processes reviewed above provides the drive. North Andros Island on the Great Bahama Bank was chosen because it has both a well-developed fresh-water lens (which may drive buoyant circulation), and the potential for reflux of elevated salinity waters from the extensive flooded banks to the west of the island (Simms, 1984). Furthermore, North Andros has a large number of inland blue holes which provide direct access to the interior of the platform and permit sampling of the saline ground waters. Preliminary results of investigations of saline discharges are presented in Whitaker and Smart (1990) and discussed in detail in Whitaker (1992).

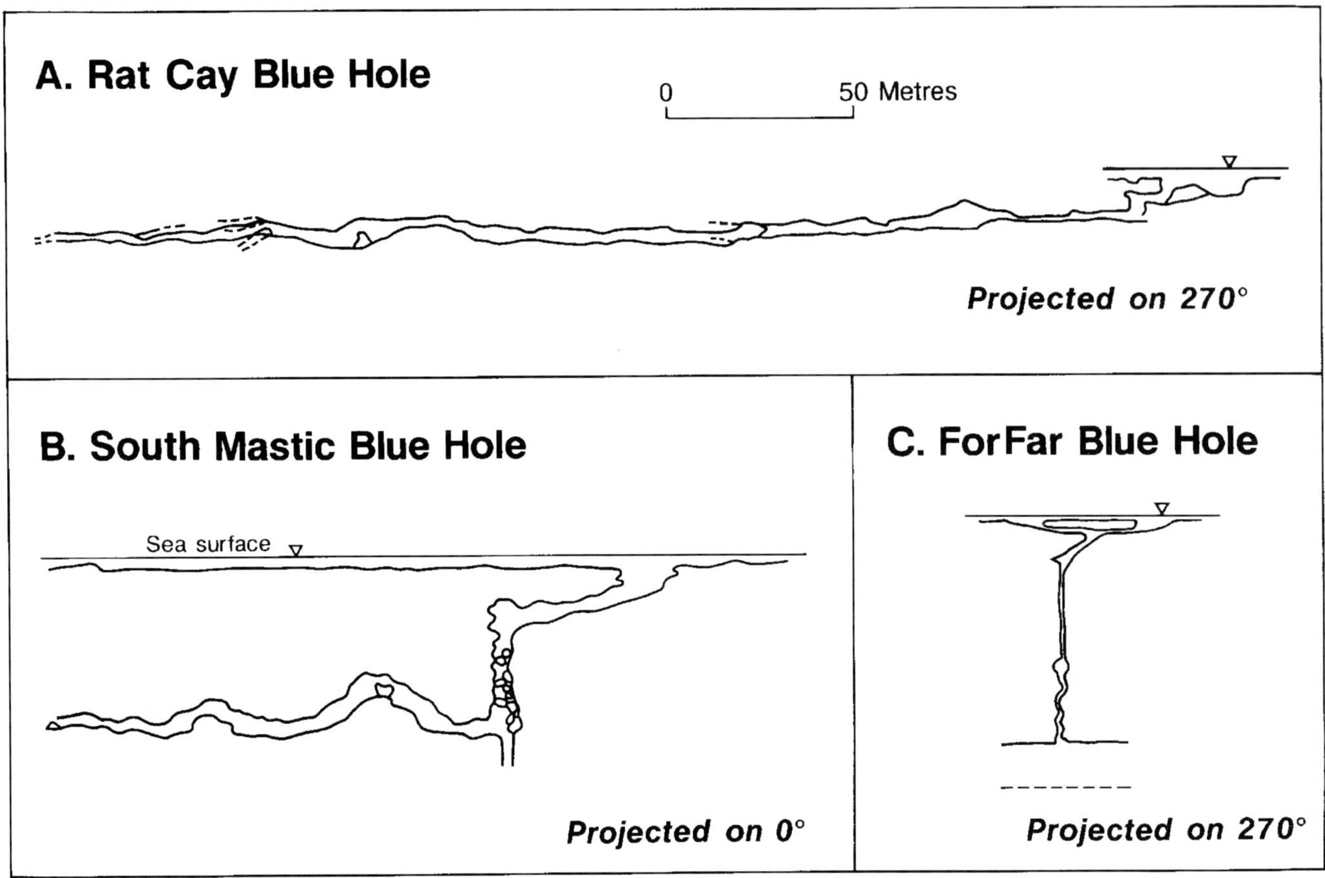

Figure 4. Surveys of oceanic blue holes on the east coast of North Andros: (A) Rat Cay Blue Hole, (B) South Mastic Blue Hole and (C) Forfar Blue Hole (projections are elevations).

The Bahamas are karstified and cavernous to depths in excess of 3000 m, as evidenced by bit drops and loss of circulation during deep drilling (Spencer, 1967). In particular the upper 100 m of the platform was exposed to meteoric diagenesis during Pleistocene lowstands, resulting in the formation of extensive cavernous porosity. Cavern collapse at depth forms vertical shafts (inland blue holes) which are flooded with platform ground waters under present highstand conditions.

The cavernous porosity provides preferred flow paths for circulating ground waters, and in oceanic blue holes which open from the shallow bank surface, strong tidally induced currents are observed. We know of ten oceanic blue holes along an 80 km stretch of the east coast of North Andros Island (Figure 3), and many other smaller openings are also present. Exploration and survey of these reveals that at some sites such as Rat Cay Blue Hole (Figure 4A) passage development is dominantly horizontal and occurs at relatively shallow depth. At other sites (such as South Mastic Blue Hole (Figure 4B) strong flows emerge from vertical fissures and rifts that—at ForFar Blue Hole (Figure 4C)—extend to depths in excess of 90 m. Many oceanic blue holes show control by the major bank-marginal fracture system which runs parallel to the drop-off into the Tongue of the Ocean (Daugherty et al., 1986; Smart et al., 1988). The fracture system appears to function as a vertical flow interceptor along the bank margin, integrating both cavernous and diffuse ground water flow beneath Andros and discharging it via oceanic blue hole outlets. On the Great Bahama Bank to the west of North Andros, higher sediment production rate, perhaps combined with the net inflow, appears to have caused choking of the karstic openings.

Magnitude of Saline Ground Water Circulation

In order to investigate the hypothesis that oceanic blue holes function as outlets for a regional-scale ground water system, the discharge from two sites (South Mastic and Rat Cay Blue Holes) was monitored using Aanderaa oceanographic recording current meters (Figure 5). These instruments record current speed and direction at pre-selected intervals (together with salinity, temperature, and local tidal head). This enables calculation of net discharge despite semi-diurnal flow reversal related to the ocean tides.

A 48-hour portion of the recording current meter record from South Mastic Blue Hole (Figure 6) shows

Figure 5. Oceanographic recording current meter in operation at Rat Cay Blue Hole.

general features of current velocity and direction common to both sites throughout the monitoring period. The semi-diurnal response to local tidal head can clearly be seen. During periods of high tide, sea surface elevation is greater than ground water elevation and inflow occurs (the inflow phase), while the converse occurs during low tides (the outflow phase). The outflow phase is longer than the inflow and achieves much higher velocities. This indicates that a net discharge of ground water must occur over the four tidal cycles shown. During the total period of observation (2.5 weeks in July 1988) at South Mastic Blue Hole, outflow occurs for an average of 7.35 ± 0.42 hours, almost 2 hours longer than the average inflow, which has a duration of 5.40 ± 1.13 hours. In addition, the average maximum velocity measured during the outflow (0.41 ± 0.02m s^{-1}), is almost twice that during inflow (0.21 ± 0.10m s^{-1}). An average of 3.75 x 10^5 m^3 of water was discharged from South Mastic Blue Hole during each outflow cycle, compared to an average of 1.71 x 10^5 m^3 drawn into the hole during each inflow phase. This is equivalent to a net discharge of 2.04 x 10^5 m^3 of water per tidal cycle, or a continuous discharge of 4.55 m^3 s^{-1} during the observation period.

A longer-term record of saline ground water discharge was obtained from Rat Cay Blue Hole, where the recording current meter operated for 29 weeks from July 1988 to January 1989 (Figure 7). During this period there was a cumulative net discharge of 6.8 x 10^6 m^3 of water (including extrapolation for periods when the current meter rotor was fouled, some 21% of the time). This is equivalent to an average net discharge of 1.71 ±1.74 x 10^4 m^3 per tidal cycle, or a continuous discharge of 0.381 ± 0.436 m^3 s^{-1}. The high standard deviation associated with the mean reflects the variability of net discharge, with high frequency (14.3 ± 2.3 days) and large amplitude (4.70 ± 2.32 x 10^4 m^3/tidal cycle) fluctuations (Figure 7A), related to the spring tides when the sun and moon are in conjunc-

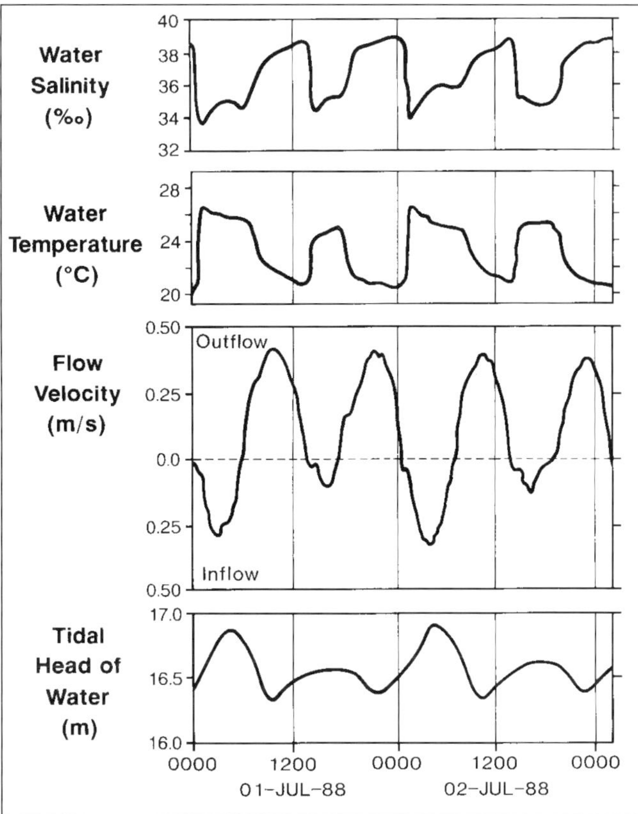

Figure 6. Current meter record from South Mastic Blue Hole, North Andros, showing variation in salinity, temperature, current velocity and direction, and local tidal head of water for 2 days in July 1988.

tion or opposition. There is a general trend of progressive increase in net discharge over the monitoring period from 2.34 x 10^4 m^3/tidal cycle during July/August (tidal cycles 0–60) to 8.51 x 10^4 m^3/tidal cycle during December/January (tidal cycles 340-400) (Figure 7B). The increased winter discharge suggests that saline ground water circulation is responding either to changing weather conditions (total rainfall, wind direction/strength or atmospheric pressure) affecting the surface of the bank, or to seasonal variations in the currents in the surrounding ocean basins.

Both oceanic blue holes monitored are thus sites of significant net discharge of saline ground water. However, the average discharge from Rat Cay is an order of magnitude less than that calculated from South Mastic during July 1988, probably due to differences in the morphology and "plumbing" of the two cave systems (compare Figures 4A and 4B). South Mastic Blue Hole is directly connected with major conduits which discharge saline ground water at depth, while at Rat Cay hydrological activity is restricted by sediment infill which appears to have partially plugged these connecting shafts. Assuming an average summer discharge of 2.0 x 10^5 $m^3 day^{-1}$ (equivalent to that of South Mastic Blue Hole) from each of the 10 known sites along the 80 km east coast of the island we can estimate a net outflow of some 49 $m^3day^{-1}m^{-1}$ of coastline. A minimum estimate based on the Rat Cay discharge would be 4.1 $m^3day^{-1}m^{-1}$ of coastline. However, the Rat Cay record (Figure 7) suggests that these estimates of discharge may increase by a factor of 3 to 4 during the winter months.

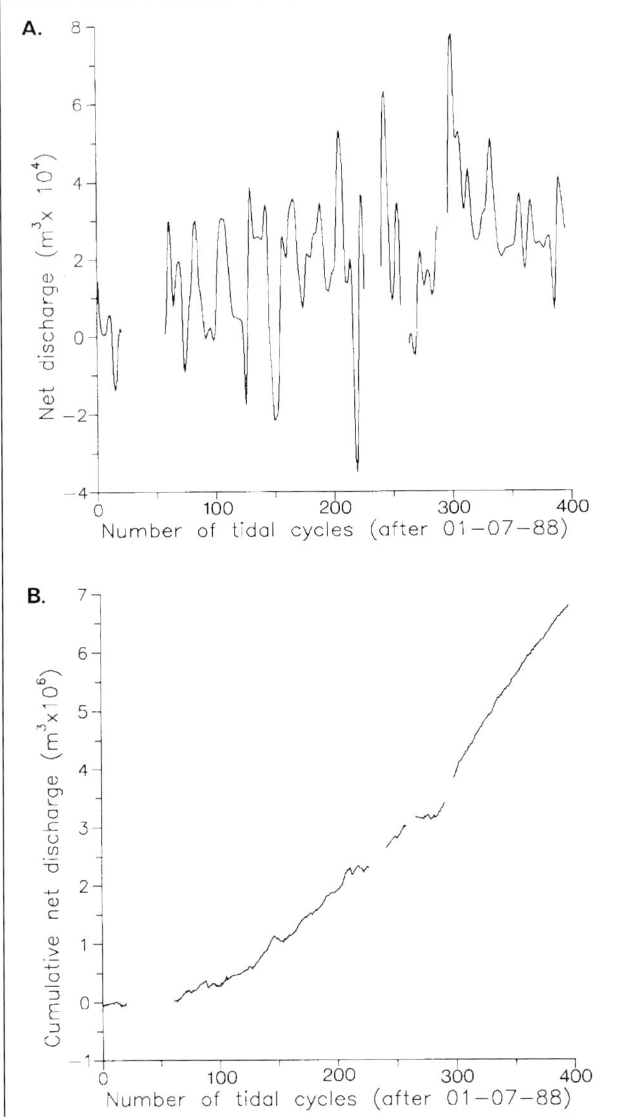

Figure 7. (A) A smoothed time-series plot of net discharge from Rat Cay Blue Hole over the period 01-07-88 to 22-01-89. Positive values indicate periods of net outflow, negative values indicate periods of net inflow, and breaks in the line are periods where rotor failure occurred. Note the 28 tidal cycle (approximately 14 days) periodicity, relating to the lunar cycle that controls tidal head, superimposed on a general increase in outflow over the monitoring period. (B) Time series plot of cumulative net discharge from Rat Cay Blue Hole over the same period. During the periods of rotor failure, which are marked by breaks in the line, discharge has been interpolated from values for adjacent periods.

Causes of Saline Ground Water Flow–Evidence from Salinity

Salinity is a conservative tracer of ground water flow that, in the absence of evaporite minerals, directly reflects sub-surface mixing between source waters. It is of particular utility in identifying waters which have undergone evaporative concentration (which may have the potential to reflux) or dilution with fresh recharge waters (indicative of buoyant circulation), and may thus aid understanding of the processes which drive the observed ground water flow. Salinity data were obtained from the recording current meters, supplemented by hand samples from nine oceanic blue holes collected at the end of the outflow phase during July/August. In situ salinity profiles were also determined in 27 inland blue holes on North Andros, mainly located in the central and eastern part of the island (Figure 3), collapse, infill and extreme inaccessibility limiting the number of sites sampled from western Andros to three.

At South Mastic Blue Hole, local sea waters have a mean salinity of 35.0‰ (Figure 6), slightly depressed relative to that of the Tongue of the Ocean due to fresh water discharge from the adjacent mainland into the restricted South Mastic Bay. In contrast the salinity of discharging ground waters is 38.7 ± 0.4‰, significantly greater than the local sea waters, and those from the Straits of Florida and Tongue of the Ocean (36.3–36.6‰). A similar pattern is observed at Rat Cay Blue Hole (average outflow salinity 37.0 ± 0.8‰) during the period July to September (unfortunately the record terminates after 186 tidal cycles due to malfunction of the salinity sensor), and in the other east coast blue holes (Figure 8). Inland ground waters on North Andros are fresh or at most brackish, even on the western side of the island. However, on the Great Bahama Bank to the west of Andros Island high evaporation rates and long residence times generate sea waters with salinities of 40.23 ± 2.0‰ in early summer, with maximum values in the immediate lee of the island in excess of 45‰ (Broecker and Takahashi, 1966). The banks are thus the only possible source of waters of elevated salinity.

A regional ground water flow must thus occur eastwards below North Andros, with discharge from the oceanic blue holes on the east coast. We would thus expect to sample these waters at depth in inland blue holes. This is in fact the case: their average bottom salinity is 38.1 ± 2.4‰ (Figure 8). The three sites from the western side of the island (marked as outliers in Figure 8) contain ground waters of 44.0 ± 0.9‰, significantly more saline than those in central and eastern Andros, and similar to the maximum salinities found on the adjacent areas of bank. Mixing calculations indicate that inland ground waters comprise between 20 and 41‰ high salinity bank water (the range of values reflecting uncertainties in the

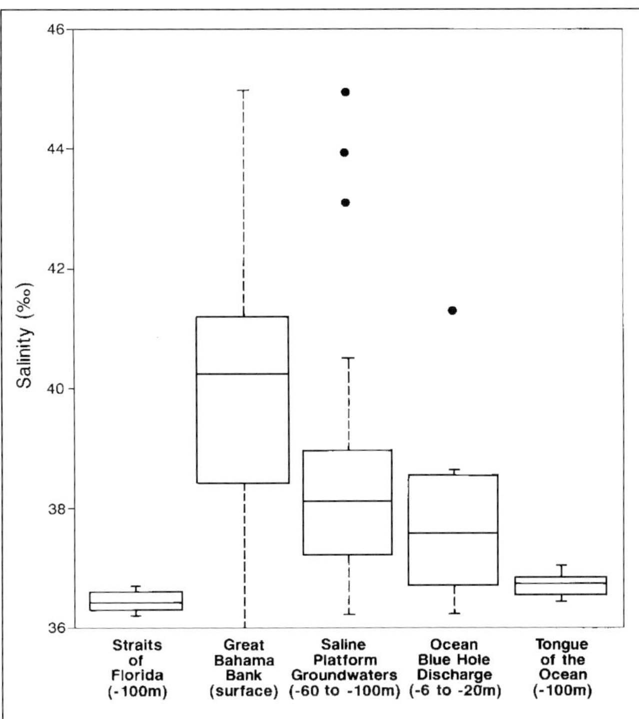

Figure 8. Box plot of salinity of ocean and bank sea waters, and platform groundwaters sampled in inland blue holes and discharging from oceanic blue holes, showing median, upper, and lower quartiles and range. Outliers are marked by black circles, and for inland blue holes comprise the three sites on the western side of the island.

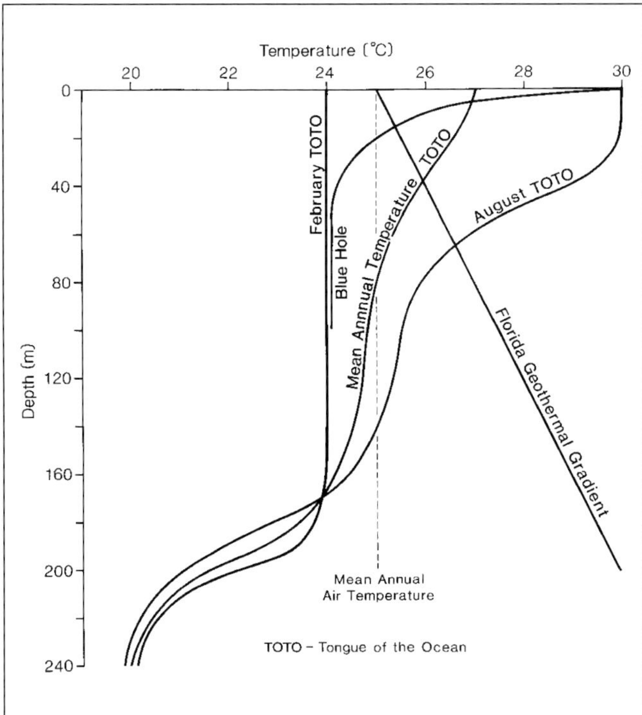

Figure 9. Comparison of temperature profiles of a typical blue hole, mean annual, February, and August profiles for the TOTO, and the Florida geothermal gradient.

salinity of source waters).

The observed flows cannot be explained by buoyant circulation in the mixing zone, which would generate salinities similar to or less than those of entrained eastern bank or Tongue of the Ocean waters. Rather these observations support the theoretical calculations of Simms (1984), and imply reflux of waters of slightly elevated salinity from the Great Bahama Bank. However, the possibility that this flow may be augmented by differences in sea-surface elevation across the island cannot be discounted and is the subject of continuing research.

Causes of Saline Ground Water Flow—Evidence from Water Temperatures

Ground water temperatures may provide both an indicator of water source and of geothermally driven circulation. In contrast to salinity however, ground water temperature is a non-conservative tracer, affected both by mixing and conductive heat transfer.

Temperature profiles of saline ground waters sampled in inland blue holes on North Andros are virtually isothermal (Figure 9), as would be expected for an open water column in which convection can occur. However, the mean bottom temperature of 24.4° C is significantly (>95% confidence interval) colder than the mean annual air temperature of 25.0° C (30-year mean, but representative of sampled years). This observation is at variance with the pattern expected for static ground waters whose temperature normally increases with depth at a rate determined by the geothermal heat flux (1°C/40m for adjacent peninsula Florida, Figure 9). The inland blue hole data therefore suggest cooling at depth. In fact the observed temperature of the isothermal section appears to be dependent on the depth of the blue hole (Figure 10A), decreasing with depth at an average rate of 1°C/69m.

These observations strongly suggest that circulation of cold ocean waters occurs at depth within the platform, the ground water temperature gradient essentially paralleling that in the adjacent Straits of Florida and Tongue of the Ocean. There is also a strong inverse relationship between ground water temperature and distance from west to east across the island (Figure 10B). This pattern suggests that cooling of the warm reflux waters from the banks occurs as they flow eastwards under Andros, either by direct mixing with cooler waters from the Straits of Florida (or Tongue of the Ocean) or by equilibration with cool waters at depth by conduction.

The temperature of discharging ground waters at South Mastic Blue Hole decreases towards the end of the outflow phase when the effect of mixing with water from the previous inflow phase is a minimum and salinities are at a maximum (Figure 6). Terminal

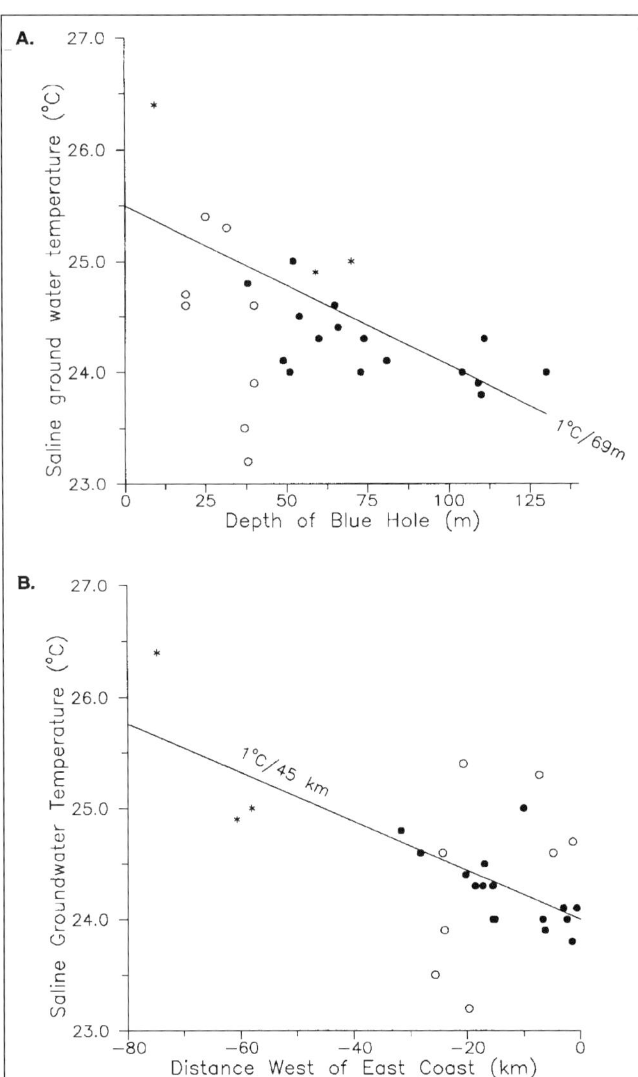

Figure 10. Relationship between the temperature of saline groundwaters and (A) the depth of the blue hole, and (B) distance from the east coast of Andros, measured on an east-west line. Three groups of sites are recognized: open circles are blue holes that penetrate <6 m of the saline zone; blue holes that fully penetrate the saline zone are represented by solid circles (sites in eastern and central Andros) and stars (sites in western Andros). Solid lines mark best-fit linear regressions lines excluding sites which penetrate < 6 m of the saline zone. In both cases the relationship is significant at >99.9% confidence interval, with t-values of -4.12 and -5.02 and R^2 of 50.0% and 59.8% for A and B, respectively.

outflow temperatures measured in the east coast oceanic blue holes in July/August are also relatively cold (average 24.7° C compared to 25–29° C for bank waters at this time), confirming the link with the cold ground waters at depth under the island. July outflow temperatures at Rat Cay (26.2 ± 0.2° C) are higher than at South Mastic, as this site discharges proportionately less deep saline ground water.

The ground water temperature data therefore strongly suggest that, in addition to the refluxing bank waters, there is a significant circulation of cold ocean water derived from depths in excess of 200 m within the platform. This is of sufficient magnitude to cool ground waters below North Andros Island to below mean annual temperature, and to reverse the geothermal gradient.

Saline Ground Water Circulation Model Through the Great Bahama Bank

The hydrological monitoring of oceanic blue holes has demonstrated a considerable net discharge of saline ground waters on the east coast of North Andros Island. These waters are of elevated salinity and thus must derive from the high salinity bank sea waters to the west of the island. Circulation is most likely to be driven by reflux of waters of only slightly elevated salinity, as predicted by Simms (1984). Reflux is at a minimum during the rainy summer and early autumn, with an increase during the subsequent dry season to a maximum in the late spring, the pattern evident in the discharge record from Rat Cay Blue Hole (Figure 7). Maximum salinities were measured at inland blue holes close to the west coast of the island. As these waters flow eastward through the platform, they mix with and become diluted by normal salinity water.

This reflux system cannot however account for the low temperatures of the saline ground waters, which appear to become colder both with increasing depth and progressively along the ground water flow path from west to east. The presence of such cold waters indicates conductive cooling or mixing with waters derived from considerable depth in the surrounding oceans. The simplest explanation of this circulation is that there is large-scale thermal convection within the platform, and these waters mix with denser elevated salinity waters derived from the bank surface (Figure 11A). Such circulation may be enhanced by the cavernous porosity associated with the bank-marginal fracture systems.

The second possibility is that there is a net flow of cold ocean water eastward from the Straits of Florida beneath Andros Island. This provides a rather better explanation of the low temperatures of saline ground waters sampled in the deeper inland blue holes. Such a circulation could be driven by an elevation potential difference across the platform generated by the Florida Current, which is confined in the shallowing straits between the Great Bahama Banks and Florida. There is a difference in sea surface elevation of as much as 66 cm across the straits between Miami and Bimini (Maul, 1986), and these waters may also flood over onto the Great Bahama Bank, creating a similar difference in elevation potential across North Andros

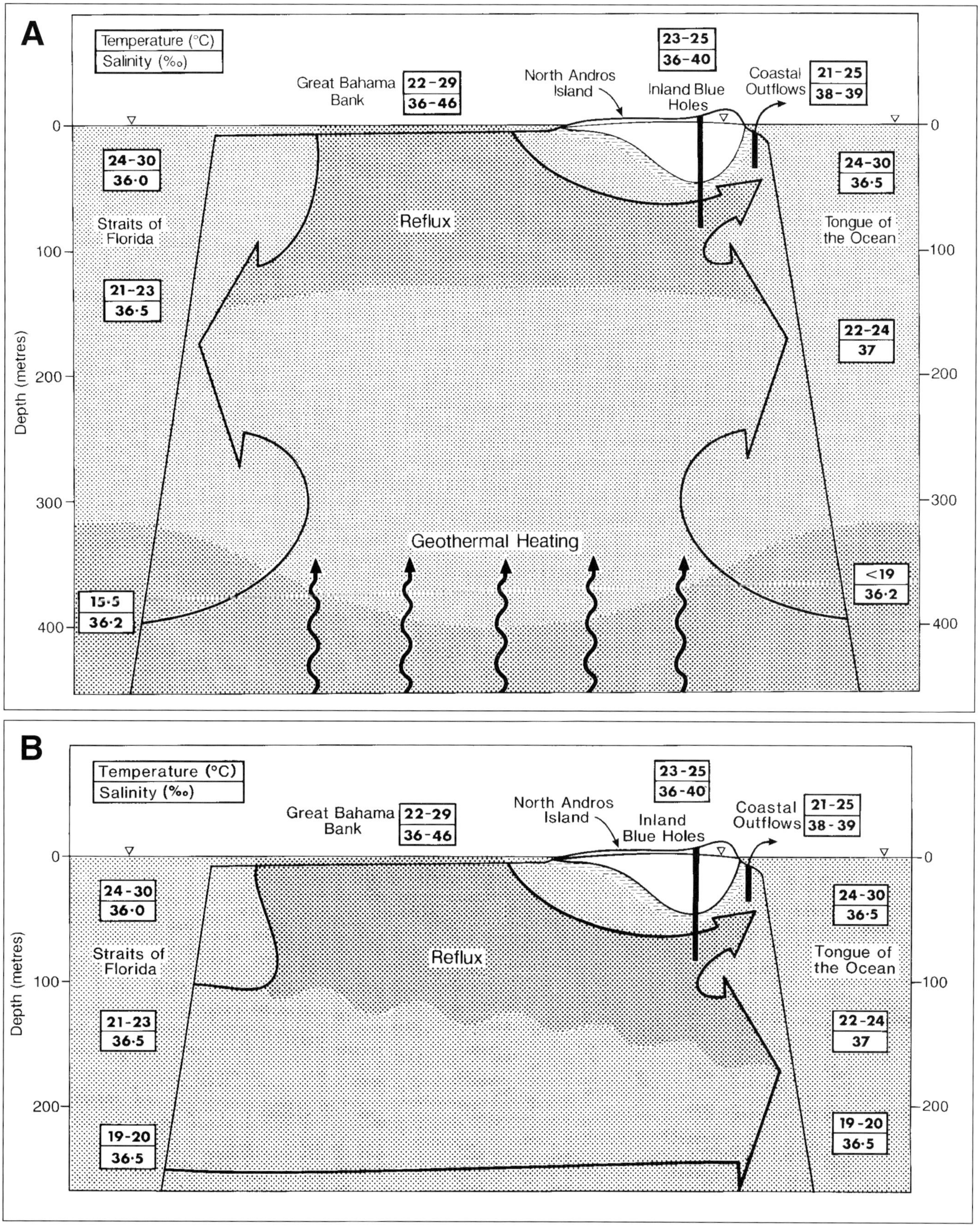

Figure 11. Summary of salinity and temperature data for ocean, bank and groundwaters, showing two possible circulation systems (open arrows), with (A) thermal and reflux drive and (B) reflux with trans-island difference in sea surface elevation. No pattern = fresh water; broken lines = mixing zone; light stipple = lower density saline water; heavy stipple = higher density saline water.

Island. This potential may be reduced during the summer and autumn when the easterly trade winds are dominant, and there is an observed reduction of ground water discharge from Rat Cay Blue Hole. This circulation (Figure 11B) is likely to be long-lived, controlled by flow of the Gulf Stream around the Caribbean and Gulf of Mexico. In contrast, the reflux-driven circulation of bank-derived waters can only occur while the banks are flooded.

The geochemistry of saline ground waters beneath Andros Island provides direct evidence of the diagenetic effects of this large-scale circulation system, and is discussed in depth by Whitaker et al. (in press). The waters involved are significantly depleted in magnesium compared to Great Bahama Bank and open ocean sea waters. In the absence of major calcium depletion, this must reflect the precipitation of modern replacement dolomite by circulating saline ground waters. In Stargate Blue Hole, South Andros, sparse replacement dolomites and dolomite cements are observed only in rock samples from the depth of the present saline zone. Stable isotope and trace element analyses suggest precipitation from a fluid of near sea water composition under slightly reducing conditions, and at temperatures in the lower range of those observed for modern saline ground waters. Strontium isotopes yield a maximum age of 0.4–0.8 Ma for the dolomites, substantially younger than that estimated for initial deposition (1.5–2.5 Ma). Order of magnitude calculations based on estimates of ground water flow combined with the observed magnesium depletion in saline ground waters, indicate dolomitization at a rate of $0.18–2.2 \times 10^{-5}\%\ yr^{-1}$. Over the long term, given the likely effect of Pleistocene sea-level fluctuations on the saline circulation system, this would be sufficient to generate 0.27–3.6% dolomite, a figure comparable to the observed abundance in the Stargate rock samples.

In conclusion, large-scale circulation of saline ground waters does occur within the upper few hundred meters of the Great Bahama Bank beneath North Andros Island, and the magnesium depletion in these waters suggests that active dolomitization is occurring. The circulation comprises saline waters input both from the bank surface and from the surrounding ocean basins, and is driven by a complex series of flow mechanisms involving differences in the vertical and horizontal distribution of salinity, ground water temperature and sea surface elevation. The complexity of this flow system, even in the relatively simple setting of an isolated carbonate platform, is in marked contrast to the simplistic models discussed above and often used to explain diagenetic patterns in ancient carbonate bodies.

DISCUSSION

We have argued in the introduction that an understanding of saline ground water movement in carbonate platforms is essential for adequate explanation of the distribution and form of shallow replacement dolomites and other early diagenetic products. Our basic tenet is that description of the ground water circulation responsible for solute transport into the system (magnesium in the case of dolomitization) will permit prediction of the three-dimensional shape of the body. Such a view is now increasingly adopted, and many studies, particularly those dealing with meteoric systems, include cartoons figuring the circulation thought to be responsible for diagenesis of particular platform sequences. However, such cartoons may prove to be at best simplistic and at worst wholly erroneous for several reasons. The contrasts in salinity and temperature driving saline ground water circulation will often be too small to be separable from normal sea water signatures using the currently available geochemical techniques, and in a water-dominated system geochemical evolution along the flow path (Banner et al., 1989) is often very small, as in the case of the Pliocene dolomites from the Little Bahama Bank studied by Vahrenkamp (1988). Thus neither the driving process nor the flow path are well constrained. Secondly, such cartoons often accord rather poorly with real flow systems because they are not adequately constrained with respect to critical hydrological parameters such as recharge (which controls the size of the freshwater lens) or permeability distribution. Thus the proposed circulation system may be physically impossible. Finally, as demonstrated in our case study, the circulation may involve not a single simple drive as generally envisaged, but a relatively complex balance of two or more processes. In some cases these may act to reinforce each other. For example, in peninsular Florida, buoyant circulation is reinforced by thermal convection, with saline inflow at depth and discharge in the mixing zone and upper part of the saline zone (Kohout et al., 1977). Conversely, two or more mechanisms may act in opposition to suppress ground water flow. In enclosed lagoons, for example, increased density induced by evaporation generates the potential for downward reflux, but it may also reduce the water surface elevation in the lagoon below that of the surrounding ocean, opposing the density potential and generating a net inflow of normal salinity water by evaporative pumping (Rougerie, 1985).

One way forward is thus to examine more present-day saline ground water circulation systems, which incorporate this complexity. However, field studies are limited by the character and configuration of the particular system studied, and may not adequately model the specific geological example of interest. Indeed, in some respects there is a general failure of the present to adequately model the past, for instance at times when glacio-eustatic effects are much less important than during the Pleistocene. A second way forward may therefore be to develop ground water flow models. These may be configured to test the sensitivity of ground water circulation to controlling parameters, which can be reliably recognized from generally available geological information. Such controls are summarized above, and include platform geography, architecture, facies distribution, and sea-level history. The potential of such general simulations has already been demonstrated, for instance by the classic paper of Garvin and Freeze (1984). There is also the potential in the longer term for coupled modeling of ground water flow and geochemical reactions, as in the recent studies of Harrison and Summa (1991) and Sanford and Konikow (1989). Such studies can also be used to predict reservoir quality (Meshri and Ostoleson 1990). Ultimately we can envisage the coupling of such diagenetic models with those describing the depositional facies and development of platform architecture over time (Bice, 1988; Sarg, 1988). Indeed, such forward modeling may be the only possible way to deal with the complex superposition and evolution of different flow systems during the development and subsequent demise of the platform (Matthews and Froelich, 1987).

ACKNOWLEDGMENTS

Financial support was provided by Amoco U.K. Exploration Company, Koninklijke/Shell Exploratie en Productie Laboratorium, and the Royal Society. We thank the Institute of Oceanographic Sciences and the Environmental Sciences Division of Wimpol Ltd., who lent, decoded and calibrated the recording current meters used on the Andros field study, and Steve Hobbs, Neil Sealey, Brad Pecel, and Rob Palmer, who provided invaluable assistance in the field. This chapter has benefited from constructive reviews by Drs. M. Esteban, A. Searl, H. Nicholson, A. Horbury and V.P. Wright.

REFERENCES CITED

Aharon, P., R.A. Socki, and L. Chan, 1987, Dolomitization of atolls by sea water convection flow: test of a hypothesis at Niue, South Pacific: Journal of Geology, v. 95, p. 187-203.

Adams, J.E., and M.L. Rhodes, 1960, Dolomitization by seepage refluxion: American Association of Petroleum Geologists Bulletin, v. 44, p. 1912-1920.

Aissaoui, D.M., D. Buigues, and B.H. Purser, 1986, Model of reef diagenesis: Mururoa Atoll, French Polynesia, in J.H. Schroeder and B.H. Purser, ed., Reef Diagenesis: Springer-Verlag, Berlin, p. 27-52.

Alfirevic, S., 1966, Hydrogeological investigations of submarine springs in the Adriatic: Hydrogeologists Memoires v. VI, Reunion de Belgrade 1963, p. 255-264.

Amdurer, M., and L.S. Land, 1982, Geochemistry, hydrology and mineralogy of the sand bulge area, Lagune Madre flats, South Texas: Journal of Sedimentary Petrology, v. 52, p. 703-716.

Atkinson T.C., and D.I. Smith, 1976, The erosion of limestones; In Ford T.D. and C.H.D. Cullingford (Eds.) The Science of Speleology, Academic Press, London, p. 151-177.

Atkinson, M., S.V. Smith, and E.D. Stroup, 1981, Circulation of Enewetak Atoll lagoon: Limnology and Oceanography, v. 26, p. 1074-1083.

Badiozamani, K., 1973, The Dorag dolomitization model–Application to the mMiddle Ordovician of Wisconsin: Journal of Sedimentary Petrology, v. 43, p. 965-984.

Baker, P.A., and S.J. Burns, 1985, Occurrence and formation of dolomite in organic-rich continental margin sediments: American Association of Petroleum Geologists Bulletin v. 69, p. 1917-1930.

Banner, J.L., G.J. Wasserburg, P.F. Dobson, A.B. Carpenter, and C.H. Moore, 1989, Trace element and isotopic constraints on the origin and evolution of saline ground waters from central Missouri: Geochimica et Cosmochimica Acta, v. 53, p. 383-398.

Bear, J., and D.K. Todd, 1960, The transition zone between fresh and saline waters in coastal aquifers: California University Water Resources Centre, Contract No. 29, 156 p.

Berner, R.A., 1971, Principles of chemical sedimentology: McGraw Hill, New York, 240 p.

Bice, D., 1988, Synthetic stratigraphy of carbonate platform and basin systems: Geology, v. 16, p. 703-706.

Bouvier, J.D., E.C.A. Gevers, and P.L. Wigley, 1990, 3-D seismic imterpretation and lateral prediction of the Amposto Marino field (Spanish Mediterranean Sea): Geologie Mijnbouw, v. 69, p. 105-120.

Broecker, W.S., and T. Takahashi, 1966, Calcium carbonate precipitation on the Great Bahama Banks: Journal of Geophysical Research, v. 71, p. 1575-1602.

Buddemeier, R.W., and J.A. Oberdorfer, 1986, Internal hydrology and geochemistry of coral reefs and atoll islands: key to diagenetic variations, in J.H. Schroeder and B.H. Purser, ed., Reef Diagenesis: Springer-Verlag, Berlin, p. 91-111.

Budd, D.A., 1988, Aragonite to calcite transformation during fresh-water diagenesis of carboantes. Bulletin of the Geogloical Society of America, v. 100, p. 1260-1270.

Budd, D.A., and H.L. Vacher, 1991, Predicting the thickness of fresh water lenses in carbonate paleo-islands: Journal of Sedimentary Petrology, v. 61, p. 43-53.

Cant, R.V., and P.S. Weech, 1986, A review of the factors affecting the development of Ghyben-

Hertzberg lenses in the Bahamas: Journal of Hydrology, v. 84, p. 333-343.

Carballo, J.D., L.S. Land, and D.E. Miser, 1987, Holocene dolomitization of supratidal sediments by active tidal pumping, Sugarloaf Key, Florida: Journal of Sedimentary Petrology, v. 57, p. 153-156.

Choquette, P.W. and R.P. Steinen, 1980, Mississippian non-supratidal dolomite, Ste. Genevieve limestone, Illinois basin: evidence for mixed water dolomitization: In Zenger D.H., J.B. Dunham and R.L. Etherington (Eds), Concepts and Models in Dolomitization. SEPM Special Publication No. 28, p. 163-196.

Cooper, H.H., 1959, A hypothesis concerning the dynamic balance of fresh water and salt water in a coastal aquifer: Journal of Geophysical Research, v. 64, p. 461-467.

Craig D.H., 1988, Caves and other features of Permain Karst in San Andres Dolomite, Yates field reservoir, West Texas: In James N.P. and P.W. Choquette, Paleokarst, Elsevier, p. 342-363.

Daugherty D.R., M.R. Boardman, and C.V. Metzler, 1986, Characteristics and origins of joints and sedimentary dikes of the Bahama islands: Proceedings of the 3rd Symposium on the Geology of the Bahamas, p. 45-56.

Davis, R.L. and C.R. Johnson, 1988, Karst hydrology of San Salvador: Proceedings of the 4th symposium on the Geology of the Bahamas, p 118-135.

Dawans, J.M., and P.K. Swart, 1988, Textural and geochemical alterations in Bahamian Pliocene dolomites: Sedimentology, v. 35, p. 385-403.

Deffeyes, K.S., F.J. Lucia, and P.K. Weyl, 1965, Dolomitization of recent and Plio-Pleistocene sediments by marine evaporite waters on Bonaire, Netherlands Antilles: Society of Economic Paleontologists and Mineralogists Special Publication No. 13, p. 71-88.

Drogue, C., 1989, Continuous inflow of sea water and outflow of brackish water in the substratum of the karstic island of Cephalonia, Greece: Journal of Hydrology, v. 106, p. 147-153.

Enos, P., 1988, Evolution of pore-space in the Poza Rica trend (Mid-Cretaceous), Mexico: Sedimentology, v. 35, p. 287-326.

Enos, P. and L.H. Sawatsky, 1981, Pre networks in Holocene carbonate sediments. Journal of Sedimentary Petrology, v.51, 961-985.

Fanning, K.A., R.H. Byrne, J.A. Breland, and P.R. Betzer, 1981, Geothermal springs of the west Florida continental shelf: evidence for dolomitization and radionuclide enrichment: Earth Science and Planetary Letters, v. 52, p. 345-354.

Folk, R.L., and L.S. Land, 1975, Mg/Ca ratio and salinity; two controls over crystallization of dolomite: American Association of Petroleum Geologists Bulletin, v. 59, p. 631-638.

Garven G., and R.A. Freeze, 1984, Theoretical analysis of the role of ground water flow in the genesis of stratabound ore deposits: Numerical and mathematical model; American Journal of Science, v. 284, 1085-1124.

Giles, M.R., 1987, Mass transfer and problems of secondary porosity creation in deeply buried hydrocarbon reservoirs: Marine and Petroleum Geology, v. 4, p. 188-204.

Given, R.K., and B.H. Wilkinson, 1987, Dolomite abundance and stratigraphic age: constraints on rates and mechanisms of Phanerozoic dolostone formation: Journal of Sedimentary Petrology, v. 57, p. 1068-1078.

Goldstein, R.H., B.P. Stephens, and D.J. Lehrmann, 1991, Fluid inclusions elucidate conditions of dolomitization in Eocene of Enewetak atoll and Mid-Cretaceous Valles platform of Mexico, in A. Bossellini et al., ed., Proceedings of Dolomieu Conference on Carbonate Platforms and Dolomitization (Abstract), p. 92-93.

Hanshaw, B.B., and W. Back, 1980, Chemical mass-wasting of the northern Yucatan peninsula by ground water dissolution: Geology, v. 8, p. 222-224.

Hanshaw, B.B., W. Back, and R Deike, 1971, AQ geochemcial hypothesis for dolomitization by ground water; Economic Geology, v.66, p. 710-724.

Hardie, L.A., 1987, Dolomitization: a critical view on some current views: Journal of Sedimentary Petrology, v. 57, p. 166-183.

Hardie, L.A., and R.M. Ginsburg, 1977, Layering: the origin and environmental significance of lamination and thin bedding, in L.A. Hardie, ed., Sedimentation on the modern carbonate tidal flats of north-west Andros Island, Bahamas: Johns Hopkins University Studies in Geology 22, John Hopkins University Press, Baltimore, p. 50-123.

Harrison, W.J. and L.I. Summa, 1991, Paleohydrology of the Gulf of Mexico Basin: American Journal of Science, v. 291, p. 109-176.

Holser, W.T., 1979, Trace elements and isotopes in evaporites, in R.G. Burns (ed): Marine minerals; reviews in Mineralogy, Mineral Society of America, v. 6, p. 301-380.

Horbury A. D. and A.E. Adams, 1989, Meteoric phreatic diagenesis in cyclic late Dinantian carbonates, northeast England: Sedimentary Geology, v. 65, p. 319-344.

Hsu, K.J., and J. Schneider, 1973, Progress report on dolomitization–hydrology of Abu Dhabi sabkhas, in B.H. Purser, ed., The Persian Gulf: Springer-Verlag, Berlin, p. 409-422.

Humphrey, J.D., 1988, Late Pleistocene mixing-zone dolomitization, West Indies; Sedimentology, v. 35, p. 327-348.

Illing, L.V., A.J. Wells, and J.C.M. Taylor, 1965, Penecontemporary dolomite in the Persian Gulf:

Society of Paleontologists and Mineralogists Special Publication No. 13, p. 89-111.

James, N.P. and Choquette, P. W., 1984, Diagnesis 9–Limestones, the meteoric diagenetic environment; Geoscience Canada, v. 11, p. 161-194.

James, N.P., and R.N. Ginsburg, 1979, The seaward margin of Belize barrier and atoll reefs: International Association of Sedimentology Special Publication No. 3, p. 1-191.

Kelts, K., and J.A. Mackenzie, 1982, Diagenetic dolomite formation in Quaternary anoxic diatomaceous muds of deep-sea drilling project leg 64, Gulf of California: Initial Reports of the Deep Sea Drilling Project, v. 64, p. 553-569.

Kendall, A.C., 1989,Brine mixing in the Middle Devonian of Western Canada and its possible significance to regional dolomitization: Sedimentary Geology 64, 271-286.

Kohout, F.A., 1960, Cyclic flow of sea water in the Biscayne Aquifer of south-eastern Florida: Journal of Geophysical Research, v. 7, p. 2133-2141.

Kohout, F.A., 1967, Groundwater flow and the geothermal regime of the Floridian Plateau: Transactions of the Gulf Coast Association Geological Society, v. 17, p. 339-354.

Kohout, F.A., and H. Kline, 1977, Effect of pulse recharge on the zone of diffusion in the Biscayne aquifer, in G. Gunay and J. Karanjac, ed., Proceedings of the Symposium on Karst Hydrogeology: DSI, Ankara, Turkey, p. 151-169.

Kohout, F.A., H.R. Henry, and J.E. Banks, 1977, Hydrogeology related to the geothermal conditions of the Floridan plateau, in D.I. Smith and G.M. Griffin, ed., The geothermal nature of the Floridan Plateau: Florida Bureau of Geology Special Publication No. 21, p. 1-40.

Land, L.S., 1980, The isotopic and trace element geochemistry of dolomite; the state of the art: Society of Economic Paleontologists and Mineralogists Special Publication No. 28, p. 87-110.

Land, L.S., 1985, The origin of massive dolomite: Journal of Geological Education, v. 33, p. 112-125.

Lippmann, F., 1973, Sedimentary Carbonate Minerals: Springer-Verlag, Berlin 228 p.

Logan, B.W., 1987, The MacLeod evaporite basin, western Australia: American Association of Petroleum Geologists Memoir, v. 44, 140 p.

Longman, M.W., 1980, Carbonate diagenetic textures from near-shore diagenetic environments; Bulletin of the American Association of Petroleum Geologists, v. 64, p. 461-487.

Lucia, F.J., 1968, Recent sediments and diagenesis of South Bonaire, Netherlands Antilles: Journal of Sedimentary Petrology, v. 38, p. 845-858.

Machel, H.G., and E.W. Mountjoy, 1986, Chemistry and environments of dolomitization–a reappraisal: Earth Science Reviews, v. 23, p. 175-222.

Marshall, J.F., 1986, Regional distribution of submarine cements within an epi-continental reef system: Central Great Barrier Reef, Australia, in J.M. Schroeder and B.H. Purser , ed., Reef Diagenesis: Springer-Verlag, Berlin, p. 8-26.

Mattes, B.W., and E.W. Mountjoy, 1980, Burial dolomitization in the Upper Devonian Miette platform, Jasper National Park, Alberta: Society of Economic Paleontologists and Mineralogists Special Publication No. 28, p. 259-297.

Matthews, R.K., 1974, A process approach to diagenesis of reefs and reef associated limestones: Society of Economic Palaeontologists and Mineralogists Special Publication No. 18, p. 234-256.

Matthews, R.K., and C. Froelich, 1987, Forward modelling of bank-margin carbonate diagenesis: Geology, v. 15, p. 673-676.

Maul, G.A., 1986, Linear correlations between Florida Current volume transport and surface speed with Miami sea-level and weather during 1964-1970: Royal Astronomical Society Geophysical Journal, v. 87, p. 55-66.

Meshri, I.D. and P.J. Ortoleva, 1990, Prediction of reservoir quality through chemical modeling, AAPG Memoir 49. 175 p.

Neild, D.A., 1968, Onset of thermohaline convection in a porous medium: Water Resources Research, v. 4, p. 553-560.

Patterson, R.J., and D.J. Kinsman, 1982, Formation of diagenetic dolomite in coastal sabkhas along the Arabian (Persian) Gulf: American Association of Petroleum Geologists Bulletin, v. 66, p. 28-43.

Phillips, O.M., 1991, Flow and reactions in permeable rocks: Cambridge University Press, Cambridge, 283 p.

Playford, P.E., 1984, Platform-margin and marginal-slope relationships in Devonian reef complexes of the Canning Basin: Proceedings of the Geological Society of Australia, Symposium, Perth, Western Australia, p. 189-214.

Pratt, B.R., and N.P. James, 1986, The St George group (lower Ordovician) of western Newfoundland: tidal flat island model for carbonate sedimentation in shallow epeiric seas: Sedimentology, v. 33, p. 313-343.

Rodgers, K.A., A.J. Easton, and C.J. Downes, 1982, The chemistry of carbonate rocks of Niue Island, South Pacific: Journal of Geology, v. 90, p. 645-662.

Rougerie, F., 1985, Main characteristics of the Takapoto lagoon and internal functioning: Proceedings of the 5th International Coral Reef Congress, v. 1, p. 363-367.

Runnels, D.D., 1969, Diagenesis, chemical sediments and mixing of natural waters. Journal of Sedimentary Petrology, v. 39, p. 1188-1201.

Saller, A.H., 1984, Petrologic and geochemical constraints on the origin of subsurface dolomite,

Enewetak Atoll: an example of dolomitization by normal sea water: Geology, v. 12, p. 217-220.

Samaden, G., P. Dallot, and R. Roche, 1985, L'atoll d'Enewetak: systeme geothermique insulaire a l'etat naturel: La Houille Blanche v. 2, p. 143-151.

Sanford, W.E., and L.F. Konikow, 1989, Simulation of calcite dissolution and porosity changes in salt water mixing zones in coastal aquifers: Water Resources Research, v. 25, p. 655-667.

Sarg, J.F., 1988, Carbonate sequence stratigraphy: Society of Economic Paleontologists and Mineralogists Special Publication No. 42, p. 155-181.

Shinn, E.A., R.N. Ginsburg, and R.M. Lloyd, 1965, Recent supratidal dolomite from Andros Island, Bahamas. In: L.C. Pray and R.C. Marray (Eds) Dolomitization and Limestone Diagenesis–a Symposium. SEPM Special Publication No. 13, p. 112-123.

Simms, M., 1984, Dolomitization by groundwater flow systems in carbonate platforms: Transactions of the Gulf Coast Association of Geological Sciences, v. 24, p. 411-420.

Simms, M.A., and L.A. Hardie, 1983, Reflux of sea water as a dolomitising mass transfer process on carbonate banks: Geological Society of America, Abstracts with Programs (Abstract), v. 15, p. 688.

Smart, P.L., and F.F. Whitaker, 1988, Controls on the rate and distribution of carbonate bedrock dissolution in the Bahamas; Proceedings of the Fourth Symposium on the Geology of the Bahamas, p. 313-322.

Smart, P.L., R.J. Palmer, F.F. Whitaker, and V.P. Wright, 1988, Neptunian dykes and fissure fills: an overview and account of some modern examples, in N.P. James and P.W. Choquette, ed., Paleokarst: Springer-Verlag, New York, p. 149-163.

Spencer, M., 1967, Bahamas Deep Test: American Association of Petroleum Geologists Bulletin, v. 51, p. 263-268.

Swartz, J.H, 1958, Geothermal measurements on Enewetak and Bimini atolls: United States Geological Survey Professional Paper 260-U, p. 711-739.

Vacher, H.L., 1978, Hydrogeology of Bermuda–significance of across-the-island variation on permeability: Journal of Hydrology, v. 39, p. 207-226.

Vahrenkamp, V.C., 1988, Constraints on the formation of platform dolomites: a geochemical study of late Tertiary dolomites from the Little Bahama Bank: Unpublished Ph.D. Thesis, University of Miami, 328 pp.

Vahrenkamp, V.C., P.K. Swart, and J. Ruiz, 1991, Episodic dolomitization of Late Cenozoic carbonates in the Bahamas, evidence from strontium isotopes: Journal of Sedimentary Petrology, v. 61, p. 1002-1014.

Von der Borch, C.C. and D. Lock, 1979, Geological significance of Coorong dolomites; Sedimentology, v. 26, p. 813-824.

Walkden G.M.. and D.O. Williams, 1991, The diagenesis of Late Dinantian Derbyshire-East Midland shelf, central England: Sedimentology, v. 38, p. 643-670.

Whitaker, F.F., 1992, Hydrology, geochemistry, and diagenesis of modern carbonate platforms in the Bahamas: unpublished Ph.D. Thesis, University of Bristol, 347 pp.

Whitaker, F.F., and P.L. Smart, 1990, Circulation of saline groundwaters through carbonate platofrms: evidence from the Great Bahama Bank: Geology, v. 18, p. 200–204.

Whitaker, F.F., P.L. Smart, V.C Vahrenkamp, R.A. Wogelius and H. Nicholson, in press, Dolomitization by near-normal sea water? Field evidence from the Bahamas, *in* Purser, B. et al. (Ed), Dolomites, a Volume in Honour of Dolomieu.

Wilson, E.N., L.A. Hardie, and O.M. Phillips, 1990, Dolomitization from geometry, fluid flow patterns and the origin of massive dolomite: the Triassic Latemar platform, northern Italy: American Journal of Science, v. 290, p. 741-797.

Wolanski, E., 1981, Aspects of physical oceanography of the Great Barrier Reef lagoon: Proceedings of the 4th International Coral Reef Symposium, v. 1, p. 375-381.

Wood, G.V., and M.J. Wolfe, 1969, Sabkha cycles in the Arab Darb formation off the Trucial Coast of Arabia: Sedimentology, v. 12, p. 165-191.

Zempolich, W.G, and L.A. Hardie, 1991, Massive burial dolomitization: the Jurassic Vajont Oolite of N.E. Italy, in A. Bosellini et al., ed., Proceedings of Dolomieu Conference on Carbonate Platforms and Dolomitization (Abstract), p. 298.

Zenger D.H., 1983, Burial dolomitization in the Lost Burro Formation (Devonian), east-central California, and the significance of late diagenetic dolomitization: Geology, v. 11, 519-522.

PART 3

Diagenesis and Faults

Chapter 10

The Influence of Fault Zone Processes and Diagenesis on Fluid Flow

R.J. Knipe
Department of Earth Sciences
The University, Leeds
Leeds, U.K.

ABSTRACT

The influence of faulting processes on the distribution of diagenetic events during basin evolution is reviewed. Faults have potentially a dual role as high permeability pathways (which enhance fluid flow) and as seals (which constrain fluid flow and generate permeability barriers). The complexities of fault array development and the fluid pressure evolution of different lithologies during burial lead to fluid flow in basins being controlled by the creation and destruction of complex 3D migration pathways. Microstructural analysis can help to identify the evolution of such pathways and provide input for the quantitative modeling of fluid flow processes. Analysis of the diagenetic histories within and outside fault zones reveals differences which can be used to establish the links between deformation mechanisms, fault zone diagenesis, and the influence of these processes on fluid flow in sedimentary basins. The combination of microstructural analysis of deformation and diagenetic sequences together with the assessment of the propagation and geometrical evolution of faults reported here offer an important route for the evaluation of (a) fault zone porosity/permeability histories; (b) the evolution and distribution of fault seals; (c) the timing of fault activity during burial and basin development; and (d) identification of fluid migration pathways and the changing drainage patterns that control the distribution and timing of open/closed compartments within the basin. A model is presented emphasizing the influence of evolving fault plane geometries, displacement patterns, tip zone processes, and fault rock evolution in controlling the juxtapositions and windows for fluid communication and diagenetic changes.

INTRODUCTION

Understanding the mechanisms, pathways, timing, and rates of fluid migration and expulsion is important to basin evolution studies, as well as to analysis of hydrocarbon reservoir development. Although the possible fluid migration pathways (bulk rock, along fracture arrays, and/or along faults) have been identified, there is still disagreement as to the exact role of faulting processes on the distribution of fluid flow and diagenetic effects within developing basins. There is little doubt about the ability of fault zones to focus fluid flow (Sibson, 1981, 1987 and 1990; Carter et al., 1990; Moore, 1989; Langseth and Moore, 1990; Knipe et al., 1991) or to act as permeability barriers which restrict fluid migration and lead to the development of compartments with different fluid pressures and chemistries (Smith, 1966, 1980; Harding,

1974; Harding and Tuminas, 1988, 1989; Hindle, 1989; Knipe, 1992). A major objective for future research concerned with basin analysis is to understand this dual behavior of faults as fluid conduits and barriers.

This chapter reviews the faulting processes that can control fluid flow and assesses the influence of fault rock evolution on flow pathways and diagenesis during basin development. The important interactions between fault rock evolution, the complexities of fault array development and diagenetic changes in different lithologies during burial are also discussed. The chapter outlines how such interactions lead to the creation and destruction of complex 3D migration pathways that can control fluid flow and thus influence diagenetic patterns within basins. Examples are presented of how the integration of microstructural analysis of deformation and diagenetic processes in fault zones, physical property measurements (porosity and permeability) of both deformed and undeformed lithologies, and the geometrical analysis of fault array evolution, are important for the identification of fluid flow pathways. The chapter contains three sections, which review and discuss three inter-related aspects of the role of faults in fluid flow:

(1) **Faulting processes and fluid pressure changes** which control the type and location of fluid migration pathways and the magnitude and distribution of fluid pressure gradients which drive fluid flow and promote precipitation/alteration processes;
(2) **Fault rock evolution** which controls the permeability changes in fault zones; and
(3) **Fault displacement patterns, activity, and array evolution** which control the distribution of flow paths and therefore drainage patterns.

FAULTING PROCESSES, FLUID PRESSURE CHANGES, AND FLUID FLOW

Slip events on faults range from earthquakes of extremely large magnitude which produce slip over tens of square kilometres on a fault plane in a few seconds, through earthquake swarms (where a cluster of quakes gradually develops and then subsides), down to micro-earthquakes and aseismic creep waves which migrate at slow rates and produce small displacements (<<5mm per event). The frequency/ magnitude distribution of earthquakes reveals that there is a simple relationship between the number and the magnitude of events; however, the relationship is not thought to hold for single faults (Sibson, 1989).

Although many of the details of fault slip are poorly understood, the fluid flow associated with each of these types of fault behavior is likely to be different. Large quakes associated with large rupture areas have a greater possibility of rapidly linking compartments with different fluid pressures, i.e., the fault acts as a valve for flow (Sibson, 1990). Where such events induce a link between overpressured compartments at depth and the surface, rapid expulsion of large volumes of water is possible (Sibson, 1981). The Imperial Valley quake (M7.5) of 1952 produced the expulsion of $10^7 m^3$ of water during a period of two months following the quake. In addition to the modification of pre-faulting fluid pressure regimes, earthquakes also produce their own fluid pressure patterns involving high fluid pressure gradients at the tip zones of the rupture area which decay with time (Nur and Booker, 1972; Booker, 1974). The smaller magnitude displacement events associated with aseismic creep (Wesson, 1988; Scholz, 1990) have the potential to slowly transfer smaller packets of fluid along the fault.

The earthquake cycle can often be divided into distinct stages, each likely to be characterised by different fluid flow behavior. The *Preseismic (α+β) stage* is generally considered to involve damage accumulation in the form of elastic strain and microfracture formation and coalescence in the source area (Scholz, 1990). The dilation associated with this process allows ingress of fluids into the damage zone and can give rise to a drop in fluid pressure in and around the fault zone (Frank 1965). That is, fluids can be drawn in and stored in the fault zone. The very transient and large dilations associated with the *main seismic (γ) phase* rupture propagation are unlikely to be preserved for long but are critical in establishing the pattern of connectivity between different fluid pressure regimes or compartments. Pockets of enhanced and/or persistent dilation are likely to be produced along the rupture surface at *dilational fault jogs* (steps at the bends in non-planar faults; Sibson, 1989) and at deformation-induced fracture concentrations at bends and in tip zones. The *postseismic (δ) phase* involves a period of decelerating displacement associated with aftershock activity and afterslip, when fluid flow occurs in response to (and to remove) the hydraulic gradients induced during the earlier stages.

The magnitude of the permeability changes induced by these different stages of faulting, and their efficiency or ability to influence flow, will depend on:

(1) the time periods when the high permeability windows can aid fluid migration;
(2) the fault zone volume affected by the event; and
(3) the magnitude and location of the dilation changes.

Each of these is influenced by the deformation mechanisms involved. The interactions between these factors, fault rock evolution, and fluid flow processes, are discussed below under "Fault Rock Evolution and Fault Zone Diagenesis."

This discussion highlights how the fluid flow potential of the fault is critically dependent on the pattern of dilation produced during the faulting event and the processes which control the decay of the

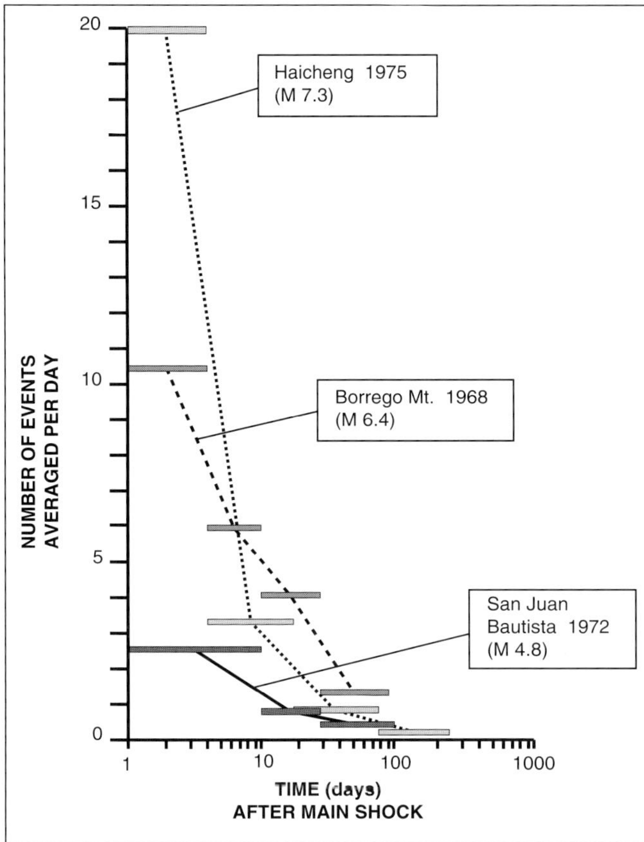

Figure 1. The aftershock decay period for earthquakes with different main event magnitudes. The horizontal time bars for each of the three earthquakes shown represent the time periods over which the number of events is averaged. Note that in each case the activity is concentrated in the 2-3 month period following the main event.

enhanced permeability (or dilation) generated in and around the fault zone. The factors that can control the modification of the initial pattern of dilation induced by a large faulting event include

(1) the distribution and magnitude of aftershocks;
(2) the time dependent processes (including afterslip) which operate in the fault zone volume during and after the aftershock activity; and
(3) the time interval before repetition of the cycle.

Aftershocks are known to cluster either in the tip zones to major ruptures or at fault bends (Sibson, 1989; Scholz, 1990). In terms of fluid migration, this is important as the locations of fluid storage volumes around the fault may change with time from an initial concentration in the source area to the dilational jogs and tip zones.

The frequency of aftershocks decreases with time (t) after the main rupture event, according to an approximately hyperbolic function, i.e., t^{-1} (Reasenburg and Ellesworth, 1982; Sibson, 1989; Scholz, 1990). Figure 1 shows the aftershock decay period for some earthquakes with different main event magnitudes. The decay period is typically 2-3 months for quakes where the magnitude is greater than four. This decay has been linked to the influx of fluid and redistribution of pore pressures after a main quake event (Booker and Nur, 1972; Booker, 1974; Li et al., 1987). However, the aftershock distribution is not a simple one, and the decay pattern can include the aftershock activity decreasing and then increasing for a time after the main event, the location of aftershocks migrating along or even away from the fault zone with time (Booker, 1974; Wesson, 1987), and significant aftershock events occurring away from the main rupturing (Scholz, 1990). In addition, given that the magnitudes of these aftershocks are smaller than the main event (the largest aftershock is usually at least one order of magnitude less than the main shock), even if the events are located on the main fault, they are likely to only affect at most approximately one-third of the initial rupture surface. This has an important influence on the possible communication history between different fluid pressure compartments. The aftershock period represents an important window for fluid migration as this is the period when communication between previously isolated lithologies or compartments with initially differing fluid pressures is most likely to be maintained and probably represents a minimum for the main time period when enhanced fluid flow paths are present in the fault zone.

Fluid flow and aftershock activity are both likely to be influenced by the interaction between fluid pressure gradients set up by the faulting itself and preexisting fluid pressure patterns modified by the propagating fault. The pattern of flow which results is likely to be asymmetric due to the volumes of extension (fluid sinks) and compression (fluid sources) at each tip (Figure 2i and ii). Flow associated with a normal fault may involve four components (Figure 2iii):

(1) sub-horizontal flow between the hanging wall and footwall at each tip (Figure 2iiia);
(2) flow from the hanging wall lower tip to hanging wall top tip where the gradient in fluid pressure will be highest due to maximum pressure in lower hanging wall tip and minimum in hanging wall upper tip (Figure 2iiib);
(3) the hanging wall lower tip compression zone may develop pore throat pinching during compression which reduces permeability. (Thus it is possible during early decay of the pressure, there may be periods when flow from the footwall lower extensional tip acts as the source for fluid flow) (Figure 2iiic);
(4) it is possible, though perhaps less likely, that if communication across the fault is prevented by seals or fault propagation is up into a high pressure horizon (see below), then down-fault flow in the footwall from the area of compression will occur (See also McCaig, 1988) (Figure 2iiid).

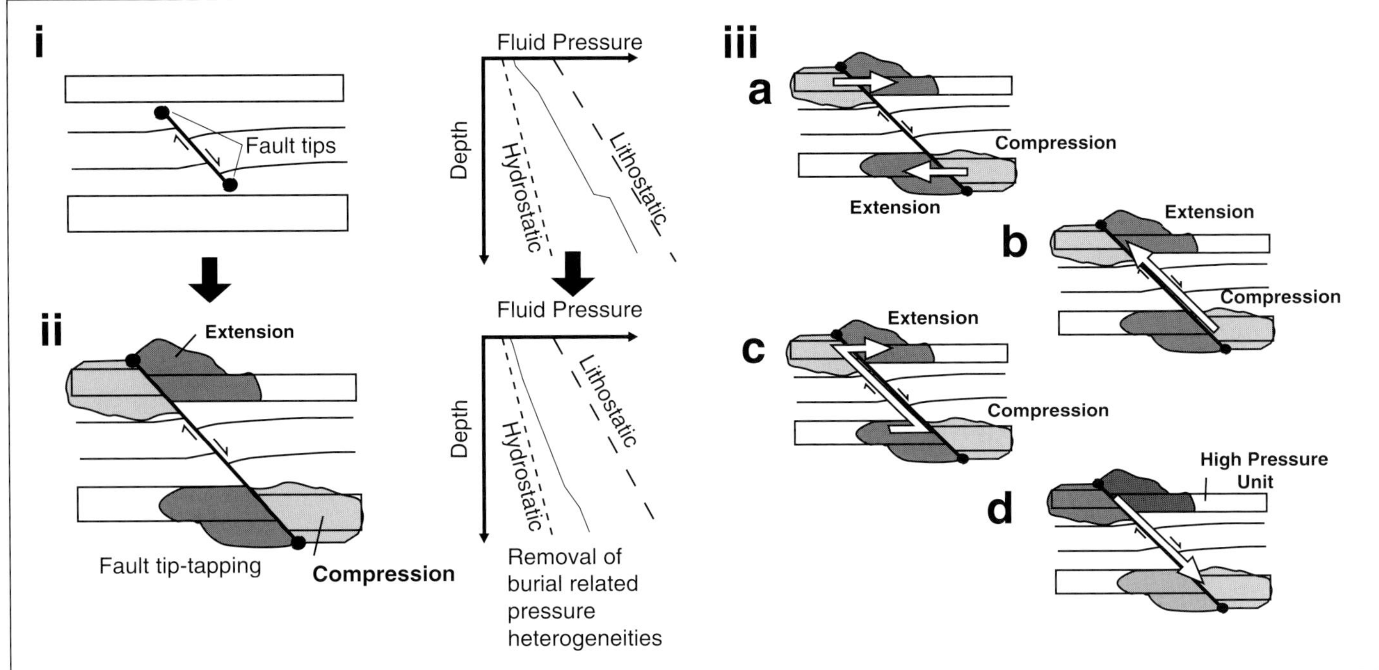

Figure 2. Fluid flow elements induced by fault activity on a normal fault. (i) and (ii) illustrate the distribution of compression and extension zones induced at the tip zones by the faulting event. Light-shaded areas are fluid sources (compression) and dark-shaded areas are fluid sinks (extension). The resulting fluid flow is likely to be asymmetric. (iii) shows the different flow patterns that may arise: (a) sub-horizontal flow between the hanging wall and footwall at each tip; (b) from the hanging wall lower tip to hanging wall top tip; (c) the hanging wall lower tip compression zone may develop pore throat pinching during compression which reduces permeability (i.e. it is possible that as pressure decays there may be periods when flow from the footwall lower extensional tip dominates; (d) down fault flow from an area of high pressure to a lower pressure unit below.

The decay of these fluid pressure gradients also will be affected by the permeability evolution of the fault rocks along the network of fractures and faults connecting pressure gradients. Some of the processes that may be involved in this permeability reduction are discussed below. A model is developed later, under "Fault Displacement Patterns, Activity, and Array Evolution," which emphasizes the interaction of these dynamic faulting processes with the exact displacement distribution, the initial permeability characteristics of the stratigraphic units affected, and the evolution of the fault rocks.

During and following the aftershock period, some fault zones are also characterized by aseismic creep. Initially, the aseismic creep or afterslip rate decreases with time and provides the background to the aftershock activity. The afterslip period usually lasts for 1–2 years for quakes with magnitudes above M4–M5 (Wesson, 1987, 1988) and the displacement decay is proportional to log t, where t is the time after the earthquake (Scholz, 1990; Wesson, 1988). Afterslip may occur as continuous slip or as discrete events where the slip rate accelerates rapidly at first and then decreases slowly. The average displacement rate for such events is usually less than 1 mm hr^{-1} and the events have a duration generally measured in hours or days. The development of episodic afterslip is thought to be associated with the presence of asperities or strong patches of fault rock in the fault zone (Wesson, 1987, 1988). Steady state or episodic aseismic slip may also dominate the displacement accumulation on faults or fault segments which are not characterised by large earthquakes. In these cases, the steady state creep or migrating creep events are thought to be a response to loading from quasi-plastic deformation on deeper parts of large fault zones (Wesson, 1988). The deformation mechanisms involved in creep events are unknown. Recorded data suggest that the fault zone material experiences a power law creep with a stress exponent of 1.2–2 (Wesson, 1988). The information on the 1975 Oroville quake in California reported in Wesson (1988) suggests that shear strain rates during such events are approximately 10^{-7} s^{-1} to 10^{-8} s^{-1}. In terms of fluid flow it is important to determine the mechanisms involved in such creep events, as packets of fluid may be able to migrate with the creep waves. Although dilation is likely during these events, a quantitative assessment of the dilation involved is not possible at present.

Between creep events or at the end of activity on a

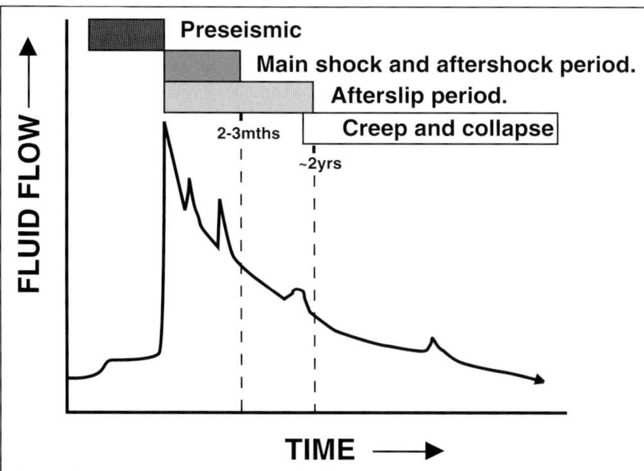

Figure 3. The idealized permeability evolution of a fault zone. Three stages can be recognized, each linked to the earthquake cycle (see text for details).

fault, the slower accumulation of displacement will be by time-dependent processes in the fault zone. These processes induce fabric modifications in the rocks generated and affect the potential for fluid flow. (A more detailed discussion, together with some examples of these aging/healing processes is given in the next section.)

In summary, the permeability evolution of the fault zone can be divided into three stages (Figure 3): a preseismic period of fluid redistribution when flow into the zone is possible; a co-seismic period when the main shock and aftershocks maintain the possibility of enhanced flow; and a postseismic period when the porosity and permeability of the fault zone can decay or collapse. Clearly, a large range of deformation processes and fluid flow behavior are possible in fault zones. The migration of fluid packets during different faulting events has been modeled recently by Stark and Stark (1991) using percolation theory. These models are important in constraining periods and general patterns of fluid flow, but cannot be used to identify particular patterns. Such identification requires markers that enable the tracing of actual fluid flow events. Fault rock analysis combined with diagenetic studies of fault zones can help to identify markers that record the history of fluid flow events.

FAULT ROCK EVOLUTION AND FAULT ZONE DIAGENESIS

The detailed understanding of fault-related fluid flow, including assessment of fluid composition and source together with the determination of the timing of flow events, requires an appreciation of the development of fault rocks. The combination of microstructural studies (to assess the evolution of the deformation fabric), and diagenetic analyses (to constrain the sequence of dissolution, precipitation, and fluid chemistries) that have affected fault zones provides a productive route for future studies. The large amount of new information (seismic, core, etc.) collected during recent oil exploration and production, particularly over the past 10 years, provides an important data base for such integrated studies. However, very few studies have attempted to apply the "standard" techniques used to assess diagenesis away from faults to the fault zones. Pioneering studies that have attempted such integrations have already highlighted the possible links between fault zones and diagenetic processes. For example, Flournoy and Ferrell (1980) have reported a concentration of mineral cements around faults. In addition, Porter and Weimer (1982), Jourdan et al. (1987), and Lee et al. (1989) present evidence for fault-controlled migration of fluids with "exotic compositions" or temperatures. A very detailed analysis of cement petrography by cathodoluminescence and fluid inclusion studies by Burley et al. (1989) provides evidence for repeated introduction of hot fluids along faults in the Tartan field of the North Sea.

Despite these advances, an exact appreciation of the interaction between fluid flow and faulting will depend upon an understanding of what controls the ability of a fault to act as a focus for fluid flow or as a barrier. This ability is determined by the evolution of the porosity and permeability of the deformed rocks in the fault zone and the adjacent volume of damaged rock. Transient or cyclic changes in the strain rate, stress variations, fluid pressure and deformation mechanisms that operate during faulting will control the evolution of the porosity and permeability in the fault zone. The permeability of the fault zone is likely to vary from a high dynamic permeability value during periods of fault activity, down to a lower value between high strain rate events. The permeability changes together with fluid pressure gradients induced by faulting will control the resulting fluid fluxes. Therefore understanding the range of deformation mechanisms which may be involved in faulting and fault rock evolution is crucial to the understanding of fluid flow.

The nature of the rocks that are generated on faults depends upon the range of deformation mechanisms which operate. Those involved in shallow level faulting can include frictional sliding, fracture, and cataclastic flow; diffusive mass transfer; and crystal plasticity (see reviews by Sibson, 1986a; Groshong, 1988, Mitra, 1988; Knipe, 1989, 1992). Although continuous displacement (steady state aseismic creep) is known in fault zones (see above), faulting usually involves transient changes in the rate of straining or the rate of displacement accumulation (Sibson, 1980; White and White, 1983; Knipe, 1989). Thus a number of deformation processes, which dominate at different strain rates and stress levels, can contribute to the development of fault rocks in individual fault zones (Knipe, 1990). Because different deformation mechanisms produce different microstructures and fabrics as well as porosity and permeability histories, the exact deformation mechanism path followed will control the fluid flow potential of faults, both in terms of flow across and along fault zones. The different faulting processes outlined above will therefore be associated with the development of different fault rocks. That dilation is involved in these processes is not questioned, but the pattern, magnitude, and preservation of dilation associated with the development of different fault rocks is poorly understood.

The **preseismic (α, β) stage** of the earthquake cycle will be associated with the development of microfracture and fracture arrays which link during the period before the main event. Such damage may form some of the fracture arrays which surround the main rupture surface, i.e some of the fractures in the fault zone may develop early but be used during later events. The **main seismic (γ) phase** results in rupture propagation and displacement and leads to the generation of frictional wear gouges and breccias where the cataclastic flow is characterised by grain rolling, frictional sliding, and comminution by brittle fracture (House and Grey, 1982; Aydin and Johnson,1983; Rutter et al., 1986; Blenkinsop and Rutter, 1986; Sammis et al.,1986). Dilation associated with the main (β) phase is instrumental in the development of implosion breccias associated with collapse into dilation sites (Sibson, 1986b). Such dilation may also result in the rapid loss of fluid pressure, the precipitation of cements and the generation of matrix (cement) supported breccias (Sibson, 1986b; Sibson et al., 1988). The aftershocks associated with the **postseismic** (δ) **phase** will involve a range of processes similar to those involved in the main rupture event.

Fault rock evolution during the displacement in the (δ + α) phases involving aseismic creep (steady state, decelerating afterslip, or discrete creep events) is responsible for the collapse of the high permeability window (i.e. the aging or sealing of the fault zone) and deserves special attention. Porosity and permeability reduction in the fault zone during the late stages of activity on a fault may be related to a range of processes (Knipe, 1992). The changing microstructures (grain size, grain contact areas) of the fault rocks during this period can also alter the rheology of the fault zone and the susceptibility to deformation by different mechanisms (Rutter and White, 1979; White and White, 1988; Knipe, 1989,1990; Lloyd and Knipe, 1992).

Porosity and permeability reduction of the fault zone after the main event are essentially the processes responsible for fault sealing. Knipe (1992) recently separated seals into three broad categories: collapse seals, cement seals, and juxtaposition seals. The first two of these are directly related to the fault rocks developed during displacement and are particularly dependent on the late stage evolution of the fault rocks. Collapse seals develop where the pore volume is decreased as a result of grain reorganization produced by grain boundary (frictional) sliding, grain deformation, fracturing, and dissolution, and include smear seals. Cement seals form where the reduced flow across or along the fault is achieved by the precipitation of minerals in and around the fault zone.

Knowledge of the pore geometry and sizes during the production of the different fault rocks described above is needed if quantification of the flow processes or entry pressures of fault zones are to be used to model drainage development or the evolution of seals (Berg, 1975; Schowalter, 1979; Downey, 1984; Watts, 1987). Very few studies of progressive changes in pore throat size, porosity, and permeability associated with the development different fault rocks have been reported (see Pittman, 1981; Knipe, 1992; Knipe et al., 1992b).

Examples of Late Stage Evolution in Fault Zones and Porosity/Permeability Characteristics of Fault Rocks

Porosity Collapse Seals Associated With Clay Smears

Figures 4a and 4b show the microstructure of a smear seal developed from an impure sandstone containing ~10–15% phyllosilicates. The fault rock evolves by the production of anastomosing zones of concentrated phyllosilicates and fractured quartz grains. The phyllosilicate-rich zones contain aligned, detrital or pre-deformation diagenetic grains and the open microporosity of the pore filling phyllosilicates from outside the fault zone is collapsed. The example also shows that smear seals can develop in the absence of shale layers and makes questionable the validity of assessing seal distribution on sandstone/shale ratios and displacement magnitudes alone.

The development of the anastomosing zones in the clay smears shown in Figure 4 are very similar to the fabrics present in faulted claystones sampled from accretionary prisms by the DSDP/ODP and the ALVIN submersible. The recent analysis of these structures (Knipe, 1986a and b; Moore et al., 1986) suggests that the associated fluid migration may be episodic and associated with the migration of dilation-displacement-collapse events along the fault zones. In these DSDP examples from accretionary prisms, high fluid pressures are known to be present along the main thrust faults and high permeability along such phyllosillicate rich faults may be associated with high fluid pressure maintaining the high dynamic porosity and permeability, which would

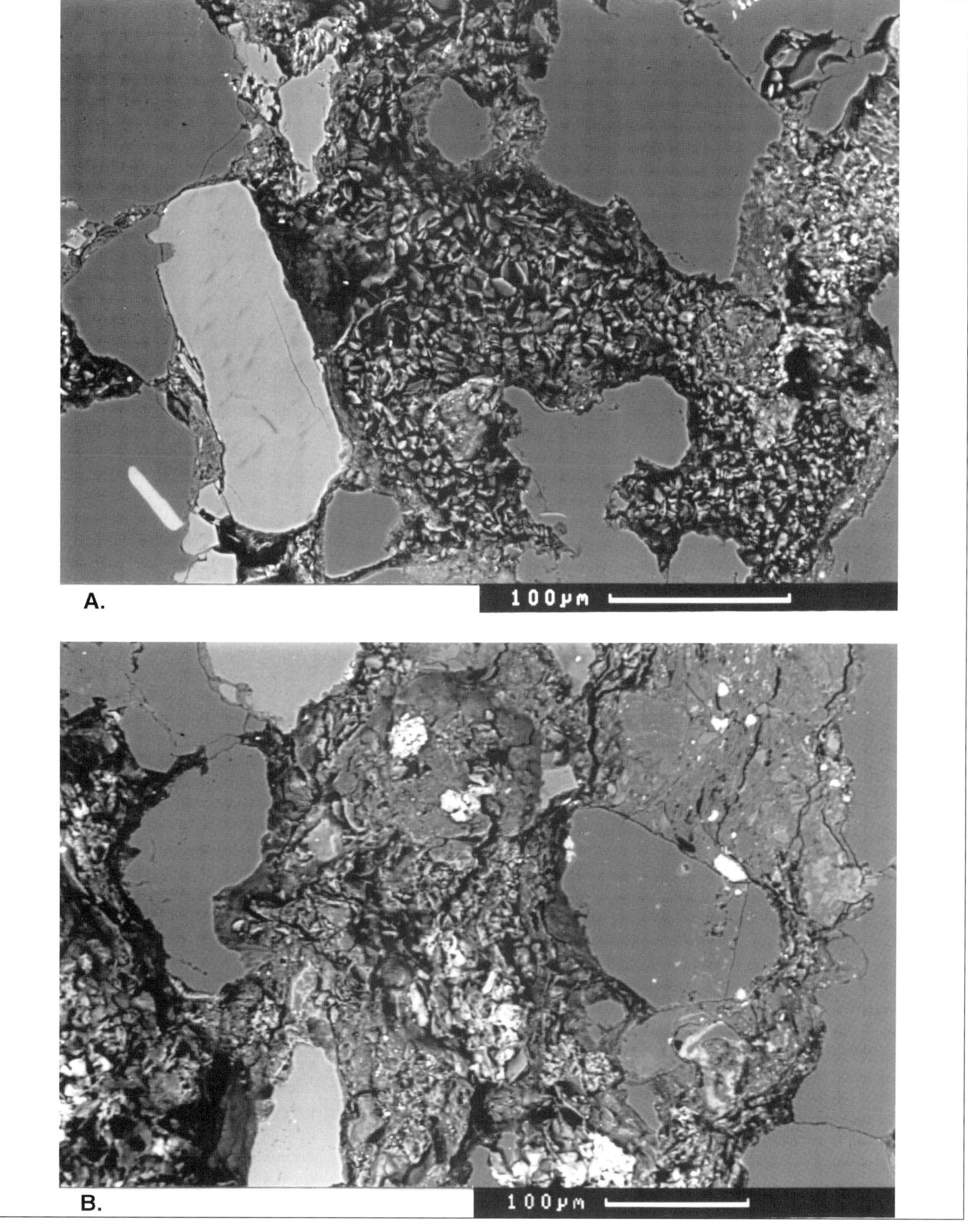

Figure 4. Microstructure of a smear seal developed from an impure sandstone initially containing ~10-15% phyllosilicates. (A) Shows the undeformed sandstone, with its characteristic open framework of clay phases in the pore space. (B) Shows the internal structure of the smear seal developed. Note the highly compacted domains parallel to the fault plane (fz).

Figure 5. Microstructures of cataclastic fault zones where late stage dissolution and precipitation processes contribute to the permeability reduction. (A) Shows the edge of a fault zone and the grain size reduction within the fault. (B) Illustrates the fine-grained nature of the fault zone, and (C) shows the new, undeformed quartz overgrowths present.

Figure 6. Example of a cemented fault zone where the carbonate cement phase forms the bulk of the fault zone volume. The dark fragments are quartz fragments. Note the concentration of fragments in a band from top right to bottom left.

decay after activity on the fault ceased (see Knipe et al., 1991, figure 4).

Seals Associated With Porosity Collapse By Dissolution

Figure 5 shows an example where the late stage collapse processes in a cataclastic fault zone are associated with dissolution and precipitation of cement phases. The importance of this diffusive mass transfer (DMT) or pressure solution mechanism (Rutter, 1976, 1983; Gratier and Guiguet, 1986; Spiers and Schutjens, 1990) to the late stage evolution of fine-grained cataclastic fault zones has been recognized by a number of workers (Rutter et al., 1979; Stel, 1985; Knipe, 1989, 1990, 1992; Lloyd and Knipe, 1992) and has been suggested as contributing to the stick-slip behavior of faults by Angevine et al. (1982). Knipe (1992) lists three situations where enhanced DMT in the fault zone may be produced by a decrease in the strain rate: (1) "rest" periods between transient faulting events; (2) in the changing pattern of fault activity on a larger scale, where during the fault array evolution, a new fault takes up the displacement and early formed faults become inactive; and (3) at the end of tectonic activity evolution in the area. In the first case, the compacted cataclasite is likely to be fractured by subsequent faulting events, while in the other cases, the compaction will continue and may also be enhanced or driven by sediment loading rather than by tectonic extension.

In this discussion, the collapse in porosity and permeability during this slow (decreasing) displacement rate can represent the final stage in the decay of the high permeability window associated with the fault. At present, a quantitative assessment of this phase is difficult given the large range of variables which influence the rate of DMT (see reviews by Gratier and Guiguet, 1986; Spiers and Schutjens, 1990). If the DMT creep rates in fine grained cataclastic quartz aggregates at temperatures of 150–200°C are approximately 10^{-10} s^{-1} to 10^{-11} s^{-1} for a grain size of ~10 µm (see Rutter, 1976; White, 1976) and shear strains of about 0.5 are needed to significantly reduce the porosity of the fractured aggregate, then periods of 100 - 1000 years are needed for this collapse. These estimates serve only to illustrate that the periods when fluid flow can use the pore system of the compacting cataclastic rock in the fault zone can be at least one or two orders of magnitude longer than those during which afterslip operates (Figure 3). It is interesting to note that the stress dependence of unity associated with DMT creep rate is lower than the 1.2–2 indicated from analysis of aseismic creep events, where the duration is in the time range 1 hr to 2 yrs (Wesson, 1988). This suggests that afterslip and associated creep events are not simply a DMT phenomenon, although they may be relaxing, with time, toward domination by such a mechanism.

The corrosion and dissolution of material in the fault zone by external fluids entering a cataclastic fault rock can have an important impact on its porosity and permeability reduction in a fault zone. Examples where widespread carbonate cement dissolution creates secondary porosity are common in the literature and the introduction of the fluid responsible often has been linked to faulting (e.g. Burley et al., 1989). Where dissolution of phases is concentrated in a fault zone, both the permeability and strength of the fault rock can be affected, and while initially the secondary porosity created may enhance permeability, large scale dissolution may weaken the rock and induce collapse. The dissolution of feldspars in fault zones may be particularly important in this context, as the clay phases produced may collapse to form smear seals during later compaction, fluid pressure cycles, fault reactivation, and/or reservoir production. Clearly the extent of such a process depends upon the initial feldspar content and the migration pathway used by corrosive fluids in the fault zone.

Seals Associated With Cement Precipitation

Precipitation of cements in fault zones may be linked to (a) changes in the fluid pressure associated with the faulting event, (b) changes in the temperatures experienced by the fluid during migration along the fault zone, (c) changes in the solubilities which arise from fault-induced mixing of fluids with different compositions, or (d) by the fluid-rock interactions involved in dissolution-precipitation processes of diffusive mass transfer. It is clear that these precipitation processes are not confined to the late stage evolution of a fault zone and some (especially a, b, and c) can be induced by the early parts of the earthquake cycle of a faulting event. The first three of the precipitation causes listed emphasize that the fault zone can act as an open system, whereas in the last case, the fault zone may be considered as a closed system and not be associated with large scale fluid movement or influx. This ability of fault zones to alternate between an open and a closed system is a consequence of the episodic or cyclic nature of the faulting processes and has become a common element in discussions of fault-fluid interactions (see reviews by Etheridge et al., 1984; Sibson et al., 1988; Carter et al., 1990).

Figure 6 shows an example of a cemented fault zone where the cement phase forms the bulk of the fault zone volume. Such breccias involve a large amount of local dilation and may initiate where wall-rock asperities and strengths cause or preserve dilation or where fluid pressure magnitudes can maintain the dilation.

Cementation events generated by fault activity are not necessarily confined to the fault zone where fault rocks have been produced. The host rock adjacent to the fault zone can also be cemented, and complex cement distributions in sandstone reservoirs have been recognised and assigned to fluid entry along faults (e.g., Flournoy and Ferrell, 1980; Burley et al., 1989). Border zone seals may develop because of a linked fracture array surrounding the fault zone or may be related to preferential fluid penetration along high permeability zones. The importance of border zone seals was discussed by Knipe (1992), who pointed to their potential strength relative to the fault rocks present along the main fault. Reactivation of the fault during later events may fail to breach the strengthened and cemented zone and maintain the seal. Border seals are also important for channeling the fluid migration associated with later events along fault zones, which can result in fluids bypassing particular sealed horizons (see below).

The examples of the late stage changes in the porosity of fault zones discussed above involve the generation of collapse seals or cement seals by different processes (grain boundary sliding, fracturing, dissolution, corrosion or alteration, and cementation). Although many seals involve a combination of these processes, it is often possible to identify one that dominates. These processes of porosity reduction will proceed at different rates and thus will each allow the high permeability window associated with faulting to remain open for different periods. Alternatively, the sealing properties of the fault can be viewed as developing at different rates. In the development of clay smears, maintaining a dilation level or porosity in the grain aggregate high enough to allow fluid flow is particularly dependent upon maintaining a high fluid pressure. Reduction in pressure is likely to cause a drastic collapse in permeability. A similar reduction in fluid pressure is likely to induce a smaller permeability reduction in a quartz aggregate because of the higher strength of the grain aggregate involved. In the case of the quartz aggregate, the permeability reduction will be more protracted, controlled by DMT processes. Sealing by cementation may either be linked to a rapid pressure drop following the faulting event, inducing a rapid permeability reduction, or it may be controlled by a slower dissolution reaction rate (Figure 7). Despite the importance of these different behaviors to both fluid flow and to assessment of the sealing properties of faults, very little information is available at present to quantify these effects.

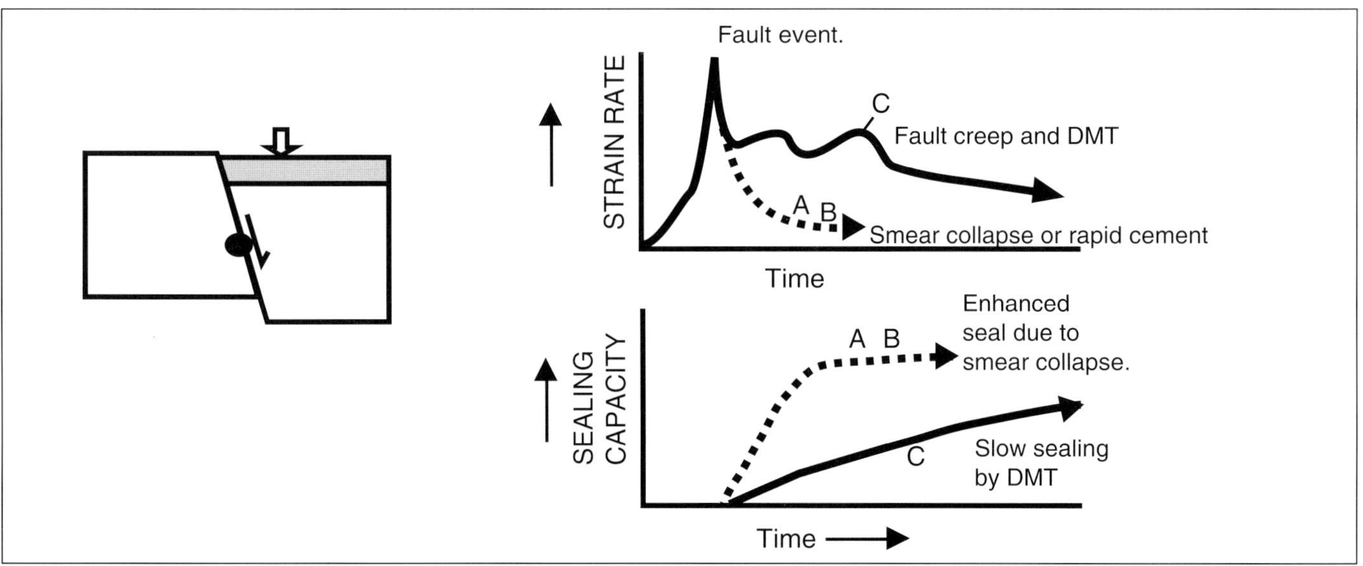

Figure 7. Relative rates of different porosity reduction processes that operate during the late stages of fault rock evolution after the faulting event shown in cartoon on the left. The diagrams compare the rate of strain rate decrease and the development of sealing capacity for three processes. The increase in the sealing capacity of the fault may be rapid when either (A) the collapse in phyllosilicate rich smears, or (B) rapid cementation takes place compared with (C) slow DMT collapse in a quartz-rich gouge. Each of these porosity/permeability collapse processes allows the high permeability window associated with faulting events to remain open for different periods.

Timing of Fault Rock Evolution, Fluid Flow, and Diagenesis

The identification of which combination of the above faulting processes are involved in the porosity/permeability development of fault rocks, as well as where and when pathways occur, are critical to fluid flow assessment. Knowledge of the timing of fault activity in a basin is a prerequisite to detailed modeling of basin hydrodynamics. Recent studies combining petrological, stable isotope, fluid inclusion, and absolute dating techniques have been used to identify timing of fluid migration in basins (Boles, 1987; Land and Fisher, 1987; Burley et al., 1989; Lee et al., 1989; Gluyas et al., this volume).

Fewer studies have attempted to test the common assumption that all faults in the array are active together to act as migration pathways during basin extension. Fault rocks contain information which can be used to evaluate the timing of activity of different faults. Detailed analysis of diagenetic sequences inside, outside, and adjacent to fault zones allows the timing of fault activity relative to the diagenetic sequence in an area. Such comparisons also allow the identification of by-pass events where the fluid flow is restricted to a fault zone and does not penetrate into particular units to induce diagenetic changes (Figure 8). Figure 9 shows diagenetic histories of fault rocks along two faults in the Morecambe Bay gas field (Irish Sea) and compares these to the general diagenetic history in the Triassic sandstone reservoir (Knipe et al. 1992a). In this example, specific faults were active at different times during the burial history of the reservoir. Assuming that all faults affecting the reservoir were active at the same time will overestimate the number of fractures available for migration. Assessment of fault population by assigning fractal numbers based on the finite number of faults present in an area may also obscure some aspects of the fault history as different fractal numbers may be characteristic of different stages of the fault array evolution.

Modeling reservoir evolution therefore requires assessing both the timing of fault activity and the type of lithologies affected by the individual events, both in terms of the tapping of developing pore fluid chemistries and of pressure compartments. Only when the right combination of fault event magnitude, fracture geometry, and fracture linking allows creation of a pathway between source and sink volumes and the production/preservation of permeable fault zone fabrics (in the fault zone and in adjacent fracture arrays) will extensive fluid flow be possible. The situation will be even more critical for hydrocarbon reservoir evolution as the additional factors of hydrocarbon maturation and the role of the buoyancy force as an aid to fault zone fluid migration or leaking needs to also be considered. In addition, influx of hydrocarbons to a fault zone can shut off diagenesis in the fault zone and postpone or prevent cement growth. This provides a mechanism of delaying fault seal and maintaining a secondary hydrocarbon migration pathway (but may cause seal by tar or asphalt formation).

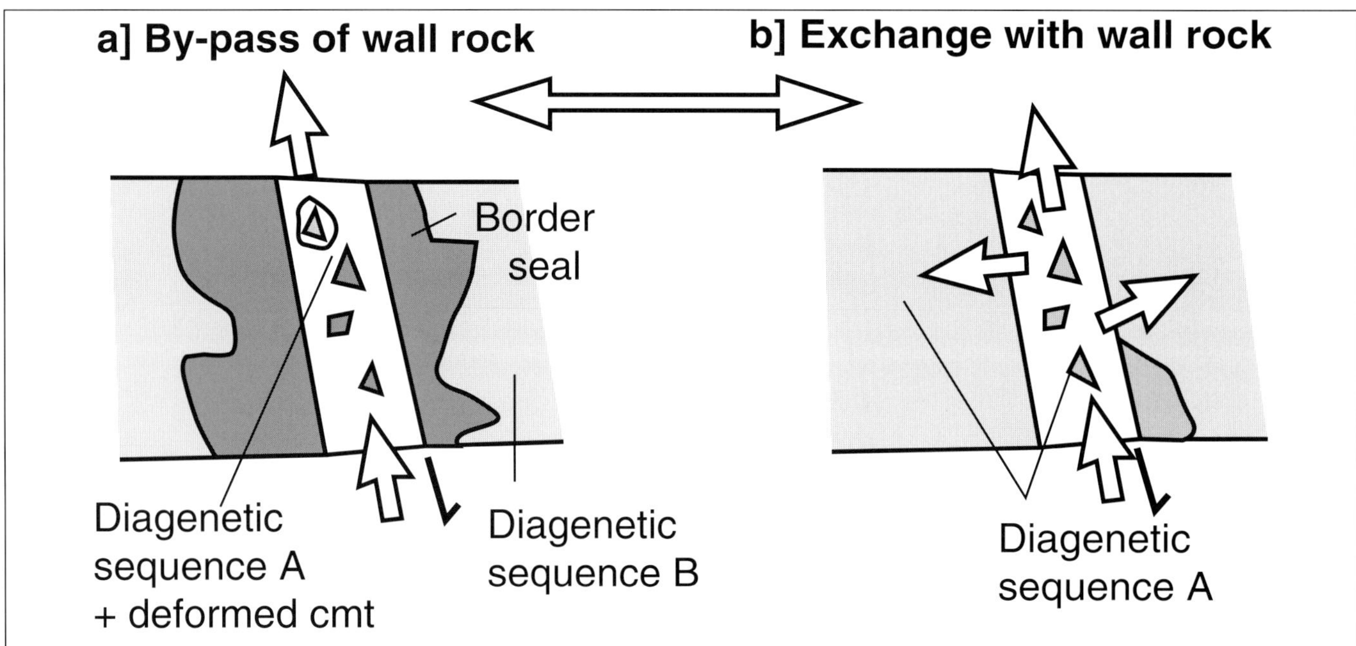

Figure 8. Review of criteria used to identify bypass events where fluid flow is restricted to a fault zone and does not penetrate along particular units/horizons to induce diagenetic changes. In the by-pass case shown in (a) deformed cements and a different diagenetic sequence within the fault zone and outside the fault zone indicate the by-pass. Where exchange between the fault zone and the wall rocks takes place, the diagenetic sequence is the same as shown in (b).

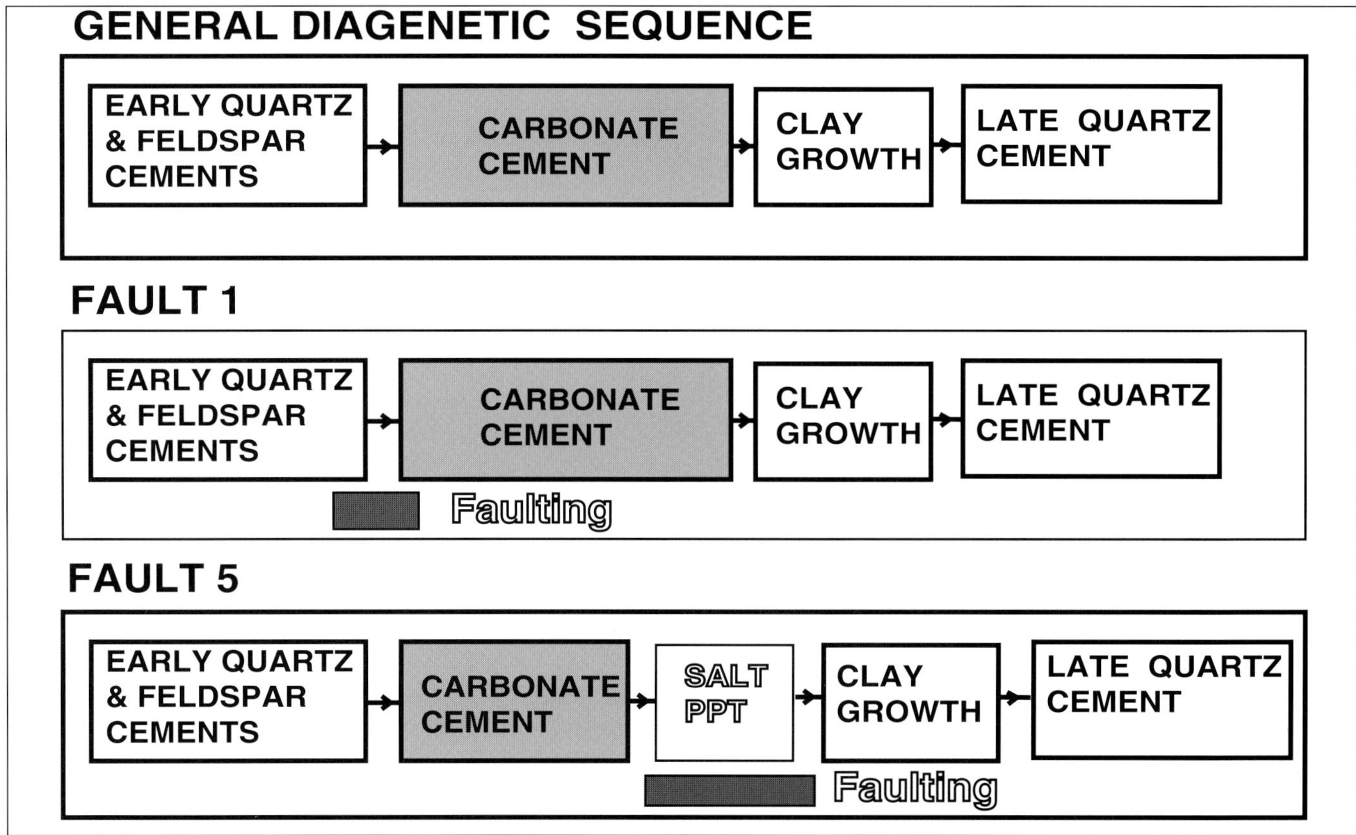

Figure 9. Diagenetic histories of a number of faults affecting Triassic Sandstones from the Morecambe Bay Gas Field (Knipe et al., 1992). In each of the faults shown the deformation associated with faulting affects a different part of the general diagenetic sequence. Activity on fault 1 is before the carbonate cement, while fault 5 is after the carbonate and associated with the invasion of salt.

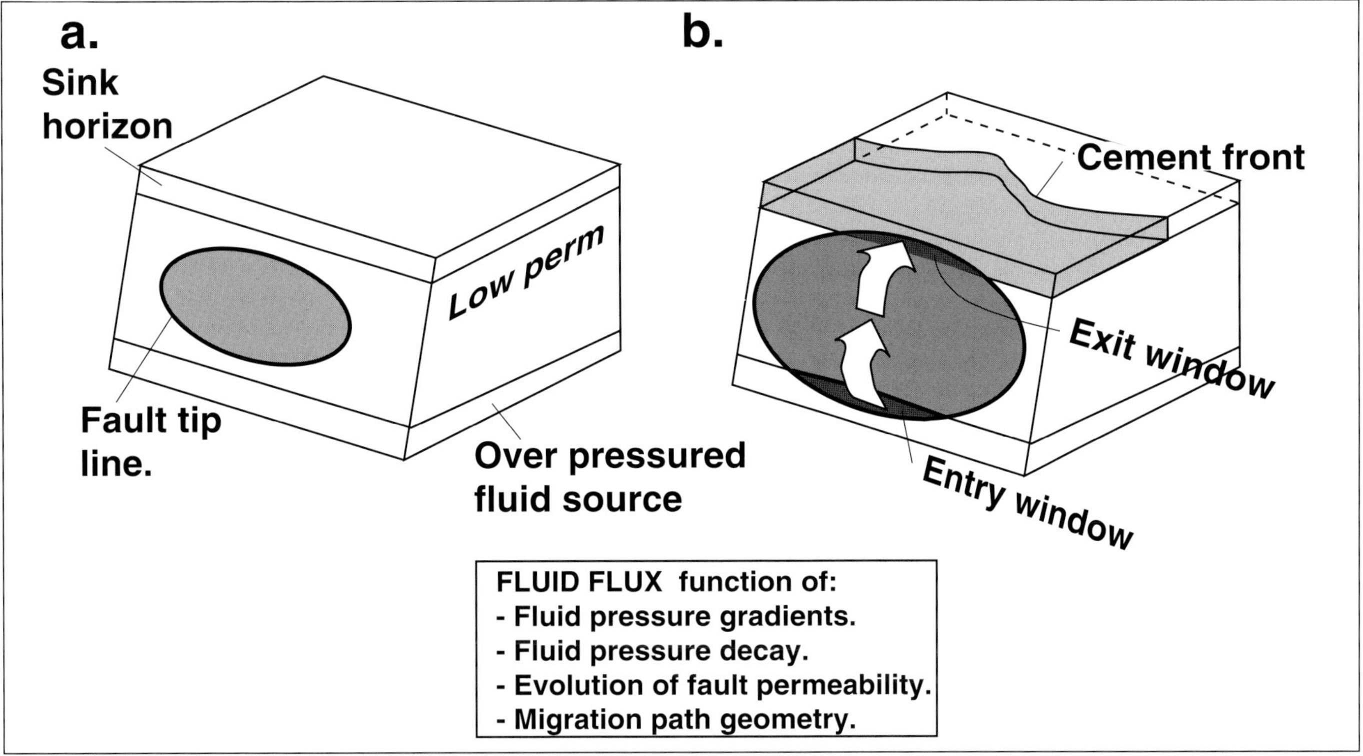

Figure 10. Review of the likely controls on the process of fault propagation and fluid flow. (a) Shows the location of a fault tip line before a faulting event where the fault is above an overpressured fluid source but below a possible sink horizon. (b) Shows the linking of the source and sink horizons after a faulting event and the development of a cemented zone in the sink horizon. The fluid flux (and cement evolution) will be dependent on (i) the pressure gradients induced by the faulting event, (ii) the rate of decay of that gradient, (iii) the permeability of the fault zone and its decay, and (iv) the location and size of source and sink areas where fluids enter and escape from the fault zone.

FAULT DISPLACEMENT PATTERNS, ACTIVITY, AND ARRAY EVOLUTION

Fault activity and the evolution of a fault array will control the distribution of flow paths (and therefore drainage patterns) by constraining the development of links and barriers between potential carrier horizons. It is therefore necessary to consider progressive changes in the communication and juxtapositions of different lithologies with different evolving pore fluid chemistry and pressures, as a fault array grows.

In a series of papers, Walsh and Waterson (1987, 1988, 1989) have presented a large dataset on the size and shape of fault planes and the associated displacement accumulation patterns. The data illustrate how isolated faults are usually elliptical in shape and how the maximum displacement in the centre of faults is related to the fault plane dimensions. More recently, Walsh and Waterson (1991) have incorporated projection of displacement data onto a single fault plane to illustrate the geometric coherence of fault displacements in volumes containing a number of faults. Allen (1988) highlighted the use of fault plane maps in the assessment of juxtapositions created by faulting. Future analysis of fluid flow requires integration of these two approaches with an evaluation of fault rock evolution patterns on the fault planes and a consideration of the fluid characteristics of lithologies affected as the faults grow and link (Knipe, 1992; Knipe et al., 1992b). In addition to the pressure gradients induced by the faulting event, the rate of decay of that gradient, and the permeability evolution of fault zones discussed above, fluid flux will also be dependent on the location and size of source and sink areas where fluids enter and escape from the fault zone. Where the entry and exit windows are produced in the tip zones of a propagating fault, fluid flow behavior is likely to be dependent on the exact location of fault tip lines with respect to the position of the source and sink horizons (Figures 10 and 11). In the ideal situation where, during a single event, the fault propagates down into an overpressured source unit and up into a high permeability sink, then given constant porosity/permeability properties along the fault and a similar array and density of tip zone fracturing, fluid flow will be relatively simple. On the other hand, where these ideal conditions are not met, a different flow behavior will result. Where propagation produces a large area in the sink horizon, rapid fluid expulsion, pressure decay, and fault locking are

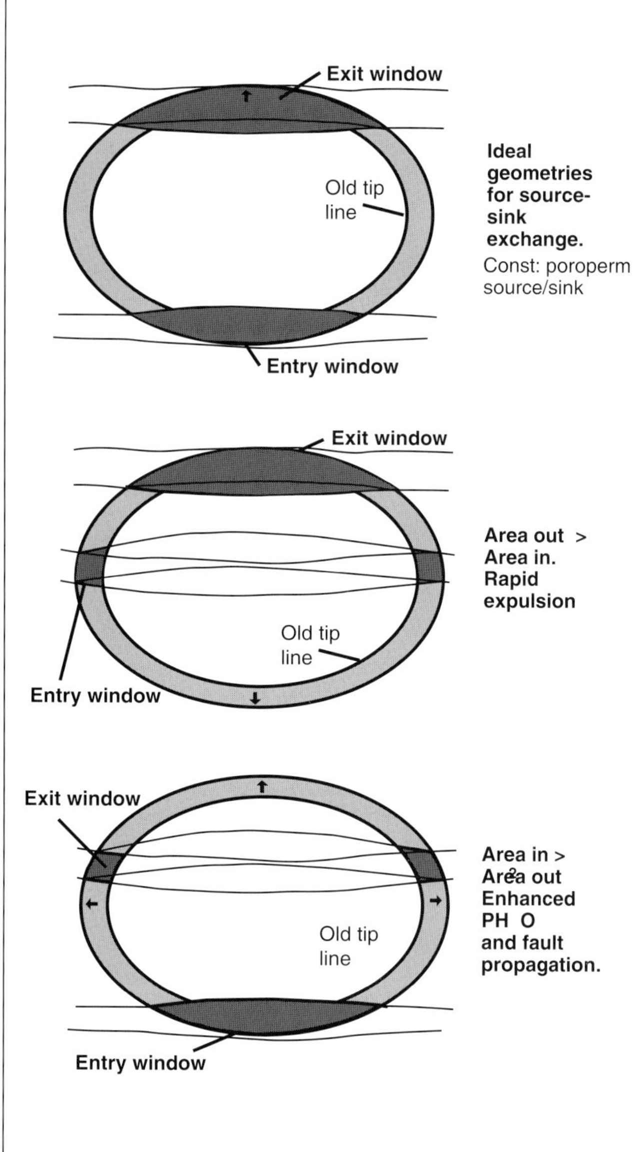

Figure 11. Fluid flow behavior and exact location of the fault tip line with respect to the position of the entry and exit windows in the source and sink horizons involved in flow. Note in the examples shown that the old fault plane is assumed not to connect with juxtaposed lithologies because of border zone cement seals or juxtaposition with low permeability horizons. (a) Illustrates a simple geometry where the entry and exit windows have similar areas, and the porosity/permeability characteristics of the source and sink are also similar. (b) Shows a situation where the entry window dimensions are much more restricted than the exit window. This may lead to a rapid loss of fluid pressure and fluid expulsion from the fault zone. (c) Illustrates a case where the entry window allows rapid access to the fault zone but expulsion from the fault plane is restricted by the small exit window. This situation may lead to the rapid reestablishment of high fluid pressure in the fault zone.

likely. If the entry window is large and the escape route or exit window small, fluid flow and pressure decay may be restricted and fault propagation and/or continued displacement enhanced (Figure 11).

Another important aspect of the exact position of the fault plane tip line, its growth, and the diagenesis associated with faults is the stratigraphy and the associated pore fluid chemistries intersected by the fault. The impact of a fault on the diagenesis of a particular horizon is often assessed from the finite displacement magnitude at that horizon level. Small faults shown on reservoir level maps are often considered to have little impact. However, if the small displacements are located on the tip of a larger fault, then horizons intersecting other parts of the fault plane may introduce fluids that can drastically affect its sealing properties (see Knipe et al., 1992b, for an example). This discussion points to the need for the assessment of the evolution of individual fault planes and consideration of the likely 3D pattern of juxtapositions, entry-exit areas, border zone cement concentrations, and bypass events during fault propagation when reservoir diagenesis models are constructed.

CONCLUSIONS

The interaction of faulting processes, fluid flow, and diagenesis presented above highlights the complex nature of fluid migration. The ability of faults to act as both high permeability pathways and as barriers has been emphasized. In addition, the importance of recognizing the different fluid flow behaviors likely during the different stages of the faulting events has been discussed. The time period during which a fault can operate as an effective migration pathway is controlled by the permeability and porosity evolution of the fault zone, which, in turn, depend on the interaction between the deformation mechanisms operating within it, the fluid pressure history, the spatial and temporal pattern of dilation, and the diagenesis induced by fluids using the fault zone. The different processes that are involved in fault rock development (and particularly those that can close the fault as a high permeability window) have been reviewed. There is a particular need to assess the mechanisms and kinetics of the late-stage porosity reduction associated with the decreasing displacement (death) of the fault zone, i.e., the sealing mechanisms of the fault, and to integrate such studies with an assessment of the geometrical evolution of the fault, in terms of the juxtapositions and intersections with different lithologies during tip-line expansion and fault zone propagation.

Despite the obvious complexities of fault-related fluid flow processes, integration of the recent advances made in the separate fields of faulting mechanisms, fault rock evolution, fault plane geometry, diagenesis, and fluid flow do provide a platform for future progress.

ACKNOWLEDGMENTS

The author would like to thank the organizers of the Geological Society Conference for their invitation and patience during the production of the manuscript. British Gas and BP are thanked for permission to publish some of the data. Tim Bevan, Stephen Hay, and Andrew Robinson are thanked for comments on the first draft.

REFERENCES CITED

Allen, U.S., 1988, Model for hydrocarbon migration and entrapment within faulted structures: AAPG Bulletin, v.73, p.803-811.

Angevine, C.L., D.L. Turcotte and M.D. Furnish, 1982, Pressure solution lithification as a mechanism for the stick-slip behavior of faults: Tectonophysics, v.1, p.151-160.

Aydin, A. and A.M. Johnson, 1983, Analysis of faulting in porous sandstones: Journal of Structural Geology, v.5, p.19-35.

Berg, R.R., 1975, Capillary pressure in stratigraphic traps: AAPG Bulletin, v.59, p.939-956.

Blenkinsop, T.G. and E.H. Rutter, 1986, Cataclastic deformation of quartzite in the Moine Thrust Zone: Journal of Structural Geology, v.8, p.669-682.

Boles, J.R., 1987, Six million year diagenetic history, North Coles Levee, San Joaquin Basin, California, *in* J.D. Marshall, ed., 1987, Diagenesis of Sedimentary Sequences, Geological Society Special Publication 36, p.191-200.

Booker, J.R., 1974, Time dependent strain following faulting of a porous medium: Journal of Geophysical Research, v.79, p.2037-3044.

Burley, S.D., J. Mullis and A. Matter, 1989, Timing of diagenesis in the Tartan Reservoir (UK North Sea): constraints from combined cathodoluminescence microscopy and fluid inclusion studies: Marine and Petroleum Geology, v.6, p.98-120.

Carter, N.L., A.K. Kronenburg, J.V. Ross and D.V. Wiltschko, 1990, Control of fluids on deformation in rocks, *in* R.J.Knipe and E.H.Rutter, eds., Deformation Mechanisms, Rheology and Tectonics. Geological Society Special Publication 54, p.1-13.

Downey, M. W., 1984, Evaluating seals for hydrocarbon accumulations: AAPG Bulletin, v.68, p.1752-1763.

Etheridge, M.A., V.J. Wall, S.F. Cox and R.H. Vernon, 1984, High fluid pressures during metamorphism and deformation: implications for mass transport and deformation mechanisms: Journal of Geophysical Research, v.89, p.4344-4358.

Flournoy, L.A. and R.E. Ferrell, 1980, Geopressure and diagenetic modifications of porosity in the Lirette field area, Terrebonne Parish, Louisiana: Gulf Coast Geological Association Transactions, v.30, p.341-345.

Frank, F.C., 1965, On dilatancy in relation to seismic sources: Revues of Geophysics, v.3, p.485-503.

Gratier, J.P. and R. Guiguet, 1986, Experimental pressure solution-deposition on quartz grains: the crucial effect of the nature of the fluid: Journal of Structural Geology, v.8, p.845-856.

Groshong, R.H., 1988, Low temperature deformation mechanisms and their interpretation: Bulletin of the Geological Society of America, v.100, p.1329-1360.

Harding, T.P., 1974, Petroleum traps associated with wrench faults: AAPG Bulletin, v.58, p.1290-1304.

Harding, T.P. and A.C. Tuminas, 1988, Interpretation of footwall (lowside) fault traps sealed by reverse faults and convergent wrench faults: AAPG Bulletin, v.72, p.738-757.

Harding, T.P. and A.C. Tuminas, 1989, Structural interpretation of hydrocarbon traps sealed by basement normal blocks and at stable flank of foredeep basins and at rift basins: AAPG Bulletin, v.73, p.812-840.

Hindle, A.D., 1989, Downthrown traps of the NW Witch Ground Graben, UK North Sea: Journal of Petroleum Geology, v.12, p.405-418.

House, W.M. and D.R. Gray, 1982, Cataclasites along the Saltville thrust, U.S.A., and their implications for thrust-sheet emplacement: Journal of Structural Geology, v.4, p.257-269.

Jourdan, A., M. Thomas, O. Brevart, P. Robson, F. Sommer and M. Sullivan, 1987, Diagenesis as the control of Brent sandstone reservoir properties in the Greater Alwyn area, East Shetland Basin: *in* J.Brooks and K.Glennie, eds., Petroleum Geology of North-West Europe, Vol. 2., Graham and Trotman, London, p.951-961.

Knipe, R.J., 1986a, Faulting mechanisms in slope sediments: examples from Deep Sea Drilling Project cores: *in* J.C. Moore, ed., Structural Fabric in Deep Sea Drilling Project Cores from Forearcs, Geological Society of America Memoir 166, p.45-54.

Knipe, R.J., 1986b, Deformation mechanism path diagrams for sediments undergoing lithification: *in* J.C. Moore, ed., Structural Fabric in Deep Sea Drilling Project Cores from Forearcs, Geological Society of America Memoir 166, p. 151-160.

Knipe, R.J., 1989, Deformation mechanisms - recognition from natural tectonites: Journal of Structural Geology, v.11, p.127-146.

Knipe, R.J., 1990, Microstructural analysis and tectonic evolution in thrust systems: examples from the Assynt region of the Moine Thrust Zone, Scotland: *in* D.J.Barber and P.G.Meredith, eds., Deformation Processes in Minerals, Ceramics and Rocks. Mineralogical Society Series, 1, p.228-258.

Knipe, R.J., 1992, Faulting processes and fault seal. *in* R.M. Larsen, ed., Structural and Tectonic Modelling and its Application to Petroleum Geology: NPF, Stavanger, p.325-342.

Knipe, R.J., S.M. Agar, and D.J. Prior, 1991, The microstructural evolution of fluid flow paths in semi-lithified sediments from subduction complexes: Philosophical Transactions of the Royal Society of London, A, 335, p.261-273.

Knipe, R.J., G.E. Lloyd, G. Cowan and I. Stuart, 1992a, Faulting and diagenesis in Sherwood Sandstones of the Morecambe Bay Gas Field: (in prep).

Knipe, R.J., A. Mitchell and S. Anderson, 1992b, Deformation processes and fault seal mechanisms: a combined microstructural analysis from the Ula Field, Central Graben, North Sea: (in prep).

Land, L.S. and R.S. Fisher, 1987, Wilcox sandstone diagenesis, Texas Gulf Coast: a regional isotopic comparison with the Frio Formation: *in* J.G.Marshall, ed., 1987, Diagenesis of Sedimentary Sequences. Geological Society Special Publication 36, p.219-235.

Langseth, M. and J.C. Moore, 1990, Introduction to special section on the role of fluids in sediment accretion, deformation, diagenesis and metamorphism in subduction zones: Journal of Geophysical Research, v.95, p.8737-8742.

Lee, M., J.L. Aronson and S.M. Savin, 1989, Timing and conditions of Permian Rotliegendes sandstone diagenesis, Southern North Sea. K/Ar and Oxygen isotope data: AAPG Bulletin, v.73, p.195-215.

Li, V.C., S.H. Seale, and T. Cao, 1987, Postseismic stress and pore readjustment and aftershock distributions: *in* R.L.Wesson, ed., Mechanics of Earthquake Faulting. Tectonophysics, v.144, p.37-54.

Lloyd, G.E. and R.J. Knipe, 1992, Deformation mechanisms accommodating faulting of quartzite under upper crustal conditions: Journal of Structural Geology, v.14, p.127-144.

McCaig, A.M., 1988, Deep fluid circulation in fault zones. Geology, 16, 867–870.

Mitra, S., 1988, Effects of deformation mechanisms on reservoir potential in Central Appalachian overthrust belt: AAPG Bulletin, v.72, p.536-554.

Moore, J.C., 1989, Tectonics and hydrogeology of accretionary prisms: role of the decollemont zone: Journal of Structural Geology, v.11, p.95-106.

Moore, J.C., S. Roeske, N. Lundberg, J. Schoonmaker, D. Cowan, E. Gonzalez and S. Lucas, 1986, Scaly fabrics from the deep sea drilling project cores from forearcs: *in* J.C.Moore, ed., Structural Fabric in Deep Sea Drilling Project Cores from Forearcs, Geological Society of America Memoir 166, p.55-74.

Nur, A. and J.R. Booker, 1972, Aftershocks caused by pore fluid flow?: Science, v.175, p.885.

Pittman, E.D., 1981, Effect of fault-related granulation on porosity and permeability of quartz sandstones, Simpson Group (Ordovician), Oklahoma: AAPG Bulletin, v.65, p.2381-2387.

Porter, K.W. and R.J. Weimer, 1982, Diagenetic sequence related to structural history and petroleum accumulation: Spindle Field, Colorado: AAPG Bulletin, v.66, p.2543-2560.

Reasenberg, P. and W.L. Ellesworth, 1982, Aftershocks of the Coyote Lake, California, Earthquake of August 6, 1979: A detailed study: Journal of Geophysical Research, v.87, B13, p.10,637-10,655.

Rutter, E.H., 1976, The kinetics of rock deformation by pressure solution: Philosophical Transactions of the Royal Society of London, A283, p.203-220.

Rutter, E.H., 1983, Pressure solution in nature, theory and experiment: Journal of the Geological Society of London, v.140, p.725-740.

Rutter, E.H. and S.H. White, 1979, The microstructures and rheology of fault gouges produced experimentally under wet and dry conditions at temperatures up to 400 degrees centigrade: Bulletin de Mineralogie, v.102, p.102-109.

Rutter, E.H., R.H. Maddock, S.H. Hall, and S.H. White, 1986, Comparative microstructures of natural and experimentally produced clay-bearing fault gouges: Pure and Applied Geophysics, v.124, p.3-30.

Sammis, C.G., R.H. Osborne, J.L. Anderson, M. Banerdt and P. White, 1986, Self-similar cataclasis in the formation of fault gouge: Pure and Applied Geophysics, v.124, p.191-213.

Schowalter, T.T., 1979, Mechanisms of secondary hydrocarbon migration and entrapment: AAPG Bulletin, v.63, p.723-760.

Scholz, C.H., 1990, The mechanics of earthquakes and faulting: Cambridge University Press, pp.439.

Sibson, R.H., 1980, Transient discontinuities in ductile shear zones: Journal of Structural Geology, v.2, p.165-171.

Sibson, R.H., 1981, Fluid flow accompanying faulting: field evidence and models: *in* D.W.Simpson and P.G.Richards, eds., Earthquake Prediction: an International Review. American Geophysical Union Maurice Ewing Series 4, p.593-603.

Sibson, R.H., 1986a, Earthquakes and rock deformation in crustal fault zones: Annual Reviews of Earth and Planetary Sciences, p.149-175.

Sibson, R.H., 1986b, Brecciation processes in fault zones - inferences from earthquake rupturing: Pure and Applied Geophysics, v.124, p.159-175.

Sibson, R.H., 1987, Earthquake rupturing as a hydrothermal mineralizing agent: Geology, v.15, p.701-704.

Sibson, R.H., 1989, Earthquakes as a structural process: Journal of Structural Geology, v.11, p.1-14.

Sibson, R.H., 1990, Conditions of fault-valve behavior, *in* R.J.Knipe and E.H.Rutter, eds., Deformation

Mechanisms, Rheology and Tectonics. Geological Society Special Publication 54, p.15-28.

Sibson, R.H., F. Robert and K.H. Poulsen, 1988, High angle reverse faults, fluid pressure cycling and mesothermal gold-quartz deposits: Geology, v.16, p.551-555.

Smith, D.A., 1966, Theoretical consideration of sealing and non-sealing faults: AAPG Bulletin, v.50, p.363-374.

Smith, D.A., 1980, Sealing and non-sealing faults in the Louisiana Gulf Coast Basin: AAPG Bulletin, v.64, p.145-172.

Spiers, C.J. and P.M.T. M. Schutjens, 1990, Densification of crystalline aggregates by fluid phase diffusional creep: *in* D.J.Barber, and P.G.Meredith, eds., Deformation Processes in Minerals, Ceramics and Rocks, Mineralogical Society Series No.1, p.334-352.

Stel, H., 1985, Crystal growth in cataclasites: diagnostic microstructures and implications: Tectonophysics, v.78, p.585-600.

Stark, C.P. and J.A. Stark, 1991, Seismic fluids and percolation theory: Journal of Geophysical Research, v.96, p.8417-8426.

Walsh, J.J. and J. Watterson, 1987, Distributions of cumulative displacements and seismic slip on a single normal fault surface: Journal of Structural Geology, v.9, p.1039-1046.

Walsh, J.J. and J. Watterson, 1988, Analysis of the relationship between displacements and dimensions of faults: Journal of Structural Geology, v.10, p.239-247.

Walsh, J.J. and J. Watterson, 1989, Displacement gradients on fault surfaces: Journal of Structural Geology, v.11, p.307-316.

Walsh, J.J. and J. Watterson, 1991, Geometric and kinematic coherence and scale effects in normal fault systems: *in* A.M.Roberts, G.Yielding and B.Freeman, eds., The geometry of normal faults. Geological Society Special Publication 56, p.193-203.

Watts, N.L., 1987, Theoretical aspects of cap-rock and fault seals for single and two phase hydrocarbon columns: Marine and Petroleum Geology, v.4. p.274-307.

Wesson, R.L., 1987, Modelling aftershock migration and afterslip of the San Juan Bautista, California, earthquake of October 3, 1972: *in* R.L.Wesson, ed., Mechanics of Earthquake Faulting. Tectonophysics, v.144, p.215-229.

Wesson, R.L., 1988, Dynamics of fault creep: Journal of Geophysical Research, v.93, p.8929-8951.

White, S.H., 1976, The effects of strain on the microstructures, fabrics and deformation mechanisms in quartzites: Philosophical Transactions of the Royal Society Series A. 283., p.69-86.

White, J.C. and S.H. White, 1983, Semi-brittle deformation within the Alpine Fault Zone, New Zealand: Journal of Structural Geology, v.5, p.579-589.

PART 4

Diagenesis, Sequence Stratigraphy, and Changes in Relative Sea Level

Chapter 11

Eustatic and Tectonic Controls on Porosity Evolution Beneath Sequence-Bounding Unconformities and Parasequence Disconformities on Carbonate Platforms

J.F. Read
Dept. of Geological Sciences
Virginia Polytechnic Institute
Blacksburg, Virginia, U.S.A.

and

Andrew D. Horbury
BP Exploration Co.
Stockley Park
Uxbridge, U.K.

ABSTRACT

Carbonate sequences and parasequences that formed under known or inferred 1 to 40 m.y. tectono-eustatic and 20 to 400 k.y. Milankovitch low to high amplitude sea level changes are shown to leave distinctive diagenetic records.

On carbonate platforms, low amplitude, high frequency sea level fluctuations typical of global green-house times form thick accumulations of meter-scale cycles with regional tidal flat caps, and there is only limited erosion of cycle tops. Humid climate cycles have early cemented, typically undolomitized intertidal fenestral caps and some supratidal laminites; aragonite fossils commonly are leached. If the climate is sufficiently arid, the cycles are dolomitized and aragonite is leached during falling sea level. Primary intergranular and moldic porosity is preserved in lower parts of cycles, and intercrystal and vuggy porosity is present in dolomitized subtidal facies and laminite caps beneath cycle top anhydrites and fine dolomite.

Moderate amplitude, high frequency fluctuations in sea level occur during times of intermediate continental glaciation. Cycles commonly have muddy facies grading up into grain-rich facies; they lack tidal flat facies and instead have small scale karstic surfaces or caliche directly over shallow subtidal facies. In humid climates, cementation plugs porosity in the upper phreatic zone, typically in grainstones just below the karstic surface. Zoned cement overgrowths in the deeper phreatic zone, although volumetrically minor, show a partial to complete record of high frequency sea level fluctuations that affected the sequence.

Primary intergranular porosity is best preserved in mid-cycle packstone-grainstones and subaerial pisolitic carbonates, with moldic porosity in leached skeletal muddy carbonates. In arid climates, there is little mineralogic stabilization or sparry calcite cementation, and dolomites can form beneath offlapping evaporatic tidal flats (particularly in highstand systems tracts). Primary porosity is generally preserved with secondary intercrystalline porosity developed in dolomites.

Large amplitude, glacio-eustatic fluctuations form erosionally bounded cycles often lacking regional tidal flat facies, because sea levelfalls off the platform faster than the tidal flats can prograde. Cycle tops commonly are highly disconformable, and can have caliche, karstic sinkholes, and caves. Large scale vertical migration of marine, mixed, meteoric phreatic and meteoric vadose zones dominate platform diagenesis. In humid climates, moldic and cavernous porosity may be extensively developed extending downward through many carbonate cycles, especially in coarse, aragonitic sediments, and sparry cements and internal sediment are associated with paleowater tables. However, aquicludes in the section may cause perching of water tables, limiting meteoric diagenesis to upper regressive parts of individual cycles where pore-rimming cements preserve primary intergranular and moldic porosity against compaction. Arid climates result in extensive caliche caps and little early spar cementation (thus much primary intergranular porosity), and dolomites can form from refluxing brines.

Low frequency, 2nd- and 3rd-order sea level falls (from one to tens of m.y. duration), associated with supersequence/sequence boundaries or tectonic uplift, are the common cause of updip porosity development and regional downdip calcite cementation under humid climates. During emergence, the sediments become mineralogically stable, chalky microporosity can form, and sediments lose intergranular porosity during cementation. Aquifers evolve from diffuse flow to conduit flow with time, as karstic passages localize flow. Earlier high frequency diagenesis is overprinted. Porosity evolves from primary intergranular and moldic to vuggy and cavernous types located in cave-fill breccias and in compaction fractured cave roofs. In arid climates, reflux dolomitization may occur if the surface is covered with hypersaline brines, but otherwise diagenesis and porosity development is proportionate to rainfall and may be very slow.

These cycle styles exert a strong influence on the development of the world's largest oil and gas fields that are reservoired in platform top carbonates. In greenhouse times, giant fields are mainly restricted to the arid cycles; during the transition into higher amplitude ice-house times, giant fields occur in both the arid and humid cycles. High amplitude, low frequency 2nd–3rd-order emergence results in giant fields mainly if the climate is humid. Reservoir and seal success are therefore critically dependent on the relationship of climate to cycle amplitude.

INTRODUCTION

Fluctuations in relative sea level strongly influence the development of carbonate sequences and their component parasequences or meter to decameter cycles (Sarg, 1988). These relative sea level changes include those tens of millions of years (2nd order) and 10 m.y. to 1 m.y. (3rd order) global sea level cycles of Vail et al. (1977), as well as tectonically induced subsidence and emergence in extensional settings or on convergent margins. Higher frequency, 400 to 100 k.y. (4th order) and less than 100 to below 20 k.y. (5th order), Milankovitch-driven sea level fluctuations, that are high amplitude during global ice-house times and low-amplitude during global green-house times, are superimposed on the longer term sea level changes (Fischer, 1964, 1982; Frakes and Francis, 1988; Koerschner and Read, 1989; Horbury, 1989; Goldhammer et al., 1990; Mitchum and Van Wagoner, 1991;

Wright, 1992). These composite sea level fluctuations form distinctive depositional sequences and sets of parasequences that should have distinctive diagenetic imprints and porosity distributions. This approach has been used widely for exploration in the high amplitude sea level Pennsylvanian and Lower Permian plays (see Wilson, 1975), but is much less prevalent in studies of rocks of other ages. Climate must also be considered as an important control on cycle diagenesis, as demonstrated by Hird and Tucker (1988).

In this chapter we test this idea, linking composite eustasy, depositional cycles, climate, and early diagenesis in order to give a better understanding and prediction of early diagenesis of specific reservoirs within a global ice-house/green-house framework. This is done by reviewing published examples of early diagenesis of carbonate sequences and parasequences, which on the basis of stratigraphic or other evidence, are known or inferred to have formed under low, moderate, or high amplitude, Milankovitch sea level changes, or during long term sea level cycles. We then examine whether the diagenetic signatures associated with high frequency Milankovitch-type sea level oscillations can be distinguished from those developed during long term sea level falls associated with 2nd/3rd order sea level cycles and tectonically generated unconformities.

Many examples cited are from well-studied surface exposures. At present there are much fewer published subsurface examples that provide sufficient details of the sequence stratigraphy, parasequences, and early diagenesis with which to assess the details of any eustatic controls, particularly at the parasequence scale. Hopefully, future examination of both failed and successful carbonate reservoir systems will allow better calibration of the proposed models. Discussion of the effects of late burial diagenesis on the early diagenetic reservoir characteristics are beyond the scope of this chapter, and for an example we refer the reader to Moore and Heydari (this volume).

DEPOSITIONAL SEQUENCES AND PARASEQUENCES OF CARBONATE PLATFORMS

Depositional sequences of shallow water carbonate platforms commonly range from 1 to 10 m.y. in duration and reflect long term changes in accommodation linked to 3rd order global eustatic cycles on cratons and passive margins. They may also reflect major extensional and/or compressive phases in tectonically active areas (Hubbard et al., 1990), although most of these sequence boundaries will coincide with eustatic sequence boundaries. Sequences are bounded updip by regional unconformities and can be subdivided into lowstand, transgressive, and highstand systems tracts (Sarg, 1988; Van Wagoner et al., 1988). Internal architecture of sequences on ramps differ significantly from those on rimmed shelves (Read et al., 1991; Burchette and Wright, 1992). Supersequences make up larger packages of sequences, all of which are resolvable on seismic records, while the sequences themselves are made up of smaller scale, typically sub-seismic resolution parasequences (meter to decameter scale depositional cycles) that may also be erosionally bounded. Most parasequences relate to high frequency, Milankovitch-driven climate changes and their associated sea level fluctuations and have periodicities in much of the Mesozoic and Cenozoic of roughly 20 k.y. (precession), 40 k.y. (obliquity or tilt), and 100 k.y. and 400 k.y. (short and long term eccentricity), although some small scale carbonate cycles may be autocyclic (Fischer, 1964; 1982; Goldhammer et al., 1990; Mitchum and Van Wagoner, 1991). Parasequences can be bundled into parasequence sets or megacycles due to high frequency sea level fluctuations riding on lower frequency sea level changes (Van Wagoner et al., 1988; Goldhammer et al., 1990). During global ice-house conditions with low CO_2 (late Precambrian, Permian-Carboniferous, and Neogene) (Figure 1), Milankovitch sea level fluctuations were generally large (up to 100 m or more), while during global green-house conditions with high CO_2 (Late Cambrian to Middle Ordovician, Silurian to Early Devonian, Late Permian to Oligocene), they were small (probably less than 10 m) (Koerschner and Read, 1989; Goldhammer et al., 1990; Berner, 1991; Wright, 1992). Although we have tentatively subdivided the high frequency, Milankovitch cycles into those formed under low, moderate, and high sea level fluctuations, we realize these form a gradational spectrum, and thus the boundaries of the classes are arbitrary. Furthermore, we doubt whether times marked greenhouse on Figure 1 were totally characterized by only low amplitude sea level fluctuations or vice versa. During times marked global green house (Figure 1), 2nd and 3rd order lowstand or transgressions, related in part to tectonically driven, low CO_2 conditions, could have relatively high amplitude, high frequency sea levels. Conversely, during times marked global ice-house conditions, 2nd and 3rd order highstands could have low amplitude, high frequency sea level changes (Ferry , 1991 pers. comm; Elrick and Read, 1991). Times of low amplitude 5th order fluctuations could have 4th order amplitudes that are much higher (Crevello, 1991). Thus during deposition of a single depositional sequence or even parasequence set, there may be a marked temporal change in amplitude of the various sea level signals.

Goldhammer et al. (1990), Kozar et al. (1990), and Hardie et al. (1991) suggest that peritidal cycles during green-house conditions could be generated by autocyclic processes. They postulate that flooding of the carbonate platform is caused by long term grad-

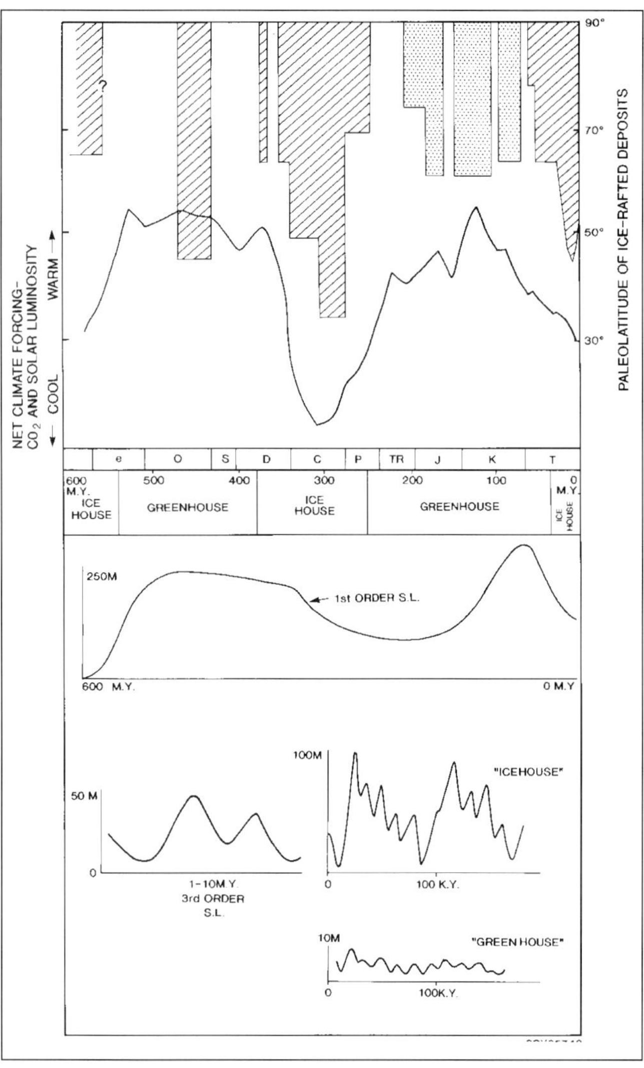

Figure 1. Relationship between ice-house and green-house conditions, eustasy, and global CO_2. Upper part of figure shows paleolatitude distributions of ice-rafted glacial deposits in continental regimes (crosshatched) and those of possible marine origin (stippled) plotted against geological time (after Frakes and Francis, 1988). The lower curve on the upper figure shows net forcing of climate due to changes in CO_2, the consequent effect on radiative forcing, and long term increase in solar luminosity (modified from Crowley and Baum, 1991 and Berner, 1991). Times of generalized ice-house vs. green-house conditions are shown on the time axis and are modified from Fischer (1982).
Lower half of the diagram shows, at the top, the first order Vail sea level curve for the Phanerozoic; compare with the CO_2/solar luminosity curve above. Bottom of diagram shows a 3rd order sea level curve with 1 to 10 m.y. periods, and schematic 4th and 5th order sea level curves which would be superimposed on the longer term curves. Low amplitude 4th/5th order sea level curves generally would be typical of times of global green-house conditions but would have been interspersed with short periods when amplitudes were substantially higher. Similarly, during ice-house times, most high frequency sea level fluctuations would have been high amplitude but probably were interspersed with times of lower amplitude sea level fluctuations.

ual subsidence during periods of non-deposition, caused by shrinkage of the subtidal carbonate factory due to progradation of tidal flats. Deposition resumes after a lag time or depth is reached and the carbonate factory restarts. Note however, that autocycles likely will lack the diagenesis associated with sea level drop off of the platform, since the autocycles will shallow to the position of static sea level.

Changes in dominant carbonate mineralogy through time will profoundly affect the response of sediments to early diagenesis, particularly to meteoric diagenesis. Wilkinson (1982) and Sandberg (1983) demonstrate that low sea level, ice-house times are times of global aragonite and high-Mg calcite production, whereas intervening high sea level greenhouse times are dominantly low-Mg calcite systems. This will result in a potential for greater diagenetic changes during subaerial exposure of the ice-house cycles, i.e. greater rates of early carbonate dissoloution and consequently greater rates of early precipitation lower in the meteoric lens. These effects will be accentuated in humid climates.

AQUIFERS AND DIAGENESIS

General reviews of early diagenetic processes affecting carbonates include Harris et al. (1985), Moore (1989), and James and Choquette (1990). Specific reviews of aquifers and karsts include James and Choquette (1988) and Wright et al. (1991), whilst Tucker (in press) also makes the link between sequence stratigraphy and carbonate diagenesis.

Tidal Flat/Sabkha Settings

During periods of arid climate and gradually falling sea level when tidal flats and sabkhas can prograde as fast as the shoreline migrates across the platform, marine or mixed meteoric marine waters/brines can reflux downward through the underlying cyclic section causing widespread regional dolomitization (Hsu and Schneider, 1973; McKenzie et al., 1980; Patterson, 1981) (Figure 2); this dolomitization is often associated with porosity development in otherwise low energy sediment. Recharge of the brine system occurs during storm flooding of the tidal flats, providing a renewed source of Mg ions. On a larger scale, brines from 3rd order highstand and lowstand evaporites can reflux down through many tens of meters of section, dolomitizing the upper parts of depositional sequences.

In humid climates, tidal flats prograding during relative sea level fall can develop a meteoric groundwater system that is thickest beneath the supratidal zone, thinning seaward (Figure 3). In tidal pond-channel belts, freshwater lenses also may develop beneath localized highs such as "palm hummocks" (Gebelein et al., 1980). Recharge for the fresh-water lens is from the surface of the tidal-supratidal flats, coupled with runoff from any surrounding higher elevations. Within the fresh-water lens, the upper part would tend to be oxidizing, along with the overlying vadose zone, passing down into increasingly reducing waters (especially where the sediments were organic rich, relatively low permeability muds).

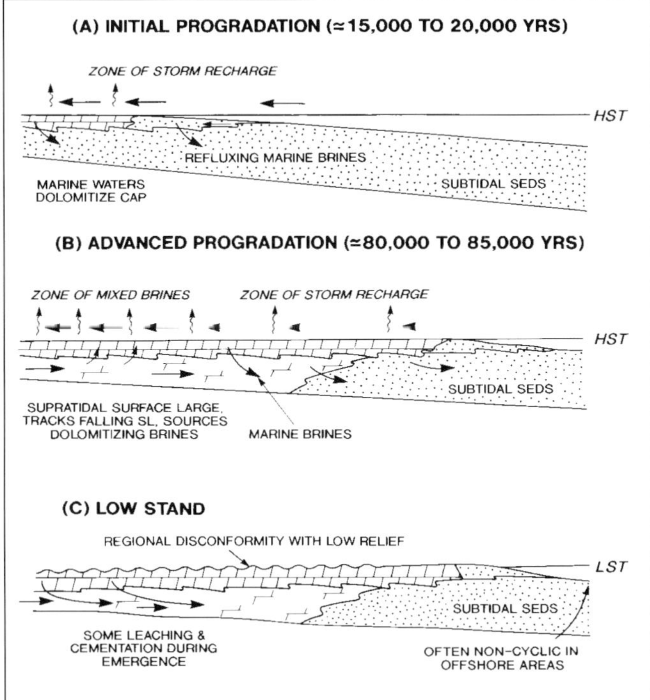

Figure 2. Schematic diagram of the hydrology of an arid tidal flat, modified from McKenzie, Hsu and Schneider (1980) and Montanez and Read (1992b). The carbonate cycle consists of a lower, subtidal unit and a capping unit of tidal flat laminite. Undolomitized limestone shown by stippled pattern; dolomite is shown by standard dolomite "brick" pattern. HST and LST are highstand sea level and lowstand sea level. (A) Highstand situation during intial progradation when dolomitizing brines are generated by evaporation of tidal and storm flooding of flats. (B) Situation during falling sea level after considerable progradation of flats. Updip, the supratidal surface becomes a disconformity, and continental waters may mix with marine brines generated by storm recharge. Much of the inner platform section becomes dolomitized by refluxing brines. (C) At the lowstand of sea level, much of the supratidal surface becomes a regional disconformity, and minor leaching of subtidal facies beneath the laminite cap may occur.

Unconformities and Aquifers

If lowered sea level during 3rd order to 5th order sea level fluctuations exposes the carbonate platform to subaerial processes, disconformities develop which act as intake areas for meteoric waters recharging regional aquifers (Meyers, 1978; Niemann and Read, 1988) (Figure 4). These aquifers may be unconfined, subject to recharge over a large area, or confined by an aquitard, in which case recharge is through the outcropping permeable limestone aquifer (White, 1969). Where the emergent area is a small island, the fresh water may form a lens from a meter to 15 m or more thick (Plummer et al., 1976; Budd, 1988; Emery and Dickson, 1989), the thickness relating to island width (Budd and Vacher, 1991). Where the meteoric water is sourced directly from rainfall onto the karst, such aquifers result in autogenic karst development (Ford and Williams, 1989). On carbonate platforms connected to large land masses (as opposed to isolated Bahamian-type platforms), meteoric waters infiltrating the unconformity surface may be supplemented by river waters from the hinterland, which on reaching the karstic coastal plain, tend to become part of the underground drainage system; such input forms allogenic karst (Ford and Williams, 1989). These waters may form regional meteoric ground water systems in which the zone of fresh water is tens of meters thick (for example 70 m in Yucatan where there is little river influx) to several hundred meters (e.g. 700 m in Florida, where there is substantial input of river water) (Back and Hanshaw, 1970). The distinction of autogenic and allogenic systems is important, since they are different in terms of hydrogeochemistry and water energy.

Marine-Meteoric Mixing Zones

The fresh water body passes downward into sea water pore fluids via a zone of diffusion or mixing (Figure 4). Potentially, these mixing zones may result in dissolution of carbonate, or in some cases, dolomitization. On small islands, the mixing zone can be as little as one or two meters to up to 15 m (Plummer et al., 1976; Budd, 1988) or can be tens of meters thick (Buddemeier and Oberdorfer, 1986). In regional aquifers with high permeability and high flow, the mixing zone may be very broad (Hanshaw et al., 1971). Movement of meteoric ground water to the sea is accompanied by circulation of water from the sea into the mixing zone and then back to the sea. The fresh water may discharge into the sea at the coastline, but if there is sufficient hydraulic head, conduit flow in karstic caverns, or confinement of the aquifer by an overlying aquitard, then fresh water may discharge considerable distances seaward of the shoreline, resulting in meteoric diagenesis beneath the water covered shelf.

Diffuse Flow vs. Conduit Flow Aquifers

During early emergence of newly deposited carbonates, given sufficient meteoric water influx, a diffuse flow aquifer develops (White, 1969) in which porosity is intergranular and water tables are relatively high. With time, the sediments become more cemented, moldic porosity develops, and the sediment is then modified by dissolution along horizontal and vertical joints to form networks of conduits, caves, and intraformational breccias that localize ground-water flow paths (average flow rates of cms/sec) to form a free flow or conduit flow aquifer (White, 1969) and lowered ground-water table (Stringfield and LeGrand, 1974). Maximum dissolution occurs at the ground water table. With more time, host sediments progres-

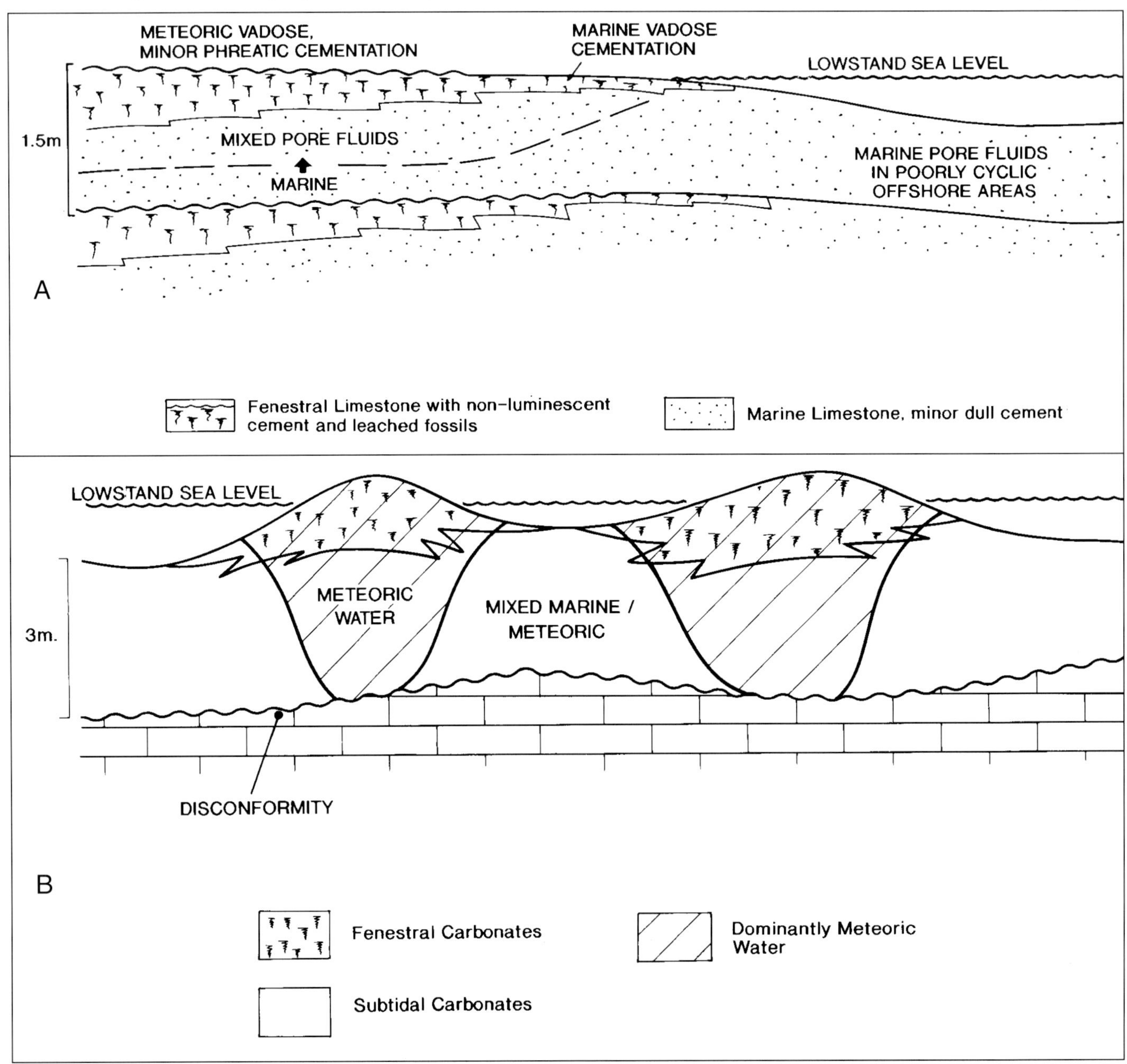

Figure 3. (A) Schematic diagram illustrating tidal flat diagenesis in non-channelled tidal flats under humid climate. A shallow meteoric vadose and phreatic zone of fresh water (less than 2 m) is generated on the inner parts of tidal flat, beneath a microkarstic surface. To seaward, tidal flats are subjected to marine vadose diagenesis. (B) Schematic diagram showing lenses of meteoric water beneath local highs ("levees and palm hummocks") of channelled tidal flats under humid climate (modified from Gebelein et al., 1980).

sively lose intergranular porosity and permeability because of cementation. Large air filled caverns may be developed beneath topographic highs, and a thick (tens of meters) vadose zone may occur. In old age, lowering of the land surface as valleys widen and flatten and caverns collapse results in plugging of the underground drainage system and formation of a shallow water table once again. Porosities and permeabilities therefore change from being relatively homogeneous, reflecting depositional fabric, to heterogeneous and compartmentalized.

Carbonate Aquifers and Calcite Cementation

Initially oxidizing aquifer waters commonly become more reducing down section, as well as down flow where aquifers are confined by aquitards (Edmunds, 1973; Champ et al., 1979; Hendry, this volume). Redox conditions strongly control the distribution of reduced ionic species incorporated into calcite, which are dominated by Fe and Mn (Oglesby, 1976; Barnaby and Rimstidt, 1989). The relative abundance of these trace elements in calcite cements may be observed with electron-stimulated light (cathodoluminescence, CL), which provides a rapid means of mapping cements regionally and relating these to position within ancient paleoaquifers (Meyers, 1974, 1978; Grover and Read, 1983; Dorobek, 1987; Niemann and Read, 1988; Horbury and Adams, 1989). In aquifer calcites, Mn is the major activator of CL, while Fe is the common quencher. It possible that other elements may also function in these roles, especially in hydrothermal systems (Machel, 1985; Machel and Burton, 1991).

Meyers (1991) reviews calcite cements in aquifers. In updip parts, where waters may be undersaturated with respect to carbonate phases, metastable aragonite is likely to dissolve to form moldic pores; the dissolved carbonate may reprecipitate as calcite cement either in place or further downdip. Calcite cements in updip, oxidizing parts of aquifers are likely to have low Mn and low Fe, and thus are dominantly nonluminescent under CL (Figure 4). These nonluminescent calcites commonly have thin bright-yellow luminescent laminae, perhaps reflecting periods of more stagnant, more reducing waters, increased Mn^{2+}/Ca^{2+} activities, or temperature fluctuations (Meyers, 1991; Machel and Burton, 1991). Not all calcite cements in updip parts of aquifers are nonluminescent; more Mn and Fe rich calcite cement can form where the carbonates are interlayered with organic rich muds or where metastable carbonates were stabilized early in the presence of organic matter undergoing oxidation, notably in the Neogene (Horbury, pers. obs. of Miocene carbonates in Southeast Asia) and in post-Miocene carbonates (Major, 1991).

Immediately downdip from the non-luminescent cements, where waters become slightly more reducing, calcite cements may have bright-yellow CL, caused by high Mn and low Fe contents (Dorobek, 1987; Emery and Dickson, 1989) (Figure 4). In many aquifers however, this regional belt dominated by bright cements is poorly defined.

In distal parts of the aquifer, waters are relatively reducing, precipitated calcites commonly are high in Fe (which quenches CL), and so are dully luminescent (dull brown to dull orange) (Frank et al., 1982; Grover and Read, 1983; Dorobek, 1987; Niemann and Read, 1988; Walkden and Williams, 1991) (Figure 4). Some of these dull cements show complex zonations related to sector zoning and differential uptake of trace elements by the various crystal faces, fluctuations in Mn^{2+}/Ca^{2+} and Fe^{2+}/Ca^{2+} activities, or fluctuations in temperature (Reeder and Grams, 1987; Machel and Burton, 1991; Hendry, this volume).

The widespread distribution of nonluminescent cements, extending over tens to 200 km in some ancient paleoaquifers, implies widespread oxidizing conditions in the paleoaquifer. Some of this may result from stratigraphic cross sections not being parallel to water flow lines thus giving a false impression of the true width of the oxidizing water zone in the aquifer; nevertheless, many of these zones are tens of kilometers across. This contrasts with some modern aquifers in mature (tightly cemented) limestones, in which the oxidizing waters are confined to a narrow (few kilometer) zone in the recharge area (Edmunds, 1973). This likely reflects the lower permeability, lower flow rates, and high organic contents of some modern aquifers compared to some of the once highly permeable ancient examples (Machel and Burton, 1991; Meyers, 1991). The wide extent of nonluminescent cement in paleoaquifers could also be due to downdip migration of the aquifer with regionally falling sea level, to form a diachronous calcite cement zone.

PARASEQUENCES AND DIAGENESIS ASSOCIATED WITH SMALL HIGH FREQUENCY SEA LEVEL CHANGES

Key Features

Carbonate platforms that formed under small amplitude (perhaps less than 10 m) sea level fluctuations ranging in frequency from 20 to 400 k.y. typically are aggraded and flat-topped to gently sloping. Facies are shallow water to peritidal, and high relief buildups such as pinnacle reefs are absent from the platform top (Koerschner and Read, 1989; Wright, 1992). The carbonate parasequences or meter-scale cycles generated tend to have "layer cake" stacking patterns, arranged into large scale transgressive onlap and regressive offlap patterns due to 1 to 10 m.y. sea level changes (Read and Goldhammer, 1988; Koerschner and Read, 1989; Read et al., 1991), although fewer cycles develop on the slowly subsiding inner

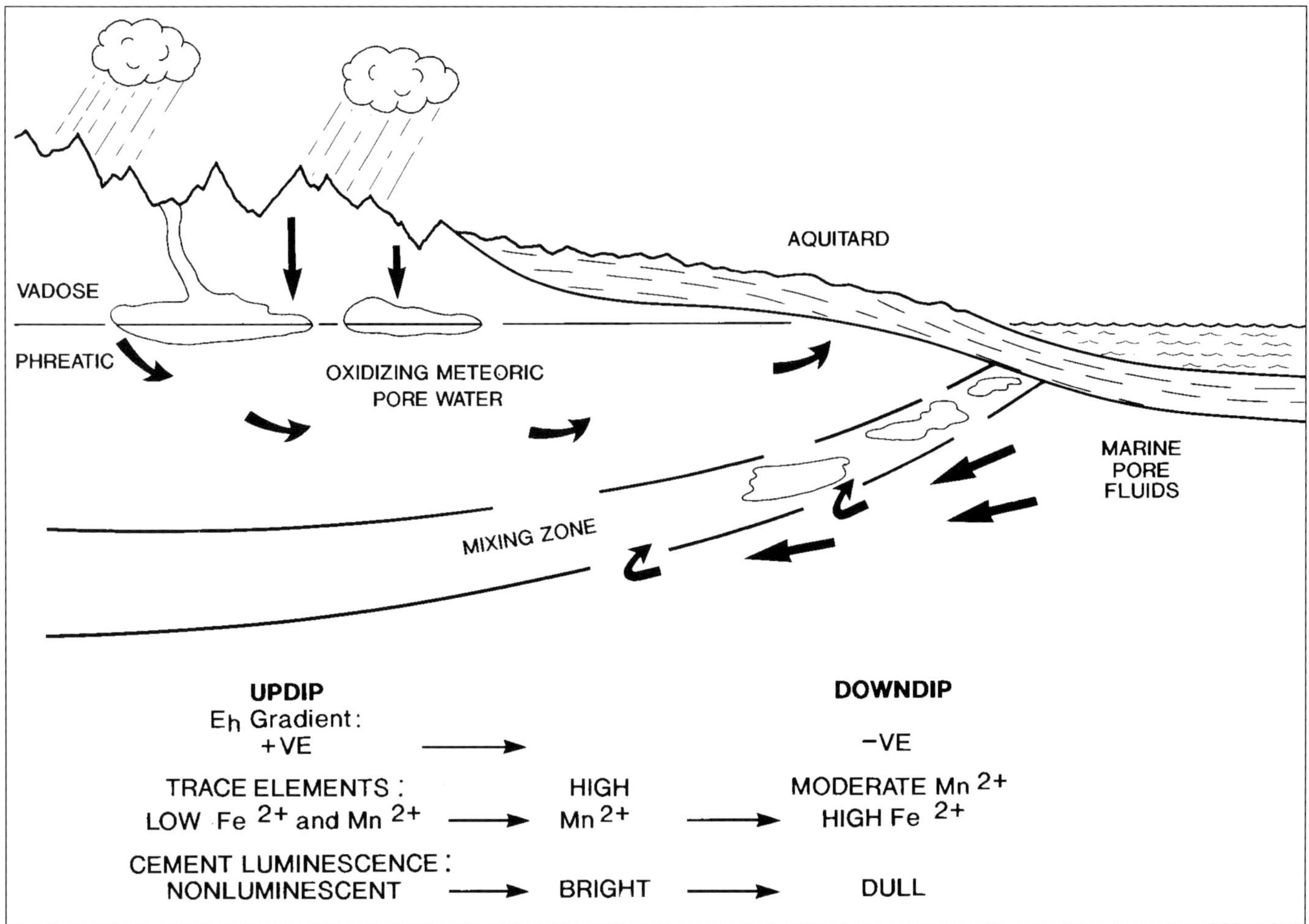

Figure 4. Schematic diagram of a carbonate aquifer that is unconfined updip and confined by aquitard downdip. The regional karstic unconformity acts as a recharge area. Numerous caves extend down to and are localized by groundwater table and by the coastal mixing zone. Regional gradients in Eh, reduced Fe and Mn, and cathodoluminescence of clear calcite cements are shown.

platform compared to the outer platform. These dominantly peritidal cycles (Figure 5) have regionally extensive tidal flat caps that cover much of the platform, because the small sea level falls allow the tidal flats to track seaward migration of the shoreline.

Disconformities are relatively poorly developed, typically on peritidal sediments, even though cycle tops may remain emergent for tens of thousands of years, because sea level only falls a short distance below the platform surface. Maximum emergence of individual cycle tops occurs during background 3rd order sea level falls when regolith, siliciclastic, or tepee capped cycles can develop (Goldhammer et al., 1990; Koerschner and Read, 1989). Tidal flat facies do not develop on many small carbonate platforms a few kilometers wide because of the high energy conditions; instead cycle tops can have dolomitic soils developed on subtidal beds (Goldhammer et al., 1990). Although platforms may be several kilometers thick, the individual cycles are sub-seismic resolution scale and are typical of the Cambrian-Ordovician of the U.S. Appalachians (Koerschner and Read, 1989; Demicco, 1985), Upper Silurian and Lower Devonian of New York (Goodwin and Anderson, 1985), many Middle to early Upper Devonian platforms (Read, 1973; Wendte and Stoakes, 1982; Elrick, personal communication, 1992), Lower Mississippian inner ramps of Northern England (Vaughan and Adams, 1984; Schofield and Adams, 1985; Adams et al., 1990), the Upper Permian of the Permian basin, U.S.A. (Borer and Harris, 1991) and larger platforms in the Tethyan Triassic (Hardie et al., 1986; Haas, 1982; Powell, 1988; A. Balog, personal communication , 1991, and Horbury, personal. observation.). Some Jurassic and Cretaceous cycles also fall into this category, although moderate amplitude 3rd/4th order processes are also important in cycle stacking and may be difficult to distinguish from most literature. Examples include the Lower Jurassic of North Africa (Crevello, 1991), Upper Jurassic of Arabia (Mitchell et al., 1988), Lower Cretaceous of the Jura Mountains (Strasser, 1988) and middle Cretaceous Edwards Formation and Stuart

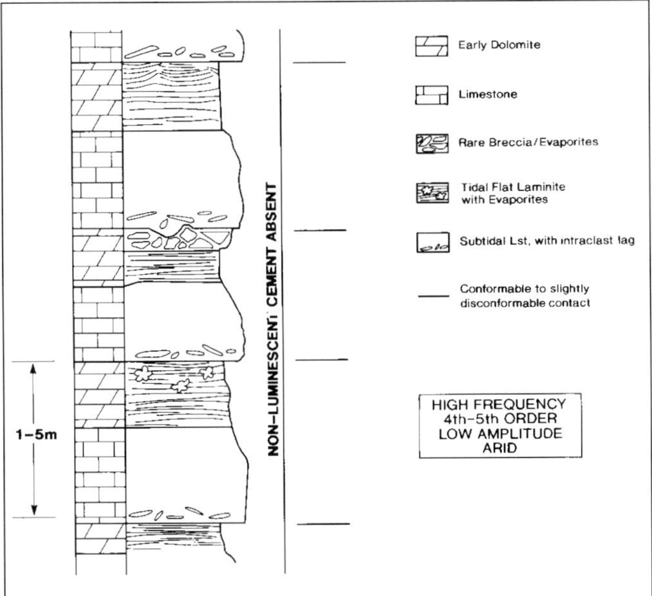

Figure 5. Idealized stratigraphic column showing carbonate cycles generated under arid climate and small sea level fluctuations. Initially, only the tidal flat laminites are dolomitized, but with repeated progradation, cycles become completely dolomitized and there may be local leaching of aragonite beneath the flats.

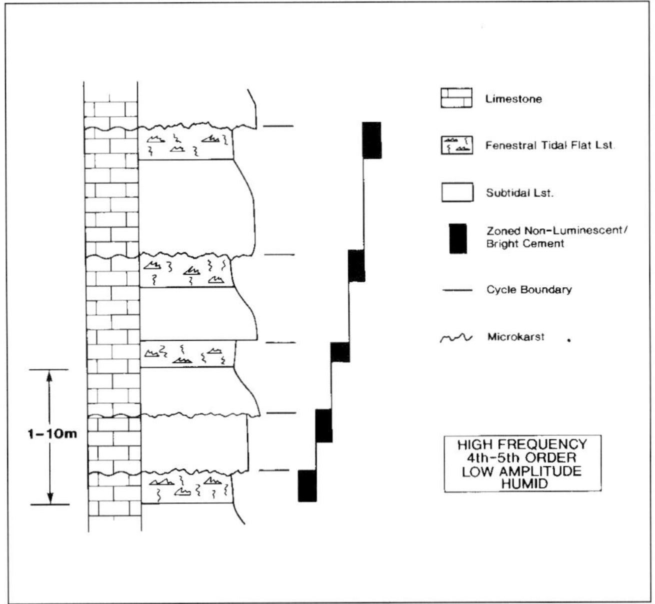

Figure 6. Idealized stratigraphic column of carbonate cycles generated during low amplitude sea level fluctuations in a humid climate. Marine vadose, meteoric vadose and possibly shallow phreatic cements are typically confined to the fenestral tidal flat cap.

City backreef of Texas (Bebout and Loucks, 1974).

Arid zone cycles have highly restricted, commonly high energy oolitic facies and low energy intertidal laminite caps; cycles are partly to completely dolomitized by brines sourced from the tidal/supratidal surface (Figure 5). Reservoir facies can occur in fine laminated dolomite caps where intercrystal porosity is preserved, in cycle capping siliciclastics, or in interparticle and intercrystal porosity in variably dolomitized, subtidal grainstones; plugging of grainstone porosity by sulfate minerals can leave only the fine dolomites permeable. Occlusion of porosity by calcite (excepting marine and beachrock cements) is rarely significant, but once emplaced, cements are rarely leached. Reservoir thickness is typically less than 3 m, and sabkha evaporites can provide extensive top and lateral updip seals, as well as completely occlude cycle top porosity in other examples. These features result in strongly stratified, layered reservoirs with potential for multiple pay zones and stratigraphic traps; where evaporite seals are present, they are usually isolated from the worst effects of burial cementation. Source rocks were typically deposited as condensed units in nearby barred or semi-barred basins which were euxinic and stratified, due to periodic restriction when sea level fell off the platform; during extreme restriction, evaporite cycles could also form in these basins. The platform cycles are evident even in highly dolomitized core, and wireline logs are useful in identifying and correlating the cycles by picking out the evaporite cap. These cycles form some of the most successful carbonate reservoirs globally, with producing fields in the Upper Permian of the Permian basin in Texas and Upper Jurassic Arab/Hith of Arabia.

Humid zone cycles commonly have very fossiliferous subtidal facies with fenestral to rare supratidal laminite caps (Figure 6). Cycle tops are planar to microkarsted and can be associated with shale. Aragonite fossils commonly are leached and fenestral porosity can be occluded by fibrous marine and vadose/phreatic sparry calcite cement. These cycles will be most porous in their lower part but cycle tops will have more primary porosity than their arid counterparts. Reservoirs will be sub-seismic scale, layered and have permeability baffles. Vertical stacking of these typically open marine cycles will tend to preclude the development of nearby source rocks integral to the depositional system. More homogeneous and better reservoirs in which there was little or no early diagenesis will often occur downdip in poorly or non-cyclic subtidal grainstone complexes. Cycles should be identifiable by modern wireline logs, particularly if cycle boundaries have a clastic component. Due to the facies and diagenetic patterns inherent to these system tracts, cycles generally do not develop good internal and updip seals and may receive significant quantities of later cements, in contrast to the equivalent anhydrite-capped arid cycles. They are therefore only successful as hydrocarbon plays due to fortuitous coincidence of later depositional events, typically where the uppermost cycles are sealed by overlying transgressive systems tract pelagic mudstone drapes with source potential, often associated with final platform demise and flooding, e.g. in the Devonian Swan Hills oil fields of Western Canada.

Diagenesis, Arid Climate, and Low Amplitude Sea Level Changes

Cambrian-Early Ordovician Cycles, U.S.A.

Extensive tidal-flat capped cycles of the U.S. Cambrian-Ordovician passive margin were probably formed under low amplitude sea level oscillations (Koerschner and Read, 1989; Osleger and Read, 1991; Montanez and Read, 1992a, b). These cycles host major petroleum and lead-zinc deposits, but their localization is controlled mainly by the overlying 2nd order unconformity. Cycles are 1 to 5 m thick with lower parts of restricted subtidal and regionally extensive caps of laminated dolomite and local, silicified evaporite nodules. Breccias and siliciclastics cap some late highstand to lowstand cycles (Read and Goldhammer, 1988; Koerschner and Read, 1989). Dolomite clasts locally are reworked across cycle boundaries into overlying subtidal limestones, suggesting early dolomitization. The dolomitization is virtually complete on the inner platform. On the outer platform, only highstand systems tracts are highly dolomitized whereas dolomite in transgressive systems tracts is restricted to laminite caps of cycles. Undolomitized subtidal parts of some cycles preserve minor evidence of marine and vadose meteoric diagenesis (aragonitic cements, meniscus cement , minor crystal silt, and some leaching of ooids). The laminated caps of cycles were dolomitized by brines beneath extensive tidal-supratidal flats (Montanez and Read, 1992a) (Figs. 2 and 5). The dolomites initially appeared to have had heavy oxygen isotope ratios and low Fe and Mn which were subsequently reset to varying degrees by replacement and overgrowth by burial dolomite (Kupecz, 1989; Montanez and Read, 1992b).

Lower Ordovician Ellenburger cycles in Texas form oil and gas reservoirs, for example the Puckett field (Loucks and Anderson, 1985), which has mean porosities of 3.5% at 4500m. Porosity is localized in early dissolution-collapse breccias after sabkha anhydrites in the Lower Ellenburger, whilst in the Upper Ellenburger most porosity is developed in packstone and grainstone filled channels. Most porosity is related to unconformity development, discussed later (Kerans, 1988).

Early Silurian Interlake Formation, Williston Basin, U.S.A.

Silurian peritidal cycles form oil reservoirs in the Williston basin (Roehl, 1985). The laminated and stromatolitic dolomites formed in arid evaporitic tidal flats and sabkhas over the ancestral Cedar Creek anticline, and are locally associated with terra rosa breccias and supratidal flat pebble layers. The succession is truncated by a Late Silurian unconformity and associated deep terra rosa. The main reservoir porosity is intercrystal in fine and microsucrosic early dolomite. These cycles contrast with the higher amplitude cyclicity in the underlying Red River Formation (see later section).

Upper Permian San Andres-Grayburg Formations, Permian Basin, U.S.A.

Upper Permian (Guadalupian) San Andres and overlying Grayburg dolomites are major oil reservoirs in the Permian basin of West Texas. On the Central Basin platform alone, these carbonates contain 32.6 billion barrels of oil in place (Major et al., 1988). The San Andres and Grayburg formations (Longacre, 1980; Chuber and Pusey, 1985; Harris and Stoudt, 1988; Hovorka et al., in press) form two or three depositional sequences which are composed of many 1 to 10 m thick parasequences that are thoroughly dolomitized. The parasequences were formed by small, high frequency fluctuations of sea level in relatively arid settings (Hovorka et al., in press). Lower parts of depositional sequences are non-cyclic, open marine carbonates, or poorly cyclic, muddy carbonates with grainstone caps. These are overlain by thick successions of cyclic restricted shelf, mollusk-peloid dolopackstone/wackestone and local tidal flat dolomites, or toward the ramp margin, by dolomitized grainstone dominated cycles (oolitic, peloidal or skeletal), some of which coarsen upwards and develop low relief over the crestal shoal. These parasequences are capped by erosion surfaces or local tidal flat facies. The upper parts of depositional sequences contain early dolomitized cyclic peritidal facies which form the superposed and marginal seals for the reservoirs. Peritidal facies are fenestral laminated, pisolitic, and commonly have tepee structures, interbedded anhydrite, and may be capped by regional eolian siliciclastics.

The San Andres-Grayburg Formations underwent early dolomitization probably from hypersaline brines from tidal flats (Major et al., 1988). The high frequency cyclicity has generated porous reservoir zones up to 15 m thick, separated by relatively impermeable seals. Some reservoirs occur in dolomitized grainstone with interparticle and intercrystal porosity (Major et al., 1988; Horkova et al., in press). These grainstones appear to have undergone diagenesis in meteoric lenses beneath local tidal flats or bar crests; meteoric lenses probably were related to high frequency sea level fluctuations that formed the parasequences (Horkova et al., in press). This caused leaching of the metastable grains, mineralogic stabilization, and partial intergranular cementation which helped preserve porosity against compaction. However, many grainstones became tight due to plugging by sulfate cement (Longacre, 1980; Chuber and Pusey,

1985). Reservoirs also commonly occur in muddy skeletal carbonates that were dolomitized and leached; porosity in these is typically in fine grained dolomite with skel-moldic and intercrystalline porosity (Longacre, 1980; Harris and Stoudt, 1988).

Upper Permian Yates Formation, Permian Basin, U.S.A.

Regressive siliciclastic units capping peritidal cycles in the Upper Permian Yates Formation, Permian basin, are interpreted to have formed under 4th order (100 to 400 k.y.) low amplitude eustasy (Borer and Harris, 1991). On the inner shelf, the cycles are dominated by anhydrite, halite, and red argillaceous siltstones. These pass into middle and outer shelf dolomite-siltstone-sandstone cycles. The carbonate parts of cycles are lagoon/tidal flat facies on the middle shelf to pisolite shoal facies on the outer shelf. The Permian carbonates were pervasively dolomitized by downward and seaward migration of brines generated from the evaporitic shelf (Adams and Rhodes, 1960). The siliciclastics relate to 4th order sea level falls and are the major hydrocarbon reservoirs in the subsurface.

Triassic Cycles of Syria and Iraq

Following the opening of Tethys (Sengor and Yilmaz, 1981), Triassic carbonate-evaporite cycles formed in Syria and Iraq. Inner shelf cycles consist of mostly interbedded varicolored clastic mudstone, occasional sandstone, subtidal ostracod/gastropod packstones and wackestones, and peritidal fenestral or algal laminated argillaceous dolomite and nodular sabkha anhydrites (Sharief, 1982). Mid- and outer shelf lithologies occur as discrete facies belts, but also often form the basal units to inner shelf cycles. They comprise locally dolomitic, massive bioclastic mudstones to grainstones with open marine benthic faunas; occasionally, small bioherms and oolitic shoals are developed (Dunnington et al., 1959; Sharief, 1982). Downdip towards interplatform basins, these facies pass into very fossiliferous, open marine slumped thin- to thick-bedded limestones and dark brown and black mudstones with pelagic bivalves and ammonoid faunas (Dunnington et al., 1959; Al-ejil and Abdu-Raheem, 1974; Sharief, 1982). Other, shallower, intrashelf basins and occasionally the interplatform basins were filled with thick laminated, locally slumped anhydrites and halite in Syria (Al-ejil and Abdu-Raheem, 1974; Lovelock, 1984) and Iraq (Buday, 1980; Sadooni and Saeed, 1987; Horbury, personal observation). The climate alternated between 1. semi-arid conditions in which cyclic sabkha-capped facies developed only on the inner platform, and 2. arid phases with major evaporite drawdown, resulting in extension of the inner platform cyclicity into intrashelf basin center evaporite cycles. In general, the poor initial reservoir quality of the generally low energy peritidal facies was little altered by dolomitization, since early evaporites plug most of the early intercrystalline porosity. Only the rare oolitic grainstone intervals show an improvement in reservoir quality, due to leaching coincident with dolomitization (Horbury, personal observation).

Lower Jurassic of North Africa

Jurassic carbonates typically are regarded as having formed during a time of global green-house conditions, although there is increasing evidence that low to moderate amplitude glacio-eustasy was important at this time (Figure 1). For example, depositional sequences on Lower Jurassic rift platforms in Morocco, described by Crevello (1991) are highly cyclic and contain numerous parasequences bundled into sets. The meter scale parasequences contain well developed marine cements and are best developed on the outer platform, where they are grainstone dominated, whereas on the peritidal inner platform, sections are less cyclic, tidal flat caps are common, and many cycles are amalgamated. The parasequences are capped by subaerial erosion surfaces, microkarst, tepees, and red argillaceous crusts and breccias that commonly developed directly on subtidal units. Below these surfaces are meniscus cements, horizontal sheet cracks, and micrite filled meteoric vadose molds. Remnant porosity was occluded by ferroan burial cements and compaction during burial. The parasequences were mainly formed by small, high frequency (20 to 100 k.y.) sea level fluctuations of less than 5 m, which did not expose the whole platform. However, larger, approximately 400 k.y. sea level drops of 10 to 30 m exposed the entire platform and formed the bounding surfaces to bundles of cycles.

Upper Jurassic of the Middle East

Ghawar, the world's largest oil field, occurs in Upper Jurassic carbonates of the Arab-D member, the lower of four regional depositional sequences of carbonate to evaporite in the Arab Formation. The Arab-D is less than 100 m thick and contains numerous small-scale upward shallowing cycles that resulted from high frequency eustasy (Mitchell et al., 1988). Fourth order cycles in the underlying Hadriya and Hanifa formations north of Ghawar form regionally traceable units marked by well developed flooding events, suggesting substantial sea level fluctuations; smaller scale cycles are less traceable (Markello, personal communication, 1992). Similarly, well developed meter scale cycles occur in sabkha facies of the Arab-Darb Formation off the Trucial Coast of Arabia (Wood and Wolfe, 1969). In the relatively low porosity, lower part of the Arab-D reservoir section at Ghawar, meter scale cycles are of skeletal-coated grain-intraclast packstone grading up into wackestone/mudstone capped by bored and burrowed hardgrounds; this facies also contains thin upward-fining, single and amalgamated storm deposits and formed below fair-weather wave base. The muddy carbonates in this part of the section partition the reservoir. The lower zone passes up into a thick succession of normal marine stromatoporoid-red algal-coral packstones and boundstones which has locally developed mounds. This is overlain by the main reservoir facies, a thick succession of skeletal-oolitic grainstone arranged in fining upward cycles of cross-bedded grainstone to burrowed packstone capped by hardgrounds. Marine fibrous to bladed cements form thin linings to intergranular voids, inhibiting over-compaction, and there was little meteoric cementation because of the arid depositional setting. Porosity is dominantly interparticle, with rare to common moldic and microporosity. Dolomite in the reservoirs has relatively low porosity and occurs as pervasive to stratiform units, especially in finer grained carbonates. The reservoir is sealed by sabkha- up into subaqueous evaporite facies which contain smaller scale cycles of dolomitized grainstone and mudstone and anhydrite facies. The high frequency eustasy has formed numerous low porosity zones (muddy carbonates) in the lower part of the reservoir, along with decreased porosity zones (packstones and dolomite) in the generally highly permeable upper part of the reservoir. Success of these arid cycles as hydrocarbon reservoirs also relates closely to their development on the prograding highstands to 3rd order cycles, at the base of which were deposited condensed organic-rich source rock shales typical of arid transgressive systems tracts in stratified intrashelf basins, such as the Hanifa in Saudi Arabia (Murris, 1980).

Lower Cretaceous Cycles, Jura Mountains

Peritidal carbonates of the Lower Cretaceous of the Jura Mountains (Central Europe) are organized into six types of 0.2-1.5m thick shallowing cycles (Strasser, 1988). Facies comprise restricted shallow subtidal bioclastic and locally oolitic limestones with beachrock cements and keystone vugs, passing up into hypersaline intertidal to supratidal flats with algal laminites and occasionally sabkha evaporites. Emergent features include birds-eye fenestrae, desiccation cracks, green Fe-smectites, and illites with charophytes, black pebbles, dissolution collapse breccias after evaporites, and sabkha dolomitization. Major bounding surfaces demonstrate fissuring with laminar calcrete, rhizoliths, and alveolar textures overlain by black pebble conglomerates with green marl matrix. Meteoric cements developed beneath all emergent surfaces are purely intracyclic. Peritidal cycles relate to 20 k.y. cyclicity, whilst larger sequences with major bounding surfaces result from 100-400 k.y. cyclicity. The climate was Mediterranean in type, with hot, dry summers and rainy winters, such that evaporite caps were not developed on all cycles and a mixture of diagenetic fabrics are superimposed, similar to that in the Lower Jurassic of Morocco.

Diagenesis, Humid Climate and Low Amplitude Sea Level Changes

Middle Ordovician Peritidal Cycles, Virginia

Regionally extensive peritidal cycles that formed under humid climate and low amplitude sea level fluctuations occur in the early transgressive and highstand systems tracts in the Virginia Appalachians of the U.S. (Read, 1980; Read and Grover, 1977; Grover and Read, 1978). More mosaic-like facies patterns on the inner ramp in Maryland may be due to autocyclic processes (Mitchell, 1985). The Virginia cycles are 1 to 10 m thick, and consist of open marine skeletal limestone, pellet limestones, and fenestral tidal flat caps topped with planar to scalloped, microkarstic surfaces. Laminated tidal flat facies are rare and limited to supratidal deposits, which include fresh water algal marsh facies and stromatolites with abundant filaments (Mitchell, 1985). Dolomite is rare and restricted to supratidal intraclastic packstone. These cycles show much vadose to shallow phreatic meteoric diagenesis limited to their upper parts reflecting development of a shallow meteoric ground-water system during emergence (Grover and Read, 1978; Read and Grover, 1977; Grover and Read, 1983). Cycle tops are microkarsted with mud-filled solution enlarged fissures. The underlying facies have leached aragonite shells and fenestral and moldic pores with pendant and pore-rimming early fibrous- and sparry calcite cement linings (zoned nonluminescent) and crystal- and pellet silt fills (Grover and Read, 1978; Dunham, 1969).

Middle to Upper Devonian Judy Creek (Swan Hills) Reef Complex, Australia

The Judy Creek platform reef complex is 67 m thick, 122 sq. km in area, and had original oil in place of 830 million barrels (Wendte and Stoakes, 1982). The complex consists of stromatoporoid reefal rim and foreslope and platform interior lagoonal facies. It contains four regionally traceable megacycles 10 to 18 m thick, each characterized by regional tidal flat caps, and backstepping of the margin. The lower three megacycles have smaller (1 to 5 m thick) cycles, most of which are traceable across the buildup, but whose tidal flat caps are less continuous. The small cycles consist of dark, muddy *Amphipora* rudstone overlain by lighter colored, mud-poor *Amphipora* rudstone with packstone/grainstone matrix, and cycle caps of fenestral and crinkly laminated pellet mudstone/packstone and beach lime sand/gravel facies and thin green shales. Although the tidal flat facies are not traceable between all wells, the flooding events mark most cycle boundaries.

The cycles were modified in a non-evaporitic setting, indicated by lack of early dolomite and evaporites. Fresh water vadose diagenesis of the lagoonal facies has produced exposure zones 30 cm to 2 m thick, along with secondary vug formation, and moldic and channel porosity that may be lined with pendant fibrous calcites; the associated green shales may be paleosols or storm transported siliciclastic mud (Walls and Burrowes, 1985; see also Read, 1973 for similar diagenetic effects in the correlative Pillara cycles, Canning Basin, Western Australia). Porosity in the Judy Creek reef complex is mainly primary and is most abundant in the grainy or reefal margin facies. Porosity of the muddy lagoonal facies tended to be occluded during stylolitization and burial cementation, but discontinuous porosity is associated with shoaling grainy facies of lagoonal cycles. The critical aspect of these cycles which makes them economic reservoirs is the presence of the overlying hemipelagic mudstones of the Waterways Formation, which acts as a seal.

The Swan Hills reefs contrast strongly with the higher energy Leduc and Rainbow reefs, which were deposited in more arid climates. In these reefs, there was major precipitation of marine cements in the bank margins. Most porosity is secondary and developed in the bank interior during later exposure of cycle tops, resulting in a strongly compartmentalized reservoir (Walls and Burrowes, 1985).

Lower Dinantian (Mississippian) Cycles, Britain

Lower Mississippian ramps in Britain were deposited in an active backarc to rift basin (Leeder, 1987; Fraser et al., 1990; Fraser and Gawthorpe, 1990), with the inner ramp environments mainly being influenced by low amplitude, high frequency (5th order) cyclicity. There are also examples of moderate amplitude, longer frequency (4th order?) events which are expressed as paleosols and paleokarsts on both inner and mid ramp facies which are discussed later. The climate varied between humid and semi-arid with seasonal rainfall (Wright, 1990), but there is only local evidence of truly arid diagenesis, such as anhydrites in graben bases. All of the cycles and emergent surfaces are sub-seismic in scale. There are also major, seismic scale, intrabasinal 2nd-3rd order unconformities which affect burial diagenesis; these are tectonic in origin and are discussed later.

Peritidal cycles 1–2 m thick developed on highs and close to the onlap margins, e.g. the upper Martin Limestone, south Cumbria (Vaughan and Adams, 1982; Adams et al., 1990); Topley Pike and Dam Dale members of the Woo Dale Limestones in Derbyshire (Schofield and Adams, 1985). Cycles comprise basal peloidal-restricted skeletal pack-grainstones while cycle tops are mudstones and wackestones with birds-eye fenestrae and channel and levee deposits. Diagenesis of peritidal cycles under humid climates resulted in complex alternations between marine and meteoric diagenesis, dissolution of molds, fissuring, and deposition of crystal silts, while evidence of more arid diagenesis is provided by desiccation cracks, tepees, and occasional evaporite pseudomorphs (Schofield and Adams, 1985; Adams et al., 1990).

Cyclicity is lost in more offshore areas where subtidal facies are massive to well bedded peloidal-skeletal grainstones (humid climate) or oolites (semi-arid climate), e.g. the Red Hill Oolite and Park Limestone in south Cumbria (Adams et al., 1990), Vincent House member of the Woo Dale Limestones in Derbyshire (Schofield and Adams, 1985), Gully Oolite and Brofiscin Oolite in South Wales (Wright, 1986; Hird and Tucker, 1988). Middle and outer ramp facies are storm-dominated argillaceous packstones and wackestones, e.g. Dalton Beds, south Cumbria (Adams et al., 1990) and the Black Rock Limestone and correlatives, southwestern Britain (Wright and Faulkner, 1990). These were not subaerially exposed in the Mississippian.

DIAGENESIS ASSOCIATED WITH MODERATE AMPLITUDE, 4TH/5TH ORDER SEA LEVEL CHANGES

High frequency, moderate amplitude fluctuations (perhaps 20 to 50 m range) in sea level should typify times when the globe is transitional between ice-house and green-house conditions. Such times might bracket the Early/Late Ordovician, Early/Late Carboniferous boundary, Early/Late Permian boundary and parts of the early Neogene. The proportion of the geological record occupied by these times is rather limited and there may be an inadequate sample database to precisely evaluate the significance of these cycles.

Key Features

Parasequences that formed under moderate fluctuations in sea level (20 to 50 m magnitudes) tend to be less layer cake than low amplitude cycles. Instead the cycles may have a moderately shingled arrangement on the platform, reflecting horizontal migration of peritidal deposystems, each deposited during 20 to 40 k.y. sea level fluctuations affected by longer term (100 to 400 k.y.) and moderate amplitude fluctuations (Read et al., 1991). Outer ramp cycles can have deeper subtidal argillaceous limestone in lower parts, beneath grain-rich shallow water caps, which are in turn succeeded by argillaceous subtidal facies, again due to rapid deepening during high frequency transgressions. These deeper water tongues die out onto the inner platform, where cycles commonly are grainstone-dominated. Tidal flat facies cap only a few cycles generally, which instead have karstic erosional tops often developed on shallow subtidal facies (Figure 7, 8) (Horbury, 1987; Horbury and Adams, 1989; Wright, 1992). Tidal flat facies can form linear belts on some platform cycles during 4th and 5th order high- and lowstands of sea level, but regional tidal flat capped peritidal facies appear to be best developed during 3rd order (1 to 10 m.y.) highstands (Elrick and Read, 1991) and close to areas of permanent emergence (Somerville, 1979a). It is not clear whether peri-

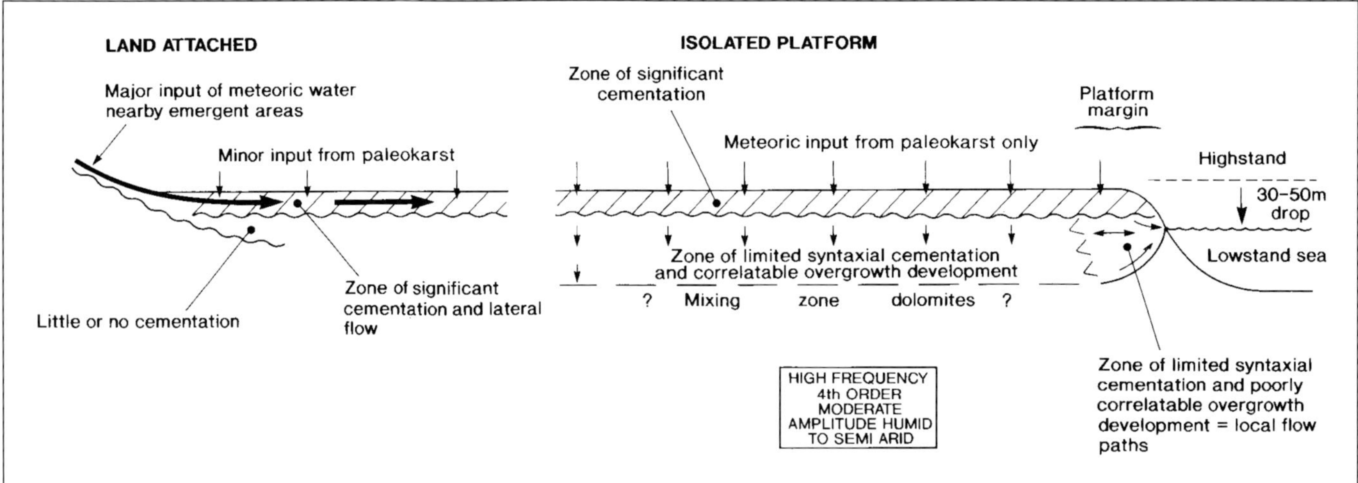

Figure 7. Schematic cross sections of the hydrologic setting of a platform cycle generated during moderate fluctuations of sea level under a humid climate. A land-attached platform is shown on the left, with considerable lateral flow of meteoric waters. An isolated platform is shown on the right, with mainly vertical infiltration of meteoric water. After Horbury (1987) and Horbury and Adams (1989).

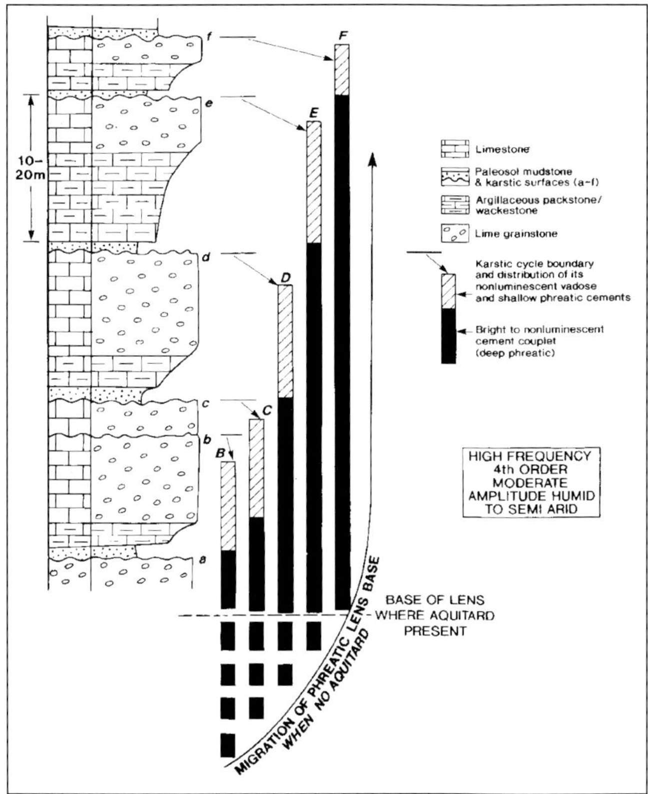

Figure 8. Idealized column of carbonate cycles generated under moderate fluctuations and a relatively humid climate. Cycles are capped by well developed karstic surfaces (a-f), each of which sourced its own meteoric zone (A-F) during lowered sea level. Each meteoric ground water zone generated its own bright to nonluminescent cement zone low in the phreatic zone, but in the shallow phreatic zone where cementation was most rapid, the cements are not easily correlated with karstic surfaces. Modified from Walkden (1987) and Horbury and Adams (1989).

tidal cyclicity at these times was due to decreased accommodation or to lowered 5th order amplitudes during 3rd order highstand deposition, compared with the transgressive systems tract, but tidal flat facies only develop well where they can nucleate off relatively positive areas of emergence.

Depositional porosities are highest in the grainy upper parts of cycles, and porous zones tend to be thicker (up to 10 m or more) and more homogeneous than the lower amplitude cycles. Also, argillaceous, muddy lower parts of cycles tend to act as partial internal seals or permeability baffles, resulting in subseismic scale stratified reservoirs and should be identifiable from cores and electric logs.

Arid zone carbonate cycles can have porosity in their tops plugged by caliche formation and vadose fibrous cements. Subtidal grainy parts of cycles lack early sparry cements and retain much primary porosity. Regionally prograding peritidal dolomites and evaporites can form seals to 3rd order influenced sequences and help to isolate the reservoir from later cementation. Arid cycles form hydrocarbon reservoirs, but not on the same scale as the low amplitude arid cycles. Oil fields are developed in the Middle to Upper Ordovician of the Williston basin and gas reservoirs in the Lower Mississippian of Wyoming and Montana, U.S.A.

Humid zone cycles (Figures 7, 8) may show some karsting and soil formation at cycle tops, while upper parts of cycles undergo some moldic and vuggy dissolution, along with plugging of primary and secondary porosity by vadose and upper phreatic sparry calcite cements. The middle or lower parts of cycles, if grainy, may be less well cemented and retain more primary porosity. Muddy cycle bases undergo stabilization of carbonate muds in meteoric waters and some moldic porosity formation. Because of development of thick ground water zones, meteoric diagenesis may extend down into underlying older cycles, if seals are poorly developed. These cycles appear to work poorly as hydrocarbon systems because of internal and external seal problems, and they may receive significant later cementation. There are possible difficulties in developing nearby source rocks where cycles stack vertically; these problems are similar those in the low amplitude humid cycles.

Diagenesis, Humid Climate, Moderate Amplitude Sea Level Changes

British Early Mississippian Cycles

Sporadic, moderate amplitude, low 4th order emergence of subtidal inner ramp oolitic and peloidal-bioclastic grainstones without tidal caps is described by Hird and Tucker (1988) and Wright (1990). The pedogenic and diagenetic fabrics developed are more mature, and thicknesses of meteoric lenses greater than in the age equivalent 5th order, low amplitude inner ramp peritidal cycles. These paleosols typically developed on inner ramp oolites which cap 3rd order scale cycles (Wright, 1986). Paleokarsts are often deeply incised and increase in number updip, to a maximum of five surfaces per 3rd order cycle in the inner ramp area (Wright, 1982; Searle, 1988). A variety of climate types are noted, although no evaporites were preserved, suggesting a broadly monsoonal climate (Wright, 1990). Dominantly humid climates are reflected by ferrous weathering in leached soils, a rubbly upper zone some 5 m thick, passing into a zone of 'pipe and fissure' karst (Wright, 1982; Robinson and Wright, 1988). Beneath this are sometimes water table caves up to 2m high, with small-scale collapse features (Wright, 1982). Diagenetic features comprise vadose grain dissolution-compaction, mold and vug development, and significant but locally extremely variable cementation down to at least 10 m in the underlying oolite, mostly of inclusion-rich and complex zoned vadose and shallow meteoric spar precipitation (Hird and Tucker, 1988; Searle, 1988). Only limited marine cements were developed initially, and there is some evidence of arid zone diagenesis (early dolomites infilling moldic porosity and associated with plant roots; Searle, 1988).

Semi-arid climate, often with strongly seasonal rainfall, is reflected by the presence of smectitic, calcrete-bearing and Rendzina type paleosols (Wright, 1990). These develop on low relief paleokarsts, sometimes with minor angular truncation (Adams and Cossey, 1981). Fabrics developed in paleosols include laminar caliche, gypsum pseudomorphs, pedogenic breccias, eolian lime sand input, rhizocretions, and needle fiber calcite (Adams and Cossey, 1981; Wright, 1982, 1983, 1988; Riding and Wright, 1981). These surfaces are sometimes associated with early marine cements, and peloidal cements occur in intraclasts, but little or no first generation meteoric or pedogenic cement occurs except close to the top of the paleokarst (Adams and Cossey, 1981; Hird and Tucker, 1988). The periodicity of these surfaces is comparable with that of the moderate amplitude surfaces described from the Lower Mississippian of Eastern U.S.A. (see next section on arid cycles).

British Upper Mississippian Cycles

Regularly stacked Upper Mississippian cycles of Britian are capped by 30 or more well developed paleokarstic surfaces and have 4th order periodicities of 100 to 200 k.y. They appear for the first time in rocks of the Asbian stage of the European Dinantian and are thought to be eustatic, coincident with the onset of the Permian-Carboniferous Gondwana glaciation (Walkden, 1987; Horbury, 1989). The cycles and paleokarsts contrast with 5th order green-house peritidal cycles that were developed on ramps, but are similar to the less frequent 4th order paleokarsts developed in the Lower Mississippian.

Upper Mississippian cycles are made up of skeletal packstones and skeletal peloidal grainstones (and rare oolite and fenestral mudstones) in upwards shoaling units on shallow open marine, normal salinity, isolated and land-attached rimmed shelves in Northern Britain (Somerville, 1979b,c; Walkden, 1987; Horbury, 1989). There is little evidence for lateral progradation within cycles, except for minor sand body bed form migration close to grainstone shoal-dominated platform margins (Gawthorpe and Gutteridge, 1990; Horbury, 1992). Regression was rapid as indicated by subaerial exposure surfaces directly on subtidal facies with evidence for erosion of up to 26m (Horbury, 1989; Walkden and Walkden, 1990). Pedogenic fabrics are complex (Davies, 1991) and reflect a dry-wet-dry cycle (S. Vanstone, personal communication, 1992), which is probably related to 5th order climatic cyclicity superimposed on 4th order climate cycles (Gray, 1981, quoted by Tucker, 1985). Paleosols usually comprise a thin, rooted rhizoconcretionary interval up to 1 m thick (semi-arid phase), cut by a smectite-covered paleokarst with up to 2 m of gently pitted relief (wet phase) (Walkden, 1987; Horbury, 1989). The surfaces may have larger erosional incisions near to the platform margins (Horbury, 1989). Superimposed later fabrics such as laminate caliche and breccia (semi-arid phase) are rare. Distinctive cements associated with these cycles may relate to the climatic cyclicity. Cement types are strongly controlled by type of platform (isolated or land-attached), by position within the meteoric lens, and by the differing sedimentation patterns on the various platforms that formed an archipelago in the basin.

In cycles on the isolated rimmed platforms in the British Carboniferous, the shallowest subsurface meteoric cements are rare pendant and ponded vadose brown cements, which have a fibrous habit and grew on all substrates including non-syntaxially on echinoderms, and brown equant cements which are very dull luminescent and are thought to be related to calcrete mottle formation in soil zones (Horbury, 1987). These represent the dry climatic phase and were short-lived precipitation events in localized diagenetic environments which were strongly controlled by local permeability pathways and sediment heterogeneity; nucleus type little influenced cement morphology, which was controlled mainly by high precipitation rates (Horbury, 1987; Horbury and Adams, 1989). Also associated with individual emergent events are shallow (5 to 15m depth) meteoric fabrics, which formed by re-precipitation of the bulk of the calcite dissolved beneath the paleokarst, probably during the wet climatic phase. They may be developed in the underlying subtidal Lower Mississippian limestones, e.g. the meteoric cements sourced from the base Upper Mississippian paleokarst noted by Schofield and Adams (1985). These cements are volumetrically abundant and constitute the pre-zone 1 type cements of Walkden and co-workers and the early meteoric cements of Horbury and Adams (1989) (Figs. 7, 8). These are clear calcites that are typically nonluminescent in CL, sometimes with well developed and very fine bright subzones which cannot be correlated for more than a few millimeters. They develop on all substrates but are thicker as syntaxial cements.

In contrast, deeper in the phreatic lenses (15 to 120 m subpaleokarst), well developed correlatable and cyclically CL sawtooth zoned clear calcite cements occur (Figures 7, 8). These are a volumetrically insignificant but important cement type that formed by slow precipitation in lenses of successively younger paleoaquifers (described by Berry (1984), Walkden and Berry (1984), and Walkden and Williams (1991) from Derbyshire, by Horbury and Adams (1989) from the Lake District, and by Solomon and Walkden (1985) from North Wales). These cements only occur below autogenic karst, where rimmed shelves are relatively isolated from probable laterally-sourced meteoric water such as onlap margins onto Caledonian basement (Figure 7). Each emergent event was recorded as a unique cementation event by growth of a couplet of bright to non-luminescent cement. Hence there are roughly the same number of cement couplets as there are eustatic cycles containing these cements (Figure 8). In the Lake District the meteoric lenses rested on a shale aquitard such that the cement couplets are most abundant low in the section and decrease upward into younger cycles (Horbury and Adams, 1989). In Derbyshire, cement couplets overlap and extend over a limited vertical range beneath the parent paleokarst surface, with younger couplets progressively appearing up section; these do not possess a basal aquitard (Walkden and Berry, 1984; Walkden, 1987). In these deeper meteoric phreatic settings, the marginally saturated waters were only capable of precipitating cement where there was strong syntaxial nucleus control influencing seeding; the cements grew slowly and are euhedral when growth went to completion.

Each isolated rimmed platform developed different cement stratigraphies because the response of each structural unit to subsidence, the facing direction, connectivity to larger oceanic water masses, and input of clastics from nearby basement terrains all influenced the detailed vertical and lateral facies mosaic on each platform. Because on an isolated platform, the availability of metastable carbonate from autogenic karst dissolution is a function of facies thicknesses deposited prior to emergence, differences between platform cements and trace elements precipitated from meteoric systems reflect the individual subsidence history of each platform, even though the mechanism that triggers cementation is eustatic. Horbury and Adams (1989) found many small scale internal variations and correlation problems with this morphology of cement where local topography influenced flow patterns in the meteoric lens, for example close to the emergent and fluvially incised platform margin during lowstands.

Cyclically exposed land-attached platforms possess many of the diagenetic features noted above, mainly the shallow meteoric cements precipitated to depths of 5 to 15m sub-paleokarst, but with some notable differences. Firstly, the regionally correlatable sawtooth-zoned syntaxial overgrowths (described above) have not been found in areas close to onlap margins. The well developed cyclic bioclastic packstones and grainstones in the lower Asbian stage lower Urswick Limestone lack these cements, and instead, possess a black/yellow 'tiger-stripe' cement with sharp zone tops and bases (Horbury and Adams, 1989) . These occur in grainstones of cycle tops and can be correlated as a general cement type between nearby localities, but not by individual zones. These cements developed on all grain types and are thickest as syntaxial overgrowths; they are interpreted as the product of locally developed meteoric lenses and are thought to possess some element of lateral flow with water input from permanently emergent areas which were being onlapped (Horbury, 1987; Horbury and Adams, 1989) (Figure 7). Allogenic karst through which such waters entered the aquifer contain many sandstone-filled pipes and incised fluvial channels close to the onlap margin (Walkden and Davies, 1983); these differ notably from the typical autogenic paleokarst found further out on the platforms. Correlative peritidal micrite/subtidal packstone 5th order cycles onlap basement in North Wales (Somerville, 1979a). These have not yielded any notable cements (e.g. Solomon and Walkden, 1985; Solomon, 1989) although they probably pass downdip into 4th order dominated systems typical of the lower Urswick Limestone. These relationships demonstrate the importance of nearby basement on facies distributions, cycle stacking patterns, and expression of cyclicity order, which ultimately influence diagenesis (c.f. Elrick and Read, 1991).

Diagenesis Under Arid Conditions and Moderate Amplitude Sea Level Changes

Middle to Upper Ordovician Red River Group, Williston Basin of Montana and North Dakota, U.S.A.

The predominantly Upper Ordovician Red River group in Montana probably formed under moderate amplitude, high frequency oscillations under arid conditions since at that time there was an important Gondwana glaciation during a green-house time of high CO_2 (Fischer, 1982; Crowley and Baum, 1991). The Red River group contains at least two to three upward shallowing cycles 10 to 20 meters thick, which are stacked in a thinning upward package. Cycles are composed of subtidal burrowed carbonate passing up into laminated dolomite and basin wide anhydrite caps (Longman et al., 1983). The laminated dolomite and evaporite facies in the various oil fields have been interpreted as subtidal shallow basinal caps of "brining-upward" cycles (Longman et al., 1983) or tidal flat/sabkha facies with solution-collapse features (Derby and Kilpatrick, 1985) on the basis of presence or absence of structures indicating emergence.

In the Cabin Creek field, cycles have disconformable, solution brecciated tops and evaporites are poorly preserved (Derby and Kilpatrick, 1985). The fine to microcrystalline dolomites are stratiform and are considered to have formed in and beneath sabkhas. Associated sulfate minerals were leached by meteoric waters to form the dominant vuggy and moldic reservoir porosity along intercrystal porosity. There is a very strong facies and intercycle control on the exact percentages of each pore type developed; the mean porosity for all cycles is 13%. Supratidal cycle tops possess vuggy porosity, intertidal facies a mixture of interparticle, intercrystalline, and moldic, and subtidal facies possess intercrystalline at the top and moldic throughout (Ruzyla and Friedman, 1985). In contrast, fields described by Longman et al. (1983) have numerous pods of porous dolomite up to 60 m thick and 1.6 km diameter which form the petroleum reservoirs below the lowest ('C') anhydrite. The regional distribution of the dolomite pods marked by concentric distribution of cryptocrystalline dolomite passing out and downward into porous fine and medium grained reservoir dolomite, and then into tight partly dolomitized burrowed limestone, suggests a model of reservoir dolomitization by brine seepage through holes in the anhydrite during or immediately after anhydrite deposition (Longman et al., 1983). This dolomite porosity probably was significantly enhanced by burial leaching (R. Perkins, personal communication, 1992).

Lower Mississippian of Wyoming-Montana

The Lower Mississippian Madison Group in the U.S. Rocky Mountains predominantly formed under high frequency, low to moderate amplitude sea level fluctuations (Elrick and Read, 1991), perhaps interspersed with times of lower amplitude fluctuations. The Madison Group consists of a supersequence, composed of basin/deep ramp to shallow ramp to tidal flat/sabkha facies and solution collapse breccia (Lodgepole-Mission Canyon-Charles evaporites). The supersequence has several 3rd order sequences with well developed 1 to 10 m thick parasequences or cycles (Dorobek, personal communication, 1991; Elrick and Read, 1991). Transgressive systems tract cycles in the Lodgepole Formation consist of commonly dolomitized skeletal wackestone/mudstone passing up into ooid grainstone with erosional tops; downslope, the cycles are deeper water, laminated or burrowed dark muds that grade up into storm-deposited skeletal/oolitic packstone-grainstone caps. Tidal flat facies are generally absent. In contrast, highstand systems tract cycles are dominated by dolomitized thin peritidal cycles of subtidal oolitic/pelletal carbonates overlain by tidal flat laminated caps. Modeling of the cycles suggested that the parasequences were formed by low to moderate (20 m) fluctuations in sea level in order to generate transgressive systems tract oolitic cycles free from tidal flat facies on the shallow ramp, while at the same time deeper water mud to storm grainstone cycles were forming on the deep ramp/slope. However, the sea level fluctuations were reduced in magnitude in order to generate the well developed tidal flat capped cycles of the highstand systems tract. This raises the likelihood of the high frequency sea level fluctuations being highest in transgressive systems tracts and decreasing into the highstand systems tracts.

The Whitney Canyon-Carter field, Wyoming, is in the Madison Group and is volumetrically the largest gas producer in the U.S. Rocky Mountains (Harris et al., 1988). Most production is from over 100 m of porous fine dolomite interbedded with fine calcitic dolomite and tight lime grainstone. This porosity zone is underlain by fine grained basinal limestone and overlain by a fractured and brecciated section of sabkha limestone and dolomite. In the main porosity zone, the best reservoirs occur in cycles of porous fine grained, dolomitized subtidal wackestone/mudstone overlain by tight crinoidal grainstone. The fine to very fine grained dolomite forms the reservoir facies and has intercrystal porosity along with skel-moldic porosity. The dolomites were probably generated by 3rd order scale processes, notably from brines from the overlying, thick prograding evaporitic sabkha facies in the upper Madison Group. These brines refluxed down section to dolomitize permeable subtidal wackestone/mudstone. The Mission Canyon dolomites were subsequently modified in meteoric or mixed water environments recharged from the overlying unconformity (Dorobek et al., in press). The tight lime grainstones may have escaped dolomitization because they were already cemented (Harris et al., 1988), although Dorobek et al. (in press) considered that much of the Mission Canyon calcite cements post-date the early dolomite. In the Little Knife field, cycle top dolomites at 3000m contain 14% porosity, most of which is located in leached intercrystalline pores (Lindsay and Kendall, 1985).

Limestones in the Frobisher-Alida zone of the upper Mission Canyon, Glenburn field, produce from stacked porosity zones. Some of this porosity occurs in carbonates previously described as marine algal and oolitic facies and reinterpreted as subaerial (Gerhard, 1985). These facies include vadose pisolite and carbonate crusts, with much vuggy interparticle and fenestral porosity. Porosity development was in part coeval with anhydrite deposition in pans and sabkhas, suggesting that the waters involved in forming the vadose fabrics were periodically hypersaline. These anhydrites form the updip and overlying seals.

Upper Mississippian, Eastern U.S.A.

Upper Mississippi cycles of the Eastern U.S.A. formed on a broad ramp and are dominated by oolitic and skeletal grainstone cycles up to 6 m thick, capped by microkarsted surfaces, local caliches, and detrital chert; caliches are best developed at the 3rd order sequence boundaries (Niemann and Read, 1988; Nelson and Read, 1989; Al-Tawil, personal communication, 1992). Lagoonal and tidal flat facies rarely cap cycles. The semi-arid climate inhibited diagenetic alteration during the moderate amplitude fluctuations because meteoric ground waters were poorly developed. Subaerial diagenesis was mainly restricted to upper parts of cycles where caliche crusts and detrital carbonate- and chert-veneered erosion surfaces formed (Niemann and Read, 1988; Harrison and Steinen, 1978). The main cements within the caliche profiles were fibrous turbid cements (needle fiber calcites), micrite coatings on grains, and rhizocretions formed by carbonate precipitation on roots and rootlets; sparry calcite cementation and leaching were relatively minor. Because there was little early sparry calcite cementation of oolitic grainstone facies during 4th/5th order sea level falls, the distribution of the hydrocarbon reservoirs is tied to the lateral and vertical distribution of grainstone bodies. Chalky microporosity is important in some of the gas reservoirs in the oolite units, although it is not clear whether this relates to later 3rd order sequence-bounding unconformities or burial fluids.

DIAGENESIS ASSOCIATED WITH HIGH AMPLITUDE 4TH/5TH ORDER SEA LEVEL CHANGES

Large, high frequency fluctuations in sea level (60 to over 100 m oscillations) appear to have characterized the Pleistocene-Miocene, the Pennsylvanian-Early Permian, and the late Precambrian during global ice-house times (Fischer, 1982)(Figure 1). Parasequences or cycles formed at these times appear to have been dominated by eccentricity (100 to 400 k.y.) rhythms (Wright, 1992). Even though sea level fluctuates with roughly 20, 40, 100, and 400 k.y. periods, only the higher sea levels riding on the eccentricity rhythm will flood shallow carbonate platforms. Consequently there are many missed beats, and cycle deposition may be interrupted by periods of emergence of tens of thousands to 100 k.y. when sea level and associated water tables lie far below the platform surface (Matthews and Frohlich, 1987; Goldhammer et al., 1990). Consequently, the parasequences commonly are bounded by major disconformities, prompting some workers to consider these as depositional sequences, in spite of the small time that they may represent.

Matthews and coworkers used numerical simulations to predict likely diagenetic facies resulting from these high amplitude, high frequency fluctuations in sea level on the accumulating sedimentary pile (Matthews and Frohlich, 1987; Quinn, 1989; Humphrey and Quinn, 1989). In their models, thickness of the vadose zone is related to position of sea level, whereas thickness of the phreatic meteoric and mixed layer are input parameters. Rates of diagenesis are input as "diagenetic clocks" (time taken to completely convert one phase to a different mineral phase). The models of Matthews and Frohlich (1987) illustrate how carbonates are subjected to repeated exposure to different diagenetic settings ranging from marine pore waters, mixing zone waters, and phreatic and vadose meteoric zones (Figure 9) to a greater extent than with lower amplitude sea level fluctuations. Unfortunately, the actual diagenetic record preserves a relatively incomplete record of these large sea level fluctuations in terms of the diagenetic stratigraphy.

Key Features

On relatively flat-topped platforms, carbonate cycles that developed under high amplitude, high-frequency sea level fluctuations form layer cake successions of 1 to 10 m thick, dominantly subtidal, 4th order cycles punctuated by subaerial, karstic disconformities (Perkins, 1977; Beach, 1982). On ramps, cycles are erosionally bounded and are highly shingled in terms of their lateral and vertical stacking pattern resulting from 20 and 40 k.y. fluctuations riding on larger amplitude 100 to 400 k.y. sea level fluctua-

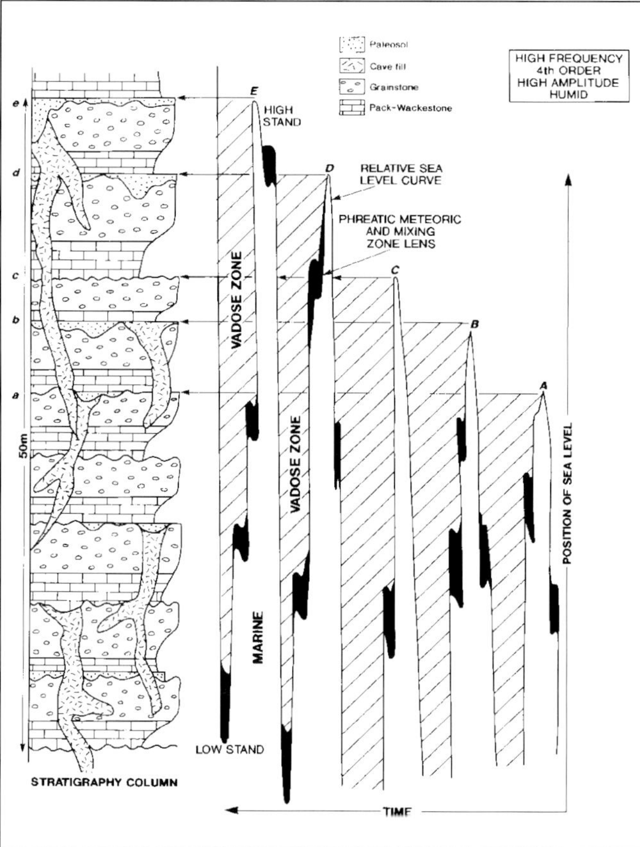

Figure 9. Idealized succession of carbonate cycles generated by large high frequency sea level fluctuations and humid or fluctuating arid to humid conditions. Each cycle is subjected to repeated exposure to different diagenetic environments (marine, mixed marine-meteoric, meteoric phreatic, and meteoric vadose) labeled on the relative sea level curve (right). Modified from Matthews and Frohlich (1987). Cycles are capped by paleokarst, which extends downward through underlying cycles as caves and passages.

tions (Read et al., 1991). Thus 100 and 400 k.y. lowstand cycles are confined to the outer ramp and have no updip equivalents. Cycles are extremely difficult to correlate without highly distinctive, continuous marker beds.

Platforms that formed under these high amplitude sea levels commonly have large relief reefs and banks due to rapid flooding of the platform, and cycles show considerable thickness variation. The cycles have well developed disconformable boundaries developed typically on subtidal facies and there may be considerable erosional relief. They generally lack tidal flat facies because sea level falls off the platform faster than tidal flats can prograde. Thus any highstand tidal flat facies and associated dolomites are likely to be in narrow bands adjacent to cratonic shorelines or in the lee of cemented islands. Cycles updip are dominated by shallow water facies capped by clastics or major disconformities. Mid-ramp or mid-shelf cycles may preserve a transgressive succession overlain by deep water shale/pelagite (maximum flooding) which shallow up into shallow water buildups or grainstone shoals, and then into clastics or a capping disconformity. Deep-shelf/ramp margin cycles can have deep water pelagics that are burrow-mixed with lowstand shallow water sediments such as ooids (Logan et al., 1969).

Significant hydrocarbon reservoirs occur in both the arid and humid cycles, contrasting with greenhouse and transitional systems where only the arid cycles form significant hydrocarbon reservoirs. Humid cycles in this case are successful as hydrocarbon traps because for the first time high amplitude flooding events resulted in greater potential for intercycle seals on muddy late Paleozoic platforms, and similarly high amplitude sea level falls resulted in significant porosity development in shelf and shelf margin bioherms.

In arid zone cycles porosity may be plugged at cycle tops by caliche, but below this zone will be mainly primary intergranular in non-dolomitized buildup and shoal-grainstone units. In dolomitized cycles related to highstand sabkhas on the inner platform or late highstand to lowstand evaporite basins whose refluxing brines dolomitize underlying facies, the porosity will be secondary intercrystal and remnant primary intergranular porosity in the dolomites, along with primary intergranular porosity in limestone. An excellent example of such a system is the Lower Permian of the Barents Sea (Stemmerik and Larsen, this volume). Although often not developed, cycle top anhydrites may seal and stratify the reservoir and protect it from effects of later cementation. There are a few examples of producing oil fields, namely in the Pennsylvanian of the Paradox basin of Utah and Colorado, U.S.A., and Miocene carbonate reservoirs of Northeast Syria, Northwest Iraq, and the Red Sea/Gulf of Suez.

Humid zone carbonate cycles can have single to multiple caliche horizons capping cycles if there is wet-dry seasonality, and numerous karstic caverns and sinkholes can extend downward through several cycles. Cycles that contain internal seals of deep water shale/carbonate can have primary porosity preserved in early cemented and leached regressive upper parts of cycles that resist compaction (Heckel, 1983) and that are relatively isolated from burial diagenesis to form highly stratified reservoirs. Coarser grained cyclic successions such as high energy reefs may lack internal seals and form poorly stratified reservoirs. In successions of high amplitude carbonate cycles with poorly developed internal seals, uppermost cycles can have primary intergranular porosity, while lower ones will show increasing moldic, vuggy, and cavernous porosity due to superimposed meteoric diagenetic events caused by large scale sea level fluctuations; the uppermost cycle is often the most porous, however. It may be difficult to identify these sub-seismic scale cycles in core or wireline logs because of the extensive diagenetic overprint and crosscutting secondary porosity and porefills. There are many examples of successful reservoirs, including the Upper Pennsylvanian Happy and Seberger fields, northwest Kansas (Ebanks and Watney, 1985), and Horseshoe "Atoll," Texas, U.S.A.

Diagenesis, Arid Climate, and High Amplitude Sea Level Changes

Middle Pennsylvanian, Paradox Basin Cycles, U.S.A.

Middle Pennsylvanian shelf cycles in the Paradox basin form major hydrocarbon reservoirs (Choquette and Traut, 1963; Pray and Wray, 1963; Wilson, 1975). The shelf carbonates contain large scale, regionally correlative cycles up to 30 m thick and possibly 100 to 400 k.y. in duration, that contain higher frequency (ave. 6m thick) cycles possibly 20 to 40 k.y. in duration (Goldhammer et al., 1991). Cycle facies include basal reworked sandstone, deep water black laminated dolomitic shale and argillaceous mudstone, spiculitic muds with packstone lenses, and small phylloid algal, mud-rich mounds 6 to 12 m thick, with non-skeletal and skeletal packstone/grainstone caps. Most caliches are only developed on cycles associated with regionally correlatable 4th order 'sequence' boundaries. Cycle-boundaries are usually marked by subaerial exposure surfaces or in downslope positions, by lowstand algal laminites and evaporites. Basin sections contain 29 correlative shale-evaporite, 4th order cycles. Meteoric diagenesis caused by major sea level falls is evident in the upper parts of shelf cycles, which show much leaching of aragonite in algal bank facies, skeletal packstone/wackestone, and non-skeletal grainstone/packstone caps (Wilson, 1975) and much non-luminescent, non-ferroan sparry cement (Dawson, 1988). The restriction of the meteoric diagenesis to upper parts of cycles, even though 4th order eustatic sea level falls of 50 to 60 m were involved, probably relates to gradual subsidence diminishing the relative sea level fall to 20 m or so (Goldhammer, et al., 1991). Shale and muddy carbonates low in the cycles also acted as aquitards, preventing deep penetration of meteoric waters. The evidence for meteoric diagenesis in these cycles is interesting in view of the arid conditions of deposition of time-equivalent evaporites in the Paradox basin. Climatic changes may have been rapid and complex, as in the Late Mississippian of Britain (see earlier section).

In contrast, Middle Pennsylvanian subsurface algal mounds of the Bug and Papoose Canyon fields of the Paradox basin, Utah and Colorado, show evidence of more typical fully arid zone diagenesis (Roylance, 1990). The mounds underwent marine aragonite cementation, followed by some leaching of capping facies above mounds, brecciation, and dolomitization; there was little sparry calcite cementation, in spite of substantial sea level fluctuations, but there was considerable porosity reduction by anhydrite and less common halite cements. Similarly, Upper Pennsylvanian arid zone cycles of the Lansing-Kansas City Group underwent little sparry cementation during emergence and development of cycle capping disconformities (Goldstein et al., 1991); these cycles apparently developed in a topographically low area of the Mid-Continent U.S.

Miocene Cycles of Iraq and Syria

Miocene cyclicity was dominated by large (greater than 60 m) eustatic sea level fluctuations of 100–1000 k.y. duration each (3rd–4th order cycles). This is suggested by detailed high resolution seismic profiling of the Miocene continental margin of the Southeastern U.S. (Snyder et al., 1992), the long term cooling trend evident in the Eocene to Pleistocene deep sea isotope record (Moore et al., 1982), and sedimentologic evidence of major ice sheet expansion from the Oligocene into the Miocene, with major glacial events in the earliest Oligocene, at the end of the Aquitanian, and from the Serravalian to the Messinian (Ingle, 1981; Fulthorpe and Schlanger, 1989; Domack and Domack, 1991) (Figure 1).

Carbonate-evaporite cycles in the Zagros foreland basin in Iraq and Syria are included in the lower Miocene Euphrates, Sericagni, and Dhiban formations, lower middle Miocene Jeribe, and upper middle Miocene lower Fars formations (Dunnington, 1959; Metwalli et al., 1974; Buday, 1980). The platform cycles have basal algal-larger foraminifer-miliolid packstones, mid-cycle fenestral wackestones and algal laminated mudstones, passing up into nodular anhydritic sabkha facies (Philip et al., 1972; Ibrahim, 1978; Shawkat and Tucker, 1978; Tucker and Shawkat, 1980). Lower Aquitainian Sericagni Formation of condensed, organic rich anoxic basin mudstones with fish debris are lateral facies equivalents of the Euphrates platform facies. Upper Aquitainian basin fill cycles of organic rich mudstone, dolomite, thin laminated anhydrite to thick halite of the Dhiban Formation developed above the Sericagni Formation (Dunnington, 1959; Hart and Hay, 1974; Metwalli et al., 1974). These were deposited in response to forced 4th-5th order glacio-eustatic desiccation cycles during an overall 3rd order scale lowstand, probably related to the late Aquitainian glacial episode. This resulted in the subaerial exposure of the top of the Euphrates platform and enhancement of reservoir quality via dissolution (Horbury, personal observation). Other comparable examples of major lowstand basinal evaporite cycles with fringing highstand carbonates are the Castille/Capitan of the Permian basin and Zechstein cycles of the North Sea.

During the middle Miocene in Syria and Iraq, cyclicity was widespread and uniform in character over wide areas. Platform cycle base packstones were often heavily dolomitized and aragonite leached to form molds due to hypersaline brine dissolution below sabkhas (Sun, 1992). This formed porosities of 17 to 28%, but porosities are lower where dolomite and anhydrite cements plug some porosity, e.g. in the Jeribe Limestone, Jebissa oil field (Philip et al., 1972). Reservoirs are strongly layered with sabkha evaporites best developed in the Serravalian glacial episode, forming good seals. The 3rd order large amplitude eustatic fluctuations were therefore related to development of Aquitainian and Serravalian ice sheets and controlled the occurrence of major evaporatic units, marine flooding events, and 4th-5th order glacio-eustatic cycle stacking patterns. Tectonics also played an important role because this large basin structurally was barred and possessed very restricted access to the Mediterranean Sea and the Indian Ocean, whilst the foldbelt to the northeast was rising continually and supplying dissolved sulfate and carbonate. The barring of and refluxing within the Zagros foreland and the similar interconnected Red Sea and Mediterranean basins may, in turn, have affected the expression and sensitivity of high-frequency glacio-eustasy by damping the amplitudes and altering rates of change of high amplitude cycles. Such damping could explain the presence of well-developed tidal flat and sabkha facies. Also, cyclic input of meteoric waters due to seasonal and longer term weather patterns in the Zagros foldbelt probably interfered with the rates of desiccation and cycle patterns, as suggested by Margaritz (1987) for Early Permian evaporite cyclicity in the Paradox basin, U.S.A. and evaporite precipitation patterns in the Dead Sea.

Pliocene-Pleistocene of Eniwetak Atoll

Diagenesis of Pliocene-Pleistocene carbonates of Eniwetak Atoll (Marshall Islands) occurred under arid to semi-arid conditions (Quinn, 1991). Shallow buried sediments (less than 100 m depth) remain aragonite-rich with a few meter-scale calcite-rich intervals at locations of Pleistocene caliche soils or thin (less than 5 m) paleophreatic lenses. More deeply buried sediments from 100 to 180 m show pervasive calcitization of aragonite, extensive moldic and vuggy dissolution, moldic porosity, and bladed and equant sparry calcite cements. Cements typically are nonluminescent and largely unzoned, reflecting the extensive development of oxic waters, low organic content, and high permeability of the reefal sediments (Quinn and Lohmann, personal communication, 1991). Calcitization occurred in superimposed meteoric phreatic zones both in response to exposure shortly after deposition and during the frequent Pleistocene sea level lowstands when sea level fell tens to over 100 m below the platform top, yet the extensive vadose environments that must have existed appear to have left little recognizable diagenetic record.

Pleistocene of Southeastern Bahamas

Pleistocene cements in Hogsty reef, southeastern Bahamas, are dominated by micritic and lesser bladed high-Mg calcite, and fibrous aragonite (Pierson and Shinn, 1985). These metastable marine components coexist with Pleistocene meteoric sparry cements, reflecting an arid climate and lack of a permanent fresh water lens.

Quaternary of Shark Bay and Lake McLeod, Western Australia

Pleistocene carbonate cycles in Shark Bay were deposited under semi-arid conditions, and are disconformity bounded with karstic surfaces, laminar and pisolitic caliches, and caliche-ooid soils (Logan, 1974). Below the laminar caliche, grainstones become plugged by mottles of fine micritic caliche cement and locally by needle spar. This mottled caliche zone passes down into relatively unlithified sediments which retain much primary porosity, as well as their metastable aragonite and high Mg calcite mineralogy due to immersion in saline groundwaters (Logan, 1974). However, even in semi-arid Shark Bay, metastable carbonates in molluscan coquina dunes in the vadose zone can be stabilized to neomorphic calcite and cemented by sparry calcite. Aragonitic sediments peripheral to ephemeral river mouths also are leached and cemented with sparry calcite .

Highstand embayments such as Shark Bay can evolve within the duration of a cycle or a few thousand years, into marginal marine evaporite basins such as Lake McLeod basin (Logan, 1989), due to barrier formation assisted by sea level fall. The initial marine transgressive facies if muddy, can function as a bottom seal for the brines and become overlain by sulfate and halite deposits. The evaporites are fed by marine influx via karstic conduits connecting the basin with the sea. Outflow of the evolved brines through karstic sinks in the basin floor can form much reflux dolomite in subjacent carbonates (Logan, 1989, Logan, personal communication, 1991), similar to that postulated for Bonaire (Deffayes et al., 1965). These dolomites can have much relict intergranular and intercrystal porosity.

Diagenesis, Humid Climate and High Amplitude Sea Level Changes

Under humid or cyclically alternating arid to humid climates, high amplitude sea level fluctuations in the 20 k.y. to 400 k.y. range are likely to leave diagenetic signatures reflecting the large scale vertical and lateral migration of diagenetic zones compared with low amplitude fluctuations of sea level. These large scale sea level fluctuations would cause the rocks to be subjected repeatedly to meteoric vadose, meteoric phreatic, mixing zone and sea water diagenetic environments (Figure 9). Any meteoric diagenesis during 4th or 5th order highstands would be restricted to eolian-generated islands or newly emergent shoals. In these local areas, sediments can undergo cementation and leaching of aragonite associated with island water tables that may be ephemeral (Budd and Land, 1990). However the bulk of the diagenesis is likely to occur during regional platform emergence, associated with lowered sea levels during 20, 40, 100 or 400 k.y. still-stands. This is because sediments reside longer in a specific diagenetic zone at these times.

Climate and Diagenesis of Pennsylvanian Cyclothems, U.S.A.

Pennsylvanian cyclothems of the U.S. experienced diagenesis under high amplitude sea level fluctuations during the major Pennsylvanian-Early Permian glaciation on Gondwana. Overall the Pennsylvanian climate of the U.S. was probably wet tropical in the east where coal-bearing deltaic facies were widely developed (Appalachians, Kansas and north central Texas) to semi-arid/arid in the west (northern Denver-, central Colorado- and Paradox basin evaporites and eolianites of the Wyoming shelf) (Driese and Dott, 1984). The Pennsylvanian climate in the U.S. also became drier with time, for example the Middle Pennsylvanian in the Mid-Continent had a relatively humid climate (indicated by blocky paleosol mudstones and overlying thick coals) compared with the drier climates of the Late Pennsylvanian to Early Permian (Heckel, 1985; Wilson, 1975).

Pennsylvanian Holder Formation, New Mexico

Goldstein (1988, 1991) describes probable humid zone diagenesis associated with Late Pennsylvanian eustatic sea level fluctuations in the cyclic Holder Formation, New Mexico, (U.S.A). There are up to 20 cycles from 6 to 30 m thick, which consist of shale/siltstone/sandstone with limestone conglomerate pebbles, overlain by wackestone, then by lime sands or bioherms, and cycle-capping paleosols. The sea level fluctuations were over 30 to 50 m in magnitude based on paleorelief of a shelf edge subaerial exposure surface (Goldstein, 1988) and probably were up to 100 m based on continent-wide facies analysis (Heckel, 1980).

Paleosol fabrics dominate the upper meter of cycles and apparently formed under freshwater, waterlogged conditions (Goldstein et al., 1991a). Most common pores are molds after aragonite fossils and few aragonite fossils were neomorphosed. The sediments show little distinguishable vadose cementation. The bulk of the calcite cementation occurred on the shelf-edge and shelf-crest associated with successive phreatic lenses fed by cycle capping disconformities. Some lenses may have been associated with perched water tables due to interlayered shale or cemented limestone aquitards which limited the thickness of the lenses from a few meters to 36 m. Some lenses show maximum early cementation just beneath the subaerial disconformity, whereas in others, maximum cementation was lower in the lens. In contrast to shelf margin facies, basin or lagoonal rocks show little early cementation. Even though the sea level fluctuations were probably greater during the Pennsylvanian than in the Late Mississippian, these cements are less correlatable than those described by Horbury and Adams (1989). This poor correlation between cements probably is due to the numerous fine shaly aquitards in the Holder Formation, which affected downward flow of water and tended to make the penetrating waters suboxic. Otherwise, the diagenesis of these two humid cycle types is broadly similar.

Pennsylvanian Cycles of the U.S. Mid-Continent

Heckel (1983) described a diagenetic model for U.S. Mid-Continent cyclothems of Middle Pennsylvanian age. The diagenetic signature has characteristics of humid conditions with widespread sparry cements, and abundant leaching, to semi-arid conditions with caliche and local evaporites. Cycles are a few meters to over ten meters thick and consist of basal near-shore shales and limestone of the transgressive systems tract, a black shale marking maximum marine flooding, followed by coarsening upward, open marine bank- and shoal water limestone facies of the highstand, and a caliche or paleosol mudstone cap marking the lowstand (Heckel, 1980; Watney et al., 1989).

Caliches, red paleosols, rhizoliths, and leach fabrics were developed on cycles that formed under more arid paleoclimates in Western Kansas, whereas karstification was more prevalent in the wetter paleoclimates of Southeastern Kansas (Heckel, 1980; Watney et al., 1989, Goldstein et al., 1991a). The transgressive limestones below the black shales show little early diagenesis and retained their aragonite mineralogy into early burial when they underwent pervasive overcompaction and neomorphism of aragonite to calcite. The black shale apparently acted as an aquiclude, shielding the underlying transgressive limestone from downward penetration of meteoric water. In contrast, the upper, regressive limestones show much early marine cement, paleocaliche and soil formation, abundant leaching of aragonitic grains, and pervasive blocky calcite cementation prior to much compaction. This moldic and vuggy porosity is important in the Pennsylvanian reservoirs whereas calcite cementation and mud infiltration have occluded porosity in non-reservoir rocks (Ebanks and Watney, 1985). Similar leached reservoirs due to large scale migration of fresh-water lenses occur in regressive bryozoan and foraminifer limestone above deeper water clayey facies in the Middle Pennsylvanian Goen Limestone cyclothems, Central Texas. These bryozoan/foraminifer facies have better porosity than the associated algal mound facies, in which early phreatic calcite cementation acted to partly occlude porosity (Marquis and Laury, 1989).

Pennsylvanian Horseshoe Atoll, Texas

The raised rim buildups on the southern and eastern margin of the Horseshoe atoll are host to billion barrel fields that developed as a result of large, high frequency sea level fluctuations and attendant fresh water diagenesis of phylloid algal buildups (Schatzinger, 1983; Reid and Tomlinson-Reid, 1991). The buildups contain numerous erosionally bounded parasequences. Tidal flat fenestral dolomitic mudstones were restricted to local shoals in the interior of the buildups on the elevated rim. Oolitic grainstones developed as belt-like shoals over the buildups along the northern margin of the rim and are also major reservoir facies. The buildup facies are bordered downslope by sponge-algal-bryozoan mounds (in water depths up to 20 m), which passed downslope into deeper shelf and slope muds. These muds form seals within and lateral to the buildups. During the repeated sea level falls, islands developed weathered zones, paleosols and eolianites, and meteoric lenses caused leaching to form oomoldic and skelmoldic porosity, some dissolution of calcite cement, recrystallization of mud and grains to microcrystalline calcite, and secondary vug- and small cave development. Repeated exposure of the phylloid algal buildups even occurred during final drowning of the complex, attesting to the large fluctuations in sea level. The leached oolite grainstones have up to 30% porosity and up to 100 md permeability. Mean porosity and permeability is lower in skeletal grainstones than in the phylloid algal buildup facies and is even lower in the leached sponge-algal-bryozoan mound facies, even though these facies may all have occasional intervals of high porosity.

Pleistocene of the Bahamas

Pleistocene platform cycles of the Bahamas are 1 to 10 m thick, disconformity bounded units (Beach, 1982). Karst features, caverns, vugs, leached peloids, and aragonitic fossils are widely developed and occur at many levels in the platform stratigraphy (Beach, 1982; Dill, 1977; Ginsburg et al., 1990) (Figure 9). Such karst features include large sea water-filled sinkholes over 100 m deep, extensive solution-enlarged joints (hundreds of meters to kilometers long) parallel and perpendicular to the bank margin, and vertical and horizontal passages commonly many meters wide (Dill, 1977; Smart et al., 1988). These are enlarged by meteoric dissolution at the surface and in the shallow subsurface by waters in the mixing zone, which is up to 10 m thick. They may have partial to complete fills of blocks and finer sediment of marine and terrestrial origin. These cavernous zones honeycomb the platform and may be represented by numerous zones of no recovery of core (Dill, 1977; Beach, 1982; Ginsburg et al., 1990).

In Bahamian platform cores (Beach, 1982), the complex interlayering of diagenetic phases as predicted by the sea level history and the models (Figure 9) can rarely be deciphered. Within each cycle, induration is greatest just beneath the disconformity surface, where primary and secondary pore space may be almost totally occluded; this induration decreases markedly downward. Also, younger cycles show little cementation overall, but induration increases into lower, older cycles, reaching a maximum a few cycles (at 10 to 15 m subsurface depth) below the present platform top where rocks are moderately well cemented and large vugs and channels are pervasive. The amount of leaching of aragonite and formation of secondary porosity appear to offset the amount of precipitated calcite cement resulting in little net loss of pore space (Beach, 1982, Harrison et al., 1984). The porosity of the nonskeletal (originally aragonitic) sediments averages 35%, whereas porosity of the skeletal (initially calcitic) sediments at depth averages 10 to 15%. Primary porosity passes downward into secondary porosity below about 10 m below the platform top. Halley and Schmoker (1983) similarly showed little net loss of porosity with fresh water diagenesis in Florida cores, whose porosity decreased only gradually and irregularly with depth over hundreds of meters.

In the Bahamian cores, two and perhaps three stages of cementation tied to Quaternary sea level falls were recognized by Beach (1982). The first cements in cycles are marine fibrous turbid cements which become recrystallized during subsequent emergence events. Sea level fall causes deposition of meteoric, intergranular sparry cements and overgrowths, with inclusion-rich rinds on cements near the disconformity, and common rhizocretions. The cements subsequently may be overlain by infiltrated marine sediment following deposition of the next marine unit. Karstic vugs and channels lined with second generation cements are ascribed to the resultant sea level fall. With the next sea level rise and fall, a third generation of meteoric cements may develop, although recognition of these is equivocal (Beach, 1982). Much of the sparry calcite cementation could be vadose even though diagenetic modification is more rapid in the phreatic than in the vadose zone (Matthews and Frohlich, 1987; Budd, 1988). However, the vadose calcites (except for caliche-related fabrics) can rarely be distinguished from the phreatic calcites (Beach, 1982; Harrison et al., 1984), although many vugs and channels are lined with coarse spar and interlayered silts, believed to be vadose.

In some beds, leaching and calcite cementation were followed by mixing-zone(?) dolomitization, while in younger units, this pre-dolomite leaching phase is absent. Even with conservative dolomite production rates, modeling suggests that substantial amounts of mixing zone dolomites in these platforms may occur below depths of about 50 m to 100 m (Matthews and Frohlich, 1987; Humphrey and Quinn, 1989). It is unclear whether the bulk of dolomites in these isolated platforms are mixing-zone (Ward and Halley, 1985; Humphrey, 1988) or related to marine water circulation through the platform (Whitaker and Smart, 1990 and this volume) due to reflux of brines generated by evaporation during more arid phases (Berner, 1965; Goldstein et al., 1991b) or due to thermal convection of cool, deeper marine waters (Saller, 1984).

DIAGENESIS ASSOCIATED WITH LONG-TERM SEA LEVEL CHANGE

Key Features

Diagenesis related to major unconformities due to regional uplift or 3rd order (1 to 10 m.y.) sea level falls in humid climates can extend tens to some hundreds of meters down section and many kilometers downdip (Figure 4). In newly deposited sediments, the water movement is via intergranular pores by diffuse flow; intergranular porosity is plugged by early meteoric phreatic cements, whose abundance decreases down section and down flow. However, development of chalky and moldic porosity may offset this porosity decrease. With increasing time and increasing cementation and leaching, water movement becomes dominated by conduit flow along widening joints and caves undergoing dissolution beneath the karst surface. These processes may be accelerated if the substrate sediments have already undergone 4th-5th order diagenesis. Beneath unconformities that are emergent for millions of years, karstic relief, breccia filled cave systems with both marine and non-marine fills, and burial induced fracturing of cave-roofs may form extremely compartmentalized reservoirs. Pleistocene mature humid karstic terrains include those documented by Purdy (1974a, b), Smart et al. (1986) and Ford and Williams (1989) from British Honduras, China, and Southeast Asia.

Many of the world's major hydrocarbon provinces are developed beneath humid karsts, particularly where platform margins with high relief are exposed. Examples occur in the Lower Ordovician of Texas; Lower Mississippian Mission Canyon of western U.S.A.; upper Paleozoic of the Volga-Ural, Timan-Pechora and North Caspian basins, Russia; Upper Jurassic Smackover and middle Cretaceous Edwards/ Stuart City Trend formations of the U.S. Gulf Coast; middle Cretaceous Golden Lane of Mexico; Lower Cretaceous Thamama and middle Cretaceous Mishrif of Arabia, Miocene isolated platforms of Southeast Asia, and Tertiary 'buried hill' reservoirs of the Far East.

Arid climate karstification declines with precipitation so that under very arid conditions the emergence does not significantly modify the character of the host limestone (Jennings, 1983; Ford and Williams, 1989). Reflux dolomitization may occur where a 2nd/3rd order cycle top is covered with hypersaline brines following transgression, although this is an extracyclic phenomena such that the dolomitization mechanism is unrelated to the actual cycle top being dolomitized. Many such cycles actually experienced major humid-type karstification prior to inundation by hypersaline brines.

In the subsurface in non-cored well data, e.g. from wireline logs, it is easier to identify the products of karstification such as porosity development/occlusion, rather than the surface itself (Esteban and Klappa, 1983). Where clastic or basinal facies overlie the unconformity, a high gamma log count due to glauconite or uranium may mark a 2nd or 3rd order maximum flooding surface. The unconformity may be evident by a marked biostratigraphic break. In seismic reflection profiles, unconformity relief will be resolved but will be difficult to distinguish from relief associated with late stage pinnacle reefs or other high relief buildups. Further detailed discussions of geological aspects of mature paleokarst are given in James and Choquette (1988), Ford and Williams (1989), and Wright et al. (1991).

Tectonically Induced Pedogenesis and Paleokarst Formation

Carboniferous Karst, Great Britain

Emergent surfaces with well developed pedogenic and paleokarst features occur in the Mississippian of Britain and appear to have developed coincident with rifting events recorded by truncation on seismic and in well data in the East Midlands (Fraser et al., 1990) and from outcrops in Northern England (Gawthorpe, 1987; Fraser and Gawthorpe, 1990). Angular unconformity and onlap relationships are commonly observed and suggest tectonic rather than eustatic origins, for example the end-Mississippian regional uplift in the East Midlands where truncated Mississippian limestones were only onlapped by the end of the Pennsylvanian (Strank, 1987). The cements precipitated below this unconformity make up 70 to 100% of the cements in both the strongly cyclic and deeper, poorly-cyclic parts of the Mississippian. Cements are dull luminescent and contain both reduced Fe and Mn; isotopic data indicate a meteoric signature with temperatures of precipitation up to 70° C (Walkden and Williams, 1991). This is a good example of how a major karst with significant erosion can completely occlude porosity in its downdip reducing zone, despite the host carbonates developing favorable porosity during 5th and 4th order cyclicity (as described previously). This emphasizes that porous shelf carbonates without good internal seals may still suffer later porosity occlusion by meteoric cement.

Late Palaeozoic Karst, Eastern Russian Craton

Open marine platform (Eastern Russian craton) rimmed shelf carbonates of Late Devonian to Early Permian age developed in three major sequences of highstand progradation (Late Devonian-earliest Mississippian, Middle Mississippian-Early Pennsylvanian and Middle Pennsylvanian-Early Permian), each separated by major erosive karstification of the platform top and subsequent maximum flooding and deposition of basinal mudstones which acted as seals (see review of Ulmishek, 1988, for details). The seal on the Permian carbonates is, however, a thick upper Lower Permian evaporite. Platform facies are predominantly limestone with dolomite interbeds and rare anhydrite (Ulmishek, 1988), suggesting a generally humid to semi-arid climate. Evidence of intraplatform emergence is indicated by the presence of many lowstand fans, particularly in late Fammenian to Early Mississippian (Grachevskiy and Ulmishek, 1966). A major karst surface which marks a 2nd order scale sequence boundary is of intra-Mississippian age and formed due to regional uplift to the west of the Russian craton in response to collision with volcanic arcs and the East Ural microcontinent chain (Ulmishek, 1988). Numerous erosional valleys up to 80m deep cross-cut the platforms and platform margins this time and are filled with coarse alluvial-fluvial clastics passing up into shallow water shales with marls and coals. Further to the west, erosion cuts down into late Frasnian carbonates. These transgressive systems tract sediments are succeeded by deep water mudstones of the next maximum flood event; facies onlap westwards. These karst/transgression relationships are also broadly correct for the Lower Pennsylvanian and Lower Permian platform tops. The extensively karstified, high relief platform margin and intraplatform algal buildups and earlier vuggy dolomites form the main oil reservoirs, e.g. West Tebuk field, Timan Pechora basin, with production rates of 7000 BOPD (Ulmishek, 1988). Slope facies to buildups are often tight and non-productive. Diagenesis of these carbonates is poorly described, but is typified by non-luminescent meteoric cements and local dolomitization (Horbury, personal observation). In the Volga-Ural province, these carbonates have produced 8 billion barrels oil from the Upper Devonian-lowest Mississippian sequence, 4 billion barrels oil from the Lower Pennsylvanian sequence and >2 billion barrels oil from the Lower Permian (Aliyev et al., 1983).

Mesozoic, Western Mediterranean

The Amposta and other oil fields in offshore Northeast Spain are associated with a major karst surface of 40 to 60 m.y. duration (Jurassic/Cretaceous to early Miocene) developed on locally dolomitic carbonates on structural paleohighs (Martinez del Olmo and Esteban, 1983). The karst surface is covered with up to 100 m of breccia with irregular pebble to boulder sized clasts, complex generations of terra rosa and speleothems, caliche, rhizoliths, and dedolomitization of clasts. The basal contact of the breccia is marked by a high gamma unit, believed to represent the top of the paleophreatic lens, whereas the top of the breccia possesses an even higher gamma peak, associated with the paleosols. Porosity, as indicated by sonic log, is best developed in the upper 40 to 50 m of breccia. An interval of lost circulation near the base of the breccia may be a zone of cavern formation associated with the paleowater table.

Mesozoic Peripheral Bulge Unconformity, Middle East

The Cenomanian Mishrif Formation of Southern Arabia has reservoir facies associated with Cenomanian-Turonian emergence due to uplift associated with the development of the peripheral bulge of the Oman Mountains at a time of rising global sea level (Burchette, 1993). This unconformity resulted in the development of wholesale leaching and doline formation with vuggy and moldic porosity in updip limestones, including rudistid shelf margin facies (Burchette and Britton, 1985). Downdip, meteoric cementation occluded much of the porosity. Along the western margin of the peripheral bulge, clastic mudstones with siliceous nodules developed at the site of a line of paleosprings, recharged from the east. This karstified shelf and shelf margin is the reservoir for the giant Fateh oil field of Dubai, located beneath the end-Cenomanian unconformity on a salt structure (Jordan et al., 1985), and the North Dome field in Qatar.

Cretaceous, Gulf of Mexico

The role of tectonic uplift in developing porosity is illustrated well by comparing the uneconomic Cretaceous Stuart City Trend reefs of Texas and the Golden Lane oil fields of the same age in Mexico. Both platform margins experienced similar massive marine cementation (Achauer, 1977; Prezbindowski, 1985; Shinn et al., 1974; Enos, 1988), with up to 21% of the total of 38% initial porosity being occluded by cement. Local synsedimentary mold development and subsequent fill by meteoric cements related to 4th-5th order cyclicity are also important in early diagenesis (Bebout and Loucks, 1974). During the 3rd order Cenomanian sea level fall, exposure of the Stuart City Trend resulted in development of some vuggy and micromoldic porosity after algae in backreef barrier beach grainstones, but the macroporosity was largely occluded and only gas reservoirs are developed (Bebout and Loucks, 1974). In the Golden Lane, the Cenomanian to Turonian unconformity represents a major phase of prolonged uplift and karstification due to Larimide tectonism. During emergence, a very deep and often undersaturated meteoric lens was established, which resulted in leaching of the earlier marine cements and development of much vuggy

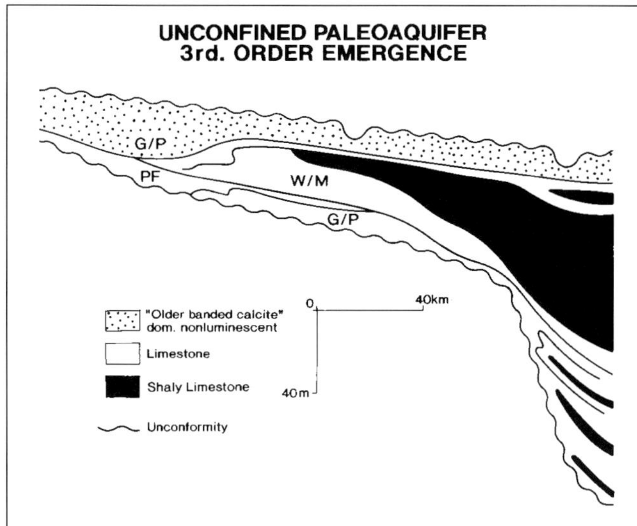

Figure 10. Stratigraphic cross section showing distribution of shallow phreatic meteoric cements in the Mississippian of New Mexico (after Meyers, 1978). "Older banded calcite" (stippled) consists of nonluminescent cement with thin luminescent bands, and was sourced from the unconformity surface on the unconfined paleoaquifer during a long-term lowered sea level. G/P is grainstone/packstone, W/M is wackestone mudstone, and PF is peritidal facies. Shaly limestone aquitards are shown in black.

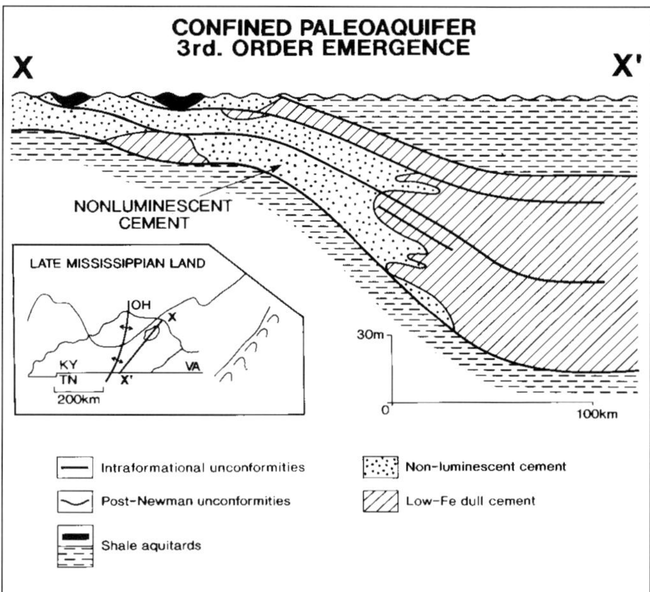

Figure 11. Distribution of dominantly nonluminescent cement in the Mississippian limestones, Kentucky (after Neimann and Read, 1988). Updip the paleoaquifer was unconfined in its upper part, but downdip, thin shales and shaly limestones within the Mississippian limestones acted as aquitards, resulting in cementation within a confined paleoaquifer. Updip early sparry cements are dominantly nonluminescent, but downdip they pass into more dully luminescent cements. Note that the section actually is oblique to depositional strike (inset figure).

and cavernous porosity (Rose, 1963; Enos, 1988). In this case, it is the length and magnitude of relative sea level fall which influences reservoir quality.

Unconformity-Related Diagenesis Associated with Diffuse Flow Aquifers

Late Mississippian Carbonates of New Mexico and the Appalachians

Upper Mississippian carbonates of New Mexico and the Appalachians of the U.S. were cemented in diffuse flow aquifers during major phases of unconformity development (Meyers, 1974, 1978; Meyers and Lohmann, 1985; Meyers, 1991; Niemann and Read, 1988; Nelson and Read, 1989). These ramp carbonates contain updip, regionally traceable shallow burial calcite cements that are non-ferroan and CL-banded, which extend down into the carbonates from a few meters to over 60 m (Figure 10). The cements in both areas can be traced over thousands of square kms and from tens to over 200 km downdip. The cement zones in both areas are (1) nonluminescent (2) luminescent and (3) nonluminescent cement, although interbasinally these zones may not correlate exactly. These "zones" commonly contain thin CL-defined cement laminae which are not regionally correlative. In the Appalachians, the updip cements pass downdip into correlative dull luminescent cements with ferroan-nonferroan-ferroan staining characteristics (Nelson and Read, 1989) (Figure 11). The early nonferroan zoned cements fill up to 80% of the pore space updip, decreasing to less than 40% 80 km to the south away from the postulated recharge area. Volumetrically, almost 60% of the New Mexico cements had been emplaced prior to karsting, and the limestones still had 10 to 20% porosity during later karst formation.

All of the updip cement zones are low in iron and could have formed from oxidizing pore waters sourced from the Upper Mississippian and Mississippian-Pennsylvanian unconformities capping the sections (Meyers, 1978; Niemann and Read, 1988). At this time the climate in the Appalachians was becoming wet and tropical, and deltas were prograding out over the carbonate ramp. The lower of these two unconformities sourced the waters for the zone 1 nonluminescent cements. The succeeding nonferroan luminescent cement formed during the subsequent Late Mississippian transgression which caused stagnation of the aquifer. The later nonluminescent zone formed during development of the Mississippian-Pennsylvanian unconformity. The nonluminescent cements (zones 1 and 3) are interpreted to have time-equivalent dull CL cements downdip where waters were more reducing. Although there are unconformities lower in the Mississippian carbonates in the Appalachians, these had little effect on cementation because they developed during arid conditions (Niemann and Read, 1988). Alternative mechanisms proposed for the New Mexico cements include (1) generation of cements from a single unconformity, with the karsting developed during the later unconformity (Meyers, 1978, 1991) or (2) formation of the cements during the latest Mississippian-pre Pennsylvanian sea level fall, which also generated the karsting (Meyers, 1988). These unconformity-sourced cements tended to plug primary porosity updip, but appear to have caused only limited plugging of pore space downdip where much of the Mississippian produces. The aquifer waters also caused metastable mineralogies to be converted to low Mg calcite updip, and could have formed chalky porosity in some of the updip ooid shoals in the Appalachians.

Note that not all shallow burial calcites are deposited from meteoric waters sourced from unconformities, but may in some cases be sourced from refluxing brines associated with evaporites higher in the section (Goldstein et al., 1991a). Similarly, other shallow burial calcite cements can be from waters sourced from tectonic highlands peripheral to foreland basins, with flow via aquifers (karstic zones, quartz sands, or carbonate sands) (Grover and Read, 1983; Dorobek, 1987).

Early Cretaceous Thamama Group, Arabian Peninsula

Chalky porosity related to formation of microrhombic calcite from ooids, lime muds, skeletal grains, and peloids, can form during long-term emergence when the sediments are subjected to meteoric pore waters. These form an important porosity type (Asquith, 1987), especially in gas reservoirs. Giant oil fields in Abu Dhabi and Eastern Saudi Arabia occur in reservoirs in Lower Cretaceous Thamama carbonates that are related at least in part to post-Thamama unconformities (Budd, 1989). These carbonates were dominantly calcitic and were recrystallized during a two-step process involving aquifers sourced from the top Thamama unconformity or from regional unconformities higher in the Cretaceous section (Budd, 1989).

Rudist buildups (Shuaiba Formation and equivalents) at the top of the Thammama Group, are the reservoirs in billion barrel fields in the Middle East. The rudist buildups formed around an intrashelf basin on the Arabian platform (Harris et al., 1984). Nearly all the reservoirs produce from moldic porosity formed by fresh water leaching of rudists and other skeletal fragments during development of the middle Aptian unconformity lasting approximately 1.5 m.y. Thick overlying transgressive systems tract shales seal the reservoirs.

Miocene of Southeast Asia

Miocene carbonate platforms throughout Southeast Asia are typically isolated and open marine in character. Oil and gas reservoirs developed in these systems include the Bombay High oil field, offshore West India (Roychoudhury and Deshpande, 1982), the Liuhua oil discovery, offshore Southeast China (Erlich et al., 1990; Turner, 1991; Tyrrell and Christian, 1992), the Luconia buildups, offshore Malaysia (Epting, 1980), and the Natuna "L" structure, offshore Northern Indonesia (Rudolph and Lehmann, 1989). Facies are typically subtidal and are divided into shallow, moderate, and deep shelf bioclast assemblages. The shallow shelf (0–15 m) is typified by green algal dominated assemblages, while the moderately deep shelf (15–40 m) is characterized by the presence of miliolids, coral fragments, imperforate foraminifera, and coralline red algae. Deep shelf (40–120 m) assemblages are typified by rhodoliths, larger benthic foraminifera, and below 80 m, by increasing abundances of planktonic foraminifera. Moderate and deep shelf assemblages dominate (>90%) most platforms. Low rates of sediment production at these water depths, absence of nearby positive basement areas off which extensive tidal flats could nucleate, and significant off-shelf sediment transport combined with rapid overall rates of relative sea level rise (Fulthorpe and Schlanger, 1989) resulted in a lack of well-developed shallow subtidal and intertidal facies (Horbury, personal observation, and S.Q. Sun, personal communication, 1992). Lowstands are marked by the development of paleokarst on all facies types, from shallow to deep shelf, and demonstrate a 3rd order periodicity of 1–2 m.y. (Rudolph and Lehmann, 1989). In other cases, mangrove and tidal flats may be developed locally during lowstands (Epting, 1980). Some 4th and 5th order cyclicity is apparent from detailed examination of electric logs and core facies, but this cyclicity appears to be purely subtidal. All of these platforms show evidence of repeated meteoric diagenesis associated with karstification (Rudolph and Lehmann, 1989; Epting, 1980; Turner, 1991; Holland et al., 1992), which rapidly stabilized the metastable carbonates that dominate these systems. Initially, mold dissolution in the vadose and upper phreatic zone and significant cavernous vuggy porosity development associated with the water table took place. Cementation of the phreatic lens progressively occluded all primary and then aragonite mold porosity as the paleokarst profile matured, and the porous vadose zone may locally have been eroded away. Primary and moldic-vuggy porosity is often only preserved in the highest (pre-demise) cycle, where it is typically 20–40%.

These sea level fluctuations must have been high amplitude such that sea level falls were rapid, with no evidence of shallowing; transgressions were equally rapid. It is likely that high-amplitude 4th order sea level falls superimposed on 3rd order sea levelfalls were responsible for these types of emergent events. During 3rd order highstands, the 4th order cycles could probably not force emergence in water depths of up to 120 m. This may explain why the emergence is characteristic of high-frequency 4th order systems, but occurs with a 3rd order periodicity where the carbonate platform was generally deposited in 15–120-m water depths. Final platform demise and backstepping often occur shortly after emergence, probably where 3rd and 4th order transgressions reinforced each other.

Diagenesis Associated with Mature Karstic Terrains, Knox-Ellenburger Unconformity, U.S.A.

The Knox-Ellenburger unconformity that caps the Cambrian-Ordovician sequence throughout much of North America (Mussman et al., 1988; Kerans, 1988; Knight et al., 1991) has localized important hydrocarbon and Pb-Zn deposits. It is up to 10 m.y. duration and is a major supersequence boundary to the "Sauk Sequence" of Sloss (1963), one of the fundamental 2nd order supercycles of North America (Vail et al., 1977). The unconformity formed during a global sea level fall and in the Appalachians during uplift associated with a subduction-related peripheral bulge. The unconformity developed during a time of increasingly humid climate, in contrast to the semi-arid conditions that typified Knox cyclic deposition, and locally has over 130 m of erosional relief. There are well developed cavern and sinkhole fills and extensive development of intraformational cave collapse breccias tens to 250 m beneath the unconformity, due to dissolution of limestone and collapse of the overlying dolomite (Mussman et al., 1988). These formed in the meteoric phreatic and mixing zone associated with a thick zone (up to 100 m) of meteoric ground water.

Nonluminescent cements occur down to 200 m below the unconformity surface (Mussman et al., 1988). Further downdip, the nonluminescent cements are absent, and the limestones are cemented by dully luminescent cement. These cements tended to plug porosity in the upper 200 m of section, so that the main porosity was karst-related vug, fissure and cavernous porosity. The initial non-luminescent cement sequence probably relates to development of a paleoaquifer whose waters were oxidizing to depths of 100 to 200 m below the unconformity, reflecting the humid setting and the highly permeable character of the carbonates. The meandering networks of caves and intraformational breccias probably carried large volumes of meteoric water into the subsurface. With stagnation of the aquifer, possibly during Middle Ordovician transgression, bright-yellow luminescent cements were deposited as the waters became more reducing. A second non-luminescent cement in the upper Knox carbonates was due to regeneration of the paleoaquifer system by upland sourced meteoric waters which caused cementation in the overlying Middle Ordovician carbonates (Grover and Read, 1983).

Paleokarstic reservoirs occur in the Lower Ordovician Ellenburger Group, Texas. These reservoirs have cumulative production of 1.4 billion barrels of oil through 1985, with low recovery efficiencies due to extreme horizontal and vertical compartmentalization of reservoirs. They have been described in detail by Kerans (1988). Most reservoirs occur in the upper 60 to 150 m of the unit, regardless of the original peritidal depositional facies, and occupy laterally extensive breccia zones (Figure 12a). They contain a lower collapse zone of chaotic dolostone clast-support breccias of highly variable thickness, averaging 15 to 30 m, with zones of unbrecciated dolomite in the lower part (Figure 12a, b). Porosity is from 1 to 15%. These are overlain by a middle zone of cave fill, siliciclastic-matrix supported chaotic breccias of Ellenburger dolomite and clasts of shale and sandstone (from the overlying Simpson Group), and intervals of crudely bedded sandstone and shale with local soft-sediment folds and faults (Figure 12a, b). This siliciclastic middle zone has a distinctive log signature and also forms a low permeability barrier. The upper zone, the cave roof, is highly fractured dolomite with fitted- to rotated clast fabrics. Porosity is from 2 to 20% and the roof zone is the major pay interval (Figure 12a, b).

The karstic cave systems developed below a stable paleoground-water table located 30 to 60 m beneath the top of the Ellenburger. The lower breccias are cave-collapse, stope-deposits whereas the middle matrix rich zone is a Middle Ordovician marine cave fill, analogous to present day blue holes. The cave-roof fracture breccias were formed during later burial compaction and fracturing of the cave roof. The breccia distribution and muddy cave fills cause the high degree of compartmentalization of these reservoirs.

Post-Mission Canyon (Mississippian) Unconformity

Mature karst associated with a supersequence boundary was developed on the Early Mississippian Mission Canyon Formation of the Madison Group, Western U.S.A., over an estimated 34 m.y., from Late Mississippian to Early Pennsylvanian time (Sando, 1988). Less well developed karst of the southern U.S. shelf is described by Meyers (1988) and Goldstein (1990). The Mission Canyon karst features described by Sando (1988) include enlarged joints (up to 30 cm wide), sinkholes (up to 15 m wide and 27 m deep), caves (up to a few tens of meters wide, extending down 60 to 105 m below the top of the Madison Group on highs, or 45 m below local base level), and two zones of evaporite solution-collapse breccias. There was 60 m of erosional relief on the mature karst surface. A complex of fault-blocks dominated the drainage pattern and many of the karstic features are filled by post-unconformity terrigenous clastics.

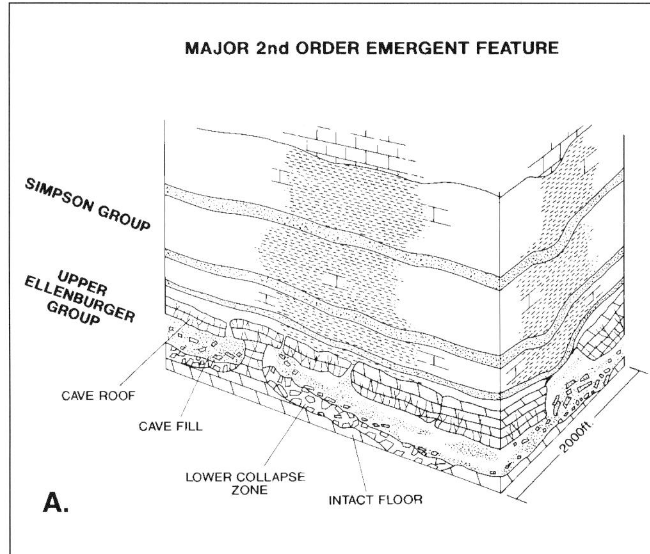

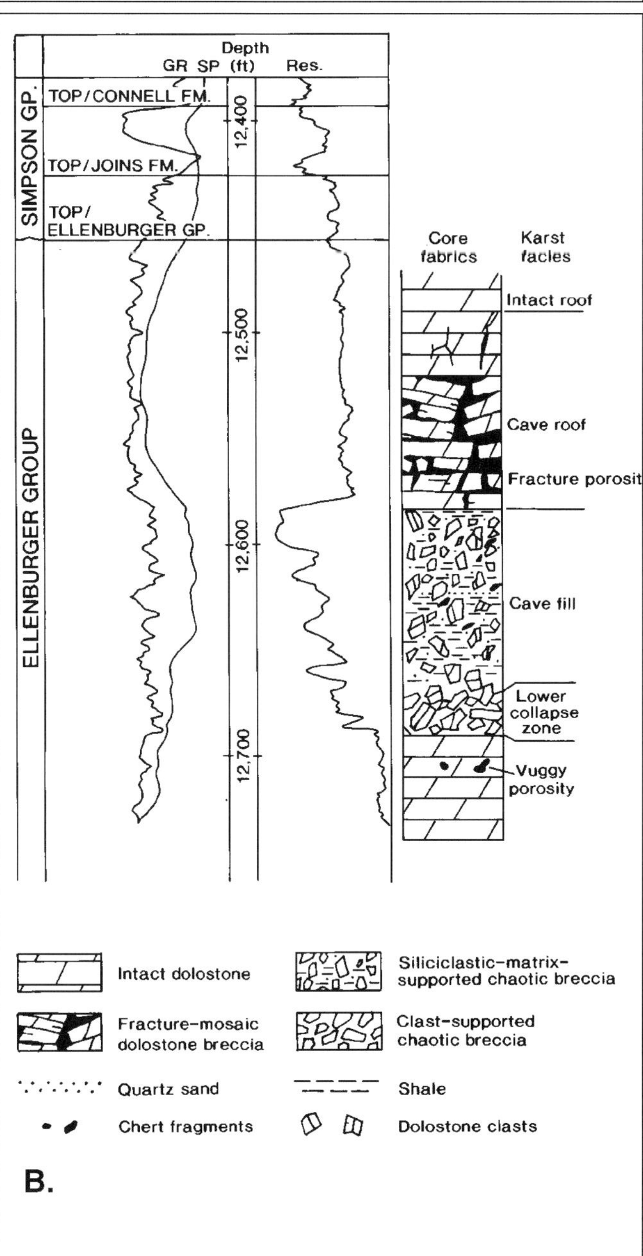

Figure 12. Karst generated during long term lowered sea level that exposed the Cambrian–Ordovician platform, U.S.A. (from Kerans, 1988). (A) Block diagram showing sub-unconformity cave system. (B) Comparison of log signature and karst facies in Gulf 000-1 TXL well, Emma Ellenburger reservoir. Much of the porosity is in the fractured cave roof.

Prior to unconformity development, the sediments were dolomitized under arid conditions and had their intergranular, fenestral and intercrystal dolomite porosities reduced by nonluminescent cements deposited during subaerial exposure of 3rd order sequence boundaries within the Mission Canyon (Smith and Dorobek, 1990). Karstic biomoldic and breccia porosity was filled by later nonluminescent cement associated with the karstic supersequence boundary. These karst related cements formed under more humid conditions and high water-rock ratios (based on stable isotopic data) associated with conduit flow than the earlier cements. Economic petroleum reserves occur in the Madison paleokarst in Northern Wyoming (Sando, 1988). The producing intervals are sealed below by a shale within the Mission Canyon, and above by a post-unconformity shale.

"Buried Hill" Paleokarstic Reservoirs

Paleokarst occurs as "buried hill" oil reservoirs in late Precambrian–early Paleozoic carbonates, onshore China (Zhai and Zha, 1982; Tong XiaoGuang and Huang Zuan, 1991). These reservoirs are typically strongly leached and karstified along pre-existing fracture trends; karstification and cavernous porosity development is due to uplift of rift shoulders and fault footwalls during the early Tertiary, producing hundreds of meters of porous buried hill (Tong XiaoGuang and Huang Zuan, 1991). Paleokarst reservoirs with significant cavernous porosity are developed in the Permian age Yates oil field of Texas, which has reserves in excess of 1 billion barrels recoverable (Craig, 1988). The porosity is best developed as water table caves in a high porosity zone some 9m thick; some caves are sited along joints, with associated sinkholes, karst towers and ?islands developed, mostly located on the platform margin grainstones of the San Andres dolomite (Craig, 1988). Dolomitization is actually a later event associated with deposition of the overlying anhydrites.

Arid Diagenesis

Where climates are truly arid (<150 mm precipitation/year) during long-lived subaerial exposure, surface-related weathering operates very slowly. Rates of 2-5mm per k.y. are quoted for the Nullarbor Plain in Australia, where porous Tertiary limestones have been exposed since the Miocene (Jennings, 1983). In autogenic karst, diagenesis comprises cementation of the upper 15m, with modest cave development beneath. Dissolution rates are higher in allogenic karst developed close to crystalline basement which occasionally protrudes through the limestones. Where limestones are already tight, karstification has little effect and only flutes develop on exposed surfaces, e.g. on Cambrian limestones exposed in the Flinders Ranges of Australia (Williams, 1978). Jennings (1983) and Ford and Williams (1989) note that the degree of karstification is proportional to precipitation. It therefore seems likely that 3rd order arid karst will little modify the inherited diagenetic and porosity textures developed prior to emergence. No examples of ancient arid karst reservoirs have been described, such that they are of limited importance compared to humid karst reservoirs.

Although dolomitization associated with refluxing evaporitic brines following 2nd-3rd order transgression cannot be considered anything other than a fortuitous extra-cyclic feature of cycle top development, there are several good examples which form hydrocarbon reservoirs. They often combine porosity generation during dolomitization with earlier, cycle top humid karstification and cavernous porosity development, with a major climate change occurring during transgression. Examples include the dolomitization of the Madison Group of the Western U.S. (Harris et al., 1988; see earlier section), dolomitization of the Yates oil field, West Texas (Craig, 1988; see above), and dolomitization of the Smackover in the Gulf of Mexico (Moore et al., 1988; Moore and Heydari, this volume). The Smackover oolites represent the highstand systems tract to a 3rd order cycle, and during the following lowstand, were initially subject to leaching of oomolds and vuggy porosity development in updip areas and partial meteoric cementation further downdip (Moore and Heydari, this volume). The climate changed from humid to arid with the establishment of the succeeding transgressive systems tract, and in the west and eastern parts of the shelf, refluxing within the Buckner evaporite lagoons resulted in dolomitization and further porosity enhancement of the underlying oolite, together with development of evaporite topseals.

SUMMARY AND CONCLUSIONS

Frequency and magnitude of sea level fluctuations, in association with climate and tectonics, influence the type of diagenetic modification on cyclic carbonate platforms.

1. On carbonate platforms formed under small, high frequency sea level fluctuations, diagenesis is mainly intracyclic. In relatively dry climates, diagenesis is dominated by early dolomitization and shallow leaching associated with prograding, regional tidal flat/evaporite facies capping parasequences (Figure 13). Depositional porosity commonly is intergranular in grainy parts of cycles or in cycle-capping siliciclastics, and intercrystal in the dolomites. Porosity loss may result from sulfate cementation of grainstones and dolomitization and leaching of muddy skeletal carbonates. Anhydrites form internal seals resulting in multiple pay zones, but will highly stratify these sub-seismic scale reservoirs (Figure 13). Under more humid conditions there is little dolomitization, and limestone caps of cycles may undergo some microkarsting, leaching, and sparry calcite cementation associated with thin fresh-water lenses. Low amplitude parasequences might be recognizable using electric logs, gamma and density-type tools recording the dense anhydrites, tight dolomites, or cycle capping siliciclastics or feldspar rich dolomite caps, or in downdip areas, the basal, possibly organic rich mudstones or argillaceous limestones.

2. Platforms that formed under moderate (few tens of meters), high frequency sea level fluctuations have more limited development of regional tidal flats and cycles are dominantly of subtidal facies that coarsen upward. Paleokarstic erosion surfaces marking 4th order emergence events cap parasequences and should be continuous over much of the shelf. Major flooding surfaces at the bases of cycles should help separate these parasequences from low amplitude types. In dry climates, there is little sparry calcite cementation or leaching, cycles have caliche at tops and metastable sediments can be preserved into the burial environment. Porosity is best developed in grainy cycles, but also can occur in dolomitized and leached muddy skeletal facies beneath prograding, evaporitic highstand systems and in subaerial pisolitic carbonates. Under wet climatic conditions, cycles are capped by karstic disconformities and diagenesis can be intercyclic, with calcite cements extending down through several cycles because of repeated establishment of successive, relatively thick groundwater bodies during lowstands (Figure 13). This results in stabilized, lithified limestone cycles with most porosity occlusion near cemented cycle tops. Because the basal facies of cycles are low-porosity muddy carbonates, the best depositional porosity will be in the grainy midparts of cycles. Reservoirs typically are 4th order, sub-seismic scale and stratified (Figure 13).

3. With large (tens to over 100 m), high frequency sea level fluctuations, diagenesis is markedly intercyclic, and sediments are subjected to rapid, large scale vertical migration of diagenetic environments through many cycles. Thus cycles are subjected to large scale (typically 4th order) alternation of diagenetic environments that include marine pore waters, mixing zone, meteoric phreatic and meteoric vadose, but they only retain a partial record of the sea level fluctuations (Figure 13). Sediments high on the platform remain in the vadose zone for much of the time. In arid climates, sediments undergo caliche formation

and some karsting, but most retain their original metastable mineralogies and are relatively poorly cemented except at cycle tops and deeper in the section. These sediments may be dolomitized by reflecting brines from highstand evaporite basins and may develop anhydrite caps which act as internal and updip seals. Under humid climates in clean carbonates, leaching of metastable carbonate in grainstones forms excellent reservoirs containing much cavernous, vuggy, and moldic porosity at many levels in the section and extending through many cycles. In highly permeable sections, cements and leach fabrics are difficult to tie to individual disconformities, and reservoirs are sub-seismic scale and weakly stratified. On the other hand, aquitards (shale or muddy carbonates) in cycles can perch ground waters, limiting meteoric diagenesis to regressive, grainy facies of cycles to form sub-seismic scale, highly stratified reservoirs (Figure 13).

4. Long-term sea level changes of 1 to 10 m.y. or more, which commonly form seismic-scale supersequence- and sequence-bounding unconformities, when coupled with humid climates may generate longstanding, regional aquifers many tens to hundreds of kilometers wide, fed from the regional unconformity surface or prograding deltaic systems (Figure 13). Aquifers in newly deposited sediments are characterized by diffuse flow and deposit cements showing a regional cement stratigraphy. Sediments in updip areas become leached to form moldic porosity, metastable carbonates are stabilized to calcite and can form chalky porosity, and cements are precipitated from oxidizing meteoric phreatic groundwaters. Further downdip or low in the section, cements decrease in abundance and are formed from more reducing phreatic groundwaters. Downdip sediments may remain as metastable mineralogies, which until they are stabilized or leached in the burial environment, can form reservoirs which are prone to over-compaction.

With time (a million to tens of millions of years), mature karst develops over the unconformity surface, and diffuse flow aquifers develop into conduit flow types as cementation plugs intergranular and moldic porosity. The karst is characterized by erosional topography, sinkholes, caves, and underground passages and most meteoric fluids move by conduit flow. Major reservoirs occur in breccias that surface the unconformity and fill caves or are localized in compaction-fractured cave roofs. These permeable zones can be separated by muddy cave fills which partition these high compartmentalized reservoirs (Figure 13).

Arid climates associated with subsequent transgression may result in extensive reflux dolomitization and deposition of evaporites. Where transgressions occur associated with humid climates, deposition of transgressive, often organic-rich shales, or condensed glauconite/phosphate mineralization may result.

5. High frequency, typically low amplitude arid cycles on a sub-seismic scale and low frequency, seismic scale humid cycles dominate the world's largest platform interior hydrocarbon reservoirs. This is because the low amplitude arid cycles either retain a high proportion of primary porosity and develop secondary porosity in metastable grainstones (e.g. oolites) or are dolomitized due to reflux in low energy facies such that intercrystalline porosity is developed. Development of anhydrite caps to individual cycles is an integral part of the systems tract. A post-carbonate regional anhydrite seal is often also developed during final basin desiccation which serves to create multiple pay zones and also often isolates the reservoirs from porosity-destructive burial diagenesis, and are best developed in the low amplitude greenhouse cycles. If there is much early marine cement, the reservoir quality can rarely be improved during later diagenesis. Source rocks are often deposited at the same time in adjacent euxinic, sediment starved basins which also suffer the effects of periodic restriction when sea levels fall off the platform.

High frequency humid zone cycles often suffer from poor reservoir quality due to repeated early meteoric cementation of cycle top high energy facies and development of isolated moldic or dolomitic intercrystal porosity in skeletal low energy facies; optimum reservoir is often developed as remnant primary porosity in the middle of cycles. Also, there is a lack of well developed or reliable internal topseals or updip lateral seals integral to the depositional systems tract due to both facies controls and because secondary porosity breaks down seal integrity; permeability baffles are mainly developed. Hydrocarbon accumulations therefore rely on later transgressive systems tract pelagic mudstone drapes, often associated with final platform demise and flooding, to seal and source the reservoir. These facies and diagenetic patterns result in stratified to well layered sub-seismic scale reservoirs. Ice-house cycles demonstrate markedly improved reservoir quality due to the more metastable primary sediment mineralogy, which improves the potential for secondary porosity development. Argillaceous seals of the transgressive systems tract are also better developed in ice-house, rather than green-house or transitional cycles.

Low frequency, 2nd and 3rd order sea level falls in humid climates are the common cause of updip macro- and chalky porosity creation and result in most of the giant, unconformity-related carbonate reservoired hydrocarbon fields. As diagenesis and cementation proceed, relatively homogeneous intergranular and early moldic porosity is lost and vuggy and cavernous porosity types result in highly compartmentalized reservoirs. These typically rely on later transgressive systems tract pelagic mudstone drapes associated with final platform demise to seal

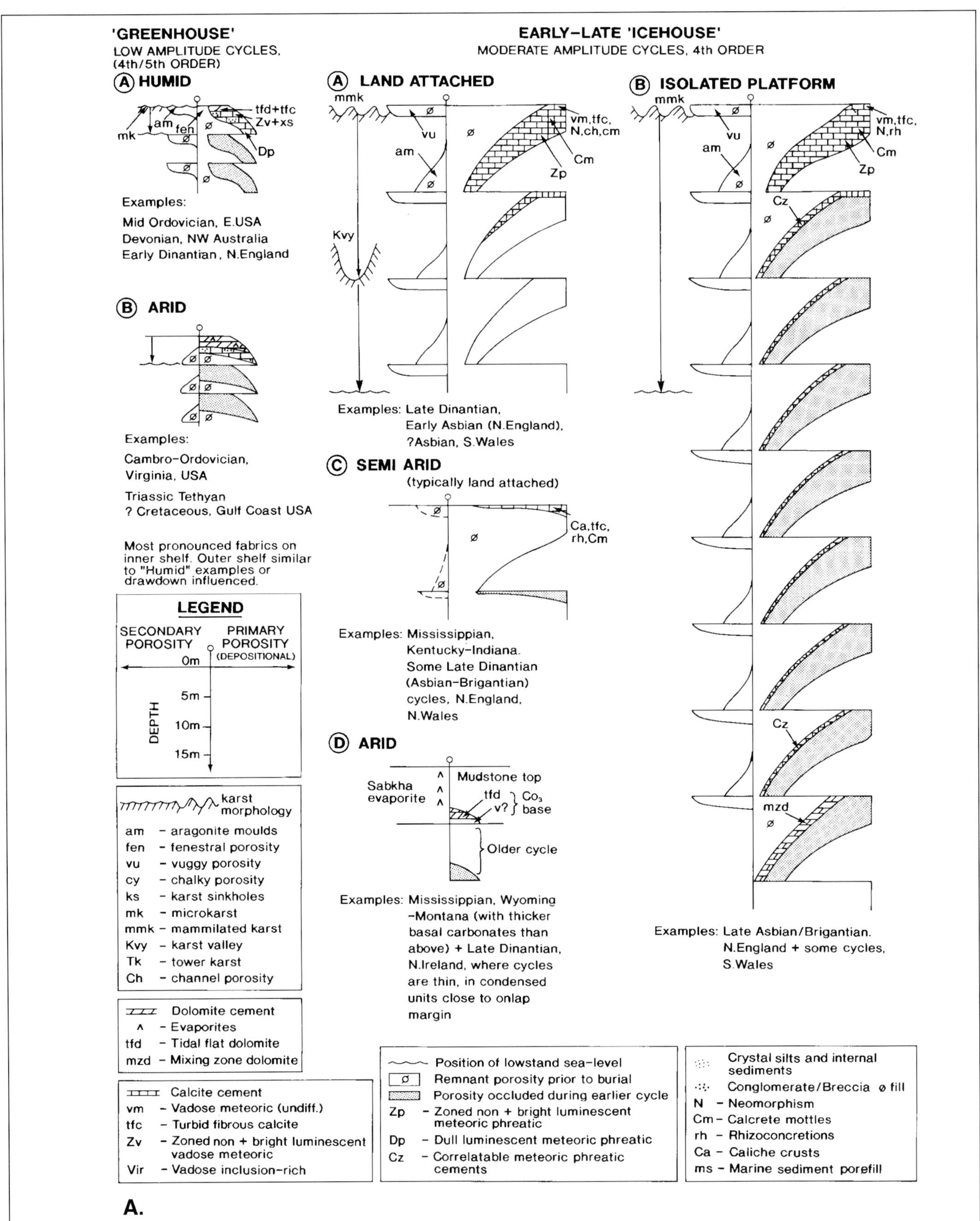

Figure 13. Summary diagram of diagenesis and porosity evolution associated with various scales of sea level fluctuation. The horizontal axes to the columns summarize primary depositional porosity and its infill by cement in each cycle (right hand side) and development and infill of secondary porosity (left hand side). Humid climates and longer term emergence (e.g. associated with 2nd and 3rd order sea level) result in reten-

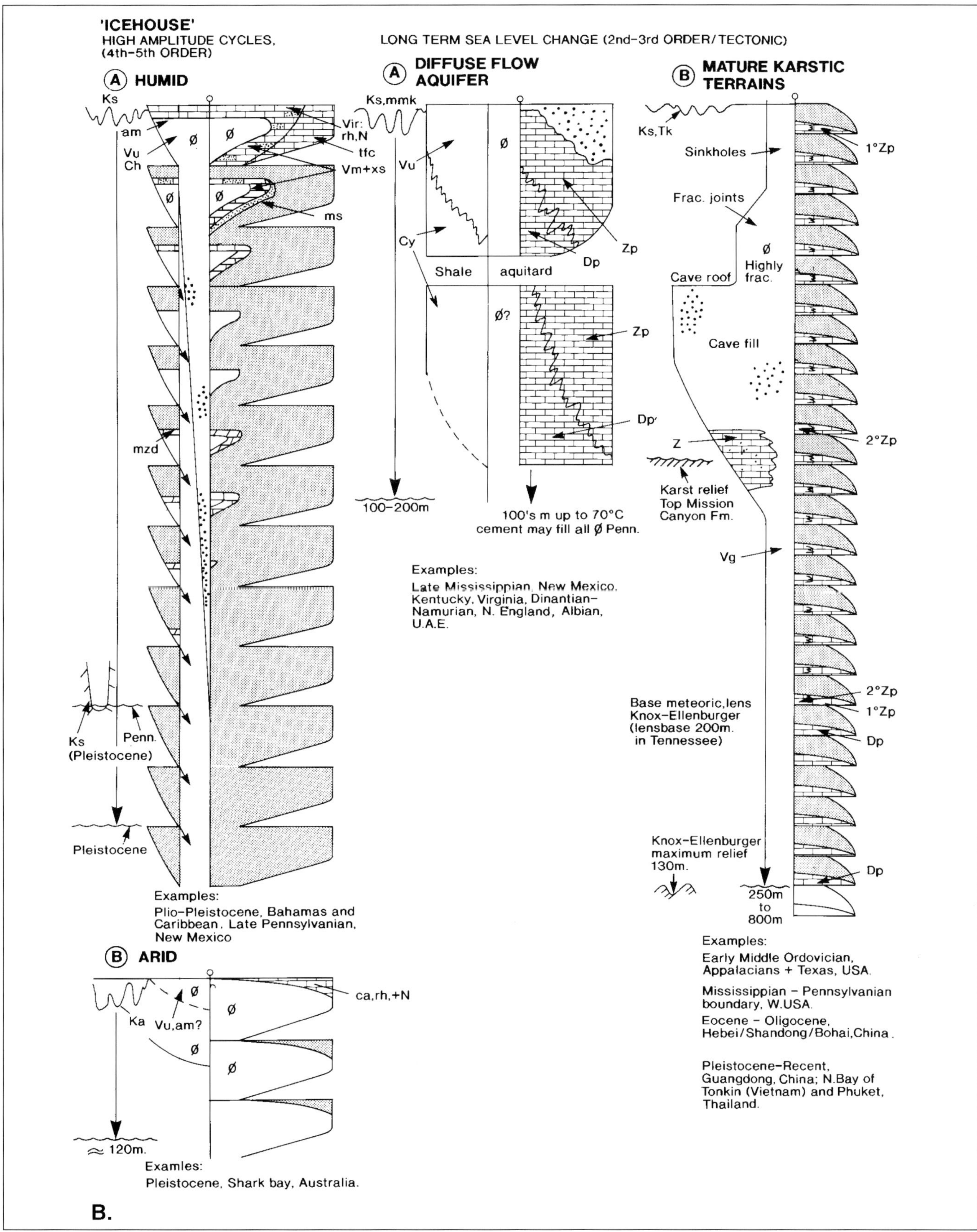

tion of less primary porosity, more cementation, and more secondary porosity development. These reservoirs are heterogenous with less facies control on diagenesis and porosity distribution than higher frequency lower amplitude fluctuations of sea level. Cycle porosity trends are stylized, with each cycle being identical for ease of illustration/modelling purposes.

and source the reservoir. Arid climate karsts are rarely successful as reservoirs in their own right since there is little diagenetic potential for modification of the host sediment. However, if the climate was arid during the following transgression, reflux dolomitization and porosity enhancement, followed by deposition of an evaporite topseal, may significantly enhance reservoir and seal capacity.

ACKNOWLEDGMENTS

J.F. Read would like to acknowledge financial support from National Science Foundation Grant No. EAR-9105558, American Chemical Society PRF21282-AC2, and Texaco, Mobil, Marathon, Chevron, Arco, Cabot, and BP oil companies. Many ideas expressed here grew from the research of past and present graduate students of the Virginia Tech carbonate research laboratory. J.F. Read also is indebted to R.N. Ginsburg, C. Kerans, J. Markello, J. Bova, R. Goldhammer, R. Goldstein, W. Dawson, T. Quinn, K.C. Lohmann, B. Logan, R. Matthews and many others for discussions, preprints and ideas. Andrew Horbury acknowledges support from British Petroleum and the National Environment Research Council, and would like to thank A.E. Adams, T.P. Burchette, D. Emery, A. Fraser, R.L. Gawthorpe, J.C. Goff, A. Gray, D. Holland, F.N.R. Sadooni and S.P. Todd for numerous discussions and ideas concerning diagenesis and sequence stratigraphy and for examples used in this paper. Typing was done by B. Pauley, and drafting courtesy of the drawing office, B.P. Exploration in London. We thank Jim Hendry and Mark Lawrence for comments on an early draft, and Paul Wright, Gordon Walkden and Andrew Robinson for helpful reviews.

REFERENCES CITED

Achauer, C.A., 1977, Constraints in cementation, dissolution and porosity development between two Lower Cretaceous reefs of Texas, *in* R.G. Loucks, and D.G. Bebout, eds., Cretaceous carbonates of Texas and Mexico: Bureau of Econ. Geol., University of Texas/Austin, Texas, v. 89, p. 127-135.

Adams, A.E., and P.J. Cossey, 1981, Calcrete development at the junction between the Martin Limestone and Red Hill Oolite (Lower Carboniferous), South Cumbria: Proceedings of the Yorkshire Geological Society, v. 43, p. 411-431.

Adams, A.E., A.D. Horbury, and A.A. Abdel Aziz, 1990, Controls on Dinantian sedimentation in South Cumbria and surrounding areas of northwest England: Proc. Geol. Assoc., v. 101 p. 19-30.

Adams, J.E., and M.L. Rhodes, 1960, Dolomitization by seepage refluxion: A.A.P.G. Bull. v. 44, p. 1912-1920.

Aliyev, I.M., G.A. Arzhevskiy, and Yu.N. Grigorenko, 1983, Petroleum provinces of the USSR. (Neftegazonosnye provintisii SSSR): Moscow, Nedra.

Al-ejil, F., and A.A.H. Abdu-Raheem, 1974, Geology of Syria: Dar-Al-fikir Publ. Co., Damascus, 266 p. (in arabic).

Asquith, G.B., 1987, Microporosity in the O'Hara Oolite Zone of the Mississippian Ste. Genevieve Limestone, Hopkins County, Kentucky, and its implications for formation evaluation: Carbonates and Evaporites, v. 1, p. 7-12.

Back, W., and B.B. Hanshaw, 1970, Comparison of chemical hydrology of the carbonate peninsula of Florida and Yucatan: Jour. Hydrology, v. 10, p. 330-368.

Barnaby, R.J., and J.D. Rimstidt, 1989, Redox conditions of calcite cementation interpreted from Mn and Fe contents of authigenic calcites: Geol. Soc. America Bull., v. 101, p. 795-804.

Beach, D.K., 1982, Depositional and diagenetic history of Pliocene-Pleistocene carbonates of northwestern Great Bahama Bank; Evolution of a carbonate platform: Ph.D. Dissertation, University of Miami, Fischer Island Station, 447 p.

Bebout, D.G., and R.G. Loucks, 1974, Stuart City trend, Lower Cretaceous, South Texas, a carbonate margin model for hydrocarbon exploration: Bureau Econ. Geol., University of Texas/Austin, Texas, v. 78, 80 p.

Berner, R.A., 1965, Dolomitization of Mid-Pacific atolls: Science, v. 147, p. 1297-1299.

Berner, R.A., 1991, A model for atmospheric CO_2 over Phanerozoic time: American Jour. Sci., v. 291, p. 339-376.

Berry, J. 1984, The sedimentology and diagenesis of the Asbian limestones of North Derbyshire: Ph.D. Thesis, University of Aberdeen (unpubl.).

Borer, J.M., and P.M. Harris, 1991, Lithofacies and cyclicity of the Yates Formation, Permian Basin: Implications for reservoir heterogeneity: A.A.P.G. Bull., v. 75, p. 726-779.

Buday, T., 1980, The Regional Geology of Iraq, Volume 1, Stratigraphy and Paleogeography, State Organization for Minerals, Baghdad, 445 p.

Budd, D.A., 1988, Aragonite-to-calcite transformation during fresh-water diagenesis of carbonates: insights from pore-water chemistry: Geol. Soc. America Bull., v. 100, p. 1260-1270.

Budd, D.A., 1989, Micro-rhombic calcite and microporosity in limestones: a geochemical study of the Lower Cretaceous Thamama Group, U.A.E.: Sedimentary Geology, v. 63, p. 293-311.

Budd, D.A., and L.S. Land, 1990, Geochemical imprint of meteoric diagenesis in Holocene ooid sands, Schooner Cays, Bahamas: Correlation of calcite cement geochemistry with extant groundwaters: Jour. Sedimentary Petrology, v. 60, p. 361-378.

Budd, D.A., and H.L. Vacher, 1991, Predicting the

thickness of fresh-water lenses in carbonate paleo-islands: Jour. Sedimentary Petrology, v. 61, p. 43-53.

Buddemeier, R.W., and J.A. Oberdorfer, 1986, Internal hydrology and geochemistry of coral reefs and atoll islands: key to diagenetic variations, *in* J.H. Schroeder and B.H. Purser, eds., Reef Diagenesis: Springer-Verlag, Berlin, p. 91-111.

Burchette, T.P., 1993, Mishrif Formation (Cenomanian-Turonian), S. Arabian Gulf: Carbonate platform growth along a cratonic basin margin: *in* T. Simo, et al., ed., Cretaceous Carbonate Platforms, A.A.P.G. Memoir 56.

Burchette, T.P., and S.R. Britton, 1985, Carbonate facies analysis in the exploration for hydrocarbons: a case study from the Cretaceous of the Middle East, *in* P.J. Brenchley, and B.P.J. Williams, eds., Sedimentology: Recent developments and applied aspects: Blackwell, Oxford, p. 311-388.

Burchette, T.P., and V.P. Wright, 1992, Carbonate ramp depositional systems. Sed. Geol., v. 79, p. 3–57.

Champ, D.R., J. Gulens, and R.E. Jackson, 1979, Oxidation-reduction sequences in ground-water flow systems: Can. Jour. Earth Science, v. 16, p. 12-23.

Choquette, P.W., and J.D. Traut, 1963, Pennsylvanian carbonate reservoirs, Ismay field, Utah and Colorado, *in* R.O. Bass, ed., Shelf Carbonates of the Paradox Basin: Four Corners Geological Society, 4th Field Conference Gdbk., p. 157-184.

Chuber, S., and W.C. Pusey, 1985, Productive Permian carbonate cycles, San Andres Formation, Reeves Field, West Texas: *in* P.O. Roehl and P.W. Choquette, Carbonate Petroleum Reservoirs, Springer-Verlag, New York, p. 289-308.

Craig, D.H., 1988, Caves and other features of Permian karst in San Andres dolomite, Yates Field reservoir, West Texas, *in*: N.P. James and P.W. Choquette, eds., Paleokarst: Springer-Verlag, New York, p. 342-363.

Crevello, P.D., 1991, High frequency carbonate cycles and stacking patterns: Interplay of orbital forcing and subsidence on Lower Jurassic rift platforms, High Atlas, Morocco, *in* E.K. Franseen, W.L. Watney, C.G.St.C. Kendall, and W.H. Ross, Sedimentary Modeling: Kansas Geol. Surv. Bull. 233, p. 207-230.

Crowley, T.J., and S.K. Baum, 1991, Toward reconciliation of Late Ordovician glaciation with very high CO_2 levels: Jour. Geophysical Research, v. 96, p. 22,597-22,610.

Davies, J.R., 1991, Karstification and pedogenesis on a Late Dinantian carbonate platform, Anglesey, North Wales: Proc. Yorkshire Geol. Soc., v. 48, p. 297-322.

Dawson, W.C., 1988, Ismay reservoirs, Paradox Basin diagenesis and porosity development: Rocky Mountain Association of Geologists: 1988 Carbonate Symposium, p. 163-174.

Deffayes, K.S., F.J. Lucia, P.K. Weyl, 1965, Dolomitization of Recent and Plio-Pleistocene sediments by marine evaporite waters on Bonaire, Netherlands Antilles: S.E.P.M. Spec. Publ. No. 13, p. 71-88.

Demicco, R.V., 1985, Patterns of platform and off-platform carbonates of the Upper Cambrian of western Maryland: Sedimentology, v. 32, p. 1-22.

Derby, J.R., and J.T. Kilpatrick, 1985, Ordovician Red River Dolomite reservoirs, Killdeer Field, North Dakota; *in* P.O. Roehl and P.W. Choquette, eds., Carbonate Petroleum Reservoirs: Springer-Verlag, New York, p. 59-70.

Dill, R.F., 1977, Blue holes - Geologically significant submerged sinkholes and caves off British Honduras and Andros, Bahama Islands, *in* Proceedings, Third International Coral Reef Symposium: Rosenstiel School of Marine and Atmospheric Science, University of Miami, Florida, p. 237-242.

Domack, E.W., and C.R. Domack, 1991, Cenozoic Glaciation, The Marine Record Established by Ocean Drilling: Joint Oceanographic Institutions, Inc. and U.S. Science Support Program, Washington D.C., 49 p.

Dorobek, S.L., 1987, Petrography, geochemistry and origin of burial diagenetic facies, Siluro-Devonian Helderberg Group (carbonate rocks), Central Appalachians (abs.): AAPG Bull., v. 71, p. 492-514.

Dorobek, S.L., T.M. Smith, and P.M. Whitsitt, in press, Microfabrics and geochemical signatures associated with meteorically altered dolomite: examples from Devonian and Mississippian carbonates of Montana and Idaho, *in* R. Rezak and D. Lavoie, eds., Carbonate Microfabrics: Frontiers in Sedimentary Geology Series: Springer-Verlag, New York.

Driese, S.G., and R.H. Dott, 1984, Model for sandstone-carbonate "cyclcothems" based on Upper Member of Morgan Formation (Middle Pennsylvanian) of Northern Utah and Colorado: A.A.P.G. Bull., v. 68, p. 574-597.

Dunham, R.J., 1969, Early vadose silt in Townsend Mound (Reef), New Mexico, *in* G.M. Friedman, ed., Depositional Environments in Carbonate Rocks, Soc. Econ. Paleontologists and Mineralogists Spec. Publ. No. 14, p. 139-181.

Dunnington, H.V., 1959, *in* R.C. Van Bellen, H.V. Dunnington, R. Wetzel and D.M. Morton: Lexique Stratigraphique International, v. 3, Asia, f.10a, Iraq. Paris International Geol. Cong. Comm. on Stratig., Centre Nat'l. Recherche, 333 p. (in English).

Ebanks, W.J. Jr. and W.L. Watney, 1985, Geology of Upper Pennsylvanian carbonate oil reservoirs, Happy and Seberger fields, northwestern Kansas, *in* P.R. Roehl and P.W. Choquette, eds., Carbonate Petroleum Reservoirs, Springer-Verlag, New York, p. 239-250.

Edmunds, W.M., 1973, Trace element variations across an oxidation-reduction barrier in a limestone aquifer, *in* E.Ingersoll., ed., Proceedings of Symposium on Hydrogeochemistry and Biogeochemistry, v.1, p. 500-526.

Elrick, M.E. and J.F. Read, 1991, Cyclic ramp-to-basin carbonate deposits, Lower Mississippian, Wyoming and Montana: a combined field and computer modelling study. Jour. Sedimentary Petrology, v. 61, p. 1194-1224.

Emery, D., and J.A.D. Dickson, 1989, A syndepositional meteoric lens in the Middle Jurassic Lincolnshire Limestone, England, U.K: Sedimentary Geology, v. 65, p. 273-284.

Enos, P., 1988, Evolution of pore space in the Poza Rica trend (Mid-Cretaceous), Mexico: Sedimentology, v. 35, p. 287-325.

Epting, M., 1980, Sedimentology of Miocene carbonate buildups, Central Luconia, offshore Sarawak: Geol. Soc. Malaysia Bull., v. 12, p. 17-30.

Erlich, R.N., S.F. Barrett, and Guo Bai Ju, 1990, Seismic and geologic characteristics of drowning events on carbonate platforms: A.A.P.G. Bull., v. 74, 1523-1537.

Esteban, M., and C.F. Klappa, 1983, Subaerial exposure environment: *in* P.A. Scholle, D.G. Bebout, and C.H. Moore, Carbonate Depositional Environments: A.A.P.G. Memoir 33, p. 1-54.

Fei Qi and Wang Xie-Pei, 1984, Significant role of structural fractures in Rengiu Buried-Hill Oil Field in eastern China: A.A.P.G. Bull., v. 68, p. 971-982.

Fischer, A.G., 1964, The Lofer cyclothems of the Alpine Triassic: Kansas State Geological Surv. Bull. 169, v. 1, p. 107-150.

Fischer, A.G., 1982, Long-term climatic oscillations recorded in stratigraphy, *in* Climate in Earth History, National Academy Press, Washington D.C., p. 97-104.

Ford, D.C. and Williams, P.W., 1989, Karst geomorphology and hydrology: Unwin Hyman, London, 601p.

Frakes, L.A., and J.E. Francis, 1988, A guide to Phanerozoic cold polar climates from high-latitude ice-rafting in the Cretaceous: Nature, v. 333, p. 547-549.

Frank, J.R., A.B. Carpenter, and T.W. Oglesby, 1982, Cathodoluminescence and composition of calcite cement in the Taum Sauk Limestone (Upper Cambrian), Southeast Missouri: Jour. Sediment. Petrology, v. 52, p. 631-638.

Fraser, A.J. and R.L. Gawthorpe, R.L., 1990, Tectono-stratigraphic development and hydrocarbon habitat of the Carboniferous in northern England, *in* R.F.P. Hardman, and J. Brooks, eds., Tectonic events responsible for Britain's oil and gas reserves: Geol. Soc. London Spec. Publ., v. 55, p. 49-86.

Fraser, A.J., D.F. Nash, R.P. Steele and C.C. Ebdon, 1990, A regional assessment of the intra-Carboniferous play of northern England, *in* J. Brooks, ed., Classic Petroleum Provinces: Spec. Publ. Geol. Soc. London, p. 417-440.

Fulthorpe, C.S. and S.O. Schlanger, 1989, Paleo-oceanographic and tectonic settings of early Miocene reefs and associated carbonates of offshore Southeast Asia: AAPG Bulletin, v. 73, p. 729-756.

Gawthorpe, R.L., 1987, Tectono-sedimentary evolution of the Bowland Basin, northern England, during the Dinantian: Jour. Geol. Soc. London, v. 144, p. 59-71.

Gawthorpe, R.L. and Gutteridge, P., 1990, Geometry and evolution of platform-margin bioclastic shoals, late Dinantian (Mississippian), Derbyshire, U.K., *in* M.E. Tucker, J.L. Wilson, P.D. Crevello, J.R. Sarg, and J.F. Read, eds., Carbonate Platforms: Spec. Publ. Int. Assoc. Sedimentologists, v. 9, p. 39-54.

Gebelein, C.D., R.P. Steinen, P. Garrett, E.J. Hoffman, J.M. Queen and L.N. Plummer, 1980, Subsurface dolomitization beneath the tidal flats of West Andros Island, Bahamas, *in* D.H. Zenger, J.B. Dunham and R.L. Ethington, Eds., Concepts and Models of Dolomitization: S.E.P.M. Special Publication, No. 28, p. 31-49.

Gerhard, L.C., 1985, Porosity development in the Mississippian pisolitic limestones of the Mission Canyon Formation, Glenburn field, Williston Basin, North Dakota, *in* P.O. Roehl and P.W. Choquette, eds., Carbonate Petroleum Reservoirs: Springer-Verlag, New York, p. 193-205.

Ginsburg, R.N., P.K. Swart, G.P. Eberli, D.F. NcNeill, and J.A.M. Kenter, 1990, Bahamas Drilling Project Phase 1, Appendix 1 and 2: Comparative Sedimentology Laboratory, Rosenstiel School of Marine and Atmospheric Science, University of Miami, 32 p.

Goldhammer, R.K., P.A. Dunn, and L.A. Hardie, 1990, Depositional cycles, composite sea levelchanges, cycle stacking patterns, and the hierarchy of stratigraphic forcing: Examples from Alpine Triassic carbonates: Geol. Soc. America Bull., v. 102, p. 535-562.

Goldhammer, R.K., E.J. Oswald, and P.A. Dunn, 1991, Heirarchy of stratigraphic forcing: Example from Middle Pennsylvanian shelf carbonates of the Paradox Basin, *in* E.K. Franseen, W.L. Watney, C. St. G. Kendall, and W. Ross, eds., Sedimentary Modelling: Computer simulations and methods for improved parameter definition: Kansas Geological Survey Bull., v. 233, p. 361-413.

Goldstein, R.H., 1988, Cement stratigraphy of Pennsylvanian Holder Formation, Sacramento Mountains, New Mexico: AAPG Bull., 72, p. 425-438.

Goldstein, R.H., 1990, Petrographic and geochemical evidence for origin of paleospeleothems, New Mexico: Implications for the application of fluid inclusions to studies of diagenesis: Jour. Sedimentary Petrology, v. 60, p. 282-292.

Goldstein, R.H., 1991, Practical aspects of cement stratigraphy with illustrations from Pennsylvanian limestone and sandstone, New Mexico and Kansas, *in* C.E. Barker and O.C. Kopp, eds., Luminescence

Microscopy and Spectroscopy: Qualitative and Quantitative Aspects: SEPM Short Course 25, p. 123-131.

Goldstein, R.H., J.E. Anderson, and M.W. Bowman, 1991a, Diagenetic responses to sea-level change: Integration of field, stable-isotope, paleosol, paleokarst, fluid inclusion, and cement stratigraphy research to determine history and magnitude of sea levelfluctuation, *in* E.K. Franseen, W.L. Watney, C. St. G. Kendall, and W. Ross, eds., Sedimentary Modelling: computer simulations and methods for improved parameter definition: Kansas Geol. Surv. Bull., v. 233, p. 139-162.

Goldstein, R.H., B.P. Stephens, and D.J. Lehrmann, 1991b, Fluid inclusions elucidate conditions of dolomitization in Eocene of Enewetak atoll and mid-Cretaceous Valles Platform of Mexico (Abstract), *in* A. Bosellini et al., eds., Dolomieu Conference on Carbonate Platforms and Dolomitization: Ortisei/St. Ulrich, Val Gardena/Grodental, The Dolomites, Italy, p. 93.

Goodwin, P.W. and E.J. Anderson, 1985, Punctuated aggradational cycles: A general hypothesis of episodic stratigraphic accumulation: Journal of Geology, v. 93, p. 515-533.

Grachevskiy, M.M., and G.F. Ulmishek, 1966, Late Paleozoic sedimentation in the North Caspian depression and the petroleum potential of its margin: Depressionnye tsikly, Moscow: VNIIOENG, p. 23-31.

Gray, D.I., 1981, Lower Carboniferous shelf paleoenvironments in North Wales: Unpubl. Ph.D. Thesis, University of Newcastle-upon-Tyne.

Grover, G.A. Jr., and J.F. Read, 1978, Fenestral and associated diagenetic fabrics of tidal flat carbonates, Middle Ordovician New Market Limestone, southwestern Virginia: Jour. Sedimentary Petrology, v. 48, p. 453-473.

Grover, G.A. Jr., and J.F. Read, 1983, Paleoaquifer and deep burial cements defined by cathodoluminescent patterns, Middle Ordovician carbonates, Virginia: A.A.P.G. Bull., v. 78, p. 1275-1303.

Haas, J., 1982, Facies analysis of the cyclic Dachstein Limestone Formation (Upper Triassic) in the Bakony Mountains, Hungary: Facies, v. 6, p. 75-84.

Halley, R.B., and J.W. Schmoker, 1983, High-porosity carbonate rocks of South Florida: progressive loss of porosity with depth: A.A.P.G. Bull. v. 67, p. 191-200.

Hanshaw, B.B., W. Back, and R.G. Deike, 1971, A geochemical hypothesis of dolomitization by ground water: Economic Geology, v. 66, p. 710-724.

Hardie, L.A., A. Bosellini, and R.K. Goldhammer, 1986, Repeated subaerial exposure of subtidal carbonate platforms, Triassic, Northern Italy: evidence for high-frequency sea level oscilations on a 10 year scale: Paleoceanography, v. 1 p. 447-457.

Hardie, L.A., P.A. Dunn, and R.K. Goldhammer, 1991, Field and modelling studies of Cambrian carbonate cycles, Virginia Appalachians - discussion: Jour. Sedimentary Petrology, v. 61, p. 636-646.

Harris, P.M., P.E. Flynn, and J.L. Sieverding, 1988, Mission Canyon (Mississippian) reservoir study, Whitney Canyon-Carter Creek field, southwestern Montana, *in* A.J. Lomando and P.M. Harris, Giant Oil and Gas Fields - A Core Workshop: SEPM Core Workshop 12, v. 2, p. 695-740.

Harris, P.M., S.H. Frost, G.A. Seiglie and N. Schneidermann, 1984, Regional unconformities and depositional cycles, Cretaceous of the Arabian Peninsula, *in* J.S. Schlee, ed., Interregional Unconformities and hydrocarbon accumulation, AAPG Mem. 36, p. 67-80.

Harris, P.M., C.G. St. C. Kendall and I. Lerche, 1985, Carbonate cements - a brief review: *in* N. Schneidermann and P.M. Harris, eds., Carbonate Cements, SEPM Spec. Publ. 36, p. 79-95.

Harris, P.M., and E.L. Stoudt, 1988, Stratigraphy and lithofacies of the San Andres Formation, C.S.Dean "A", XIT, and SW Levelland units of Levelland-Slaughter field, Permian Basin: in A.J. Lomando and P.M. Harris, Giant Oil and Gas Fields, A Core Workshop: SEPM Core Workshop No. 12, p. 649-694.

Harrison, R.S., L.D. Cooper, and M. Coniglio, 1984, Late Pleistocene carbonates of the Florida Keys: *in* Carbonates in Subsurface and Outcrop: 1984 CSPG Core Conference, Canadian Society of Petroleum Geologists, Calgary, Alberta, Canada, p. 291-306.

Harrison, R.S., and R.P. Steinen, 1978, Subaerial crusts, caliche profiles and breccia horizons: comparison of some Holocene and Mississippian exposure surfaces, Barbados and Kentucky: Geol. Soc. America Bull., v. 89, p. 389-396.

Hart, E., and J.T.C. Hay, 1974, Structure of Ain Zalah Field, Northern Iraq: A.A.P.G. Bull., v. 58, p. 973-981.

Heckel, P.H., 1980, Paleogeography of eustatic model for deposition of midcontinent Upper Pennsylvanian cyclothems, *in* T.D. Fouch and E.R. Magathan, eds., Paleozoic Paleogeography of west-central United States: West-Central United States Paleogeography Symposium 1, Rocky Mountain Section SEPM, p. 197-215.

Heckel, P.H., 1983, Diagenetic model for carbonate rocks in Mid-continent Pennsylvanian eustatic cyclothems: Jour. Sedimentary Petrology, v. 53, p. 733-759.

Heckel, P.H., 1985, Recent interpretations of Late Paleozoic cyclothems, *in* Proceedings of the Third Annual Field Conference: Mid-Continent Section, SEPM, Lawrence, Kansas, p. 1-22.

Hendry, J.P., 1993, Geological controls on regional subsurface carbonate cementation: an isotopic-paleohydrological investigation of Middle Jurassic limestones in central England: (this volume).

Hird, K. and M.E. Tucker, 1988, Contrasting diagenesis of two Carboniferous oolites from South Wales; a tale of climatic influence: Sedimentology, v35, p. 587-602.

Holland, D.C., J.S. Dickens, and A.D. Horbury, 1992, A case of drowning- the death of a carbonate platform in the South China Sea, *in* Exploration frontiers in Asia and the western Pacific: 1992 AAPG International Conference and Exhibition, Sydney, Australia (Abs.). p. 56-57.

Horbury, A.D., 1987, Sedimentology of the Urswick Limestone Formation in South Cumbria and North Lancashire. Unpubl. Ph.D. Thesis, University of Manchester, 668 p.

Horbury, A.D., 1989, The relative roles of tectonism and eustacy in the deposition of the Urswick Limestone Formation in South Cumbria and north Lancashire, *in* R.S. Arthurton, P. Gutteridge and S.C. Nolan, eds., The Role of Tectonics in Devonian and Carboniferous Sedimentation in the British Isles: Yorkshire Geol. Soc. Occ. Publ. no. 6, p. 153-169.

Horbury, A.D., 1992, A Late Dinantian peloid cementstone-paleaeoberesellid buildup from North Lancashire, England: Sedimentary Geology, v. 76, in press.

Horbury, A.D., and A.E. Adams, 1989, Meteoric phreatic diagenesis in cyclic late Dinantian carbonates, northwest England: Sedimentary Geology, v. 65, p. 319-344.

Hovorka, S.D., H.S. Nance and C. Kerans, in press, Parasequence geometry as a control on permeability evolution: examples from the San Andres and Grayburg Formations in the Guadalupe Mountains, New Mexico *in* R. Loucks and J. F. Sarg, eds., Carbonate Sequence Stratigraphy, AAPG Memoir 57.

Hubbard, R.J., S.P. Edrich, and R.P. Rattey, 1990, Geological evolution and hydrocarbon habitat of the 'Arctic Alaska microplate', in: J. Brooks (Ed), Classic Petroleum Provinces: Geol. Soc. London Spec. Publ. v. 50, p. 143-187.

Humphrey, J.D., 1988, Late Pleistocene mixing zone dolomitization, southeastern Barbados, West Indies: Sedimentology, v. 35, 327-348.

Humphrey, J.D., and T.M. Quinn, 1989, Coastal mixing zone dolomite, forward modelling, and massive dolomitization of platform-margin carbonates: Jour. Sedimentary Petrology, v. 59, 438-454.

Hsu, K.J., and J. Schneider, 1973, Progress report on dolomitization - hydrology of Abu Dhabi sabkhas, Arabian Gulf, *in* B.H. Purser, ed., The Persian Gulf; Holocene Carbonate Sedimentation and Diagenesis in a Shallow Epicontinental Sea: New York, Springer-Verlag, p. 409-422.

Ibrahim, G.M.S., 1978, Sabkha sediments of the Lower Fars Formation, North Iraq: Msc. thesis, University of Bristol, (unpubl.), 174p.

Ingle, J.C., 1981, Origin of Neogene diatomites around the north Pacific rim, *in* R.E. Garrison, R.G. Douglas, K.E. Pisciotte, C.E. Isaacs and J.C. Ingle, eds., The Monterey Formation and related siliceous rocks of California: Spec. Publ. Pacific Section, SEPM. p. 157-179.

James, N.P. and P.W. Choquette, eds., 1988, Paleokarst: Springer-Verlag, New York, 416 p.

James, N.P. and P.W. Choquette, 1990, Limestones - the meteoric diagenetic environment, *in* I.A. McIlreath and D.W. Morrow, eds., Diagenesis: Geoscience Canada Reprint Series v. 4, p. 35-73.

Jennings, J.N, 1983, The disregarded karst of the arid and semiarid domain: Karstologica, v.1, p. 61-73.

Jordan, C.F. Jr., T.C. Connally, Jr. and H.A. Vest, 1985, Upper Cretaceous carbonates of the Mishrif Formation, Fateh field, Dubai, U.A.E., *in* P.O. Roehl, and P.W. Choquette, eds., Carbonate Petroleum Reservoirs: Springer-Verlag, New York, p. 428-442.

Kerans, C., 1988, Karst-controlled reservoir heterogeneity in Ellenburger Group carbonates of West Texas: A.A.P.G. Bull., v. 72, p. 1160-1173.

Knight, I., N.P. James, and T.E. Lane, 1991, The Ordovician St. George unconformity, northern Appalachians: The relationship of plate convergence at the St. Lawrence Promontory to the Sauk/Tippecanoe sequence boundary: Geological Society of America Bull., v. 103, p. 1200- 1225.

Koerschner, W.F., and J.F. Read, 1989, Field and modelling studies of Cambrian carbonate cycles, Virginia Appalachians: Jour. Sed. Petrology, v. 59, p. 654-687.

Kozar, M.G., L.J. Weber, and K.R. Walker, 1990, Field and modelling studies of Cambrian carbonate cycles, Virginia Appalachians - Discussion: Jour. Sedimentary Petrology, v. 60, p. 790-794.

Kupecz, J.A., 1989, Petrographic and geochemical characterization of the Lower Ordovician Ellenburger Group, west Texas: Ph.D. Dissertation, University of Texas, Austin, 157 p.

Leeder, M, 1987, Tectonic and palaeogeographic models for Lower Carboniferous Europe, *in* J. Miller, A.E. Adams and V.P. Wright, eds., European Dinantian Environments: Wiley, Chichester, p. 1-19.

Logan, B.W., 1974, Inventory of diagenesis in Holocene-Recent carbonate sediments, Shark Bay, Western Australia, *in* B.W. Logan et al., Evolution and diagenesis of Quaternary carbonate sediments, Shark Bay, Western Australia: A.A.P.G. Memoir 22, p. 195-250.

Logan, B.W. , 1989, The Lake McLeod evaporite basin: AAPG Memoir 44, 140 p.

Logan, B.W., J.L. Harding, W.H. Ahr, J.D. Williams, and R.G. Snead, 1969, Carbonate sediments and reefs, Yucatan Shelf, Mexico: AAPG Memoir 11, p. 1-196.

Longacre, S.A., 1980, Dolomite reservoirs from Permian biomicrites, *in* R.B. Halley and R.G. Loucks, eds., Carbonate Reservoir Rocks, SEPM Core Workshop No. 1, p. 105-117.

Longman, M.W., T.G. Fertal, and J.S. Glennie, 1983, Origin and geometry of Red River Dolomite reservoirs, western Williston Basin: A.A.P.G. Bull. 67, p. 744-771.

Loucks, R.G., and J.H. Anderson, 1985, Depositional facies, diagenetic terrains, and porosity development in the Lower Ordovician Ellenberger dolomite, Puckett field, West Texas, *in* P.O. Roehl, and P.W. Choquette, eds, Carbonate Petroleum Reservoirs: Springer, New York, p. 19-37.

Lovelock, P., 1984, A review of the tectonics of the northern Middle East region: Geol. Magazine, v. 121, p. 577-587.

Machel, H.G., 1985, Cathodoluminescence in calcite and dolomite and its chemical interpretation: Geo. Sci. Can., 12: p. 139-147.

Machel, H.G., and E.A. Burton, 1991, Factors governing cathodoluminescence in calcite and dolomite and their implications for studies of carbonate diagenesis, *in* C.E. Barker and O.C. Kopp, eds., Luminescence microscopy and spectroscopy: qualitative and quantitative applications: S.E.P.M. Short Course 25, Dallas, Texas, p. 37- 57.

Magaritz, M., 1987, A new explanation for cyclic deposition in marine evaporite basins: meteoric water input: Chem. Geol.: v. 62, p. 239-250.

Major, R.P., 1991, Cathodoluminescence in post-Miocene carbonates, *in* C.E. Baker, and O.C. Kopp, eds., Luminescence Microscopy and spectroscopy: Qualitative and quantitative applications: S.E.P.M. Short Course 25, Dallas, Texas, p. 149-153.

Major, R.P., D.G. Bebout, and F.J. Lucia, 1988, Depositional facies and porosity distribution, Permian (Guadalupian) San Andres and Grayburg Formations, P.J.W.D.M. field complex, Central Basin Platform, West Texas, *in* A.J. Lomando, and P.M. Harris, eds., Giant Oil and Gas Fields, A Core Workshop: S.E.P.M. Core Workshop 12, v. 2, p. 615-648.

Marquis, S.A. Jr., and R.L. Laury, 1989, Glacio-eustasy, depositional environments, and reservoir character of Goen Limestone cyclothem (Desmoinesian), Concho Platform, Central Texas: A.A.P.G. Bull. 73, p. 166-181.

Martinez del Olmo, W., and M. Esteban, 1983, Paleokarst development, *in* P.A. Scholle, D.G. Bebout and C.H. Moore, eds., Carbonate Depositional Environments: A.A.P.G. Memoir 33, p. 93-95.

Matthews, R.K., and C. Frohlich, 1987, Forward modelling of bank-margin carbonate diagenesis: Geology, v. 15, p. 673-676.

McKenzie, J.A., K.J. Hsu, and J.F. Schneider, 1980, Movement of subsurface waters under the sabkha, Abu Dhabi, U.A.E., and its relation to evaporative dolomite genesis: S.E.P.M. Spec. Publ. No. 28, p. 11-30.

Metwalli, M.H., G. Philip, and M.M. Moussly, 1974, Petroleum bearing formations in northeastern Syria and northern Iraq: A.A.P.G. Bull., v. 58, p. 1781-1796.

Meyers, W.J., 1974, Carbonate cement stratigraphy of the Lake Valley Formation (Mississippian), Sacremento Mountains, New Mexico: Jour. Sediment. Petrology, v. 44, p. 837-861.

Meyers, W.J., 1978, Carbonate cements: their regional distribution and interpretation in Mississippian limestones of southwestern New Mexico: Sedimentology, v. 25, p. 371-400.

Meyers, W.J., 1988, Paleokarstic features in Mississippian limestones, New Mexico, *in* N.P. James and P.W. Choquette, Paleokarst: Springer- Verlag, New York, p. 306-328.

Meyers, W.J., 1991, Calcite cement stratigraphy: an overview, *in* C.E. Barker and O.C. Kopp, eds., Luminescence Microscopy and Spectroscopy: Qualitative and Quatitative Applications: SEPM Short Course 25, p. 133-148.

Meyers, W.J., and K.C. Lohmann, 1985, Isotope geochemistry of regionally extensive calcite cement zones and marine components in Mississippian Limestones, New Mexico, *in* Schneiderman, N. and Harris, P.M., eds., Carbonate Cements: S.E.P.M. Special Publication No 36, p. 223-239.

Mitchell, J.C., P.J. Lehmann, D.L. Cantrell, I.A. Al-Jallal, and M.A.R. Al-Thagafy, 1988, Lithofacies, diagenesis and depositional sequence; Arab-D Member, Ghawar Field, Saudi Arabia, in SEPM Core Workshop No. 12, p. 459-514.

Mitchell, R.W., 1985, Comparative sedimentology of shelf carbonates of the Middle Ordovician St. Paul Group, Central Appalachians: Sedimentary Geology, v. 43, p. 1-41.

Mitchum, R.M., Jr., and J.C. Van Wagoner, 1991, High frequency sequences and their stacking patterns: sequence-stratigraphic evidence of high-frequency eustatic cycles: Sedimentary Geology, v. 70, p. 131-160.

Montanez, I.P., and J.F. Read, 1992a, Fluid-rock interaction history during stabilization of early dolomites, Upper Knox Group (Lower Ordovician), U.S. Appalachians: Jour. Sedimentary Petrology, in press.

Montanez, I.P., and J.F. Read, 1992b, Eustatic control on dolomitization of cyclic peritidal carbonates: Evidence from the Early Ordovician Knox Group, Appalachians: Geol. Soc. America Bull., in press.

Moore, C.H., 1989, Carbonate Diagenesis and Porosity , Elsevier, Amsterdam, 338 p.

Moore, C.H. and E. Heydari, 1993, Burial diagenesis and hydrocarbon migration in platform limestones: a conceptual model based on the upper Jurassic of the Gulf Coast of the U.S.A. (this volume).

Moore, C.H., A. Chowdhury, and L. Chan, 1988, Upper Jurassic Smackover platform dolomitization, northwestern Gulf of Mexico: a tale of two waters, in V. Shukla and P.A. Baker, eds., Sedimentology and Geochemistry of dolostones: S.E.P.M. Spec. Publ., v. 43, p. 175-189.

Moore, T.C., N.G. Pisias, and I.D. Keigwin, Jr., 1982, Cenozoic variability of oxygen isotopes in benthic forminifera, *in* Climate in Earth History: National Academy Press, Washington D.C., p. 172-182.

Murris, R.J., 1980, Middle East: stratigraphic evolution and oil habitat. A.A.P.G. Bull., p. 597-618.

Mussman, W.J., I.P. Montanez, and J.F. Read, 1988, Ordovician Knox paleokarst unconformity, Appalachians, *in* N.P. James, and P.W. Choquette, eds., Paleokarst: Springer-Verlag, New York, p. 211-228.

Niemann, J.C., and J.F. Read, 1988, Regional cementation from unconformity-recharged aquifer and burial fluids, Mississippian Newman Limestone, Kentucky: Jour. Sedimentary Petrology, v. 58, p. 688-705.

Nelson, W.A., and J.F. Read, 1989, Updip to downdip cementation and dolomitization patterns in a Mississippian aquifer, Appalachians: Jour. Sedimentary Petrology, v. 60, p. 379-396.

Oglesby, T.W., 1976, A model for the distribution of manganese, iron and magnesium in authigenic calcite and dolomite cements in the Upper Smackover Formation in eastern Mississippi: M.S. Thesis, University of Missouri, 122 p.

Osleger, D.A., and J.F. Read, 1991, Relation of eustasy to stacking patterns of meter-scale carbonate cycles, Late Cambrian, U.S.A.: Journal of Sedimentary Petrology, v. 61, p. 1225-1252.

Patterson, R.J., 1981, Hydrologic framework of a sabkha along Arabian Gulf: AAPG Bull., v. 65, p. 1457-1475.

Perkins, R.D., 1977, Depositional framework of Pleistocene rocks in South Florida, *in* P. Enos, and R. Perkins, Quaternary Sedimentation in South Florida: Geol. Soc. America, Mem. 147, p. 131-198.

Philip, G., M.H. Metwali, and M.M. Moussly, 1972, Petrographic characteristics of the oil-bearing Jeribe Formation, Jebissa oil field (Syrian Arab Republic): 8th Arab Petroleum Congress, Algiers, Paper 67 (B-3), 20 p.

Pierson, B.J. and E.A. Shinn, 1985, Cement distribution and carbonate mineral stabilization in Pleistocene limestones of Hogsty reef, Bahamas, *in* N. Schneiderman and P.M. Harris, Carbonate Cements: S.E.P.M. Spec. Publ. No. 36, p. 153-168.

Plummer, L.N., H.L. Vacher, F.T. McKenzie, O.P. Bricker, and L.S. Land, 1976, Hydrogeochemistry of Bermuda: A case history of ground-water diagenesis of biocalcarenites: Geol. Soc. America Bull., v. 87, p. 1301-1316.

Powell, J.H., 1988, Stratigraphy and sedimentation of the Phanerozoic rocks in central and southern Jordan, Part B., Kurnuk, Ajlan, and Belga groups: The Hashemite Kingdom of Jordan, Ministry of Energy and Natural Resources Authority, Geology Directorate, Geological Mapping Division.

Pray, L.C., and J.L. Wray, 1963, Porous algal facies (Pennsylvanian) Honaker Trail, San Juan Canyon, *in* R.O. Bass, ed., 4th Field Conf. Gdbk.: Four Corners Geological Society, p. 204-234.

Prezbendowski, D.R., 1985, Burial cementation- is it important?. A case study, Stuart City Trend, southcentral Texas, *in* N. Schneidermann and P.M. Harris , eds, Carbonate Cements: S.E.P.M. Spec. Publ. No 36, p. 241-264.

Purdy, E.G., 1974a, Karst-determined facies patterns in British Honduras: Holocene carbonate sedimentation model: A.A.P.G. Bull., v. 58, p. 825-855.

Purdy, E.G., 1974b, Reef configurations: cause and effect, *in* L.E. Laporte, ed., S.E.P.M. Spec.Publ. No 18, p. 9-76.

Quinn, T.M., 1989, Forward modelling of bank-margin carbonate diagenesis: Results of sensitivity tests and initial applications, *in* T.A. Cross, ed., Quantitative Dynamic Stratigraphy: Prentice-Hall, p. 445-455.

Quinn, T.M., 1991, Meteoric diagenesis of Plio-Pleistocene limestones at Enewetak Atoll: Jour. Sedimentary Petrology, v. 61, p. 681-703.

Read, J.F., 1973, Paleo-environments and paleogeography, Pillara Formation (Devonian), Western Australia: Bull. Can. Petrol. Geol., v. 21, p. 344-394.

Read, J.F., 1980, Carbonate ramp-to-basin transitions and foreland basin evolution, Middle Ordovician, Virginia Appalchians: A.A.P.G. Bull.,v. 64, p. 1575-1612.

Read, J.F., and R.K. Goldhammer, 1988, Use of Fischer plots to define 3rd order sea levelcurves in peritidal cyclic carbonates, Early Ordovician, Appalchians: Geology, v. 6, p. 895-899.

Read, J.F., and G.A. Grover, Jr., 1977, Scalloped and planar erosion surfaces, Middle Ordovician limestones, Virginia: analogues of Holocene exposed karst or tidal rock platforms: Jour. Sedimentary Petrology, v. 47, p. 956-972.

Read, J.F., D.A. Osleger, and M.E. Elrick, 1991, Two-dimensional modelling of carbonate ramp sequences and component cycles, *in* E.K. Franseen, W.L. Watney, C. St. G. Kendall, and W. Ross, eds., Sedimentary Modelling: Computer simulations and methods for improved parameter definition: Kansas Geol. Surv. Bull., v. 233, p. 473-488.

Reeder, R.J., and J.C. Grams, 1987, Sector zoning in calcite cement crystals: implications for trace element distribution coefficients in carbonates: Geochim. et Cosmochim. Acta, v. 51, p. 187-194.

Reid, A.M., and S.A. Tomlinson-Reid, 1991, The Cogdell field study, Kent and Scurry Counties, Texas: A post-mortem, *in* M.P. Candelaria, ed., Permian Basin Plays - Tomorrow's Technology Today: West Texas Geol. Soc. Publ. No. 91-89, p. 39-66.

Riding, R. and V.P. Wright, 1981. Paleosols and tidal flat/lagoonal sequences on a Carboniferous carbonate shelf: Jour. Sedimentary Petrol.ogy, v. 51, p. 1323-1339.

Robinson, D, and V.P. Wright, 1988, Ferrolytic weathering in Lower Carboniferous paleosols: evidence of shifts in monsoonal patterns, *in* Abstracts, British Sedimentological Research Group Meeting, Cambridge, 1988: British Antarctic Survey, Cambridge.

Roehl, P.O., 1985, Depositional and diagenetic controls on reservoir rock development and petrophysics in Silurian Tidalites, Interlake Formation, Cabin Creek field area, Montana, *in* P.O. Roehl and P.W. Choquette, eds., Carbonate Petroleum Reservoirs: Springer-Verlag, New York, p. 85-106.

Rose, P.R., 1963, Comparison of type El Abra of Mexico with "Edwards Reef Trend" of south-central Texas, *in* Geology of Peregrina Canyon and Sierra de El Abra, Mexico: Geol. Soc. America, Ann. Fld. trip guidebook, Corpus Christi , p. 57-64.

Roychoudhury, S.C., and Deshpande, S.V., 1982, Regional distribution of carbonate facies, Bombay offshore region, India: A.A.P.G. Bull., v. 66, p. 1483-1496.

Roylance, M.H., 1990, Depositional and diagenetic history of a Pennsylvanian algal-mound complex: Bug and Papoose Canyon fields, Paradox Basin, Utah and Colorado: A.A.P.G. Bull. 74, p. 1087-1099.

Rudolph, K.W. and P.J. Lehmann, 1989, Platform evolution and sequence stratigraphy of the Natuna Platform, South China Sea, *in* P.D. Crevello, J.L. Wilson, J.F. Sarg, and J.F. Read, eds., Controls on Carbonate Platform and Basin Development, Spec. Publ. Soc. Econ. Paleont. Mineral., v. 44, p. 353-361.

Ruzyla, K., and G.M. Friedman, 1985, Factors controlling porosity in dolomite reservoirs of the Ordovician Red River Formation, Cabin Creek Field, Montana, in P.O. Roehl, and P.W. Choquette, eds, Carbonate Petroleum Reservoirs: Springer, New York, p. 39-58.

Sadooni, F.N.R. and S. Saeed, 1987, Upper Triassic carbonate-evaporite sediments of NW Iraq. Abstracts, 8th Int. Ass. Sedimentologists Regional Meeting, Tunisia.

Saller, A.H., 1984, Petrologic and geochemical constraints on the origin of subsurface dolomite, Enewetak Atoll: an example of dolomitization by normal seawater: Geology, v. 12, p. 217-220.

Sandberg, P.A., 1983, An oscillating trend in Phanerozoic non-skeletal carbonate mineralogy: Nature, v. 305, p. 19-22.

Sando, W.J., 1988, Madison Limestone (Mississippian) paleokarst: A geologic synthesis, *in* N.P. James and P.W. Choquette, eds., Paleokarst: Springer-Verlag, p. 256-277.

Sarg, J.F., 1988, Carbonate Sequence Stratigraphy, *in* C.K. Wilgus and others, eds., Sea- Level Changes: An Integrated Approach: S.E.P.M. Spec. Publ. 42, p. 156-181.

Schatzinger, R.A., 1983, Phylloid algal and sponge-bryozoan mound-to-basin transition: A Late Paleozoic facies tract from the Kelly-Snyder field, West Texas, *in* P.M. Harris, ed., Carbonate Buildups—A Core Workshop: SEPM Core Workshop No. 4, p. 244-303.

Schofield, K., and A.E. Adams, 1985, Stratigraphy and depositional environments of the Woo Dale Limestones Formation (Dinantian), Derbyshire: Proc. Yorkshire Geol. Soc., v. 45, p. 225-23.

Searle, A., 1988, The limitations of cement stratigraphy as revealed in some Lower Carboniferous oolites from South Wales: Sedimentary Geology, v. 57, p. 171-183.

Sengor, A.M.C., and Y. Yilmaz, 1981, Tethyan evolution of Turkey: a plate tectonic approach: Tectonophysics, v. 75, p. 181-241.

Shawkat, M.G., and M.E. Tucker, 1978, Stromatolites and sabkha cycles from the Lower Fars Formation (Miocene) of Iraq: Geol. Rundschau, v. 67, p. 1-14.

Sharief, F.A., 1982, Lithofacies distribution of the Permian-Triassic rocks in the Middle East: J. Petrol. Geol., v. 4, p. 299-310.

Shinn, E.A., W.E. Bloxsom, and R.M. Lloyd, 1974, Recognition of submarine cements in Cretaceous reef limestones from Texas and Mexico (abs): A.A.P.G. Bull., v. 58, p. 82-83.

Sloss, L.L., 1963, Sequences in the cratonic interior of North America: Geol. Soc. America Bull., v. 74, p. 93-113.

Smart, P.L., T. Waltham, M. Yang, and Y. Zhang, 1986, Karst geomorphology of western Guizhou, China: Trans. Brit. Cave Res. Ass., v. 13, p. 89-103.

Smart, P.L., R.J. Palmer, F. Whitaker, and V. P. Wright, 1988, Neptunian dikes and fissure fills: an overview and account of some modern examples, *in* N.P. James and P.W. Choquette, eds., Paleokarst: Springer-Verlag, New York, p. 149-163.

Smith, T.M., and S.L. Dorobek, 1990, Meteoric calcite cements from the Mission Canyon Formation, Montana (abs.): Geol. Soc. America Abs. with Prog., v. 22, p. 335.

Snyder, S.W. , A.C. Hine, C.W. Snyder, and S.R. Riggs, 1992, Miocene glacioeustasy below the resolution of multiple, planktic biostratigraphic correlation schemes (abs): A.A.P.G. Annual Conv., Calgary, Canada, p. 123.

Solomon, S.T., 1989, The early diagenetic origin of Lower Carboniferous mottled limestones (pseudo breccias): Sedimentology, v. 36, p. 399-418.

Solomon, S.T., and G.M. Walkden, 1985, The application of cathodoluminescence to interpreting the diagenesis of an ancient calcrete profile: Sedimentology, v. 32, p. 877-896.

Somerville, I.D., 1979a, A sedimentary cyclicity in early Asbian (Lower D1) limestones in the Llangollen district of North Wales: Proceedings of the Yorkshire Geological Society, v. 42, p. 397-404.

Somerville, I.D., 1979b, Minor sedimentary cyclicity in late Asbian (upper D1) limestones in the Llangollen district of North Wales: Proc. Yorkshire Geol. Soc., v. 42, p. 317-341.

Somerville, I.D., 1979c, A cyclicity in early Brigantian (D2) limestones east of the Clwydian Range, North Wales, and its use in correlation: Geological Journal, v. 14, p. 69-86.

Stemmerik, L., and G.B. Larsen, 1993, Diagenesis and porosity evolution of Lower Permian *Palaeoaplysina* build-ups, Bjornoya, Barents Sea: an example of diagenetic response to high frequency sea level fluctuations in an arid climate (this volume).

Strank, A.R.E., 1987, The stratigraphy and structure of Dinantian strata in the East Midlands, U.K., *in* J. Miller, A.E. Adams and V.P. Wright, eds., European Dinantian Environments: Wiley, Chichester, p. 157-175.

Strasser, A., 1988, Shallowing-up sequences in Purbeckian peritidal carbonates (lowermost Cretaceous, Swiss and French Jura Mountains): Sedimentology, v. 35, p. 369-383.

Stringfield, V.T., and H.E. LeGrand, 1974, Karst hydrology of northern Yucatan Peninsula, Mexico, *in* Field Seminar on Water and Carbonate Rocks of the Yucatan Peninsula, Mexico: New Orleans Geol. Soc., Field Trip 2 Ann. Mtg., Miami, Geol. Soc. Amer., p. 26-44.

Sun, S.Q., 1992, Aragonite dissolution from hypersaline seawater: A hypothesis. Sed. Geol., v. 77, p. 249-257.

Tong XiaoGuang and Huang Zuan, 1991, Buried hill discoveries of the Dinantian Depression in north China: A.A.P.G. Bull., v . 75, 780-794.

Tucker, M.E., 1985, Shallow-marine carbonate facies and facies models, *in* P.J. Benchley and B.P.J. Williams, Sedimentology: Recent Developments and Applied Aspects: Geol. Soc. London Spec. Publication 18, p. 139-161.

Tucker, M.E. and M.G. Shawkat, 1980, The Miocene Gachsaran Formation (formerly Lower Fars) of the Mesopotanian Basin, Iraq: Sabkha cycles and Depositional Controls: J. Geol. Soc. Iraq, v. 13, p. 261-267.

Tucker, M.E, in press, Carbonate diagenesis in a Sequence Stratigraphic Framework: Sed. Review.

Turner, N.L., 1991, The Lower Miocene Liuhua carbonate reservoir, Pearl River Mouth Basin, offshore Peoples Republic of China: 23rd Offshore Technology Conference, Houston, p. 113-123.

Tyrrell, W.W., and H.E. Christian, 1992, Exploration history of the Liuhua 11-1 field, Pearl River Mouth Basin, China: A.A.P.G. Bull., v. 76, p. 1209-1223.

Ulmishek, G.F., 1988. Upper Devonian-Tournaisian facies and oil resources of the Russian craton's eastern margin, *in* N.J. McMillan, A.F. Embry, and D.J. Glass, eds, Devonian of the World: Can. Soc. Petrol. Geol., Calgary, p 527-549.

Vail, P.R., R.M. Mitchum, and S. Thompson, III, 1977, Seismic stratigraphy and global changes of sea level, Part 4: Global Cycles of relative changes of sea level, *in* C.A. Payton, Seismic Stratigraphy - Applications to hydrocarbon exploration: A.A.P.G. Memoir 26, p. 83-97.

Van Wagoner, J.C., H.W. Posamentier, R.M. Mitchum, P.R. Vail, J.F. Sarg, T.S. Loutit and J. Hardenbol, 1988, An overview of the fundamentals of sequence stratigraphy and key definitions; in C.K. Wilgus, B.S. Hastings, C. G. St. C. Kendall, H.W. Posamentier, C.A. Ross, and J.C. Van Wagoner, eds., Sea-Level Changes - An Integrated Approach: S.E.P.M. Spec. Publ. 42, p. 39-46.

Vaughan, R.D. and A.E. Adams, 1984, Chadian and Arundrin sedimentation in the Furness and Millom areas, South Cumbria: European Dinantian Environments, 1st Meeting, Manchester, 1984 (Abs): Open University, U.K., p. 120-123.

Walkden, G.M., 1987, Sedimentary and diagenetic styles in late Dinantian carbonates of Britain, *in* J. Miller, A.E. Adams and V.P. Wright, eds., European Dinantian Environments: Wiley, Chichester, p.131-155.

Walkden, G.M., and J. Berry, 1984, Syntaxial overgrowths in muddy crinoidal limestones: cathodoluminescence sheds new light on an old problem: Sedimentology, v. 31, p. 251-267.

Walkden, G.M. and J.R. Davies, 1983, Polyphase erosion of subaerial omission surfaces in the Late Dinantian of Anglesey, North Wales: Sedimentology, v. 30, p. 861-878.

Walkden, G.M. and G.D. Walkden , 1990, Cyclic sedimentation in carbonate and mixed carbonate-clastic environments: four simulation programs for a desktop computer, *in* M.E. Tucker, J.L. Wilson, P.D. Crevello, J.R. Sarg, and J.F. Read, eds., Carbonate Platforms, Facies Sequences and Evolution: International Assoc. of Sedimentologists, Spec. Publ. No. 9, p. 55-78.

Walkden, G.M., and D.O. Williams, 1991, The diagenesis of the Late Dinantian Derbyshire-East Midland carbonate shelf, central England: Sedimentology, v. 38, p. 643- 670.

Walls, R.A. and G. Burrowes, 1985, The role of cementation in the diagenetic history of Devonian reefs, western Canada, *in* N. Schneidermann and P.M. Harris, eds., Carbonate Cements: S.E.P.M. Spec. Publ. 36, p. 185-220.

Ward, W.C., and R.B. Halley, 1985, Dolomitization in a mixing zone of near-seawater composition, Late Pleistocene, northeastern Yucatan Peninsula: Jour.

Sedimentary Petrology, v. 55, p. 407-420.

Watney, W.L., J. French, and E.K. Franseen, 1989, Sequence stratigraphic interpretations and modelling of cyclothems in the Upper Pennsylvanian (Missourian) Lansing and Kansas City groups in eastern Kansas: Kansas Geological Society 41st Annual Field Trip Guidebook, Lawrence, Kansas, 211 p.

Wendte, J.C. and F.A. Stoakes, 1982, Evolution and porosity devlopment of the Judy Creek reef complex, Upper Devonian, Central Alberta: in W.G. Cutler, Canada's Giant Hydrocarbon Reservoirs: Can. Soc. Petrol. Geol., Calgary, Alberta, p. 63-81.

Whitaker, F.F. and P.L. Smart, 1990, Active circulation of saline ground waters in carbonate platforms: Evidence from the Great Bahama Bank: Geology, v. 18, p. 200-203.

Whitaker, F.F., and P.L. Smart, 1993, Circulation of saline ground water in carbonate platforms; a review and case study from the Bahamas (this volume).

White, W.B., 1969, Conceptual models for carbonate aquifers: Ground Water, v.7, p. 15-21.

Wilkinson, B.H., 1982, Cyclic cratonic carbonates and Phanerozoic calcite seas: Jour. Geological Education, v. 30, p. 189-203.

Williams, P.W., 1978, Interpretations of Australian karsts, *in* J.L. Davies and M.A.J. Williams, eds, Landform evolution in Australia: ANU Press, Canberra, p. 259-286.

Wilson, J.L., 1975, Carbonate facies in geologic history: Springer-Verlag, New York, 471 p.

Wood, G.V., and M.J. Wolfe, 1969, Sabkha cycles in the Arab/Darb Formation off the Trucial Coast of Arabia: Sedimentology, v. 12, p. 165-191.

Wright, V.P., 1982, The recognition and interpretation of paleokarsts: two examples from the Lower Carboniferous of South Wales: Jour. Sedimentary Petrology, v. 52, p. 83-94.

Wright, V.P., 1983, A rendzina from the Lower Carboniferous of South Wales: Sedimentology, v. 30, p. 159-179.

Wright, V.P., 1986, Facies sequences on a carbonate ramp: the Carboniferous Limestone of South Wales: Sedimentology, v. 33, p. 221-241.

Wright, V.P., 1988, Palaeokarsts and palaeosols as indicators of palaeoclimate and porosity evolution: a case study from the Carboniferous of South Wales, *in* N.P. James and P.W. Choquette (Eds.), Palaeokarst: Springer, New York, p. 329-341.

Wright, V.P., 1990, Equatorial aridity and climatic oscillations during the early Carboniferous, southern Britain. Jour. Geol. Soc. London, v. 147, p. 359-363.

Wright, V.P., 1992, Speculations on the controls on cyclic peritidal carbonates: ice-house versus greenhouse eustatic controls: Sed. Geol., v. 76, p. 1-5.

Wright, V.P., and T.J. Faulkner, 1990, Sediment dynamics of early Carboniferous ramps: Geol. Jour., v. 25, p. 139-144.

Wright, V.P., M. Esteban and P. Smart, 1991, Palaeokarst and palaeokarstic reservoirs: University of Reading, 158 p.

Zhai Guangming and Zha Quanheng, 1982, Buried hill oil and gas pools in the North China Basin: AAPG Mem. 32, p. 317-335.

Chapter 12

Diagenesis and Porosity Evolution of Lower Permian *Palaeoaplysina* Buildups, Bjørnøya, Barents Sea: An Example of Diagenetic Response to High Frequency Sea Level Fluctuations in an Arid Climate

Lars Stemmerik
Grønlands Geologiske Undesøgelse
Copenhagen, Denmark

and

Geir B. Larssen
Oljedirektoratet
Harstad, Norway

ABSTRACT

The lower Kapp Duner Formation on Bjørnøya consists of two Palaeoaplysina buildup complexes separated by a karst surface, believed to represent a third order sequence boundary. The buildup complexes are composed of 3–10 m-thick, stacked, shallowing upward cycles of restricted shelf wackestone and Palaeoaplysina wackestone and packstone. Each cycle is terminated by an event of subaerial exposure related to fourth order fluctuations in sea level. The upper complex passes laterally into restricted shelf wackestone and mudstone facies. Diagenesis started with early dolomite replacement of carbonate mud, dissolution of aragonitic material, and precipitation of anhydrite cement in secondary pores. These took place in an arid climate during repeated subaerial exposure related to fourth order fluctuation in sea level, and were associated with evaporite sedimentation in the adjacent basin. They were followed by dolomite replacement of anhydrite and precipitation of limpid dolomite rhombs and ankeritic dolomite. These processes are related to prolonged subaerial exposure in a more humid climate during the middle and Late Permian. Post-Permian diagenesis involved minor quartz and calcite cementation.

Seismically identified mounds in the Early Permian succession of the offshore areas of the Barents Sea were probably not subjected to prolonged middle and Late Permian exposure. Their porosity development was controlled mainly by early, intra-depositional processes, and they are predicted to form potential but highly heterogeneous reservoirs with porosities in the range of 10–20%.

INTRODUCTION

The upper Paleozoic carbonates in the Barents Sea area form a major, almost untested exploration target in the Norwegian sector (Figure 1). Most of the information about these carbonates has come from seismic studies (Gerard and Buhrig, 1990), and predictions concerning facies development and diagenesis have to rely on studies of time-equivalent rocks on Spitsbergen, Bjørnøya, Greenland and Arctic Canada (e.g. Bruce and Toomey, 1993; Cecchi, 1993; Nilsen et al., 1993). The most promising reservoir unit in the succession appears to be Upper Carboniferous–Lower Permian (Gzhelian-Asselian) *Palaeoaplysina* buildups and associated sediments. The buildups have not been drilled but restricted shelf sediments of this age are dolomitized over large areas and secondary porosity has been created, apparently as the result of dolomite replacement of carbonate mud and dissolution of carbonate grains (Lønøy, 1988; Stemmerik and Larssen, in press).

This chapter outlines the controls on diagenesis of dolomitized lower Asselian (Lower Permian) *Palaeoaplysina* buildups of the Kapp Duner Formation on Bjørnøya and discusses the implications for seismically identified mounds of the same age in the subsurface. Most porosity in the Bjørnøya buildups was created during dolomite replacement of carbonate mud and dissolution of aragonite grains during subaerial exposure in an arid climate. The exposure events were related to fourth order fluctuations in relative sea level and the dolomite replacement was associated with deposition of subaqueous evaporites in the basinal areas. Later, porosity destructive processes started with anhydrite cementation (now replaced by dolomite) during subaerial exposure. This was followed by precipitation of limpid dolomite during events of tectonically controlled, prolonged subaerial exposure in the latest Permian, and calcite cementation during burial. Up to 25% secondary porosity was created as a result of dolomite replacement and aragonite dissolution. Porosity was reduced to 10–20% as the result of anhydrite cementation, and further reduction took place during later dolomite and calcite cementation.

The late Paleozoic was a period with high amplitude fluctuations in sea level due to glaciations in the southern hemisphere (Veevers and Powell, 1987). The Barents Sea carbonate platforms were also affected by these sea level fluctuations (e.g. Stemmerik and Worsley, 1989), and it is therefore likely that the sea level controlled diagenetic alterations seen in the Bjørnøya buildups also affected buildups in the offshore areas. However, later phases of dolomite precipitation in the Bjørnøya buildups appear to be related to local tectonism and did not have a regional effect. Furthermore, calcite cementation post-dates hydrocarbon migration in the Bjørnøya buildups (Lønøy, 1988) and may not have affected reservoir properties in the subsurface. It is therefore possible that similar *Palaeoaplysina* buildups in the subsurface have good reservoir potential.

GEOLOGICAL BACKGROUND

Together with the marginal parts of North Greenland, the Barents Sea area formed an extensive mosaic of inter-connected intracratonic basins during the late Paleozoic. The limits of this huge depositional area are fairly well established to the southeast and southwest but the northern basinal margin is poorly understood (Figure 1). Eastwards, the basin continued to Franz Joseph Land and Novaya Zemlya and was connected to the Timan-Pechora basin. The marine connection southwards to the East Greenland and Zechstein basins was not established until the mid-Permian (Stemmerik and Worsley, 1989; and in press).

Following Early Carboniferous rifting with deposition in narrow and semi-isolated half grabens, most of the Barents Sea area became a relatively stable and gently subsiding platform during the latest Carboniferous. This pattern continued into the Permian with some downwarping and increased subsidence over sites of pre-existing grabens such as the Nordkapp basin. However, the Stappen high (on which Bjørnøya is located), southwestern Spitsbergen and Loppa high were subjected to renewed tectonic activity and structural inversion during the Late Carboniferous to mid-Permian (Worsley et al., 1990). The sedimentary succession on Bjørnøya is therefore thinner and stratigraphically less complete than that recorded elsewhere in the region (Steel and Worsley, 1984). Sedimentation started in the latest Devonian and approximately 600m of continental siliciclastic sediments were deposited during the latest Devonian and Early Carboniferous (Gjelberg, 1981). These sediments are overlain by 250–300 m of shallow marine carbonates and sandstones of Late Carboniferous age. A marked fall in sea level during the latest Carboniferous led to exposure and erosion in many places throughout the region. A subsequent, earliest Asselian sea level rise led to re-establishment of relative uniform carbonate sedimentation over large areas and also led to flooding of Bjørnøya and deposition of the Kapp Duner Formation (Figure 2). Deposition was followed by a major break in sedimentation during the Sakmarian, and other than during two brief depositional events in the latest Early Permian, Bjørnøya was subaerially exposed during the Permian. Renewed sedimentation started in the earliest Triassic and approximately 200m of Triassic shales are preserved. The Stappen high was inverted during the Tertiary, and post-Triassic sediments are not preserved on Bjørnøya. A maximum overburden of 1.5 to 2.0 km of sediment is suggested from geochemical data.

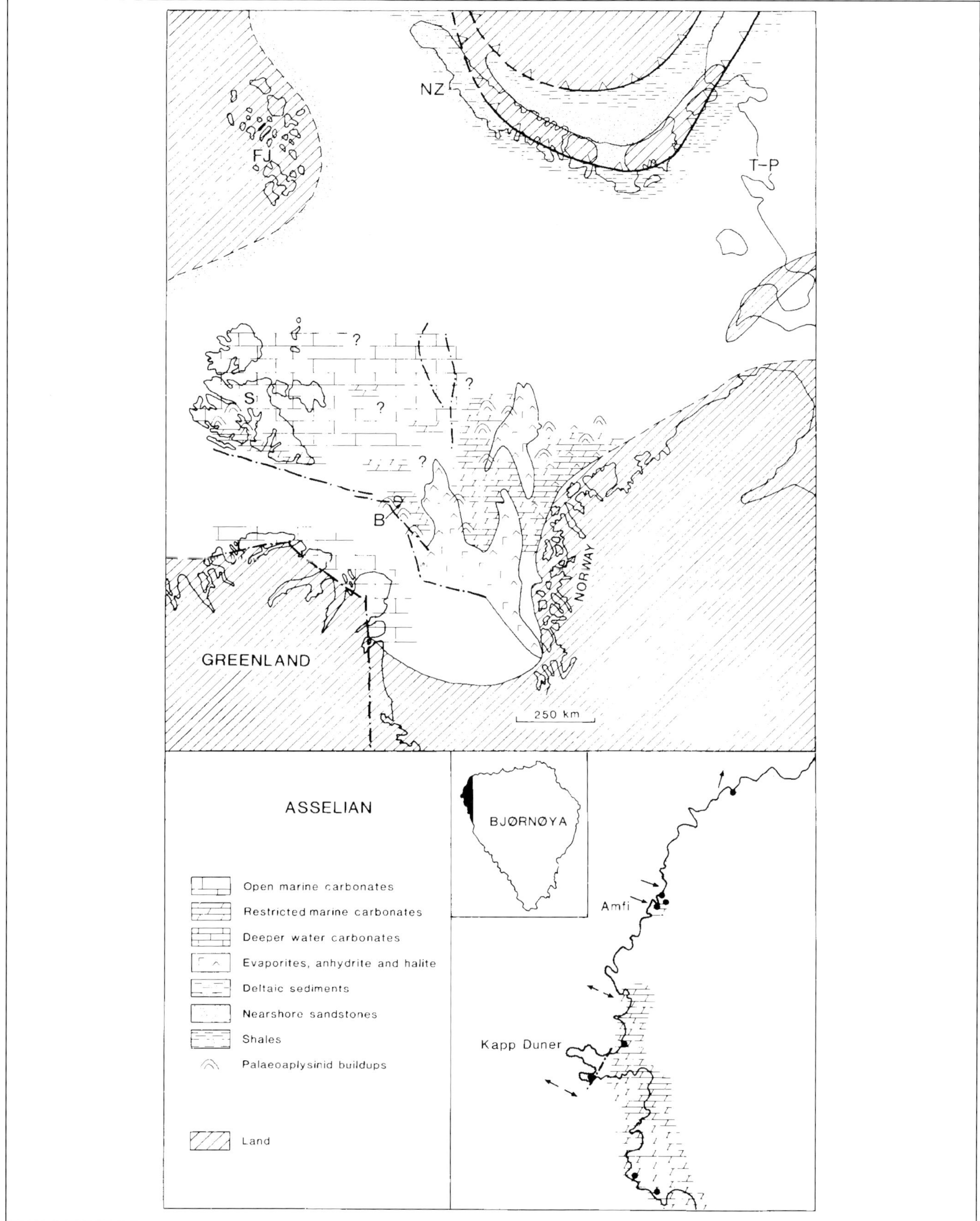

Figure 1. Pre-drift location map with a reconstruction of the Barents Sea area showing Asselian depositional environments. Locations mentioned in the text: B: Bjørnøya, S: Spitsbergen, FJ: Franz Joseph Land, NZ: Novaya Zemlya, T-P: Timan-Pechora basin. Inset map shows outcrop of Kapp Duner Formation; arrows indicate direction of migration of the *Palaeoaplysina* buildups. Modified from Stemmerik and Worsley (in press).

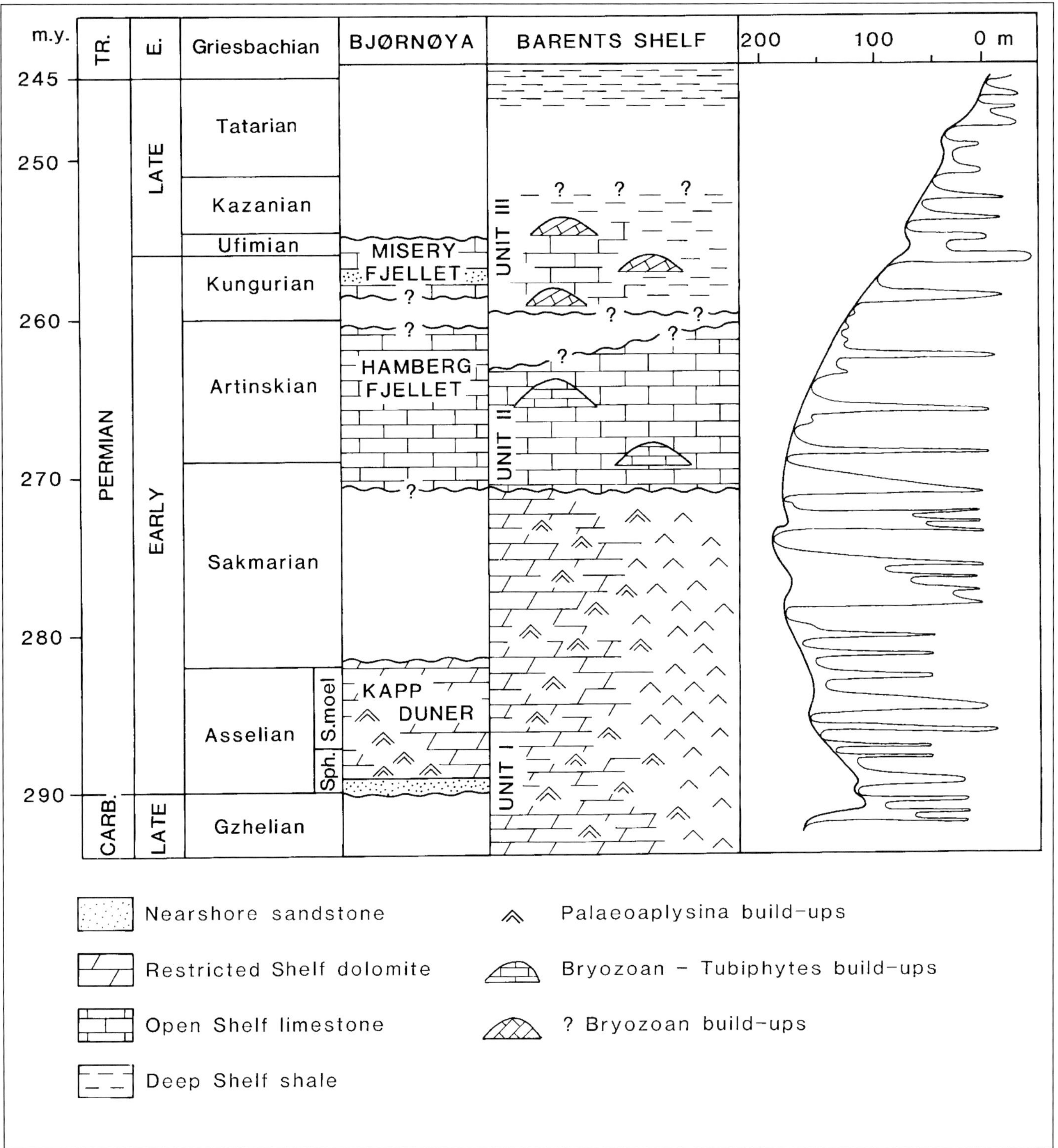

Figure 2. Correlation of major depositional units between Bjørnøya and offshore Barents Sea. Sea level curve (far right column) from Ross and Ross (1987).

PALEOGEOGRAPHY AND STRATIGRAPHY

The Barents Sea area was a mosaic of isolated carbonate platforms, carbonate ramps and shallow evaporite basins during the Asselian (Figure 1). The sediments form a single seismic-stratigraphic unit, the Asselian sequence set P1-1 of Cecchi (1993). This contains thick accumulations of *Palaeoaplysina* buildups along the margins of the Nordkapp and Bjørnøya basins and possibly also west of Bjørnøya. These buildups appear to have been controlled by former structural lineaments. Restricted marine and lagoonal carbonates, with isolated *Palaeoaplysina* mounds were

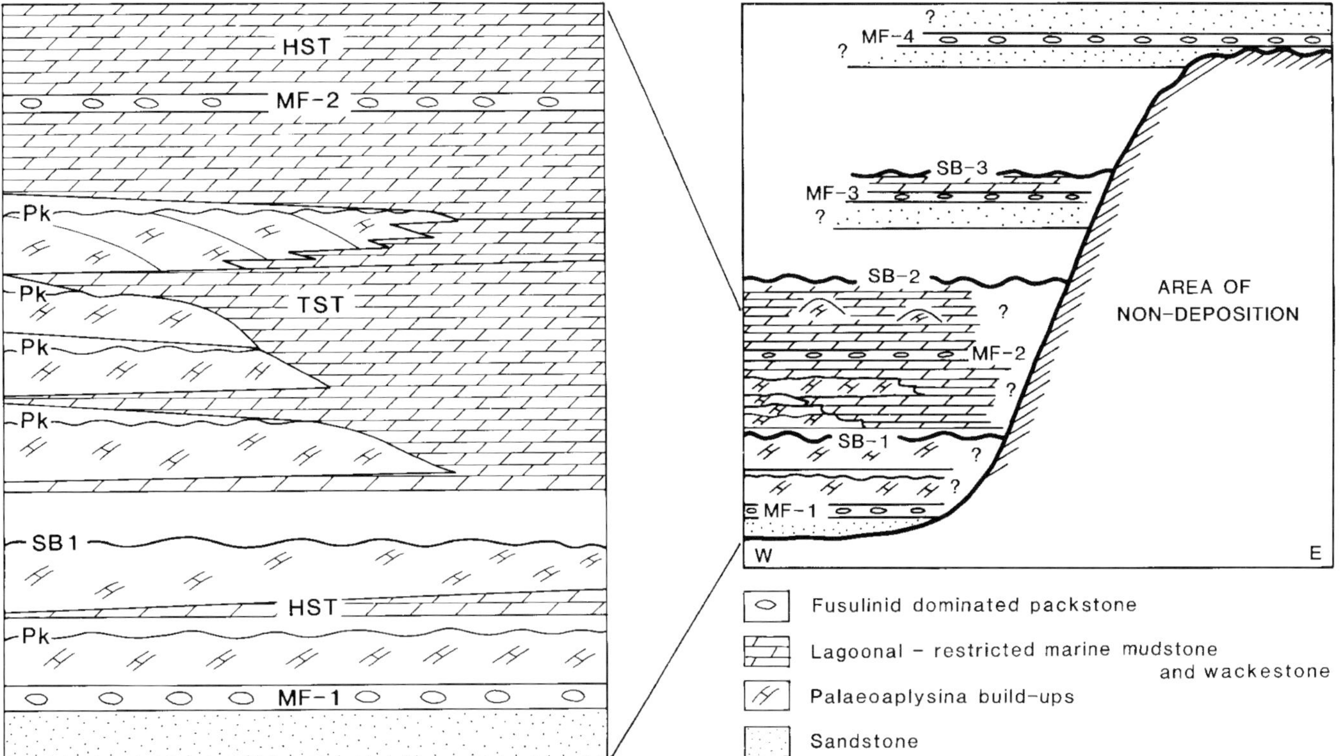

Figure 3. Stratigraphy of the Kapp Duner Formation on Bjørnøya. SB: sequence boundary, MF: maximum flooding surface, HST: highstand systems tract, TST: transgressive systems tract. All sequence stratigraphic nomenclature relates to the third order sequences. Pk: paleokarst surface related to fourth order fluctuations in relative sea level.

deposited over most of the platform areas including eastern Bjørnøya, and locally, thick accumulations of sabkha deposits occur (Figure 1). Halite and gypsum were deposited in the basinal areas.

The Asselian deposits on Bjørnøya may be viewed as a transgressive succession related to a second order rise in relative sea level. They consist of four depositional sequences related to third order fluctuations in sea level, and the two lower sequences are composed of stacked cycles related to fourth order fluctuations (Figure 3). The third order sequences can be identified in offshore wells but are below seismic resolution. On Bjørnøya, the tectonic setting complicates the sequence pattern so that the two youngest third order sequences are highly condensed and dominated by sandstones. *Palaeoaplysina* buildups are confined to the two oldest sequences (Figure 3). Offshore in contrast, *Palaeoaplysina* buildups are expected throughout the Asselian.

DEPOSITIONAL ENVIRONMENTS

Two distinctive sedimentary environments can be distinguished in the lower part of the Kapp Duner Formation on Bjørnøya: *Palaeoaplysina* buildups and lagoonal-restricted marine carbonates with isolated *Palaeoaplysina* mounds and subordinate sandstone beds (Figure 4). The *Palaeoaplysina* buildups form mud-dominated, massive to poorly bedded tabular and lenticular bodies up to 10 m thick and consist of *Palaeoaplysina* wackestone and packstone (Figure 5). Large colonial corals occur frequently and other faunal elements include rare bryozoans, brachiopods, ostracods, fusulinids, and phylloid algae. Internal, inclined stratification due to lateral accretion is common. The top surfaces of the buildups are undulating and evidence of subaerial exposure can be found. The buildups were dolomitized and lithified, aragonitic material was dissolved, and dissolution enlarged fractures were formed prior to deposition of the overlying sediment. Fractures and vugs were partly filled with anhydrite cement (now replaced by dolomite) and internal, laminated sediment. The most prominent paleokarst surface within the succession, probably a third order sequence boundary (SB-1 in Figure 3), has relief of up to 2 m with incised sandstone- and conglomerate-filled channels and sand-filled fractures and vugs (Figure 6). The internal sediment is brownish and easily distinguished from internal sediment related to later, mid-Permian exposure events. Pedogenic structures and *in situ Microcodium* (as described from *Palaeoaplysina* buildups on Spitsbergen; Skaug et al., 1982) have not been observed. This may be due to the pervasive dolomitization which has destroyed most primary textures.

The lagoonal-restricted marine deposits are domi-

Figure 4. Stacked *Palaeoaplysina* buildups at location Amfi in Figure 1. Note the lateral transition into lagoonal and restricted marine deposits towards the east (right). Person on top of the exposure for scale.

Figure 5. *Palaeoaplysina* packstone. Note the densely packed plates and well-developed vuggy porosity.

nated by well-bedded 10–30 cm thick biogenic wackestones and mudstones with a very sparse fauna of ostracods and fusulinids. Three distinctive, up-to-50-cm-thick fusulinid packstone beds occur within the lagoonal-restricted marine deposits. They contain a diverse fauna of crinoids, bryozoans, brachiopods, algae, and small foraminifers, in addition to the fusulinids (Simonsen, 1987). These beds record temporary events of fully marine conditions on the carbonate platform and are interpreted as deposited during maximum flooding. Each has a distinctive fusulinid fauna indicating that considerable time spans are represented between depositional events. These beds record flooding events related to third order fluctuations in relative sea level (Ross and Ross, 1988).

The *Palaeoaplysina* buildups are divided into two complexes separated by a distinctive paleokarst surface. Both consist of shoaling upward cycles of subtidal deposits punctuated by subaerial exposure surfaces. This is typical for cycles formed under high amplitude sea level changes (Wright, 1992; Read and Horbury, 1993). The lower buildup complex is ca. 20 m thick and consists of two shoaling upward cycles of restricted marine wackestone and *Palaeoaplysina* buildups. The buildups are tabular and these cycles can be traced for more than 4 km in the north-south direction; they are also laterally consistent in the east-west direction. The lower buildup complex forms the highstand systems tract of the Nordvestbukta sequence (Figure 3). It overlies a fusulinid packstone bed and is terminated by a karstic surface, interpreted as a sequence boundary (SB-1). The buildups migrated from north to south and probably represent rapidly migrating east-west oriented barriers that adjusted to falling sea level. The upper buildup complex is ~25 m thick and forms part of the transgressive systems tract of the following Grytvika sequence. It overlies sequence boundary 1 and is overlain by the Grytvika fusulinid bed, recording maximum flooding (Figure 3). The buildup complex consists of five shoaling upward cycles of restricted marine wackestones and *Palaeoaplysina* buildups. Individual cycles are up to 8 m thick and are terminated by a subaerial exposure surface. They are laterally restricted in the east-west direction and towards the east pass laterally into a succession of lagoonal and restricted marine carbonate mudstones and wackestones.

DIAGENESIS

Introduction and Methods

Deposits of the Kapp Duner Formation have undergone severe and complex diagenetic alterations. The diagenesis of the *Palaeoaplysina* buildups is more complex and extensive than that seen in the nearby lagoonal deposits, and it appears that there was a strong facies control. Petrographic studies have been supplemented by stable isotope and chemical analyses. The latter were carried out for Ca, Mg, Fe, Mn, Sr, and Na on a JEOL 733 Superprobe automated by a Tracor Northern TN 2000 X-ray analyzer. Sr and Na contents are generally low ($<< 500$ ppm) and usually undetectable using this method. $\delta^{13}C$ and $\delta^{18}O$ are expressed in ‰ relative to the PDB standard. The diagenetic sequence outlined below differs slightly from that described by Folk and Siedlecka (1974), Siedlecka (1975), and Lønøy (1988) in earlier accounts on the diagenesis of the Kapp Duner Formation.

Replacement Dolomite

The majority of the dolomite found in the *Palaeoaplysina* buildups and the associated lagoonal-restricted shelf deposits is a replacement of lime mud and calcite fossils, like fusulinids and brachiopods. Fossil fragments and most lime mud is replaced by dolomite which is brownish in color and consists of

Figure 6. Paleokarst surface on top of a *Palaeoaplysina* buildup.

anhedral to subhedral crystals generally less than 0.05 mm in size (Figures 7 A and B). Locally, a more coarsely crystalline fabric of subhedral to anhedral, often somewhat elongated crystals occurs associated with the microcrystalline dolomite. These crystals are cloudy due to numerous inclusions of microcrystalline dolomite. The crystal size is between 0.05 mm and 0.1 mm. The average Ca/Mg is 51.9/47.9 and the average iron content is less than 0.2 % in the replacement dolomite. $\delta^{13}C$ and $\delta^{18}O$ average 6.0‰ and 0.1‰ respectively. These values are comparable with results from sabkha dolomites (e.g. McKenzie, 1981; Wallace, 1990).

Dolomite Cements

Cement A: Elongated Dolomite

Cement A, elongated dolomite, is the earliest cement generation. It consists of elongated dolomite crystals, up to 0.4 mm long and 0.1 mm wide, which are rich in inclusions and occasionally display a relict fibrous texture. The crystals form a pore-lining cement in most solution fractures and vugs and in partly dissolved *Palaeoaplysina* plates (Figure 7 B). They typically occur immediately beneath the subaerially exposed surfaces of the *Palaeoaplysina* buildups. This dolomite corresponds to dolomite 2 of Lønøy (1988), who suggested on the basis of textural observations that it replaces primary anhydrite cement. Elongated dolomite is slightly more Ca-rich (Ca/Mg: 52.7/47.1) than replacement dolomite and the limpid dolomite cement B (as expected if it is replacing anhydrite).

Anhydrite cementation and replacement by elongated dolomite post-date replacement dolomite and a phase of aragonite and calcite dissolution and associated fracturing. The precursor anhydrite pre-dates infilling of marine fossils of the following sea level cycle.

Cement B: Non-Ferroan Limpid Dolomite

Cement B, non-ferroan limpid dolomite, occurs as limpid, equant and subhedral to euhedral crystals, often with a outer, brownish ferroan zone of ankeritic composition (Figure 7 C,D). Crystal size ranges upwards from 0.1 mm. Many crystals have a core of elongated dolomite crystals (cement A) or an inclusion-rich rhombic core (Figure 7B). This dolomite corresponds to dolomite 3 of Lønøy (1988) and the limpid dolomite of Folk and Siedlecka (1974) and Siedlecka (1975). It is geochemically very similar to the replacement dolomite (Fe <0.2 %; Ca/Mg: 52.2/47.6; $\delta^{13}C$ = 4.0‰ and $\delta^{18}O$ = –3.7‰).

Minor dissolution of the non-ferroan limpid dolomite took place prior to precipitation of the ankeritic dolomite. More extensive dissolution took place later, and limpid crystals with dissolved cores or cores filled by calcite are common. Cement B postdates dolomite cement A. It forms pore-lining and pore-filling cement in vugs, fractures and other solution-enlarged pores (Figure 7 B,C,D). The relationships to vadose silt, filled into the pores during mid-Permian exposure, are more complicated. Some small inclusion-rich crystals pre-date the vadose silt while larger limpid crystals post-date this event. The occurrence together with vadose silt suggests that the limpid dolomite is precipitated from surface-derived water. This is confirmed by the isotopic composition of the crystals. The low iron content indicates precipitation in an oxic environment.

Cement C: Ankeritic Dolomite

Cement C, Fe-rich dolomite with an ankeritic composition, is widespread in partly filled pores, vugs and fractures (Figures 7C,D, and 8A–D). It also fill very thin (< 0.01 mm wide) fractures. The ankeritic dolomite forms an outer rim on many limpid dolomite crystals and, in rare cases, (>1 mm) rhombs. More finely crystalline ankeritic dolomite fills fractures and in some cases dissolved parts of limpid dolomite rhombs. The ankeritic dolomite has an average composition of Mg/Fe: 22.2/23.2 and Ca/Mg+Fe: 53.3/45.4. Furthermore, it contains up to 3.2% Mn (average 2.0%). $\delta^{13}C$ and $\delta^{18}O$ average 3.4‰ and –4.7‰, respectively.

The ankeritic dolomite is the youngest dolomite seen in the *Palaeoaplysina* buildups. It post-dates an episode of fracturing and minor dolomite dissolution. The isotopic composition indicates that it precipitated from surface-derived water, and the abundance of iron and manganese indicates a reducing environment.

Silica Cement

Silica occurs as quartz and chalcedony. It is volumetrically insignificant but has been found as cement in *Palaeoaplysina* molds and other solution vugs in the buildups and also as isolated nodules in the lagoonal-restricted shelf sediments. The nodules consist of length-slow chalcedony and megacrystalline quartz.

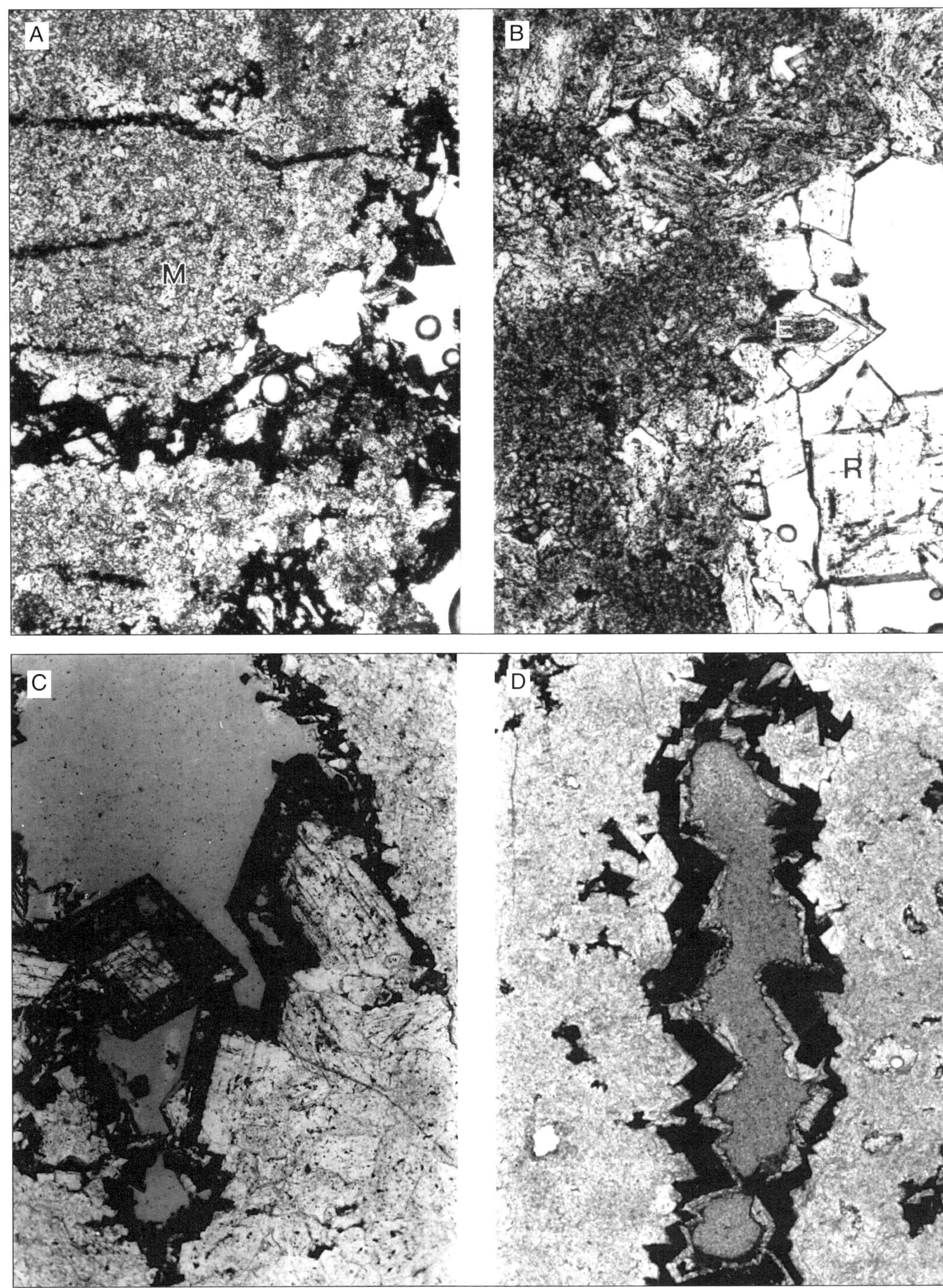
A
M
B
E
R
C
D

Figure 7. Dolomite cements. (A). Microphoto of fractures in microcrystalline dolomite (M). The largest fracture is partly filled by ankeritic dolomite and rhombic dolomite. Long axis equals 1.5 mm. (B). Microphoto of rhombic dolomite (R) filling vug in sediment preserved as microcrystalline dolomite. Note lining of elongated dolomite crystals; in some rhombic dolomite crystals there is a core of elongated dolomite (E). Long axis equals 1.5 mm. (C). Microphoto of rhombic dolomite with a rim of ankeritic dolomite. Note partial dissolution of the ankeritic dolomite. Long axis equals 1.3 mm. (D). Pore partly filled by rhombic dolomite, ankeritic dolomite, and late calcite. Long axis equals 10 mm.

The chalcedony often forms spherulites and contains small inclusions of sulfate (Siedlecka, 1972; Folk and Siedlecka, 1974). Megacrystalline quartz also occurs as idiomorphic crystals together with vadose silt in solution enlarged vugs. Silica cement post-dates precipitation of ankeritic dolomite and is probably a replacement of gypsum (cf. Siedlecka, 1972, 1975).

Calcite Cement

An iron-poor to moderately iron-rich (red to purple in stained thin sections) calcite spar forms the youngest cement in many pores (Figure 7D). It is zoned under cathodoluminescence with one to three zones with bright yellow luminescence. Calcite with similar luminescence also fills very thin (<< 0.01 mm wide) fractures and pores in dolomite cements B and C. The crystals are very coarsely crystalline (up to 3 mm across). In many cases, the calcite is replacive and contains numerous inclusions of dolomite.

Calcite post-dates all other authigenic minerals. Lønøy (1988) reports $\delta^{18}O$ values between –11‰ and –12‰ (which suggest precipitation at elevated temperatures) and $\delta^{13}C$ values in the range –2.7‰ to +0.5‰. A burial origin is confirmed by the presence of traces of hydrocarbons between the dolomites and the calcite (Lønøy, 1988).

Timing Of Diagenetic Events

Dolomitization and lithification were early and essentially syndepositional (prior to or during aragonite dissolution) and were completed before deposition of the next cycle of marine sediments (see figure 9 and observations (i) and (ii) in figure caption). Anhydrite (or gypsum) was precipitated after dissolution of the aragonite but prior to deposition of the overlying sediments (iii). The vadose silt is associated with a pronounced latest Sakmarian-earliest Kungurian karst event (Lønøy, 1988). Dolomite B cementation began prior to this event but the main phase was later (v and vi). Most of the dolomite cements B and C post-date deposition of the Kungurian Miseryfjellet Formation (vii) and were precipitated before calcite cement (viii). Thin calcite-filled fractures cross-cut vugs filled with dolomite cements A, B and C and have a trend different from that of fractures filled with ankeritic dolomite. These relationships indicate that precipitation of dolomite cements B and C and of calcite cement were two separate events, not closely associated as suggested by Folk and Siedlecka (1974) and Siedlecka (1975). Replacement of anhydrite by elongated dolomite took place before or as dolomite cement B was precipitated (iv) and prior to the latest Sakmarian-earliest Kungurian karst event (v).

Controls On Diagenesis

Diagenesis of the *Palaeoaplysina* buildups in the Kapp Duner Formation took place as four separate events, three of which took place before any significant burial (Figure 9).

(1) Alteration of micritic mud to microcrystalline dolomite, accompanied by aragonite dissolution and fracturing and followed by precipitation of anhydrite in the newly created fractures and vugs. These processes were associated with subaerial exposure of the buildups related to fourth order fluctuations in sea level. During the last few years, it has been increasingly evident that important diagenetic modifications are related to both high frequency and longer term third order fluctuations in relative sea level (e.g. Humphrey, 1988; Humphrey and Quinn, 1989; Read and Horbury, 1993; Scholle et al., in press), and that these alterations greatly influence later burial diagenesis (Tucker, 1992). It is also evident that the diagenetic signal related to these sea level cycles is climate-dependent and controlled by the amplitude of the sea level cycle. Early diagenesis of the Bjørnøya buildups took place during a time dominated by high frequency, high amplitude fluctuations in sea level. The climate was arid as indicated by precipitation of halite and anhydrite/gypsum in the nearby basins (see Figure 1). Early dolomitization is known from low amplitude cyclic deposits in arid climates where it is associated with sabkha progradation (e.g. Koerschner and Read, 1989). However, sabkhas are not likely to form during high amplitude fluctuations in sea level and sabkha deposits have not been found in the Asselian on Bjørnøya. More probably dolomitization was associated with reflux of hypersaline water during sea level low stand, a process described from other carbonate platforms associated with major evaporite basins (e.g. Tucker, 1991; Kendall, 1989). Dolomitization and evaporite cementation are also common in other carbonate platform successions surrounding evaporite basins (e.g. Stentoft, 1990; Tucker and Hollingworth, 1986; Tucker, 1991).

(2) During mid-Permian time, there was a change

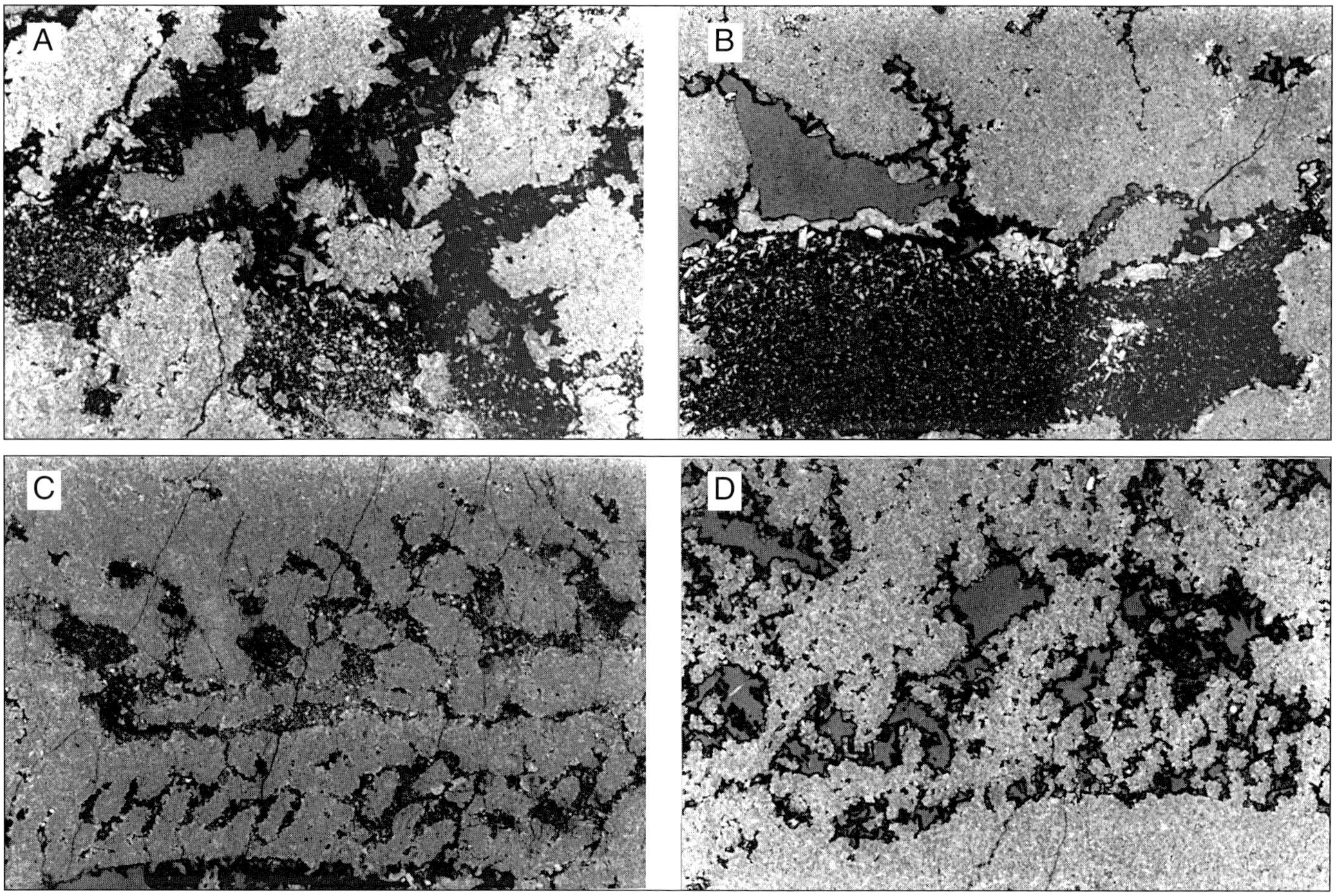

Figure 8. Other cements. (A). Secondary vug filled partly by vadose silt (V) and ankeritic dolomite. Note abundant pore lining of elongate dolomite. Long axis equals 1.7 cm. (B). Secondary vug partly filled by vadose silt. Note pore lining of ankeritic dolomite followed by calcite. Long axis equals 3.4 cm. (C). Two *Palaeoaplysina* plates. The original skeleton has been dissolved and is partly filled by vadose silt and ankeritic dolomite (dark). Channels were filled by carbonate mud that was replaced by dolomite (light) before dissolution of the skeletons. Long axis equals 1.7 cm. (D). Dissolved *Palaeoaplysina* plate with pore space partly filled by elongate dolomite and ankeritic dolomite. Long axis equals 1 cm.

toward more temperate and humid climatic conditions in the region (Steel and Worsley, 1984), as recorded both by a change in biota and by a regional shift from dolomite to limestone deposition (Stemmerik and Worley, in press). The more humid conditions are also recorded in the type of diagenesis affecting the sediments on Bjørnøya. Extensive karstification took place during mid- and Late Permian subaerial exposure (contrast with the lack of karstification associated with Asselian exposure events). Dissolution mainly affected the post-Kapp Duner limestone deposits where sediment-filled fractures tens of meters deep were formed. The more stable dolomite of the Kapp Duner Formation suffered only minor dissolution during these events. The first diagenetic alterations related to this karstification are probably replacement of anhydrite by elongated dolomite cement A and precipitation of small dolomite rhombs. These processes took place during early Sakmarian subaerial exposure (Figure 9).

(3) The third diagenetic phase is also related to subaerial exposure and karstification. It consists of precipitation of dolomite cements B and C and is associated with minor dissolution and fracturing which post-date deposition of the late Kungurian-(?)Ufimian Miseryfjellet Formation. They are most likely related to a prolonged event of subaerial exposure during the late Permian. The latest stages of precipitation took place in a reducing environment and include the ankeritic dolomite and minor amounts of pyrite. This change towards reducing pore water is probably controlled by the Early Triassic transgression of the area and deposition of organic-rich shales.

(4) Post-Permian burial diagenesis of the buildups includes minor silica cementation and precipitation of sparry calcite. Calcite cementation post-dates hydrocarbon migration into the buildups and is associated with a new generation of fractures, perhaps related to Tertiary uplift of the Stappen high.

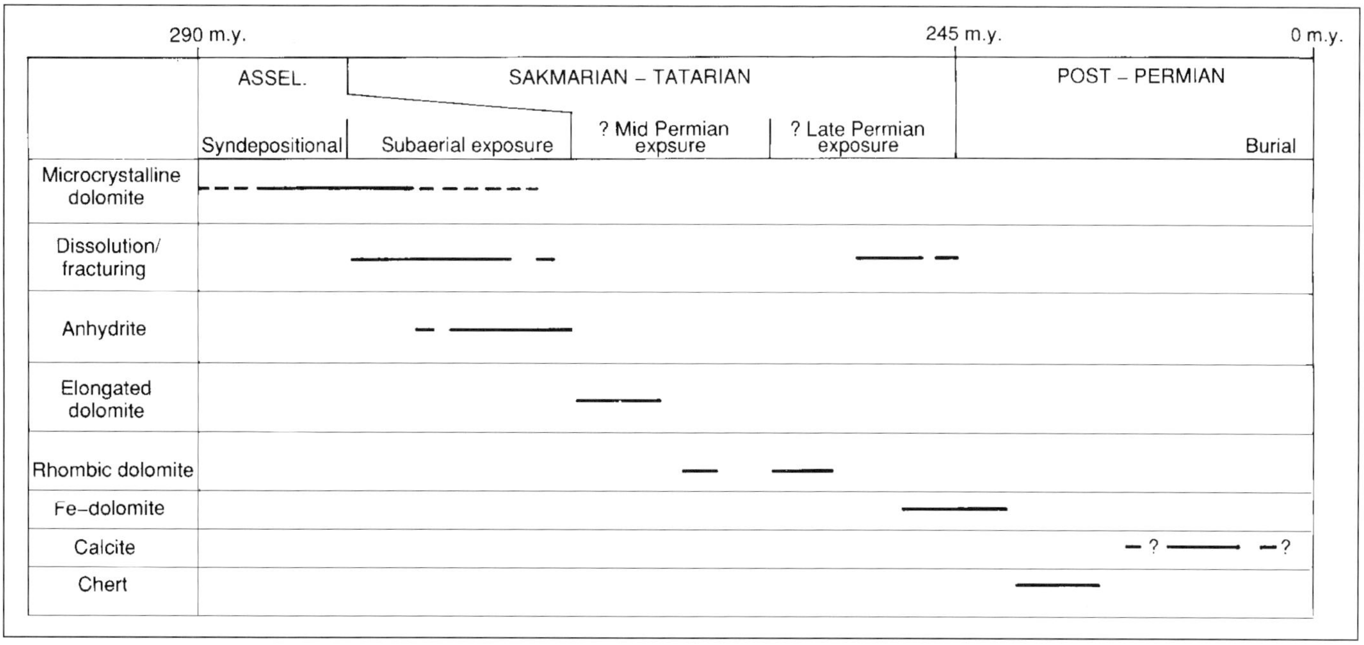

290 m.y.
245 m.y.
0 m.y.
ASSEL.
SAKMARIAN – TATARIAN
POST – PERMIAN
Syndepositional
Subaerial exposure
? Mid Permian exposure
? Late Permian exposure
Burial
Microcrystalline dolomite
Dissolution/ fracturing
Anhydrite
Elongated dolomite
Rhombic dolomite
Fe–dolomite
Calcite
Chert

Figure 9. Relative timing of diagenetic events in the *Palaeoaplysina* buildups. Dashed lines indicate uncertainty. The following evidence was used to interpret the timing of the dolomite replacement and precipitation of anhydrite, later replaced by dolomite cement A : (i) Mud-filled channels within *Palaeoaplysina* plates were dolomitized before or as the remaining part of the aragonitic plates were dissolved (Figure 8 C,D); (ii) Solution fractures and solution enlarged vug within sediments preserved as microcrystalline dolomite occasionally contain marine fossils; and (iii) Fractures cemented by anhydrite (dolomite cement A) occasionally contain marine fossils. The timing of dolomite cements A, B, and C are constrained by the following observations (Figure 9): (iv) Dolomite cement B is in optic continuity with the elongated dolomite A crystal it nucleated on; (v) Middle Permian vadose silt postdates dolomite cement A and some small inclusion-rich dolomite rhombs of cement type B (Figure 8 A,B); (vi) Most limpid dolomite (B) and all ankeritic dolomite (C) post-date vadose silt (Figure 8 A,B); (vii) Limpid and ankeritic dolomite (B and C) occur also in overlying Kungurian sediments (Siedlecka, 1975); and (viii) All dolomite pre-dates the calcite cement.

POROSITY EVOLUTION

The porosity found in the *Palaeoaplysina* buildups exposed on Bjørnøya is secondary vuggy and fracture porosity (c.f. Choquette and Pray, 1970). The vast majority of these secondary pores was formed as the result of dissolution of aragonite grains and possibly dolomitization of carbonate mud during early exposure. Later diagenetic modifications are porosity destructive, although some porosity was created during late (?) Tertiary fracturing.

The initial replacement of carbonate mud by microcrystalline dolomite produced no significant porosity. The associated dissolution and fracturing of the buildups, particularly of *Palaeoaplysina* skeletons during intra-depositional exposure events, created significant secondary porosity (20-25%). However, part of this was filled by anhydrite (later replaced by elongate dolomite) so that the net increase in porosity during the exposure events was probably 10-20%. The most porous intervals are found in the grain-rich parts of individual buildups (see Figure 8 C,D). Porosity within the buildups was reduced further by precipitation of dolomite cements B and C during mid- and Late Permian tectonically controlled exposure events. A decrease in porosity of between 1% and 5% is estimated to have occurred at this time, although some new fractures and vugs were formed. Minor additional reductions in porosity were related to precipitation of sparry calcite during burial.

IMPLICATIONS FOR EXPLORATION OFFSHORE

The outcrop studies of the *Palaeoaplysina* buildups on Bjørnøya indicate that subsurface equivalent buildups are likely to be dolomitized and dominated by secondary vuggy and fracture porosity. The Asselian successions on Bjørnøya and in the subsurface show similar sequence stacking patterns, and high frequency cycles such as those seen in the outcrop succession are also recognizable in wells (Cecchi, 1993). This means that the subsurface buildups were subjected to the same cyclic fluctuations in sea level as were those on Bjørnøya, and that they probably followed a similar early diagenetic pattern of dolomite replacement, dissolution of aragonitic grains, and anhydrite cementation. However, the later diagenetic alterations related to local uplift and prolonged subaerial exposure of Bjørnøya are not likely to be regionally widespread features. They include anhydrite replacement by elongated dolomite (A) and precipitation of limpid and ankeritic dolomite (B and C).

The diagenesis of the Bjørnøya buildups differs significantly from that described in age equivalent buildups from the Sverdrup basin in Arctic Canada by Beauchamp et al. (1989). Those in the Sverdrup basin were deposited in more agitated environments along the margins of non-evaporitic basins and their diagenesis is dominated by pervasive, early marine cementation. The buildups are preserved as calcite, and regionally widespread dolomitization did not occur during this interval in the Sverdrup basin. *Palaeoaplysina* buildups dominated by early marine cement are also likely to be found in the subsurface in the Barents Sea area along the platform margins but are likely to have poorer reservoir quality than those on Bjørnøya.

SUMMARY AND CONCLUSIONS

The *Palaeoaplysina* buildups on Bjørnøya were deposited during an early Asselian sea level high stand and the subsequent transgression. Individual buildups are 2–10 m thick and are stacked along the east coast of the island to form a northeast–southwest-trending barrier. East of this barrier, lagoonal and restricted marine wackestones were deposited. Individual *Palaeoaplysina* buildups were terminated by paleokarst surfaces related to short term, possibly glacio-eustatic, fluctuations in sea level.

Diagenesis is mainly early and related to repeated subaerial exposure during third and fourth order fluctuations in sea level and later Permian tectonic uplift. Diagenetic alteration includes syndepositional dolomitization of carbonate mud, formation of secondary vuggy porosity and precipitation of anhydrite, followed by anhydrite replacement by elongated dolomite (A) and precipitation of non-ferroan limpid dolomite rhombs (B) and ankeritic dolomite (C). Diagenesis related to burial and late uplift is very limited. It includes formation of small volumes of coarsely crystalline calcite and quartz. The porosity development is therefore dominated by processes related to early, intra-depositional exposure of the buildup complex. Most porosity is secondary vuggy and fracture porosity and the best reservoir facies are those dominated by *Palaeoaplysina* skeletons.

Seismically identified mounds in the Early Permian succession of the offshore areas of the Barents Sea were probably not subjected to prolonged mid- and Late Permian exposure. Their porosity development was therefore controlled mainly by early, intra-depositional processes and they are predicted to form potential but highly heterogeneous reservoirs with porosities in the range 10 to 20%. Silica cement, which is a major concern in younger parts of the succession, appears not to be a problem in the *Palaeoaplysina* buildups.

ACKNOWLEDGMENTS

This project was financially supported by the Norwegian Petroleum Directorate (NPD). Atle Mørk, IKU is gratefully acknowledged for introducing us to the geology of Bjørnøya. The paper greatly benefited from improvements suggested by P. Gutteridge, A. Horbury, and an anonymous AAPG reviewer. Permission to publish was granted by the Geological Survey of Greenland (LS) and NPD (GBL).

REFERENCES CITED

Beauchamp, B., Davies, G. R. and Nassichuk, W. W., 1989, Upper Carboniferous to Lower Permian *Palaeoaplysina*–phylloid algal buildups, Canadian Arctic Archipelago. *in*: Reefs—Canada and adjacent areas, Geldsetzer, H. H. J., N.P. James and G.E. Tebbutt, eds., Canadian Society Petroleum Geologists Memoir 13, p. 590-599.

Bruce, J. R. and Toomey, D. F., 1993, Late Paleozoic bioherm occurrences of the Finnmark Shelf, Norwegian Barents Sea: analogues and regional significance. *in*: Arctic geology and petroleum potential, Vorren , T. O., et al. , eds., NPF Special Publication No. 2, Elsevier, p. 377–392.

Cecchi, M., 1993, Carbonate sequence stratigraphy: application to the determination of play models in the Upper Paleozoic succession of the Barents Sea. *in* Arctic geology and petroleum potential, Vorren, T. O., et al., eds., NPF Special Publication No. 2, Elsevier, p. 419–438.

Choquette, P. W. and Pray, L. C., 1970, Geological nomenclature and classification of porosity in sedimentary carbonates. AAPG Bulletin, v. 54, p 207-250.

Folk, R.L. and Siedlecka, A., 1974, The "schizohaline" environment: its sedimentary and diagenetic fabric as exemplified by Late Paleozoic rocks of Bear Island, Svalbard. Sedimentary Geology, v. 11, p. 1-15.

Gerard, J. and Buhrig, C., 1990, Seismic facies of the Permian section of the Barents Sea: analysis and interpretation. Marine and Petroleum Geology, v.7, p. 234-252.

Gjelberg, J.G., 1981, Upper Devonian (Fammenian)–Middle Carboniferous succession of Bjørnøya. Norsk Polarinstitutt Skrifter, 174, p. 1-67.

Humphrey, J.D., 1988, Late Pleistocene mixing zone dolomitization, southeastern Barbados, West Indies. Sedimentology, v. 35, p. 327-348.

Humphrey, J.D. and Quinn, T.M., 1989, Coastal mixing zone dolomite, forward modelling, and massive dolomitization of platform-margin carbonates. Journal of Sedimentary Petrology, v. 59, p. 438-454.

Kendall, A.C., 1989, Brine mixing in the Middle Devonian of western Canada and its possible significance to regional dolomitization. Sedimentary Geology, v. 64, p. 271-285.

Koerschner, W.F. and Read, J.F., 1989, Field and modelling studies of Cambrian carbonate cycles, Virginia Appalachians. Journal Sedimentary Petrology, v. 59, p. 654-687.

Lønøy, A., 1988, Environmental setting and diagenesis of Lower Permian *Palaeoaplysina* buildups and associated sediments from Bjørnøya. Implications for exploration of the Barents Sea. Journal of Petroleum Geology, v. 11, p. 141-156.

McKenzie, J., 1981, Holocene dolomitization of calcium carbonate sediments from the coastal sabkhas of Abu Dhabi, U.A.E.: a stable isotope study. Journal of Geology, v. 89, p. 185-198.

Nilsen, K. T., Henriksen, E., and Larssen, G. B., 1993, Exploration of the Late Paleozoic carbonates in the southern Barents Sea—a seismic stratigraphic study, *in* Arctic geology and petroleum potential, Vorren, T. O., et al., eds., NPF Special Publication No. 2, Elsevier, p. 393–404.

Read, J.F. and Horbury, A.D, 1993, Eustatic and tectonic controls on porosity evolution beneath sequence-bounding unconformities and parasequence disconformities on carbonate platforms. This volume.

Ross, C. A. and Ross, J. R. P., 1987, Late Paleozoic sea-levels and depositional sequences. Cushman Foundation Foraminiferal Research Special Publication 24, 137-149.

Ross, C. A. and Ross, J. R. P., 1988, Late Paleozoic transgressive-regressive deposition.*in*: Sea level changes—an integrated approach, Wilgus, , Ed., Society Economic Paleontologists Mineralogists Memoir 42, 227-247.

Scholle, P.A., Stemmerik, L., Ulmer, D.S., Di Liegro, G. and Henk, F.H., in press, Palaeokarst-influenced depositional and diagenetic patterns in Upper Permian carbonates, Karstryggen area, central East Greenland. Sedimentology.

Siedlecka, A., 1972, Length-slow chalcedony and relict sulphates—evidence of evaporitic sulphates in the Upper Carboniferous and Permian beds of Bjørnøya, Svalbard. Journal of Sedimentary Petrology, v.42, p. 812-816.

Siedlecka, A., 1975, The petrology of some Carboniferous and Permian rocks from Bjørnøya, Svalbard. Norsk Polarinstitutt Skrifter 1973, p. 53-72.

Simonsen, B. T., 1987, Upper Paleozoic fusulinds of Bjørnøya. IKU Report 88.090, 67 p.

Skaug, M., Dons, C. E., Lauritzen, Ø. and Worsley, D., 1982, Lower Permian *Palaeoaplysina* bioherms and associated sediments from central Spitsbergen. Polar Research, v. 2, p. 57-75.

Steel, R. J. and Worsley, D. 1984, Svalbard`s Post-Caledonian strata: An atlas of sedimentational patterns and palaeogeographic evolution, *in* Petroleum geology of the North European margin, Spencer, A. M., et al., eds., Graham and Trotman for the Norwegian Petroleum Society, p. 109-135.

Stemmerik, L. and Larssen, G. B. in press, Upper Paleozoic carbonates in the Loppa High area, western Barents Sea.

Stemmerik, L. and Worsley, D., 1989, Late Palaeozoic sequence correlations, North Greenland, Svalbard and the Barents Sea. *in* Correlation in hydrocarbon exploration, Collinson, J., ed., Graham and Trotman for the Norwegian Petroleum Society, p. 99-111.

Stemmerik, L. and Worsley, D., in press, Permian history of the Barents Sea area *in*: Permian of the northern continents, Scholle, P. A. and T.M. Peryt, eds., Berlin, Springer Verlag.

Stentoft, N., 1990, Diagenesis of the Zechstein Ca-2 carbonate from the Løgumkloster-1 well, Denmark. Danmarks Geologiske Undersøgelse Serie B, v.12, 42 p.

Tucker, M. E., 1991, Sequence stratigraphy of carbonate-evaporite basins: models and applications to the Upper Permian (Zechstein) of Northeast England and adjoining North Sea. Journal of the Geological Society of London, v.148, p. 1019-1036.

Tucker, M.E., 1992, Carbonate diagenesis and sequence stratigraphy, *in* Wright, V.P., ed., Sementology Review. Blackwell, London. v. 1, p. 51–72.

Tucker, M.E. and Hollingworth, N.T.J., 1986, The Upper Permian reef complex (EZ1) of North East England: diagenesis in a marine to evaporitic setting. *In* Schröder, J.H. and B.H. Purser, eds, Reef Diagenesis. Springer Verlag, Berlin, p. 270-290.

Veevers, J. J. and Powell, C. M. C. A., 1987, Late Paleozoic glacial episodes in Gondwanaland reflected in transgressive-regressive depositional sequences in Euramerica. Geological Society of America Bulletin 98, p. 475-487.

Wallace, M. W., 1990, Origin of dolomitization on the Babwire Terrace, Canning Basin, Western Australia. Sedimentology, v. 37, p. 105-122.

Worsley, D., Agdestein, T., Gjelberg, J. and Steel, R. J., 1990, Late Palaeozoic basinal evolution of Bjørnøya: Implications for the Barents Sea. Geonytt, v. 1, p. 124-125.

Wright, V.P., 1992, Speculations on the controls on cyclic peritidal carbonates: ice-house versus green house eustatic controls. Sedimentary Geology, v. 76, p. 1–5.

Chapter 13

Burial Diagenesis and Hydrocarbon Migration in Platform Limestones: A Conceptual Model Based on the Upper Jurassic of the Gulf Coast of the USA

Clyde H. Moore and Ezat Heydari
Applied Carbonate Research Program
Department of Geology/Geophysics
Louisiana State University
Baton Rouge, Louisiana, USA

ABSTRACT

Subsurface Upper Jurassic Smackover limestones of the Gulf Coast of the U.S.A. are an ideal laboratory for the study of the burial diagenesis of platform carbonates because of the enormous quantity of material available occasioned by extensive hydrocarbon production, the wide range of burial depths over which they occur, and the relative simplicity of their depositional setting. The model presented is based on several long term studies of these limestones. Pathways for burial diagenetic fluids were established early on during the sedimentological evolution of the platform and were driven by relative sea level fluctuations. Regionally extensive high porosity trends, that subsequently acted as conduits for diagenetic fluid flow, were generally restricted to siliciclastic lowstand fans and wedges, and highstand blanket carbonate sands. Transgressive systems tracts are generally muddy and act as aquicludes, or in some cases, hydrocarbon source rocks. Relative sea level lowstands can dramatically influence later diagenetic fluid flow by modifying original porosity distribution and permeability characteristics through dissolution and cementation. Carbonate mineralogies are generally stabilized during this early, surface-related diagenetic phase.

Burial diagenesis in platform limestones proceeds in two main phases: pre-hydrocarbon migration diagenesis and post-hydrocarbon migration diagenesis. Pre-hydrocarbon migration diagenesis is driven by chemical compaction in an initially open system, dominated by marine connate and mixed marine connate and meteoric water. This hydrologic system became compartmentalized during hydrocarbon trap formation. New pore fluids, derived from underlying siliciclastics below and from the dewatering of adjacent basinal deposits with faults as conduits, moved into porous platform carbonates and mixed with evolved connate waters. Calcite cements were predominantly sourced from pressure solution of host limestones. However, their composition also reflects the importation of Fe, Mn, radiogenic Sr, Pb, Ba, etc., from subjacent lowstand siliciclastic sequences and adjacent basinal deposits. Hydrocarbon precursors, such as H_2S and organic acids were also introduced, leading to sulfide formation and minor secondary porosity generation. Migration and trapping of hydrocarbons arrested further diagenesis within the carbonate reservoir sequence.

Post-hydrocarbon migration diagenesis is driven by thermal destruction of reservoir hydrocarbons under the influence of high temperatures. The resulting high H_2S environment led to metal fixation by sulfide precipitation, thermochemical sulfate reduction, late calcite cementation, and sulfur-driven methane destruction. Progressive loss of porosity through compaction as well as the emplacement of the by-products of late diagenesis, such as calcite cements and pyrobitumen, resulted in progressive closure of the diagenetic system. Trace element and isotopic compositions of late, post-hydrocarbon calcite cements reflect these processes.

This model should be applicable to hydrocarbon and sulfate-bearing carbonate shelf-ramp depositional sequences featuring high energy, highstand systems tracts, developed in response to 3rd order relative sea level fluctuations.

INTRODUCTION

Burial diagenesis associated with carbonate rock sequences has received a great deal of attention during the last decade. Many of the important diagenetic processes active in the subsurface, such as chemical compaction (Scholle, 1977; Choquette and James 1987), subsurface cementation (Moore, 1985), stylolitization (Koepnick, 1985), and thermo-chemical sulfate reduction (Heydari and Moore, 1989) have been thoroughly discussed in specific as well as conceptual terms. In addition there have been a number of general survey papers on the burial diagenetic environment that have attempted to assess the importance of processes responsible for subsurface diagenetic rock modification and porosity evolution (Scholle and Halley, 1985; Choquette and James, 1987; Moore, 1989).

To date, most important regional carbonate diagenetic studies stressing porosity evolution, such as Meyer's work in the Mississippian (1974; 1978; 1980; 1988), have involved stable cratonic sequences that have suffered repeated exposure to subaerial conditions. In this setting, intensive early meteoric diagenesis usually destroys available porosity long before hydrocarbon migration, limiting the study's value to the petroleum industry as a predictive model or tool.

In contrast, the economically important subsurface carbonates of the Upper Jurassic Smackover Formation of the U.S.A. Gulf of Mexico were deposited under conditions of rapid subsidence driven by extensional tectonics associated with the formation of the Gulf of Mexico by rifting (Moore, 1985). Exposure of Smackover carbonate reservoir facies to porosity modification by meteoric diagenetic conditions was therefore intermittent and short-lived, resulting in the retention of large porosity volumes during its rapid and permanent burial into the subsurface diagenetic regime (Moore and Druckman, 1981). This porosity provided the necessary access for subsurface fluids to interact with the rock during burial as well as a pathway for the migration of hydrocarbons forming in adjacent source rocks.

The present chapter is based on the results of a series of continuing diagenetic studies (Moore and Druckman, 1981; Klosterman, 1981; Moore, 1985; Moore et al., 1986, 1988; Heydari and Moore, 1989), stratigraphic studies (Moore, 1984; McGillis, 1984; Stewart, 1984; Wilkinson, 1984; Meensden et al., 1987; Moore and Druckman, 1991) and organic geochemical studies (Sassen et al., 1987; Sassen and Moore, 1988) carried out on a large Smackover subsurface data set under the auspices of the L.S.U. Applied Carbonate Research Program. The integration of these regional diagenetic, stratigraphic, and organic geochemical trends within a single stratigraphic sequence, extending across an entire geological province, under conditions of shallow (1600 m or 5500 ft) to deep (6300 m or 21,000 ft) burial, allows the development of a well-constrained model of the burial diagenesis and porosity evolution of a porous carbonate platform sequence as it is buried into the deep subsurface. This model should be a useful predictive tool for future exploration and exploitation efforts in deeper, frontier carbonate provinces.

In the sections that follow, we will trace the diagenetic and hydrologic evolution of the Smackover carbonate aquifer as it is deposited, buried into the subsurface, and interacts with evolving pore fluids under the influence of increasing temperatures and pressures, and we will witness its ultimate destruction as an effective regional aquifer as it is compartmentalized by structure and its pore system destroyed by compaction and cementation.

Subsurface Data And Analytical Methods

The subsurface data set used for this study is enormous. The basic stratigraphic, diagenetic, and organic geochemical framework outlined here, detailed in the studies mentioned in the introductory remarks above, is based on some 2000 geophysical well logs, over 500 sample logs, nearly 300 well cores, and four modern regional seismic lines.

Petrographic observations were made on polished and stained thin sections. Cathodoluminescent observations were performed with a standard Technosyn luminoscope. Operating conditions were approximately 15 KV gun potential, focused beam diameter of about 1 cm, 0.5 millitorr vacuum, and 300-400 micro amp beam current. Trace element analyses were conducted by dissolving 50 mg of powdered sample in 0.5 N HCl and analyzing with an Inductively Coupled Plasma-Atomic Emission Spectrometer (ICP-AES). Electron probe analyses were performed on a JOEL-733 using an accelerating voltage of 15 KV, beam current of 30 nannoamps, and beam size of 20 microns. Oxygen and carbon isotope analyses were performed using standard techniques. All data are reported relative to the PDB standard in per mil, with a precision of 0.2‰. Sr extraction for isotope analyses was done by the common ion exchange method. Precision of the analyses is ±0.00005.

Structural Setting

The Upper Jurassic of the U.S. Gulf Coast was deposited on a passive margin, shortly after the opening of the gulf by rifting (Salvador, 1987). Its regional structural framework is dominated, therefore, by extensional tectonics, featuring the development of a series of distinct syn-rift interior salt basins. These salt basins, the South Texas, East Texas, North Louisiana, and Mississippi, formed in areas of modified continental crust (Figure 1). Each is bounded to the south by positive continental blocks that separate the salt basins from the main Gulf of Mexico Basin (Figure 1). The interior salt basins were sites of rapid subsidence during deposition of the Louann Salt and exhibit high rise salt structures today.

A zone of more stable crust (less subsidence) occurs just north of each salt basin. The Louann Salt onlaps and wedges out to the north onto this ramp-like surface (Moore, 1984). This stable zone overlying the Louann Salt wedge, and north of the Louann

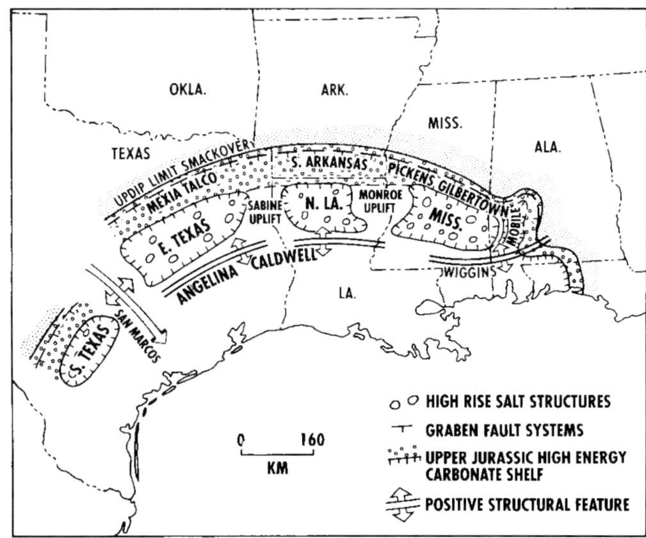

Figure 1. Regional structural setting for the Upper Jurassic of the central Gulf of Mexico. Modified from Moore, 1984.

depocenters in the interior salt basins, was the location for the development of the Upper Jurassic Louark carbonate shelf during a major Oxfordian sea level rise (Moore, 1984; Moore and Druckman, 1991; Goldhammer et al., 1991).

As the Smackover shelf developed, differential tilting of the shelf toward the basin, coupled with sediment loading, triggered small-scale salt tectonics along the basin margins, resulting in the formation of subtle salt anticlines and turtle structures (Jackson and Seni, 1983). During subsequent deposition of Smackover and post-Smackover Upper Jurassic units, sliding of the shelf sediment wedge over the salt toward the basin, coupled with further salt movement, led to the development of a regional, graben feature coincident with the updip limits of the Louann Salt. This graben feature is called the Mexia-Talco fault zone in East Texas, the South Arkansas graben in the central Gulf and the Pickens-Gilbertown-Mobile fault zone to the east (Figure 1). The majority of this early fault activity took place immediately following the deposition of the Smackover, reached its peak during deposition of the Buckner and Haynesville formations (Late Oxfordian), and was completed by the time of deposition of the Cotton Valley (Kimmeridgian) (Figure 2) (Moore, 1984). This early faulting and salt movement played a major role in the development of Upper Jurassic Smackover hydrocarbon traps and had a strong impact on the hydrologic continuity of upper Smackover aquifers around the Gulf rim prior to hydrocarbon migration.

Sequence Stratigraphic Framework

The Upper Jurassic of the U.S. Gulf rim is only encountered in the subsurface. The sequence has been

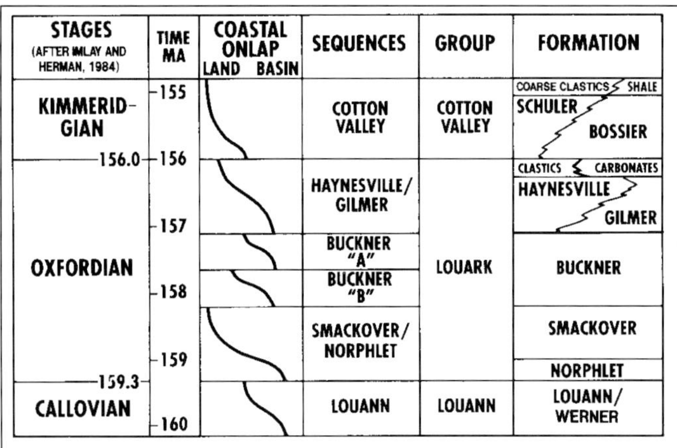

Figure 2. Stratigraphic framework of the Upper Jurassic of the central Gulf of Mexico. Compiled from Moore and Druckman, 1991.

divided into three rock stratigraphic groups: the Louann group, consisting of the Louann and Werner evaporites at the base, followed by the mixed carbonate-clastic-evaporite Louark group, which is capped by the siliciclastic Cotton Valley Group (Figure 2). The Louark consists of four formations: the siliciclastic Norphlet at the base; the carbonate dominated Smackover, the main subject of this chapter; the mixed siliciclastic-evaporite-carbonate Buckner; and the mixed siliciclastic-carbonate Haynesville at the top (Figure 2) (Moore and Druckman, 1991).

Ammonite fossil-based chronostratigraphic data on the Upper Jurassic sequence in the central Gulf (Imlay and Herman, 1984) suggest that the Louark group should be placed into the Oxfordian stage, that the overlying Cotton Valley Group, into the Kimmeridgian, and that the underlying Louann group, into the Callovian (Figure 2). These age assignments, based on more current paleontologic data, however, run counter to the earlier chronostratigraphic framework of Todd and Mitchem, 1977, Haq et al., 1988, and Sarg, 1988, who restrict the Oxfordian stage to the Smackover Formation, and assign post-Smackover units to Kimmeridgian-Portlandian stages. In this chapter we will use the chronostratigraphic assignments of Imlay and Herman (1984).

Moore and Druckman (1991) recognized four third order sequences in the Oxfordian of the central Gulf: the Smackover-Norphlet, the Buckner B, the Buckner A, and the Haynesville-Gilmer (Figure 2). Each of these four sequences commenced with siliciclastic lowstand systems tracts, and ended with high energy, oolitic carbonate highstand systems tracts.

Figure 3 shows the log, lithology, depositional environments, and stratigraphic interpretation of a continuous core through the entire Smackover sequence in the Humble McKean #12 well in Columbia County, Arkansas. This well can be considered typical of the Smackover-Norphlet sequence of the central Gulf of Mexico, and seems to be representative of the Smackover-Norphlet sequence to the east and to the west along the Upper Jurassic trend.

The terrestrial red desert sands and shales of the Norphlet lie unconformably on the Louann Salt and represent the lowstand systems tract of the Smackover sequence. The upper 2 m of the Norphlet are gray, re-worked, and represent the transgressive surface of the Smackover transgressive systems tract. A 50 m sequence of organic rich, varved, marine lime mudstones occurs immediately above this surface. These mudstones are thought to be relatively deep water anoxic basin deposits associated with salinity stratification established soon after the Smackover flooding event. This varved sequence occurs across the entire trend of the Smackover from Texas to Florida and acts as the source rock for Jurassic hydrocarbons (Sassen and Moore, 1988).

At the top of the varved sequence, gray lime muds showing soft sediment slump structures are thought to represent the base of the slope and the beginning of the Smackover highstand systems tract. The top of the varved sequence then, would be a downlap surface, and represents the maximum flooding surface for the Norphlet-Smackover sequence.

The highstand systems tract, in the center of the Louark shelf, is a classic continuous shoaling upward sequence representing slope, shelf, sand shoal environments, and finally a shoreline beach complex, and has been termed a prograding carbonate ramp by Ahr (1973) and by Budd and Loucks (1981). Moore (1984), Stewart (1984), Wilkinson (1984), and McGillis (1984) suggested, instead, that the Smackover, in the western part of the trend, represented a well-defined grainstone rimmed carbonate shelf. A thick but laterally restricted buildup of ooid grainstones (up to 200 m thick at the basin margin, and seldom over 35 m thick in the interior of the shelf) in the Smackover occurs downdip along the basin margin and interfingers updip with restricted lagoonal carbonate mudstones in East Texas. This facies architecture suggests a shelf, rather than a ramp depositional setting. This well-defined Smackover carbonate shelf margin can be mapped on regional seismic throughout East Texas (Moore, 1984).

The Smackover upper sequence boundary in this well is marked by a distinct unconformity exhibiting a soil profile, soil breccias, and caliches strongly suggesting exposure during the relative sea level lowstand separating the Buckner and Smackover sequences. The regional extent of this unconformity, from interior shelf across the shelf margin in the western part of the Smackover trend (Texas, Arkansas, and Louisiana), would suggest a type 1 sequence boundary.

The subsequent Buckner flooding event is marked by high energy shoaling upward oolitic carbonate

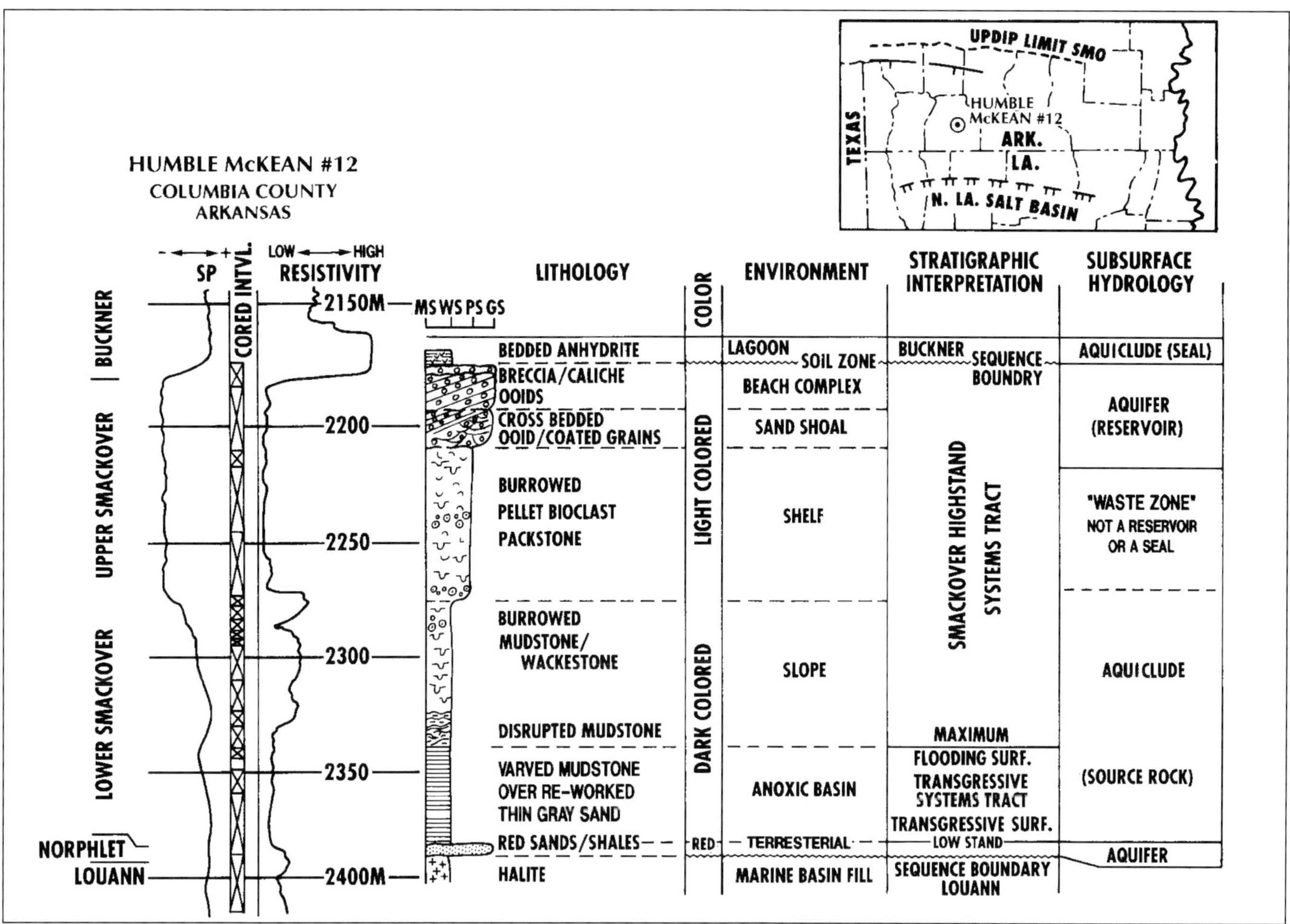

Figure 3. Core description, environmental and stratigraphic interpretation of the Louann to Buckner formations in the Humble McKean #12 well located in the Haynesville field, central Columbia County, Arkansas. Those sequences that should act as aquifers or aquicludes in the subsurface are noted. Shell Oil Company, Houston Texas made the core available to the investigators.

sequences along the basin margin interfingering with bedded anhydrite to the north. This subaqueous bedded anhydrite was deposited in a shelfal lagoon behind a Buckner shelf margin carbonate barrier (Dickinson, 1968; Moore, 1984). These lagoonal anhydrites apparently represent the Buckner transgressive systems tract on the shelf. Moore and Druckman (1991) documented the onlap of the anhydrite onto the unconformity surface of the Smackover sequence boundary across Arkansas and Texas by seismic and stratigraphic analysis.

The sequence in the Humble McKean well represents the classic response of carbonate sedimentation to a third order relative sea level cycle. While Smackover sequences elsewhere are remarkably similar to the McKean sequence, as one might expect, present lack of chronostratigraphic control (particularly to the east in Mississippi, Alabama and Florida) precludes one from making a definitive correlation of these eastern shoaling upward sequences with the Smackover in Arkansas. Nevertheless, the facies architecture, and response to relative sea level fluctuations of these sequences, are so similar that given similar burial histories, eogenetic diagenetic patterns, aquifer development, and subsequent response to burial history should be almost identical.

The next section details the influence of this facies architecture in defining the ultimate pathways available for the movement of burial diagenetic fluids and the impact of eogenetic diagenesis on these aquifer systems as a result of the significant relative sea level fall that occurred at the end of the Smackover.

AQUIFER DEVELOPMENT AND IMPACT OF DIAGENETIC OVERPRINT AT THE TIME OF BURIAL

Aquifer Development

Smackover facies architecture, established during a third order sea level cycle, resulted in the development of two major, regional aquifer systems. The siliciclastic lowstand systems tract at the base (the Norphlet) is a regional, highly porous blanket siliciclastic sand that had the capacity to permit movement of significant volumes of water throughout the burial history of the Smackover. The efficiency of the Norphlet aquifer is particularly enhanced if it is intersected and truncated by a later, regional unconformity such as the updip truncation suffered by the Upper Jurassic along the mid-Cretaceous unconformity in Arkansas (Murray, 1961). Rock-water interaction in these siliciclastic sands, coupled with solution of the subjacent Louann evaporites during burial, may furnish Mn-rich, Fe-rich, and radiogenic Sr-rich brines to the pure carbonate sequence above, if conduits are available (Moore and Druckman, 1981; Moore, 1985; Moore 1989).

The uppermost, shallow water, high energy phase of the Smackover highstand systems tract (beach, ooid shoal facies) is generally dominated by thick (up to 35 m) ooid grainstones. It is uncertain at this point whether these grainstones are aggradational (Moore, 1984) or progradational (Budd and Loucks, 1981). Hydrologically the result is the same in that the entire high energy facies complex acts as a regional, interconnected, highly porous, carbonate aquifer. This porous aquifer system forms the major Upper Jurassic hydrocarbon reservoir when modified by structural dislocation. The onlapping evaporites or shales associated with the initial Buckner flooding event often act as seals.

The organic-rich varved lime mudstones of the transgressive systems tract and the lower slope mudstones of the highstand systems tract should act as aquicludes, effectively isolating the two major aquifers of the sequence: the Norphlet lowstand systems tract and the upper Smackover highstand systems tract. In addition, the Smackover source rock (varved lime muds) is also initially separated from the Smackover reservoir rock (ooid grainstones). Vertical conduits, therefore, must be established across this aquiclude if Norphlet-related waters and lower Smackover hydrocarbons are to migrate into, and interact with, the reservoir rocks of the upper Smackover aquifer.

The most realistic way to develop vertical conduits is by a series of faults that cut the entire sequence, from the Norphlet through the upper Smackover. It is clear from stratigraphic, structural, and seismic analysis that such conduits were established during the development of the post-Smackover regional graben systems and in conjunction with the initiation of salt tectonics (Moore and Druckman, 1981; Moore, 1985; Heydari, 1990).

Impact Of Early Diagenetic Overprint

Smackover highstand deposits were generally dominated by high-energy ooid grainstones, the mineralogy of which varied from aragonite in the Smackover shelf interior to calcite along the shelf margin (Figure 4) (Moore et al., 1986; Swirydczuk, 1988; Moore, 1989). This pattern of ooid mineralogical variation, which was first described for the Louisiana-Arkansas area, seems to have extended eastward through the Mississippi Salt basin (Heydari, 1990) and westward, across East Texas (Moore, 1989). The Smackover shelf was exposed to subaerial conditions, and an extensive meteoric water system was developed (Moore and Druckman, 1981) during the sea level lowstand marking the end of the Smackover-Norphlet sequence (Moore and Druckman, 1991; see Figures 3 and 4 and discussion above). Those areas dominated by aragonite ooids (shelf interior) underwent extensive moldic porosity development (up to 50%) with concurrent precipitation of large volumes of early pre-compaction calcite cement in intergranular pores (up to 40%; Moore and Druckman, 1981) (Figure 4A). Along the shelf margins, where calcite ooid mineralogy was dominant, little moldic porosity was developed (less than 1%), and only minor early pre-compaction intergranular calcite cements were precipitated (less than 1%), sourced by the dissolution of minor aragonitic bioclasts such as mollusk fragments (Figure 4A). Primary intergranular porosity at the time of burial for these shelf edge sequences averaged over 35%. This distinctive and predictable diagenetic and pore system gradient in the upper Smackover was first recognized by Moore and Druckman in 1981 in Arkansas and Louisiana and was later extended westward and eastward across most of the Smackover trend (Moore, 1984; Wilkinson, 1984; Stewart, 1984; McGillis, 1984; Meensden et al., 1987; Moore, 1989; Heydari, 1990).

During the subsequent Buckner relative sea level rise, the Smackover shelf was flooded by marine waters. In westernmost Arkansas and East Texas, Buckner carbonate shelf margin barriers were responsible for the development of extensive back-barrier evaporative lagoons overlying the Smackover shelf (Moore, 1984; Moore et al., 1988) (Figure 4B). The subsequent reflux of Buckner evaporative waters through the porous, oomoldic grainstones of the upper Smackover highstand deposits resulted in extensive platform dolomitization across East Texas and into westernmost Arkansas (Moore et al., 1988). This dolomitization event significantly enhanced the porosity-permeability characteristics of the Smackover reservoirs in this region (Moore, 1984) (Figure 4B). Undolomitized Smackover oomoldic grainstones commonly exhibit porosities of 30–35% with permeabilities less than 0.01 md. Dolomitized oomoldic Smackover grainstones have porosities of 35–45% and permeabilities in excess of 50 md. To the east across central Arkansas and into the Mississippi Salt basin, extensive siliciclastic influx precluded the development of widespread Buckner evaporites, and the upper Smackover generally retained its limestone lithology (Moore, 1984; Moore et al., 1988). The upper Smackover in Alabama and Florida, however, is also overlain by extensive Buckner evaporites, and is also dolomitized, after an earlier

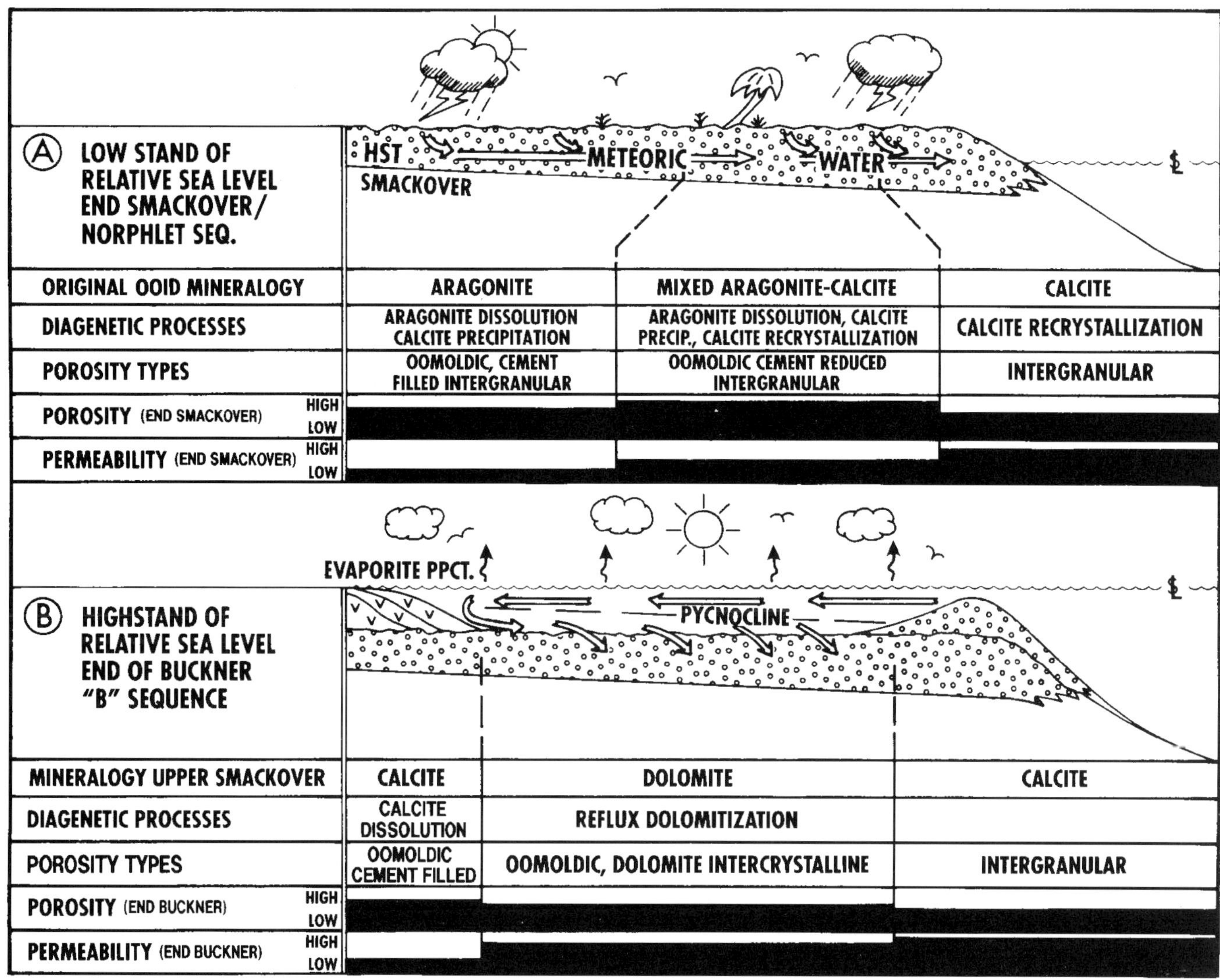

Figure 4. Cartoon illustrating the early diagenetic history of the upper Smackover ooid grainstones in response to: A. the relative low sea level stand at the end of the Smackover/Norphlet sequence; and B. the dolomitization associated with the relative highstand of sea level during the Buckner "B" sequence. Range of porosity values shown on both figures: low <10%, high 40% or greater. Range of permeability values shown on both figures: low <0.01 md; high 500 md or greater.

meteoric diagenetic event (Moore, 1984). Lack of chronostratigraphic control in this area precludes correlating these evaporites and related dolomitization events to the same sea level cycle.

At the time of burial into the subsurface, then, the hydrologic continuity and porosity-permeability characteristics of the upper Smackover aquifer had been significantly modified by early diagenetic events associated with changes in relative sea level as summarized below and in Figure 4.

In areas unaffected by evaporite-related dolomitization, hydrologic continuity of the aquifer was affected by the distribution of early pre-compaction intergranular cements sourced by the dissolution of aragonite ooids (Moore and Druckman, 1981; Moore, 1989). In those areas where these cements were concentrated, permeability was generally destroyed, while porosity remained high, effectively forming a barrier to free fluid migration during later burial. Ooid grainstones at the shelf margin, however, were buried into the subsurface with hydrologic characteristics relatively unaffected by early diagenetic overprint (Figure 4A).

In areas affected by the regional reflux-dolomitization events associated with relative sea level rise at the start of the Buckner sequence, regional hydrologic continuity was reestablished due to porosity-permeability enhancement associated with dolomitization (Moore, 1989). Pore system characteristics are dramatically modified, however, since these rocks now exhibit secondary moldic, intercrystalline dolomite porosity, rather than primary intergranular porosity or moldic porosity combined with cement filled intergranular porosity (Figure 4B).

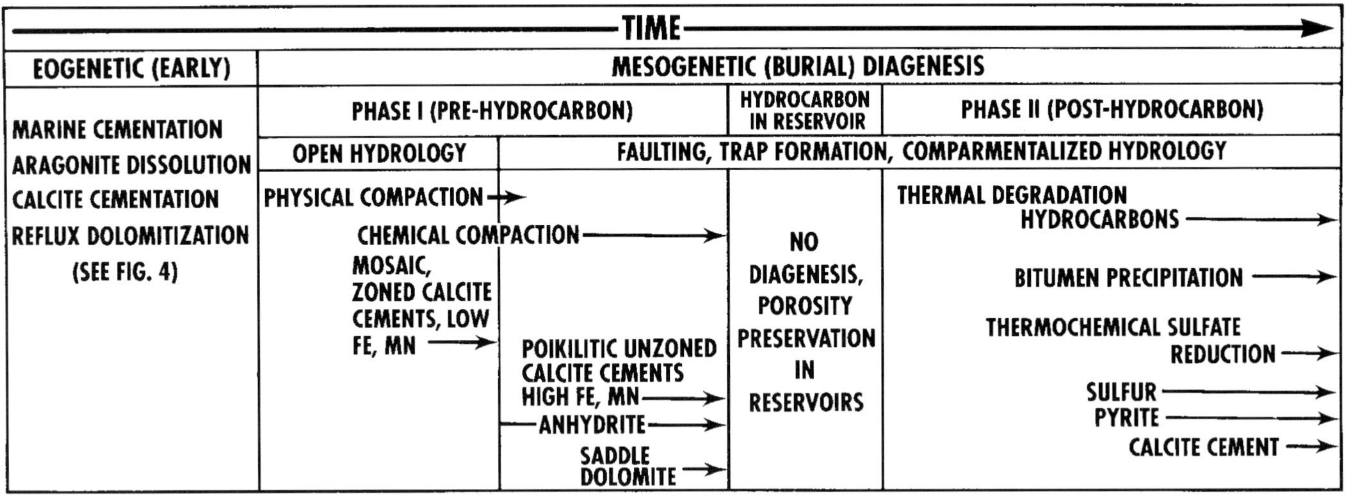

Figure 5. Relative paragenetic sequence of upper Smackover ooid grainstones. Phase I burial diagenesis and early diagenesis is based on Arkansas-Louisiana studies (Moore and Druckman, 1981; Moore, 1985) and the definitive work of Heydari (1990) on the Mississippi Clarke County shelf. Phase II burial diagenesis is from the Mississippi Salt basin (Heydari and Moore, 1989).

BURIAL DIAGENESIS OF THE SMACKOVER

Introduction

The burial diagenesis of the Smackover will be considered in two phases: Phase I, diagenesis before hydrocarbon migration; and Phase II, diagenesis after hydrocarbon arrival in the reservoir (Figure 5). The first phase is dominated by pressure solution, cementation related to pressure solution, and the modification of Smackover aquifers by warping and faulting leading to formation of hydrocarbon traps. It is in this phase of diagenesis that hydrologic conduits were established between deeper siliciclastic aquifers and Smackover source rocks leading to dramatic changes in the chemical composition of Smackover pore fluids and the migration of oil into upper Smackover reservoirs (Figure 5). The second phase of burial diagenesis commences when the oil filled reservoirs have reached burial temperatures high enough to cause thermochemical cracking of the trapped hydrocarbons with concomitant production of CO_2 and H_2S. The resulting aggressive pore fluids drive the processes of deep burial diagenesis, including: thermochemical sulfate reduction, thermochemical methane oxidation, calcite dissolution, and cementation (Figure 5).

(Phase I) Pre-Hydrocarbon Burial Diagenesis of the Upper Smackover

As the Smackover was buried, the surficial meteoric water system that had so intensely affected the upper Smackover highstand sequence tract was terminated, and a hydrologic regime dominated by marine-connate or marine-connate-meteoric water mixtures was established. Since these pore waters had no direct access to significant volumes of CO_2, they became equilibrated with the rock, in this case, calcite and dolomite. As burial continued, however, these equilibrium conditions were affected by increasing lithostatic pressure at grain-to-grain contacts and increasingly higher temperatures. The solubility of calcite increases with pressure, and the calcite grains tended to dissolve, initially at the grain contacts, and later along solution seams and stylolites in the Smackover (Moore, 1989). Increasing temperature, however, decreases the solubility of calcite, and the tendency is, therefore, for the calcite dissolved at grain contacts to be precipitated as calcite cement in available pore spaces (Moore, 1989). This process of grain-to-grain pressure solution and concurrent cementation has been termed chemical compaction (Choquette and James, 1987) and is one of the major carbonate diagenetic processes operating during burial (Moore, 1989).

The process of chemical compaction in the upper Smackover, is reflected by the common occurrence of coarse crystalline, mosaic, to poikilotopic burial cements (Figure 6). The burial origin of these upper Smackover cements has been well established previously (Moore and Druckman, 1981; Moore, 1985). Evidence developed in these earlier papers includes: petrographic relationships clearly indicating that these cements were precipitated after grain breakage and grain-to-grain pressure solution (Figure 6), relatively light oxygen isotopic compositions suggesting precipitation under elevated temperatures; and invariant carbon isotopic composition indicating that cement carbon was sourced from the ooid grainstones of the upper Smackover (Figure 7). These cements were emplaced prior to the arrival of hydrocarbons in upper Smackover reservoirs as constrained by the consistent occurrence of bitumen and oil stains on the

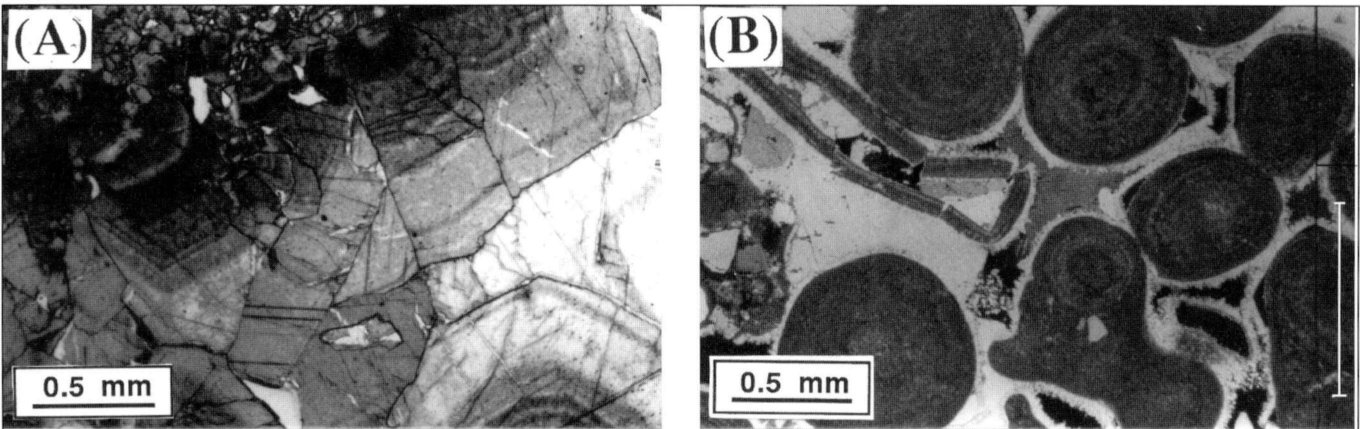

Figure 6. Photomicrographs of Phase I burial calcite cements. Stained with alizarin red-S and potassium ferrocyanide. This is the material from which the stable isotopic analysis shown on Figure 8 is taken. A. Fe-zoned, pore fill mosaic calcite cement from the Lear Clements #1, Miller County, Arkansas. Plain light. B. Ooid grainstone from the Arco Bodcaw #1, Columbia County, Arkansas. Two generations of cement. An early Fe-free marine crust, and a later Fe-rich poikilitic calcite which heals the fractures of the crushed ooid in the left center of the field. Crossed polars.

outer crystal faces, with no oil stain or bitumen between grain and crystal (Moore and Druckman, 1981; Moore, 1985). Obviously, when hydrocarbons arrived in upper Smackover reservoirs, pressure solution and cementation ceased, resulting in the preservation of porosity. Outside the reservoirs, however, chemical compaction could have continued with further degradation of effective porosity. These basic differences in chemical compaction between oil-wet reservoir and water-wet sections have not, to date, been well documented.

The chemical and isotopic composition of these burial cements should reflect the chemical evolution of Smackover pore fluid, and by extension, the changing hydrological conditions of the Smackover aquifer during burial.

Data

In the western Gulf of Mexico, two basic types of burial cement are recognized:

1. a strongly Fe-zoned mosaic calcite with average Fe composition < 500 ppm and Mn 143 ppm (Figures 6 and 8).
2. an unzoned, more Fe-rich poikilotopic calcite with average Fe composition about 2000 ppm and Mn levels averaging close to 500 ppm (Figures 6 and 8).

While occurring together in a single thin section, these two cement types can be petrographically separated, with the mosaic calcite being clearly the oldest cementation event (Moore, 1985). When cross plotted, the oxygen and carbon stable isotopic composition of each of these two cement types clearly overlap (Figure 7), with the mosaic calcites actually exhibiting both the lightest and the heaviest $\delta^{18}O$ (PDB) values. As seen in Figure 7, when analyzed by zone, the $\delta^{18}O$ composition of the zoned mosaic calcite shows progressively lighter values toward the later zones, with the very last zone showing a distinct reversal of the trend, toward heavier values. The carbon isotopic values of both cement types are similar, averaging about +3‰ $\delta^{13}C$ (PDB).

The strontium isotopic composition ($^{87}Sr/^{86}Sr$) of the unzoned Fe-rich burial cements is generally elevated above that expected for Upper Jurassic Oxfordian sea water (0.7070). A single value obtained for the zoned mosaic calcite cement in Arkansas, exhibited an Upper Jurassic sea water value (Moore, 1985).

Both mosaic zoned and poikilotopic unzoned calcite cements contain two phase fluid inclusions. The zoned mosaic calcites have a mean homogenization temperature of 107°C, and the unzoned calcites a mean homogenization temperature of 127°C. Previous fluid inclusion studies on these materials, however, suggest that step-wise re-equilibration may a problem in the Smackover (Moore, 1985).

Discussion

The relatively invariant carbon isotopic data strongly suggests that carbon was sourced from the rock mass itself, and hence supports the supposition that most of these Phase I burial cements are related to pressure solution. The stable oxygen isotopic data, however, presents an interesting dilemma since the earlier zoned cements exhibit lighter $\delta^{18}O$ values than the younger Fe-rich cements. Since the Fe-rich cements were precipitated after the mosaic calcites in the subsurface, one would expect these cements to have a lighter $\delta^{18}O$ composition than the mosaic zoned cements because of higher temperatures encountered later in the burial history of the sequence. This would also presume that the pore fluid composition remained the same during the precipitation of both

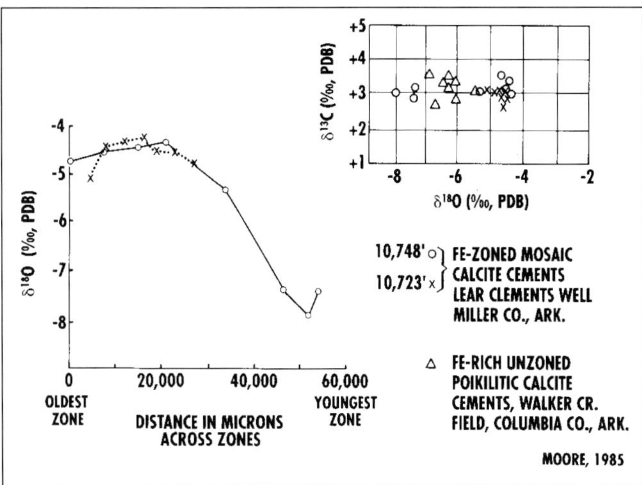

Figure 7. Stable isotopic composition of Phase I burial cements taken from the upper Smackover from the Lear Clements #1 and Arco Bodcaw #1 wells in Miller and Columbia counties, Arkansas, respectively. The cements from the Lear well were analyzed by zone, and their oxygen isotopic compositions are plotted against distance from the oldest zone to the youngest zone. The stable isotopic composition of Phase I cements from both the Lear and Arco wells are cross plotted in the upper right hand corner of the diagram.

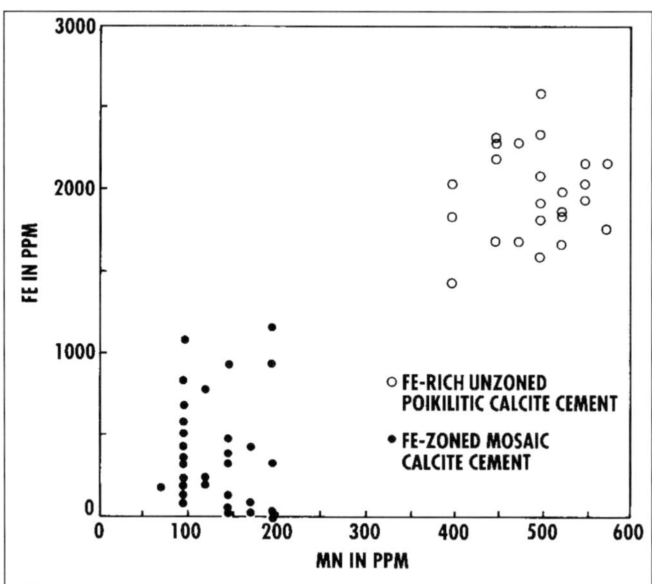

Figure 8. Fe-Mn composition of Phase I burial calcite cements taken from the upper Smackover in southern Arkansas. Zoned cements from the Lear Clements #1, a wildcat located in Miller County Arkansas. The unzoned cements taken from the Arco Bodcaw #1, Walker Creek field, Columbia County, Arkansas.

cements. The major differences between the Fe and Mn compositions of these two cements strongly suggests that, indeed, the pore fluid composition changed dramatically from one cement type to the other. If pore fluids with elevated salinities (such as are now seen in the Smackover, Moore and Druckman, 1981) were involved in the precipitation of the unzoned cements, while the zoned cements precipitated from an earlier meteoric water or marine connate water, one would expect just such a reversal of the oxygen isotopic trend as is observed in these data. This change to more elevated salinities in the later Fe-rich cements may well signal the onset of salt tectonics and the breaching of the Norphlet by faulting.

The elevated Fe and Mn in the unzoned cements is most likely to have been sourced from siliciclastics containing feldspars and other Fe, Mn-bearing minerals. This theory is supported by the elevated $^{87}Sr/^{86}Sr$ compositions of the Fe-rich poikilotopic cements. The Smackover highstand systems tract in the western part of the trend is basically siliciclastic-free and as such cannot be used to explain the regional development of this cement type. In addition, the upper Smackover is blanketed by evaporites over most of this area, leaving the Norphlet as the only logical source for the Fe, Mn, and radiogenic Sr as incorporated into the unzoned poikilotopic cements. The Louann, below the Norphlet, could contribute heavy $\delta^{18}O$, and raise the total dissolved solids of the pore fluid, as noted above, if water moving through the Norphlet aquifer interacted with and dissolved sulfates and salts associated with the Louann. The presence of late stage (contemporaneous with or post-poikilotopic calcite) replacement anhydrites described by Moore and Druckman (1981) in the upper Smackover, and discussed later in this paper under post-hydrocarbon diagenesis, would certainly support this scenario.

Fault conduits associated with early salt movement and withdrawal, probably were fluid conduits from the Norphlet aquifer up through the lower Smackover aquiclude into the upper Smackover aquifer. At this point, the Fe, Mn, radiogenic Sr, and heavy $\delta^{18}O$ were incorporated into the poikilotopic cements forming by chemical compaction. As burial continued and temperatures rose, the organic-rich transgressive systems tract of the lower Smackover reached the oil window and maturation and migration commenced (Figure 9). These same fault conduits could then be used to charge the upper Smackover traps that had subsequently formed during early salt movement. Hydrocarbon inclusions within the poikilotopic cements suggest that these cements formed shortly before or during hydrocarbon emplacement.

The Smackover aquifer had been hydrologically compartmentalized by faulting and the trap formation had commenced at the point when pore fluid composition changed and the poikilotopic cements began to precipitate. These traps were subsequently

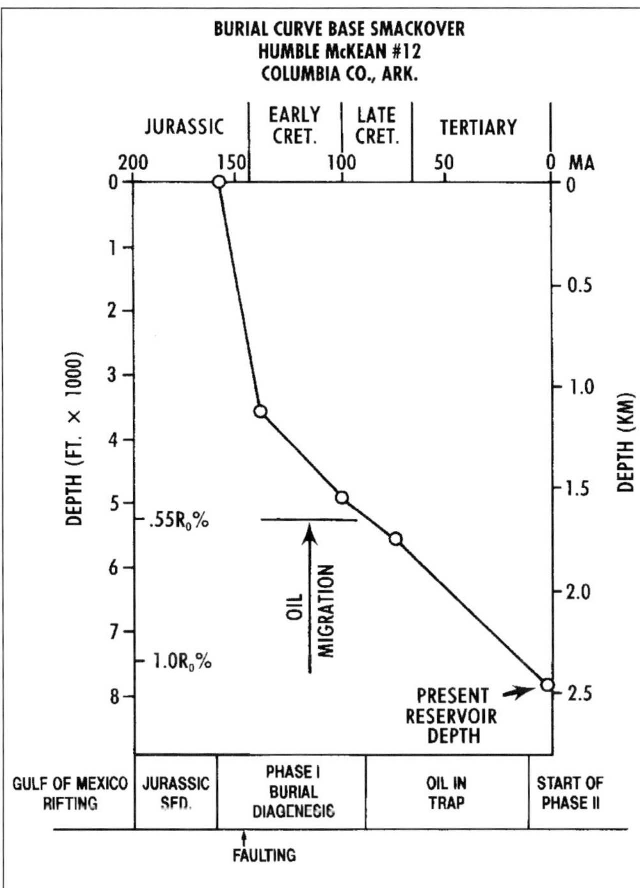

Figure 9. Burial curve of the Humble McKean #12 well. The curve was constructed from log derived chronostratigraphic unit thickness. The curve was not corrected for compaction nor for unconformity section loss. R_o values were calculated.

charged with hydrocarbons.

Data developed for similar upper Smackover prehydrocarbon burial cements to the east, in the Mississippi Salt basin (Heydari, 1990), strongly parallel the scenario developed above for the western portion of the Smackover trend. Similar strong Fe, Mn, radiogenic Sr, and stable isotope trends are present in these cements precipitated during compartmentalization of the upper Smackover aquifer by salt-associated faulting and the initial charging of upper Smackover traps with hydrocarbons through these same faults.

The following discussion of post-hydrocarbon burial diagenesis will center on a deeply buried (6 km) upper Smackover sequence found in the Black Creek field located in the southern part of the Mississippi Salt basin (Figures 10 and 11).

(Phase II) Post-Hydrocarbon Burial Diagenesis of the Upper Smackover

Figure 10 shows the general geological setting of the Mississippi Salt basin. Figure 11 illustrates an idealized cross section across the basin, showing the location of Black Creek field. The sequence stratigraphic setting of the Smackover at Black Creek field is similar to that described earlier for the western and central part of the trend, with the lower Smackover laminated source rock facies representing a transgressive sequence tract, while the upper Smackover ooid grainstone reservoir is developed in a highstand systems tract. There is no Norphlet lowstand systems tract represented at Black Creek field. Finally, it is uncertain whether the Smackover sequence at Black Creek is of the same sea level cycle as the Smackover sequence in the Arkansas-Louisiana-east Texas area described above.

Black Creek field is a two-well noncommercial field that produced gas consisting of 78% H_2S (Parker, 1974). It is located in an area of the Mississippi Salt basin deemed overmature for liquids (Sassen and Moore, 1988). The upper Smackover is at 6060 m at Black Creek with adjusted bottom hole temperatures of 205°C. Equivalent vitrinite reflectance for the upper Smackover at this site probably exceeds 2.5% (Sassen and Moore, 1988) (Figure 12). The late stage, post-hydrocarbon (Phase II) paragenetic sequence for the upper Smackover at Black Creek is shown in Figure 5. Prehydrocarbon diagenesis at Black Creek is essentially identical to that described earlier for the western trend and for shallower sequences located along the margins of the Mississippi Salt basin to the north (Heydari, 1990). Fe, Mn, and radiogenic Sr trends, however, in these Phase I calcites are not developed at Black Creek, probably because the clastic Norphlet Formation is missing at this site. This further supports the Norphlet source for the Fe/Mn/Sr model developed in the last section. Hydrocarbon migration filled the reservoir at Black Creek during the Cretaceous at temperatures near 100°C (Sassen and Moore, 1988), essentially shutting down chemical compaction within the reservoir and preserving the porosity present at the time of migration (up to 15%, Heydari, 1990) (Figure 12).

With further burial and increased temperatures (between 120 and 200°C), liquid hydrocarbons began to crack, with the production of large volumes of methane, H_2S, and solid bitumen (Sassen and Moore, 1988) (Figures 5 and 12). During this time, the last two diagenetic phases precipitated. The late stage replacement anhydrite began to be replaced by calcite, with the liberation of native sulfur. Finally, poikilotopic calcite cement precipitated directly over the reservoir bitumen (Figures 5 and 13) (Heydari and Moore, 1989).

Data

Petrographically, the calcite in Figure 13B is clearly post-bitumen. The bitumen cracks as it loses volatiles, and the calcite is precipitated within these cracks. In addition, devolatilization of bitumen results in an intricate, highly irregular outer surface, facing the pore. The pore filling calcites nucleate on and engulf

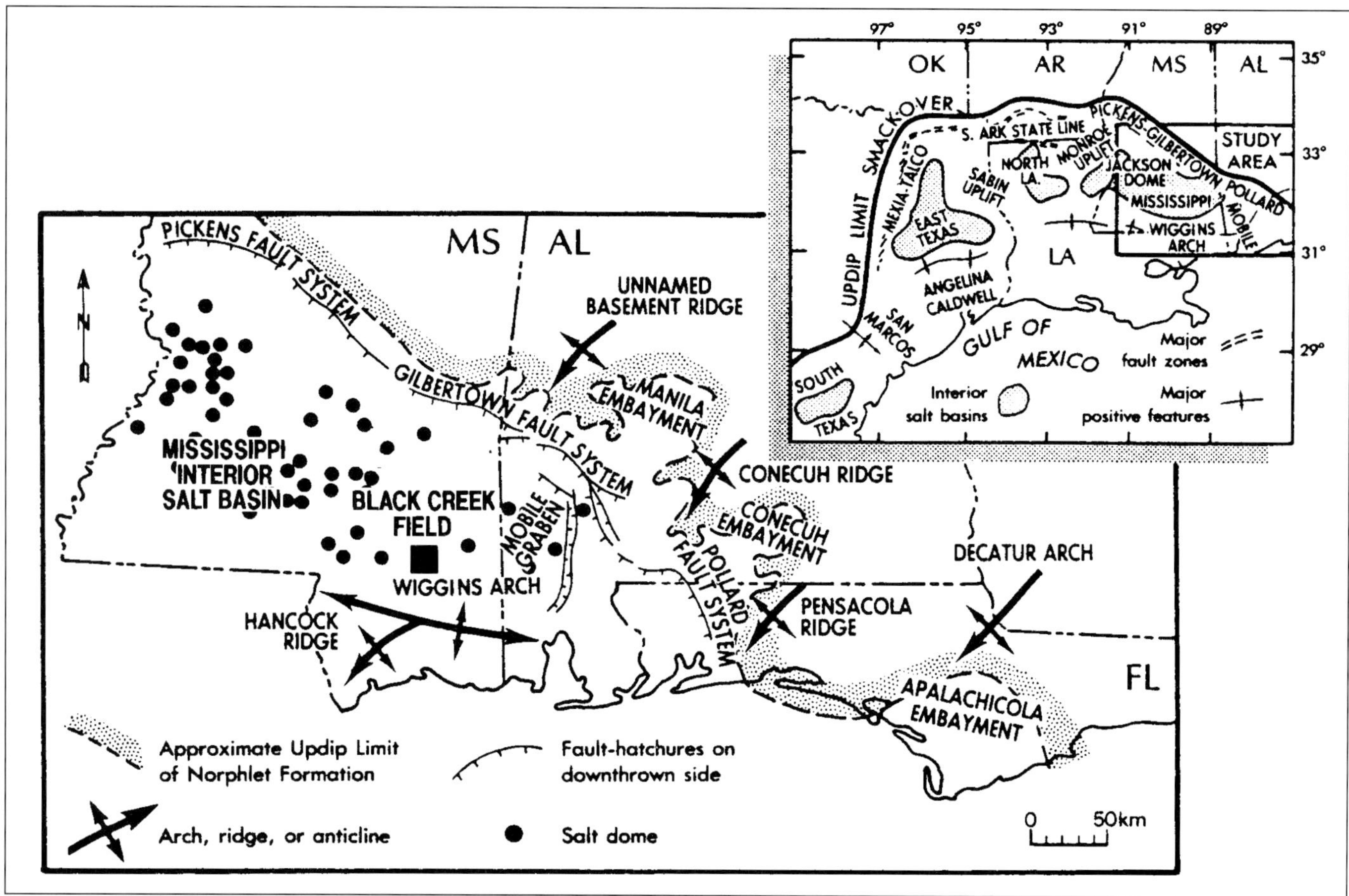

Figure 10. Structural setting for the Upper Jurassic, eastern Gulf of Mexico. Taken from Sassen and Moore (1988).

this surface (Figure 13). Finally, Klosterman (1981) reported two phase fluid inclusion homogenization temperatures near 200°C from these calcites.

Figure 14 illustrates an oxygen-carbon isotope cross plot of all calcite phases found at Black Creek as well as the ooid allochems that make up the bulk of the upper Smackover. The pre-hydrocarbon calcites occupy a trend (-4.5 to -8.5‰ $\delta^{18}O$, +4.0 to 5.0‰ $\delta^{13}C$) almost identical to that of the pre-hydrocarbon calcites described from the western trend earlier (see Figure 7). The calcite that replaces anhydrite has a range of oxygen isotopic values from -3 to -4‰, while their carbon isotopic compositions range from -2.0 to +4.2‰ (Figure 14). The oxygen isotopic composition of the post-bitumen calcites range from -3.5 to -5.6‰, while their carbon isotopic values range from -1.6 to -16.3‰ (Figure 14). The strontium isotopic composition of all late phase calcites at Black Creek is centered on Upper Jurassic sea water (0.7070) (Heydari, 1990).

Discussion

The isotopic compositions of the pre-hydrocarbon calcite phases at Black Creek field show a clear trend toward elevated temperatures, with carbon being sourced from the rock itself. These cements are obviously the product of pressure solution during progressive burial into the subsurface. Heydari and Moore (1989) ascribed the anhydrite replacement by calcite to thermochemical sulfate reduction, because of its close association with native sulfur, and the similarity of its sulfur isotopic composition to that of the anhydrite cement, Jurassic sedimentary anhydrite, and the included native sulfur. Heydari and Moore (1989) noted that experimental work suggested that reaction rates for this reaction (reaction 2) are prohibitively slow below 200°C and concluded that the calcites must have replaced the anhydrite at a temperature near 200°C. The relatively heavy carbon isotopic composition of these calcites suggest that reduction of sulfate to native sulfur occurred predominantly by reaction with H_2S rather than involving massive oxidation of hydrocarbon gases, which would have resulted in much lighter carbon isotopic compositions for these calcites.

The pore fill bitumen at Black Creek was analyzed by pyrolysis by Sassen and Moore (1988). They concluded that the HI values of the bitumen (33 to 126 mg/g) were consistent with the late stages of crude oil destruction and suggested that significant bitumen precipitation commenced at 1% vitrinite reflectance (see also Sassen, 1988), or at a temperature near 200° C at Black Creek. These data imply that the calcite

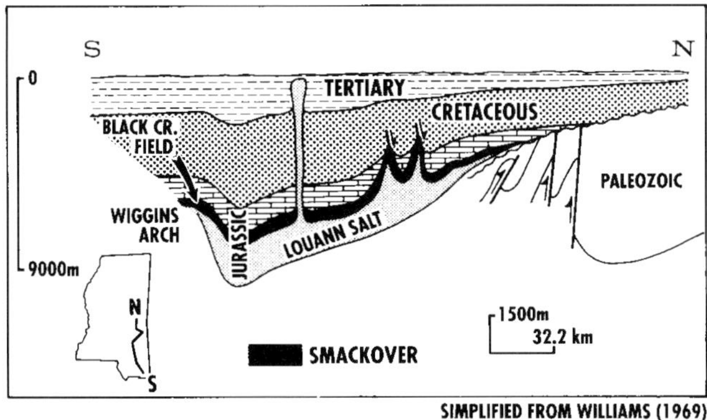

Figure 11. Simplified cross section across the Mississippi Salt basin showing the relative location of Black Creek field in Perry County, Mississippi.

cements that were subsequently precipitated on the bitumen must have formed at similar temperatures. These conclusions are consistent with the available two phase fluid inclusion data and clearly constrain the temperature of formation for the post-bitumen calcite to near 200°C.

Both post-bitumen calcite and calcite replacing anhydrite show relatively heavy oxygen isotopic values. Both calcites, however, formed at temperatures near 200°C. These data suggest that the available oxygen in the pore fluid at the time of precipitation was enriched in ^{18}O by extensive rock-water interaction under conditions of decreasing water-to-rock ratios (progressively reduced porosity) and increasing temperature. Heydari and Moore (1989) calculated that the upper Smackover pore fluids at Black Creek may well have had a $\delta^{18}O$ composition of +15‰ at the time of precipitation of the last calcite phase. These data support the trend to heavier oxygen isotopic brine compositions toward the basin center outlined by Kharaka et al., (1987) for the Mississippi Salt basin. Land and Prezbendowski (1981) reported oxygen isotopic values for subsurface brines produced from Mesozoic limestones in the South Texas salt basin in excess of +18 ‰.

The light carbon trend (down to -16.4‰) exhibited by post-bitumen calcite cements suggests the involvement of some CO_2 derived from the thermal oxidation of methane. There is little doubt that some heavier carbon is also being contributed by the limestone because these cements would have much lighter carbon isotopic compositions if carbon was only being contributed by thermal methane oxidation (Hudson, 1977).

The thermal oxidation of methane could well be driven by reaction of methane under elevated temperatures with native sulfur derived from the thermochemical reduction of sulfate (Heydari and Moore, 1989) (reaction 1).

$$CH_4 + 4S^o + 2H_2O \rightarrow 4H_2S + CO_2 \quad (1)$$

The products of this reaction are H_2S and CO_2. The

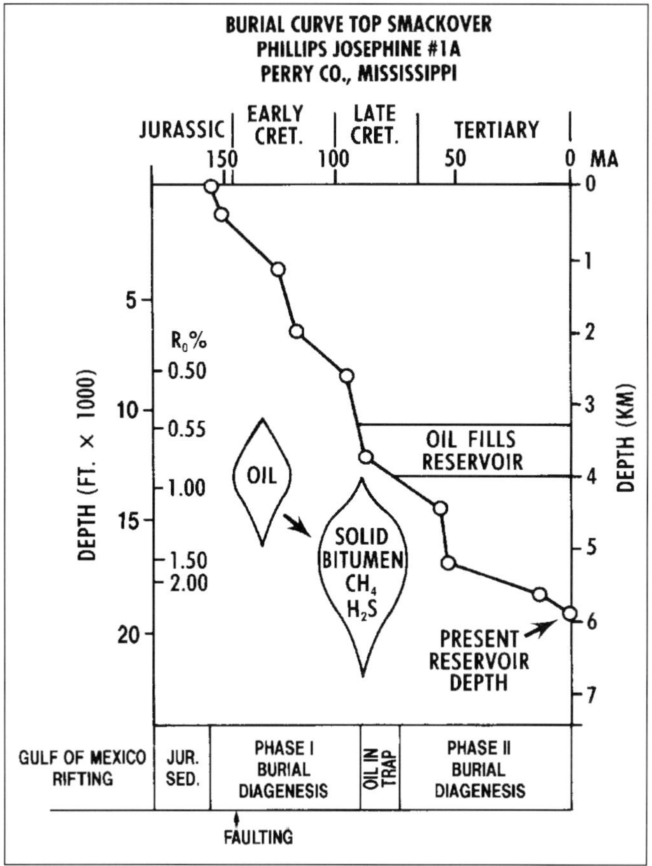

Figure 12. Burial curve for the Phillips Josephine #1A, Perry County Mississippi at the Black Creek field. The curve was constructed from log-derived chronostratigraphic unit thicknesses without correction for compaction or section loss at unconformities. R_0 values from Sassen and Moore (1988). Oil and bitumen windows are shown relative to R_0 values, not to time.

H_2S, in turn, will tend to fuel the thermochemical reduction of sulfate (reaction 2):

$$CaSO_4 + 3H_2S + CO_2 \rightarrow CaCO_3 + 4S^o + 3H_2O \quad (2)$$

thus linking together two diagenetic reactions that could ultimately lead to the destruction of available methane in a sulfate-rich, Fe-poor system, such as the upper Smackover at Black Creek (Sassen and Moore, 1988; Heydari and Moore, 1989).

By the time the post-bitumen calcite had precipitated, almost all available pore space in the reservoir had been filled, either with bitumen or with late calcites, and one ends with a closed system that is totally rock dominated and with only limited volumes of water available for diagenetic processes.

APPLICABILITY OF THE SMACKOVER MODEL TO OTHER SEQUENCES

The pre-hydrocarbon migration diagenetic model for the Smackover should be widely applicable to

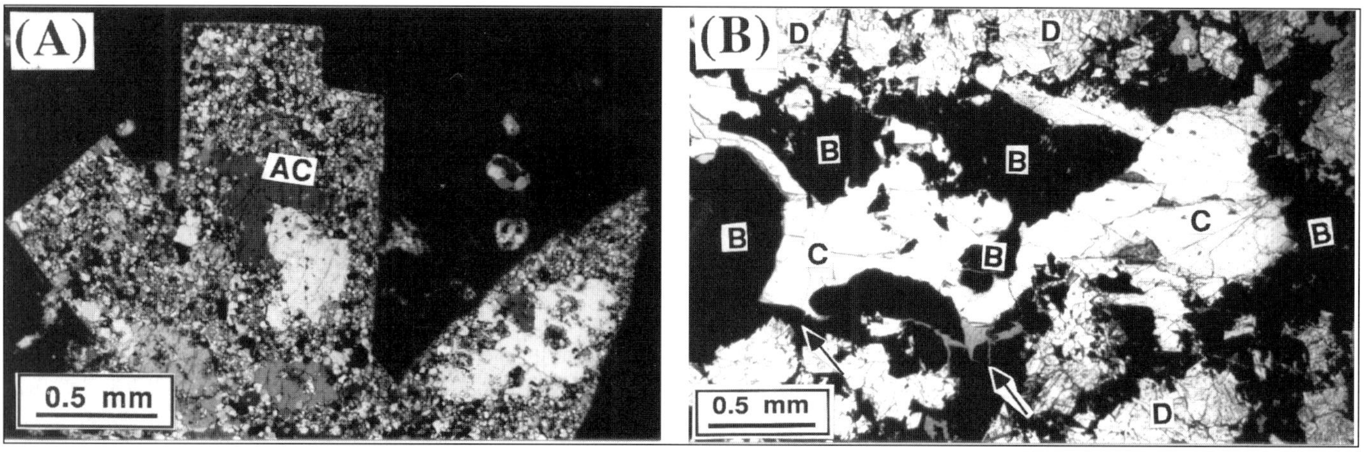

Figure 13. Photomicrographs of Phase II diagenetic products from the Phillips Josephine #1A, Perry County Mississippi. A. Mosaic calcite (AC) after anhydrite. B. Poikilitic calcite (C) precipitated in void over bitumen. This calcite is the post-bitumen calcite shown in the stable isotope cross plot on Figure 14. Host is dolomitized (D) ooid grainstone.

shelf-ramp depositional sequences featuring high energy, highstand systems tracts developed in response to third order relative sea level fluctuations. This setting results in the formation of a hydrologic framework, facies architecture, and early diagenetic history similar to the Smackover.

The sequence most directly comparable to the Smackover, in terms of similar facies development and burial history (up to hydrocarbon migration) is the Jurassic of northern Arabia (Wilson, 1975; Murris, 1980). Upper Jurassic reservoirs of the Middle East are developed in ooid-bearing highstand systems tracts, are sourced from contiguous transgressive systems tracts, and exhibit an early meteoric diagenetic overprint, oomoldic porosity, and reflux dolomitization associated with sequence boundaries. These sequences are subsequently rapidly buried and suffer porosity loss, presumably by pressure solution which is terminated by hydrocarbon migration (Murris, 1980; Koepnick, 1985.). The Upper Jurassic of west Central Europe (Wilson, 1975; Purser, 1978; Sellwood et al., 1985) and west-central Argentina (Legarreta, 1991) and the Silurian-Devonian Helderberg Group of the central Appalachians, U.S.A. (Dorobek, 1987) show similar sequence architecture and seem to exhibit similar diagenetic histories.

The post-hydrocarbon migration diagenetic model for the Smackover should be widely applicable to hydrocarbon-sulfate-bearing carbonate rock sequences that have reached thermal maturities high enough to trigger the thermal destruction of the associated hydrocarbons. Carbonate sequences that fulfill these conditions and that exhibit late diagenetic histories similar to those described for the Smackover include: the Lower Cretaceous of South Texas (Woronick and Land, 1985); the Lower Permian of West Texas (Mazzullo, 1986); the Devonian of the Alberta sedimentary basin (Mattes and Mountjoy, 1980; Krouse et. al., 1988); the Ordovician Ellenburger Formation of West Texas (Kupecz and Land, 1991); and the Cambrian Bonneterre Formation of Missouri (Gregg and Shelton, 1989). It should be noted, that these sequences often host Mississippi Valley type (MVT) sulfide ore deposits (Kyle, 1983).

CONCLUSIONS

Early Diagenesis

1. Ultimate diagenetic pathways are established early, under the influence of changing relative sea level.
2. Lowstand siliciclastic fans, and highstand carbonate systems tracts tend to develop into regional aquifer systems that focus hydrologic flow, concentrate diagenetic processes, and provide migrational pathways for hydrocarbons.
3. Transgressive systems tracts are often aquicludes and may well contain the source rocks for later hydrocarbon formation.
4. Due to this strong stratigraphic control, vertical conduits, such as faults, connecting aquifer systems and source-rocks to reservoirs are important components of the subsurface hydrological regime. Burial diagenesis is therefore closely linked to structural history.
5. Aquifer systems may be modified by early diagenetic overprint, such as solution, precipitation, and dolomitization associated with relative sea level fluctuations.
6. The geometries of high porosity-permeability units and their early diagenesis are most strongly controlled by the depositional sequence stratigraphy. These heterogeneities, including porosity distribution, are then inherited through into burial diagenesis. Porosities as high as 40% may be present at the time of burial.

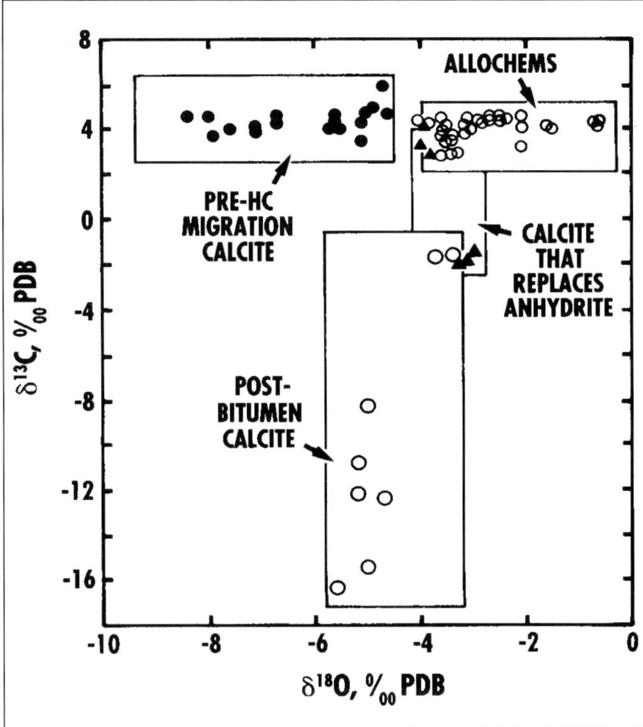

Figure 14. Stable isotope cross plot of burial calcites and ooids from the Phillips Josephine #1A, Perry County, Mississippi.

Burial Diagenesis

1. Pre-hydrocarbon diagenesis is driven by chemical compaction in an initially open, marine connate to mixed meteoric-marine water dominated system. The hydrologic system becomes compartmentalized by structural trap formation and development of fault-related vertical conduits providing for the movement of basinal related, metal bearing brines and ultimately of hydrocarbons into the reservoirs of the aquifer system. This phase of diagenesis is terminated by arrival of hydrocarbons into the reservoirs.
2. The high porosities present at the time of burial (up to 40%) are rapidly lost in limestone sequences by chemical compaction. Porosities encountered in Smackover limestone hydrocarbon reservoirs averages 15%. Dolomite hydrocarbon reservoirs show much higher porosities (up to 40%) because the rate of chemical compaction for the less soluble dolomites is much slower.
3. The pre-hydrocarbon migration burial diagenetic model for the Smackover should be widely applicable to shelf-ramp depositional sequences featuring high energy, highstand systems tracts developed in response to third order relative sea level fluctuations. The Upper Jurassic of northern Arabia, west Central Europe, west-central Argentina and the Silurian-Devonian of the Eastern U.S. are cited as examples.
4. Post-hydrocarbon diagenesis is driven by the thermal cracking of reservoired hydrocarbons in a closed, rock dominated system. Important processes include thermochemical sulfate reduction and thermal oxidation of methane. The two processes seem to be cross-linked by H_2S and native sulfur in sulfate-rich, Fe-poor sequences such as the Smackover, leading to the relatively early destruction of methane.
5. Hydrocarbons preserve porosity in a continuing burial situation. Porosity occlusion in a reservoir where reservoir hydrocarbons are undergoing thermal destruction is by the precipitation of bitumen and calcite. Final porosities under these conditions are generally less than 5%.
6. The post-hydrocarbon migration diagenetic model for the Smackover should be widely applicable to hydrocarbon-sulfate-bearing carbonate rock sequences that have reached thermal maturities high enough to trigger the thermal destruction of the associated hydrocarbons. The Lower Cretaceous of South Texas, the Lower Permian of West Texas, the Devonian of Alberta, Canada, the Ordovician of West Texas, and the Cambrian of Missouri are cited as examples.

ACKNOWLEDGMENTS

The authors are grateful for the long term support, and confidence, that the associates have given the program over the years. David Stoudt of Mossbacher Energy, and George Herman of Tulane University provided much needed material and data. Lynton Land of the University of Texas and Texaco Inc. provided stable isotope analyses. Roger Sassen kept us honest with the organics and Bill Wade was our local expert on thermochemical sulfate reduction and H_2S. Barbara Delaville, Pat, Pam, all provided much needed support. M.L. Eggert and C. Duplechain prepared the illustrations. West Yellowstone Mt. provided much needed inspiration. Reviews by Andy Horbury, Alistair Gray (both of BP, London), and Frances Abbots (Shell, KSEPL, The Hague) significantly enhanced the manuscript.

REFERENCES CITED

Ahr, W., 1973, The carbonate ramp, an alternative to the shelf model: Transactions Gulf Coast Association Geological Societies, v. 23, p. 221-225.

Budd, D. A. and R. G. Loucks, 1981, Smackover and lower Buckner formations, south Texas: Depositional systems on a Jurassic carbonate ramp: Bureau of Economic Geology, Report of Investigations #112, 34p.

Choquette, P. W., and N. P. James, 1987, Diagenesis #12. Diagenesis in limestone-3. The deep burial environment: Geoscience Canada, v. 14, p. 3-35.

Dickinson, K. A., 1968, The Upper Jurassic stratigraphy of some adjacent parts of Texas, Louisiana, and Arkansas: USGS Professional Paper 549E, 25 p.

Dorobek, S.L., 1987, Petrology, geochemistry and origin of burial diagenetic facies, Siluro-Devonian Helderberg Group (carbonate rocks), central Appalachians: AAPG Bulletin, v. 71, p. 492-514.

Goldhammer, R. K., P. J. Lehmann, R. G. Todd, J. L. Wilson, W. C. Ward, and C. R. Johnson, 1991, Sequence stratigraphy and cyclostratigraphy of the Mesozoic of the Sierra Madre Oriental, northeast Mexico-A Field Guidebook: Gulf Coast Section Society of Economic Paleontologists and Mineralogists Foundation, 85 p.

Gregg, J. M., and K. L. Shelton, 1989, Minor and trace element distributions in the Bonneterre dolomite (Cambrian), southeast Missouri: evidence for possible multiple-basin fluid sources and pathways during lead-zinc mineralization: Geological Society of America Bulletin, v. 101, p. 221-230.

Haq, B. U., J. Hardenbol, and P. R. Vail, 1988, Mesozoic and Cenozoic chronostratigraphy and cycles of sea-level changes: *in* C. K. Wilgus,. B. S. Hastings, C. G. S. C. Kendall, H. W. Posamentier, C. A. Ross, and J. C. van Wagoner, eds., Sea-level changes: an integrated approach: Tulsa, SEPM Special Publications #42, p.71-108.

Heydari, E., 1990, Burial diagenesis of the Smackover Formation, southeast Mississippi Salt basin: Ph. D. thesis, Louisiana State University, 268 p.

Heydari, E. and C. H. Moore, 1989, Burial diagenesis and thermochemical sulfate reduction, Smackover Formation, southeastern Mississippi Salt basin: Geology, v. 17, p. 1080-1084.

Hudson, J. D., 1977, Stable isotopes and limestone lithification: Journal Geological Society London, v. 133, p. 637-660.

Imlay, R. W., and G. Herman, 1984, Upper Jurassic Ammonites from the subsurface of Texas, Louisiana, and Mississippi, *in* W. P. S. Ventress, D. G. Bebout, B. F. Perkins, and C. H. Moore eds., The Jurassic of the Gulf Rim: Austin, Proceedings of Third SEPM Gulf Coast Section Research Conference, p. 149-170.

Jackson, M. P. A., and S. J. Seni, 1983, Geometry and evolution of salt structures in a marginal rift basin of the Gulf of Mexico, east Texas: Geology, v. 11, p. 131-135.

Kharaka, Y. K., A. S. Maest, W. W. Carothers, L. M. Law, P. J. Lamothe, and T. L. Fries, 1987, Geochemistry of metal-rich brines from central Mississippi Salt Dome Basin: Applied Geochemistry, v. 2, p. 543-561.

Klosterman, M. J., 1981, Application of fluid inclusion techniques to burial diagenesis in carbonate rocks sequences: Louisiana State University, Applied Carbonate Research Group, Contribution #7, 101p.

Koepnick, R. B., 1985, Distribution and permeability of stylolite-bearing horizons within a Lower Cretaceous carbonate reservoir in the Middle East: Society of Petroleum Engineers 14171, p. 1-7.

Krouse, H. R., C. A. Viau, L. S. Eliuk, A. Ueda, and S. Halas, 1988, Chemical and isotopic evidence of thermochemical sulphate reduction by light hydrocarbon gases in deep carbonate reservoirs: Nature, v. 333, p. 415-419.

Kupecz, J. A. and L. S. Land, 1991, Late stage dolomitization of the Lower Ordovician Ellenburger Group, west Texas: Journal of Sedimentary Petrology, v. 61, p. 551-574.

Kyle, R. J., 1983, Economic aspects of subaerial carbonates, *in* P.A. Scholle, D. G. Bebout, and C. H. Moore, eds., Carbonate Depositional Environments, AAPG Memoir 33, p.73-92.

Legarreta, L., 1991, Evolution of a Callovian-Oxfordian carbonate margin in the Neuquen Basin of west central Argentina: facies, architecture, depositional sequences and global sea-level changes: Sedimentary Geology, v. 70, p. 209-240.

Land, L. S., and P. R. Prezbindowski, 1981, The origin and evolution of saline formation water, Lower Cretaceous carbonates, south-central Texas, U.S.A., Journal Hydrology, v. 54, p. 51-74.

Mattes, B. W. and E. W. Mountjoy, 1980, Burial dolomitization of the Upper Devonian Miette buildup, Jasper National Park, Alberta, *in* D. H. Zenger, J. B. Dunham and R. L. Ethington eds., Concepts and Models of Dolomitization, SEPM Special Publication No. 28, p. 259-297

Mazzullo, S. J., 1986, Mississippi Valley-type sulfides in Lower Permian dolomites, Delaware Basin, Texas: implications for basin evolution: AAPG Bulletin, v. 73, p. 388.

McGillis, K.A., 1984, Upper Jurassic stratigraphy and carbonate facies, northeastern east Texas basin, *in*, M. W. Presley, ed., The Jurassic of East Texas, East Texas Geological Society, Special Publication, Tyler Texas, p. 63-66.

Meensden, F. C., C. H. Moore, E. Heydari, and R. Sassen, 1987, Upper Jurassic depositional systems and hydrocarbon potential of southeast Mississippi, Transactions Gulf Coast Association Geological Societies, v. 37, p. 161-174.

Meyers, W. J., 1974, Carbonate cement stratigraphy of the Lake Valley Formation (Mississippian) Sacramento Mountains, New Mexico: Journal of Sedimentary Petrology, v. 44, p. 837-861.

Meyers, W. J., 1978, Carbonate cements: their regional distribution and interpretation in Mississippian limestones of southwestern New Mexico: Sedimentology, v. 25, p. 371-400.

Meyers, W. J., 1980, Compaction in Mississippian skeletal limestones southwestern New Mexico: Journal of Sedimentary Petrology, v. 50, p. 457-474.

Meyers, W. J., 1988, Paleokarst features in Mississippian skeletal limestones New Mexico, *in* N. P. James and P. W. Choquette eds., Paleokarst: New York, Springer-Verlag, p. 306-328.

Moore, C. H., 1984, The upper Smackover of the Gulf rim: depositional systems, diagenesis, porosity evolution and hydrocarbon production, *in* W. P. S. Ventress, D. G. Bebout, B. F. Perkins, and C. H. Moore eds., The Jurassic of the Gulf Rim: Austin, Proceedings of Third SEPM Gulf Coast Section Research Conference, p. 283-307.

Moore, C. H., 1985, Upper Jurassic subsurface cements-a case history, *in* N. Schneidermann, and P. M. Harris eds., Carbonate cements: Tulsa, SEPM Special Publication #36, p. 291-308.

Moore, C. H., 1989, Carbonate diagenesis and porosity: New York, Elsevier, 338p.

Moore, C. H., and Y. Druckman, 1981, Burial diagenesis and porosity evolution, Upper Jurassic Smackover, Arkansas and Louisiana: AAPG Bulletin, v.65, p. 597-628.

Moore, C. H., and Y. Druckman, 1991, Sequence stratigraphic framework of the Upper Jurassic Smackover and related units, western Gulf of Mexico: AAPG Bulletin, v. 75, p. 639.

Moore, C. H., A. Chowdhury, and E. Heydari, 1986, Variation of ooid mineralogy in Jurassic Smackover limestone as control of ultimate diagenetic potential: AAPG Bulletin, v. 70, p. 622-623.

Moore, C. H., A. Chowdhury, and L. Chan, 1988, Upper Jurassic Smackover platform dolomitization northwestern Gulf of Mexico: a tale of two waters, *in* V. Shukla and P. A. Baker, eds., Sedimentology and geochemistry of dolostones: Tulsa, SEPM Special Publication 43, p.175-189.

Murray, G. E., 1961, Geology of the Atlantic and Gulf coastal province of North America: New York, Harper and Brothers, 692 p.

Murris, R. J., 1980, Middle East: stratigraphic evolution and oil habitat: AAPG Bulletin, v. 64, p. 597-618.

Parker, C. A., 1974, Geopressures and secondary porosity in the deep Jurassic of Mississippi: Gulf Coast Association of Geological Societies Transactions, v. 24, p. 69-80.

Purser, B. H., 1978, Early diagenesis and the preservation of porosity in Jurassic limestones. Journal Petroleum Geology, v. 1, p. 83-94.

Salvador, A., 1987, Late Triassic-Jurassic paleogeography and origin of Gulf of Mexico basin: AAPG Bulletin, v. 71, p. 419-451.

Sarg, J. F., 1988, Carbonate sequence stratigraphy *in* C. K. Wilgus, B. S. Hastings, C. G. S. C. Kendall, H. W. Posamentier, C. A. Ross, and J. C. van Wagoner, eds., Sea-level changes: an integrated approach: Tulsa, SEPM Special Publications #42, p. 155-181.

Sassen, R., 1988, Geochemical and carbon isotopic studies of crude oil destruction, bitumen precipitation, and sulfate reduction in the deep Smackover Formation: Organic Geochemistry, v. 12, p. 351-361.

Sassen, R., C.H. Moore, J. A. Nunn, F. C. Meendsen, and E. Heydari, 1987, Geochemical studies of crude oil generation, migration, and destruction in the Mississippi Salt basin: Gulf Coast Association of Geological Societies Transactions, v. 37, p. 217-224.

Sassen R., and C. H. Moore, 1988, Framework of hydrocarbon generation and destruction in eastern Smackover trend: AAPG Bulletin, v. 72, p. 649-663.

Sellwood, B. W., J. Scott, P. Mikkelsen, and P. Akroyd, 1985, Stratigraphy and sedimentology of the Great Oolite Group in the Humbly Grove oilfield, Hampshire, S. England: Marine Petroleum Geology, v. 2, p. 44-55.

Scholle, P. A., 1977, Chalk diagenesis of deep-water carbonate turbidites, upper Cretaceous Monte Antola flysch, northern Appennies, Italy: Journal of Sedimentary Petrology, v. 41, p. 233-250.

Scholle, P. A., R. B. Halley, 1985, Burial diagenesis: out of sight, out of mind, *in* N. Schneidermann, and P. M. Harris eds., Carbonate cements: Tulsa, SEPM Special Publication #36, p. 309-334.

Stewart, S. K., 1984, Smackover and Haynesville facies relationships in north-central east Texas, *in*, M. W. Presley, ed., The Jurassic of East Texas, East Texas Geological Society Special Publication, Tyler, Texas, p. 56-62.

Swirydczuk, K., 1988, Mineralogical control on porosity type in Upper Jurassic Smackover ooid grainstones, southern Arkansas and northern Louisiana: Journal of Sedimentary Petrology, v. 58, p. 339-347.

Todd, R. G., and R. M., Mitchum, Jr., 1977, Identification of upper Triassic, Jurassic, and Lower Cretaceous seismic sequences in Gulf of Mexico and offshore west Africa, *in*, C. E. Payton, ed., Seismic stratigraphy-applications to hydrocarbon exploration: Tulsa, AAPG Memoir 26, p. 145-163.

Wilkinson, S., 1984, Upper Jurassic facies relationships and their interdependence on salt tectonism in Rains, Van Zandt, and adjacent counties, east Texas: *in*, M. W. Presley ed., The Jurassic of East Texas, East Texas Geological Society Special Publication, Tyler, Texas, p. 153-156.

Williams, C.H., Jr., 1969, Cross section from Mississippi-Tennessee state line to Horn Island in the Gulf of Mexico: Jackson, Mississippi, Mississippi Geological Survey.

Woronick, R.E. and L.S. Land, 1985, Late burial diagenesis, Lower Cretaceous Perasall and Lower Glen Rose formations, south Texas, *in* N. Schneifermann and P.M. Harris, eds., Carbonate Cements, SEPM Special Publication #36, p. 265–275.

Chapter 14

Geological Controls on Regional Subsurface Carbonate Cementation: An Isotopic-Paleohydrologic Investigation of Middle Jurassic Limestones in Central England

James P. Hendry
Department of Geology and Petroleum Geology
University of Aberdeen, King's College
Aberdeen, U.K.

ABSTRACT

Bathonian (Middle Jurassic) carbonate sediments deposited on the northern margins of the Wessex Basin and over the Central Midlands Platform were pervasively cemented by a single generation of sparry calcite cement. The nature and evolution of the hydrologic system responsible for the cementation has been investigated through integrating petrographic and isotopic data with regional tectonic, stratigraphic, and sedimentological information.

The spar cements precipitated in a confined paleoaquifer, sourced by Early Cretaceous meteoric recharge, when tectonic uplift of the basin margins and concomitant fall in relative sea level generated the late Cimmerian unconformity. Isotopic data show that variable degrees of mixing of recharging groundwaters and marine-like pore fluids expelled updip from thick dewatering shales in the Wessex Basin took place during cementation. Consideration of the tectono-stratigraphic history of the basin and surrounding regions, together with diagenetic information from other carbonate formations, indicates that the overall hydrologic system lasted for a maximum of 20 m.y. and extended through much of the Middle and Late Jurassic over a large geographical area. In contrast, earlier, syndepositional emergence did not provide the necessary conditions for development of extensive paleoaquifers, and did not generate pore filling cements.

This study emphasizes links between basin evolution, basin geometry, subsurface hydrology, and cementation characteristics. Comparisons are drawn with published examples to highlight the importance of an integrated approach to regional diagenesis.

INTRODUCTION

Regional occurrence of a pore filling cement in a sedimentary unit can be evidence that a major hydrologic system was developed during the evolution of the host sedimentary basin. Some regional cementation can be a product of solute diffusion in static pore fluids (see Aplin et al., this volume); however the scale of mass transfer involved in the precipitation of large volumes of low-solubility carbonate and silicate cement phases generally indicates that considerable subsurface fluid flow was involved. Geological condi-

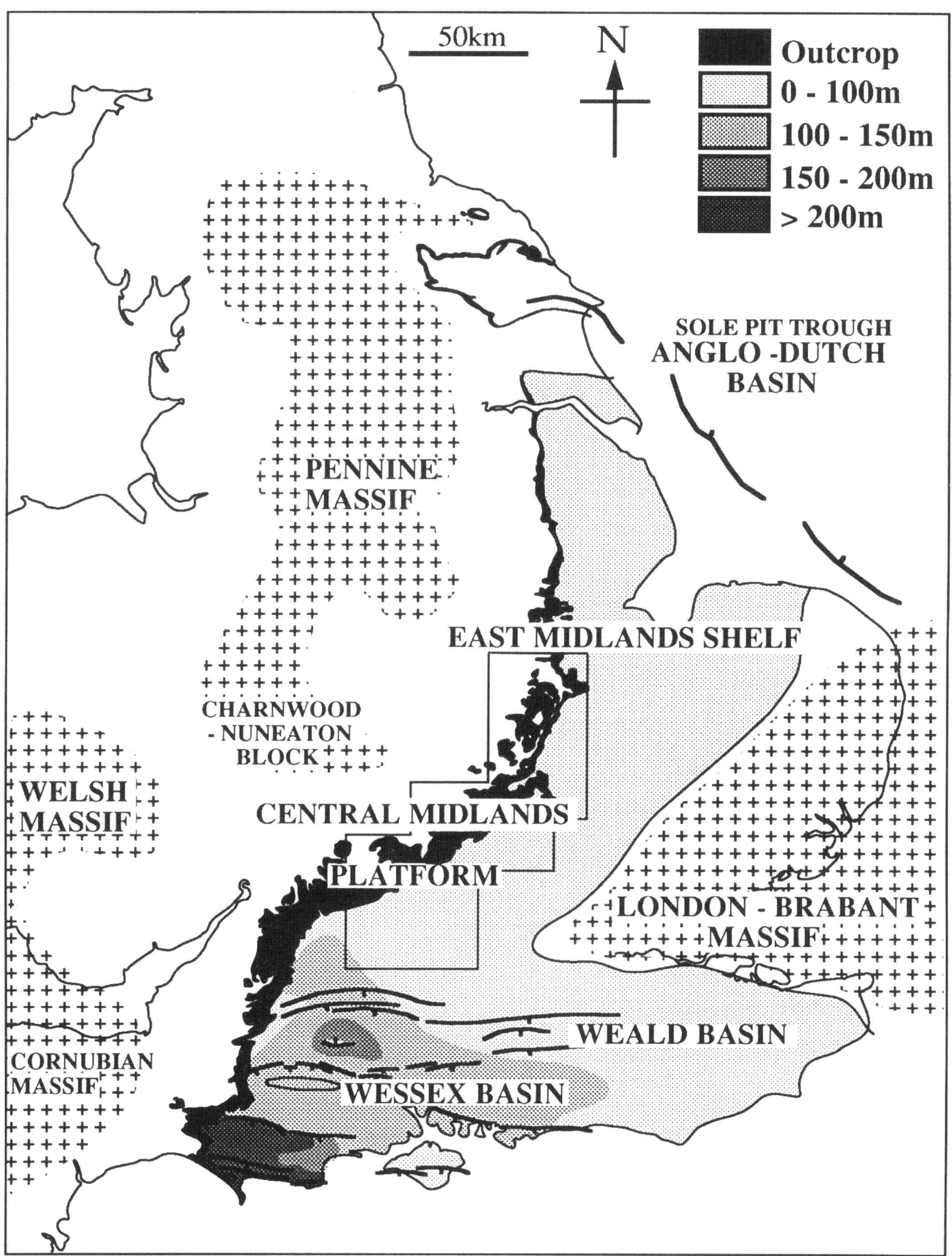

Figure 1. Tectonic setting and Bathonian isopach map for England. Emergent Paleozoic massifs during the Bathonian are indicated by cross-hatching. The study area is the box in the center of the map (see Figure 2).

tions required to produce suitable hydraulic gradients for such large scale fluid movement are fundamentally related to the tectonic and sedimentological evolution of the host basins (Galloway, 1984; Bethke, 1989). The purpose of this chapter is to show how the nature and evolution of such paleohydrologic systems can be studied via analyses of regional patterns of petrography and geochemical tracers (particularly oxygen and carbon isotopes) in pore-filling calcite cements, using as an example the Bathonian (Middle Jurassic) White Limestone Formation of central England. It will be shown that interpretation of otherwise equivocal diagenetic data can yield valuable results if it is carried out within a framework of the regional geological and tectonic history, combined with a consideration of basic hydrologic concepts. Although the cements in question display most of the characteristics of "burial cements" described in the literature, they will be shown to be related to the development of a major meteoric paleoaquifer sourced from the Early Cretaceous late Cimmerian unconformity. This approach to paleohydrologic modeling of an extensive sedimentary unit demonstrates the significance of basin development in controlling the timing and extent of regional cementation, and the results have implications for porosity prediction in ancient limestones.

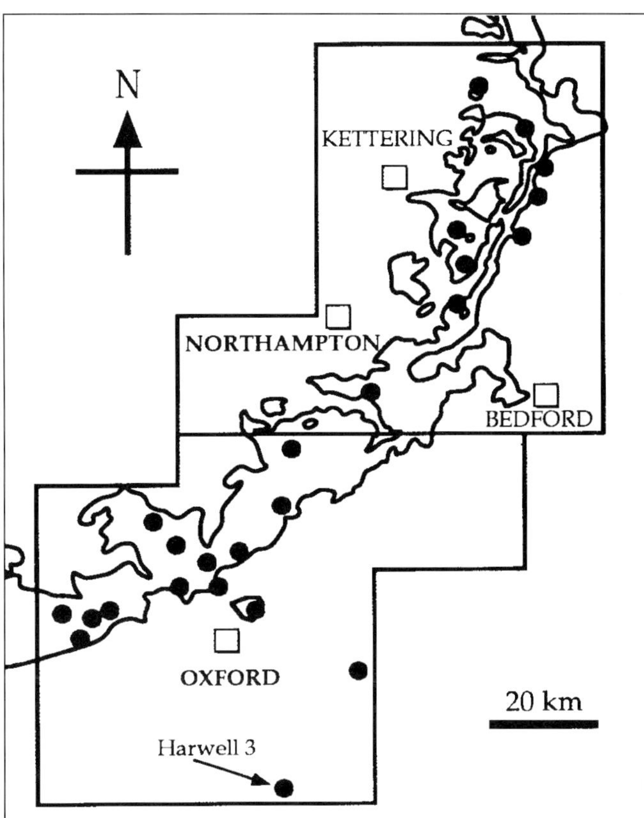

Figure 2. Map of study area showing location of sampled outcrops and boreholes. The horizontal line separates the Oxfordshire (southern) and Northamptonshire (northern) localities. Grid references in Hendry (1990).

GEOLOGICAL FRAMEWORK

Regional Tectonic Setting

The Mesozoic and Tertiary tectonic history of central and southern England has been extensively studied in recent years (e.g. Chadwick, 1985a; Lake and Karner, 1987; Karner et al., 1987). A simplified tectonic and isopach map is shown in Figure 1 and the study area of Oxfordshire to Northamptonshire (Figure 2) is highlighted. The study area encompasses a transition from the northern edge of the Wessex Basin onto the Central Midlands Platform, a more tectonically stable area which underwent relatively slow subsidence through the Jurassic. The Central Midlands Platform passes northeastwards into the East Midlands Shelf and eventually the Anglo-Dutch Basin. It is surrounded to the east and northwest by Paleozoic basement blocks of the London-Brabant Massif and Pennine–Charnwood Massifs. These Paleozoic massifs remained relatively positive features throughout most of the Mesozoic.

The faults bounding the Wessex-Weald and Anglo-Dutch basins were active through the Mesozoic. The Wessex Basin underwent intermittent, rapid, and differential subsidence through the Early and Middle Jurassic. More gradual regional subsidence took place in the Cretaceous corresponding to post-rift thermal relaxation, and compaction of the predominantly argillaceous Jurassic strata in the depocenter (Karner et al., 1987). Regional lithospheric readjustment to rejuvenated crustal extension caused a major phase of basement fault reactivation and uplift of Mesozoic basin margins across northern Europe in the latest Jurassic and Early Cretaceous (Lake and Karner, 1987). This was accompanied by gradual global sea level fall through the Early Cretaceous (Haq et al., 1987), and much of the Jurassic sequence onlapping the London–Brabant Massif was uplifted and subsequently truncated by Early Cretaceous erosion (the "late Cimmerian unconformity"; Chadwick, 1985b). Following renewed eustatic sea level rise in the Aptian, most of southern England was covered by pelagic marine sedimentation. Regional inversion of the Mesozoic basins was initiated at the end of the Cretaceous, in association with the onset of Alpine compression. Continuing uplift in the Tertiary resulted in emergence and erosion of much of England, but subsidence of the London-Brabant Massif (Chadwick, 1985c; Lake and Karner, 1987).

Middle–Late Jurassic Sedimentation

The Jurassic deposits of central England comprise a sequence of predominantly marine shelf sediments, dominated by mudstones but with important carbon-

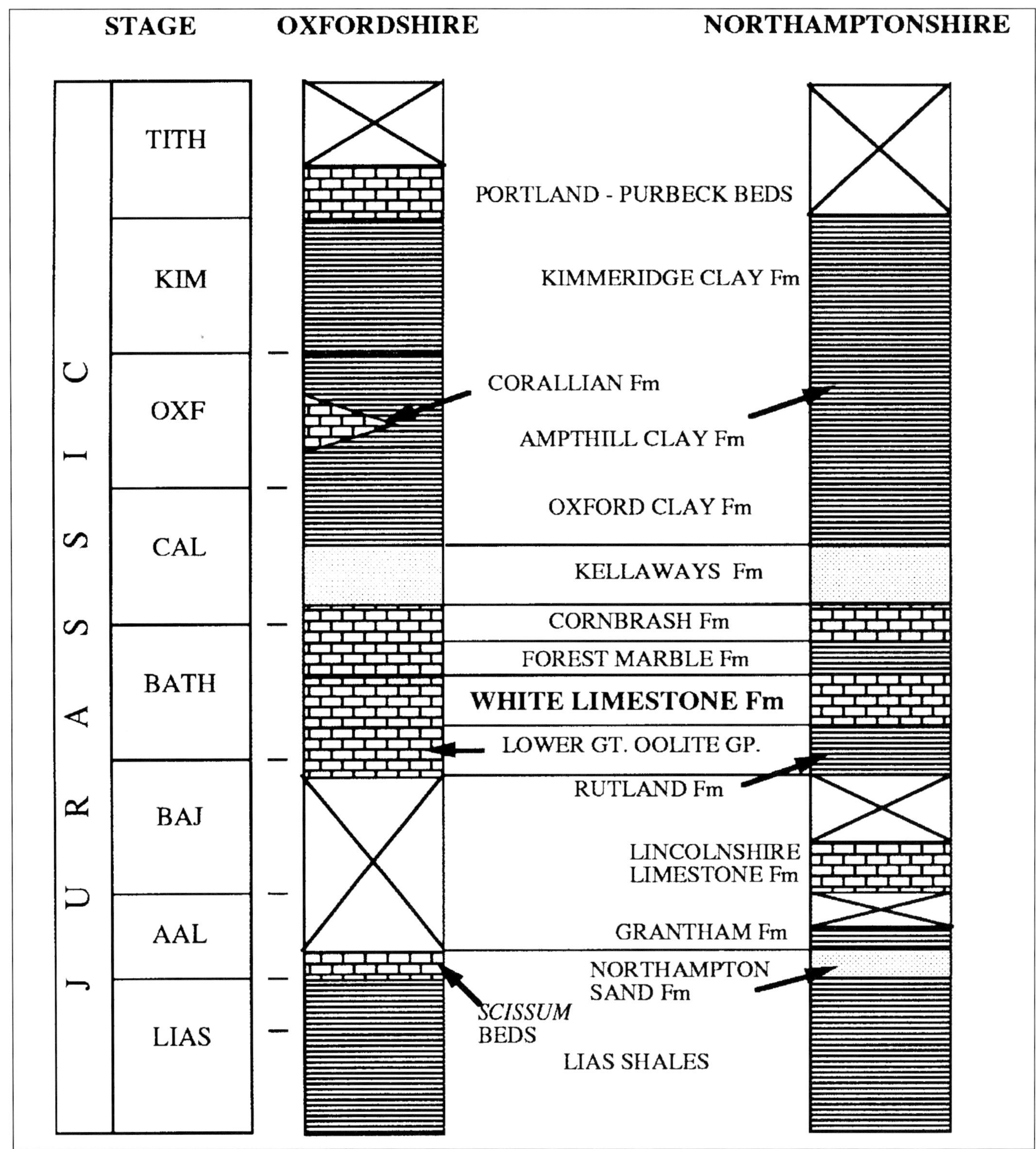

Figure 3. Simplified stratigraphic columns for the Jurassic of Oxfordshire and Northamptonshire.

ate facies developed particularly in the Middle Jurassic (Figure 3). Bajocian–Bathonian sedimentation was controlled by three competing factors: (1) progressive eustatic sea level rise, (2) intermittent and relatively minor subsidence of the Central Midlands Platform, especially in the southwest (probably related to continuing deepening of the Wessex Basin), and (3) progradation of paralic to alluvial, argillaceous, siliciclastic sediments from the emergent basement massifs (Anderton et al., 1979). These factors resulted in complex, cyclic deposition in the study area (Cripps, 1986). Transgressions (mainly from the offshore Wessex Basin area) spread carbonate deposition northeastwards over the Central Midlands Platform. Shallowing

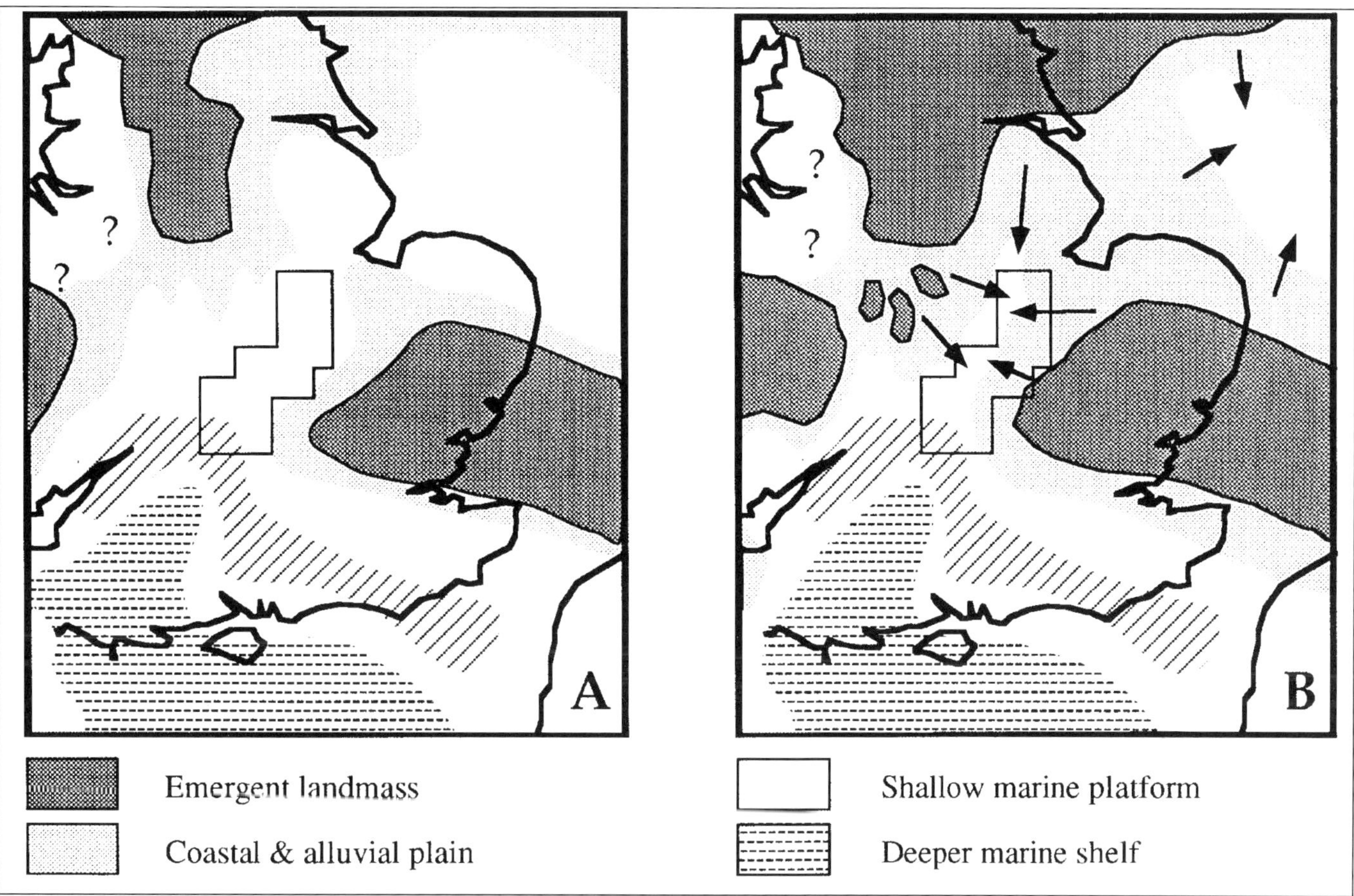

Figure 4. Generalized Bathonian paleogeography of central and southern England at times of (A) maximum transgression and (B) during coastal progradation from the emergent land masses. Diagonal hatching indicates the approximate position of the major oolitic barrier separating the study area from the offshore facies in the central and southern Wessex Basin. Adapted from Cripps (1986). Arrows indicate direction of coastal progradation.

of the depositional environment as sedimentation outpaced subsidence led to deposition of peritidal carbonates and progradation of paralic and non-marine sediments from the land masses to the north and east. As a result, local to widespread emergence occurred at the tops of individual cycles in the central and northeastern parts of the study area (Figure 4), although the tops of cycles in the southwestern parts of the study area generally remained submerged. Minor uplift took place in the Oxfordshire area during the early Middle Jurassic. This restricted Bajocian carbonate sedimentation to shoal and lagoonal environments in the Wessex Basin area (Anderton et al., 1979), and to a separate barrier island-lagoon complex in northern Northamptonshire, which was established in response to transgressions from the Anglo-Dutch Basin (Ashton, 1977).

Lower Bathonian transgressions from the Wessex Basin did not result in carbonate deposition further northeast than central Oxfordshire. However in the middle to late Bathonian, carbonates of the White Limestone Formation ("WLF") were deposited across the study area by a sequence of five more extensive transgressions (Figure 4). One of these may have temporarily linked the Wessex and Anglo-Dutch basins (Cripps, 1986), but this is difficult to establish since the WLF is thin and very argillaceous over the East Midlands Shelf. The overall depositional environment of the study area by the Upper Bathonian represented a shallow inner platform setting, with maximum water depths of the order of 10–15 m. It was separated from the open marine shelf of the central and southern Wessex Basin area by a major oolitic barrier, part of which encroaches into southern Oxfordshire (Palmer, 1979; Cripps, 1986) and continues into the Weald Basin (Sellwood et al., 1987). South of the barrier, Bathonian strata are dominated by open marine shales of the Fuller's Earth Formation (Anderton et al., 1979).

The thickness of the WLF in Oxfordshire reaches a maximum of about 18 m and it gradually thins northeastward into Northamptonshire, farther away from the Wessex Basin. Individual cycles range from less than a meter to several meters thick. Carbonate facies comprise backshoal oolitic, peloidal and bioclastic grainstones and packstones, with more argillaceous and/or faunally restricted, subtidal to peritidal wackestones relatively more common northeastward.

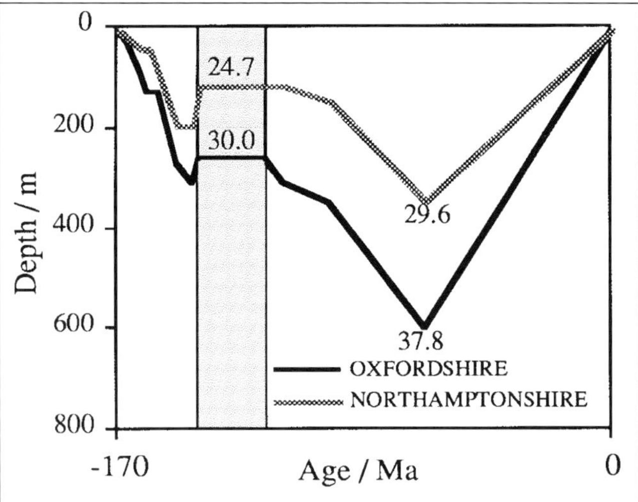

Figure 5. Burial histories of the WLF in Oxfordshire and Northamptonshire with burial temperatures at selected intervals. From data in Table 1. Note that the WLF was not exposed within the study area at the time of the late Cimmerian unconformity (shaded box).

Molluscan/peloidal packstones and grainstones are the most abundant lithologies overall. Terrigenous shales occur interbedded with the carbonates, particularly in Northamptonshire, and brackish-water shales and marls with rootlets are prominent above peritidal beds at the tops of some sedimentary cycles throughout the study area. Where these shales are absent in the more offshore (southwestern) parts of the study area, the tops of cycles are sometimes marked by burrowed and encrusted hardgrounds.

Major transgression, initiated in the Callovian, resulted in shallow marine carbonate and siliciclastic deposition across the Central Midlands Platform. Eustatic sea level rise accompanied by gradual regional subsidence continued until the Kimmeridgian. This led to sustained marine mudstone sedimentation in the study area (Anderton et al., 1979), except for nearshore carbonate and sandstone deposition fringing the London-Brabant massif in the Oxfordian (de Wet, 1987). Central England remained inundated by marine waters until the end of the Jurassic. After the uplift and truncation of Middle and Upper Jurassic strata on the margins of the London-Brabant Massif at the late Cimmerian unconformity, renewed transgression in the Aptian blanketed the study area and surrounding basins with middle–Upper Cretaceous pelagic sediments. Marine conditions persisted until Early Tertiary inversion and eustatic sea level fall, which rendered the study area emergent until the present day.

Burial and Thermal Histories

Approximate burial histories for the WLF in Oxfordshire and Northamptonshire are shown in Figure 5. Average estimated overburden thicknesses were compiled from information in Anderton et al. (1979) and Cope et al. (1980) and are summarized in Table 1. No attempt has been made to decompact argillaceous units, and the detailed history of Late Jurassic/Early Cretaceous and Tertiary uplift is poorly constrained. Nevertheless it is clear that there was differential burial in the study area, reflecting the relative proximity of the southwestern part (Oxfordshire) to the central Wessex Basin. Figure 5 also shows estimated burial temperatures at selected intervals during the subsidence histories. Although ancient heat flows are not known, the relatively stable tectonic setting of the study area from the Bathonian to Recent allows the assumption of a constant continental heat flow. Astin and Scotchman (1988) suggest that thermal effects associated with Early Cretaceous rifting increased heat flow in eastern England. However, these are likely to have been significant in basinal areas, where crustal thinning was concentrated, and not in the stable Central Midlands Platform.

CEMENTATION OF THE WHITE LIMESTONE FORMATION

Early Diagenesis

WLF carbonates of diverse facies were sampled extensively from outcrop and from two borehole cores, in a 120 km long belt across Oxfordshire and Northamptonshire (Figure 2). A vertical suite of specimens was obtained from each locality so that stratigraphic and geographical variation in cementation could be documented. Slabbed and polished specimens were studied using stained acetate peels (e.g., Miller, 1988). Approximately 100 polished thin sections were examined in transmitted light, and in cathodoluminescence (CL). Stable isotopic and ICPAES (Inductively Coupled Plasma Atomic Emission Spectrophotometry) analyses of cements are summarized in Table 2.

Syndepositional marine diagenesis in the WLF featured extensive microboring and subsequent partial cementation of microporosity, especially in the micritized components (Hendry, 1990). Extensive burrowing and internal sediment deposition were important at hardgrounds. Macroscopic marine cements are very rare, reflecting the low energy of the inner platform setting. Pyrite occurs throughout the formation, and minor, early calcite cement fringes in some Northamptonshire samples were probably precipitated during bacterial degradation of organic matter.

Evidence of temporary emergence and desiccation of peritidal beds at the tops of some sedimentary cycles includes fenestral fabrics, shrinkage cracks, and rare cryptalgal lamination (Palmer and Jenkyns, 1975). Where marls and shales with rootlets occur, emergence of the cycle tops possibly was more prolonged. However, no vadose or other cements specifically related to the emergence surfaces have been observed. Dissolution of bioclastic aragonite is pervasive through the formation, but there is no unequivocal evidence for its near-surface origin. The only possible exceptions are rare examples at hardgrounds, where the sporadic occurrence of clay-filled biomolds may indicate localized synsedimentary *submarine* aragonite dissolution (cf. Palmer et al., 1988).

Regional Cementation

Primary and biomoldic porosity of the WLF throughout the study area is cemented by a single generation of medium to coarse grained, equant calcite spar cement (Figures 6A, B). The spar is ubiquitous in all the carbonate facies and occurs throughout the vertical extent of the formation. Intergranular porosity is almost always tightly cemented, but large primary shelter and secondary biomoldic pores (such as coral dissolution voids) are incompletely occluded (e.g. Figure 6A). Evidence of mechanical compaction is not abundant in the limestones, but it pre-dates spar cementation. Chemical compaction (dissolution seam formation) also generally pre-dated spar cementation. The spars are very similar in texture and occurrence throughout Oxfordshire and Northamptonshire. However, there are some subtle petrographic and compositional variations, which are detailed below.

Oxfordshire

The spars display an overall poorly to moderately ferroan stain and moderate to dull orange-brown

Table 1. Data used in constructing the burial histories for the WLF in Oxfordshire and Northamptonshire. Cumulative depth to the WLF for each stratigraphic division is calculated from averaged thicknesses of the strata, estimated from the literature but not decompacted. Uplift values are poorly constrained but estimated values for the late Cimmerian unconformity are included and the Tertiary uplift rate is assumed to be constant. Burial temperatures at the middle of the WLF are calculated for each interval by estimating the steady state heat transfer by conduction in a homogenous flat crust (Andrews-Speed et al., 1984; Emery et al., 1988) where the temperature T (°C) at any depth Z (m) is given by

$$T_z = T_s + q\sum_{i=1}^{n}\left[\frac{z_i}{k_i}\right]$$

where T_s is the surface (or sea floor) temperature in °C, q is the conductive heat flow in Wm^{-2} and k_i is the thermal conductivity in $Wm^{-1}\,°C^{-1}$ of the *i* th layer of thickness z_i where $\Sigma\,[z_n] = Z$. q is estimated to have been about 55 mWm^{-2} in the study area from the onshore data in Wilson and Luheshi (1987) and Richardson and Oxburgh (1979). In the absence of a detailed record of the thermal conductivities of the strata in Oxfordshire and Northamptonshire the following values were extrapolated from data for the East Midlands shelf quoted in Emery (1991), but with a higher value for the Chalk, comparable to those reported in Andrews-Speed et al. (1984) and Wilson and Luheshi (1987):

Clays = 1.3 $Wm^{-1}\,°C^{-1}$
Sandstones = 1.7 $Wm^{-1}\,°C^{-1}$
Limestones = 2.2 $Wm^{-1}\,°C^{-1}$
Chalk = 3.0 $Wm^{-1}\,°C^{-1}$

An average value of 20 °C is estimated for T_s in the Jurassic and Cretaceous, reflecting generally warm climatic conditions and shallow epeiric seas.

OXFORDSHIRE

Stratigraphic Unit	Age At Top (m.y.)	Depth To Middle Of WLF (m)	Temperature At Middle Of WLF (°C)
WLF	170	8	20.2
Forest Marble Fm.	169	15	20.4
Cornbrash Fm.	168	19	20.5
Kellaways Clay	167.5	23	20.6
Kellaways Sand	167	26	20.7
Oxford Clay Fm.	162	86	23.3
Corallian Beds	160	126	24.3
Nondeposition	156	126	24.3
Kimmeridge Clay Fm.	150	266	30.2
Portland & Purbeck Strata	144	306	31.2
Late Cimmerian Unconformity	142	261	30.0
Nondeposition	119	261	30.0
Aptian Strata	113	306	31.5
Albian Strata	97.5	346	33.2
Chalk	65	596	37.8
Tertiary Uplift	0	8	

NORTHAMPTONSHIRE

Stratigraphic Unit	Age At Top (m.y.)	Depth To Middle Of WLF (m)	Temperature At Middle Of WLF (°C)
WLF	170	4	20.1
Thrapston Clay Fm.	169	10	20.4
Cornbrash Fm.	168	12	20.4
Kellaways Clay	167.5	15	20.5
Kellaways Sand	167	19	20.7
Oxford Clay Fm.	160	44	21.7
Non Deposition	159	44	21.7
Ampthill Clay Fm.	156	94	23.8
Kimmeridge Clay Fm.	150	194	28.1
Late Cimmerian Unconformity	142	114	24.7
Nondeposition	113	114	24.7
Albian Strata	97.5	144	25.9
Chalk	65	344	29.6
Tertiary Uplift	0	4	

luminescence. Spar crystals display striking and complex compositional zonation patterns (Figure 6C) (Hendry and Marshall, 1991a). Concentric zoning is minor and restricted to the earliest crystals of pore-fills. The bulk of the spar throughout Oxfordshire displays sector zonation with many crystals containing additional zones restricted to certain sectors ("sector-specific zonation"). The spars contain 537–8868 ppm Fe (mean 3547 ppm), 210–824 ppm Mn (mean 487 ppm), 361–1926 ppm Mg (mean 755 ppm) and 58–253 ppm Sr (mean 131 ppm) (Table 2).

Northamptonshire

The pore filling spar in Northamptonshire gives a moderate to strongly ferroan stain and has a dull to very dull orange-brown luminescence. Spar crystals show sector and sector-specific zonation, but whereas the zoning patterns are generally uniform in Oxfordshire (Hendry and Marshall, 1991a), there is a wider variety of fabrics in the Northamptonshire spars. Limited concentric zoning is sometimes combined with the sector zoning (Figure 6D), but is quite subtle and is not correlatable between localities. The spars contain 4273–9293 ppm Fe (mean 6727 ppm), 459–897 ppm Mn (mean 620 ppm), 432–1943 ppm Mg (mean 843 ppm) and 65–400 ppm Sr (mean 182 ppm) (Table 2).

ISOTOPIC RESULTS

Stable isotopic ratios from the Oxfordshire WLF spars are plotted in Figure 7A. Also shown are analyses from non-luminescent oyster (*Praeexogyra* sp.) calcites and from brachiopod (*Epithyris* sp.) calcites. Although largely non-luminescent, the latter contain minor early diagenetic ^{13}C-depleted calcite and pyrite in punctae, which could not be avoided during sampling. The majority of the spars occupy a relatively tight field on Figure 7A. Overall isotopic compositions range between -3.7 and -7.4‰ PDB for oxygen and –1.5 and +2.1‰ PDB for carbon. Values from individual localities fall in smaller clusters, generally $\leq$ 2.0‰ for $\delta^{18}O$ and $\leq$ 2.5‰ for $\delta^{13}C$ (see full data tables in Hendry, 1990). Where it was possible to extract two samples from large void fills, the central sample is generally slightly ($\leq$ 1.5‰) more depleted in ^{18}O than the outer one. All of the $\delta^{18}O$ values are depleted with respect to Bathonian marine calcite.

Stable isotope analyses of oysters, brachiopods and spar cements from the WLF in Northamptonshire are plotted in Figure 7B. The four brachiopod analyses with positive $\delta^{13}C$ values are from impunctate *Kallirhynchia* sp. while the other two (*Epithyris* sp.) contain early ^{13}C-depleted cement in the punctae. The

Table 2. Summarized isotopic and ICPAES analyses for the WLF spars in Oxfordshire and Northamptonshire. Microsamples of cements and other components were extracted from polished slabs using a hand-held PCB drill with a tungsten carbide bit. Between 10 and 15 mg of powder were extracted for combined isotopic and trace element analysis of each sample. Stable isotope samples were analysed on a VG SIRA 12 mass spectrometer at Liverpool University Stable Isotope Laboratory and raw data were corrected following standard procedures using an acid fractionation factor of 1.01025 (Friedman and O'Neil, 1977). Trace element contents were measured by ICPAES at Royal Holloway and Bedford New College, London. Average analytical precisions for stable isotope analyses (measured on repeat unknowns) were better than 0.11 ‰ for $\delta^{18}O$ and 0.06 ‰ for $\delta^{13}C$. Precision and accuracy for ICPAES were determined from repeat analyses of two calcites of known composition, with the following averaged results (in relative %): Fe 2.6 and 1.2, Mg 1.1 and 1.7, Mn 0.9 and 6.8, Sr 1.0 and 3.7. Precisions are all calculated at 95% confidence level. Full data tables can be found in Hendry (1990). n = number of analyses; s.d. = standard deviation (1σ).

OXFORDSHIRE						
	$\delta^{18}O$ (‰PDB)	$\delta^{13}C$ (‰PDB)	Fe (ppm)	Mn (ppm)	Mg (ppm)	Sr (ppm)
Range	–7.4 to –3.7	–1.5 to + 2.1	537 to 8868	210 to 824	361 to 1926	58 to 253
Mean	–5.8	0.2	3547	487	755	131
S.D.	0.7	0.7	1541	106	338	44
n	183	183	108	108	108	108

NORTHAMPTONSHIRE						
	$\delta^{18}O$ (‰PDB)	$\delta^{13}C$ (‰PDB)	Fe (ppm)	Mn (ppm)	Mg (ppm)	Sr (ppm)
Range	–11.2 to – 6.1	–4.2 to + 1.0	4273 to 9293	459 to 897	432 to 1491	65 to 400
Mean	–7.9	–0.6	6727	620	843	182
S.D.	1.0	1.1	1372	100	261	99
n	66	66	26	26	26	26

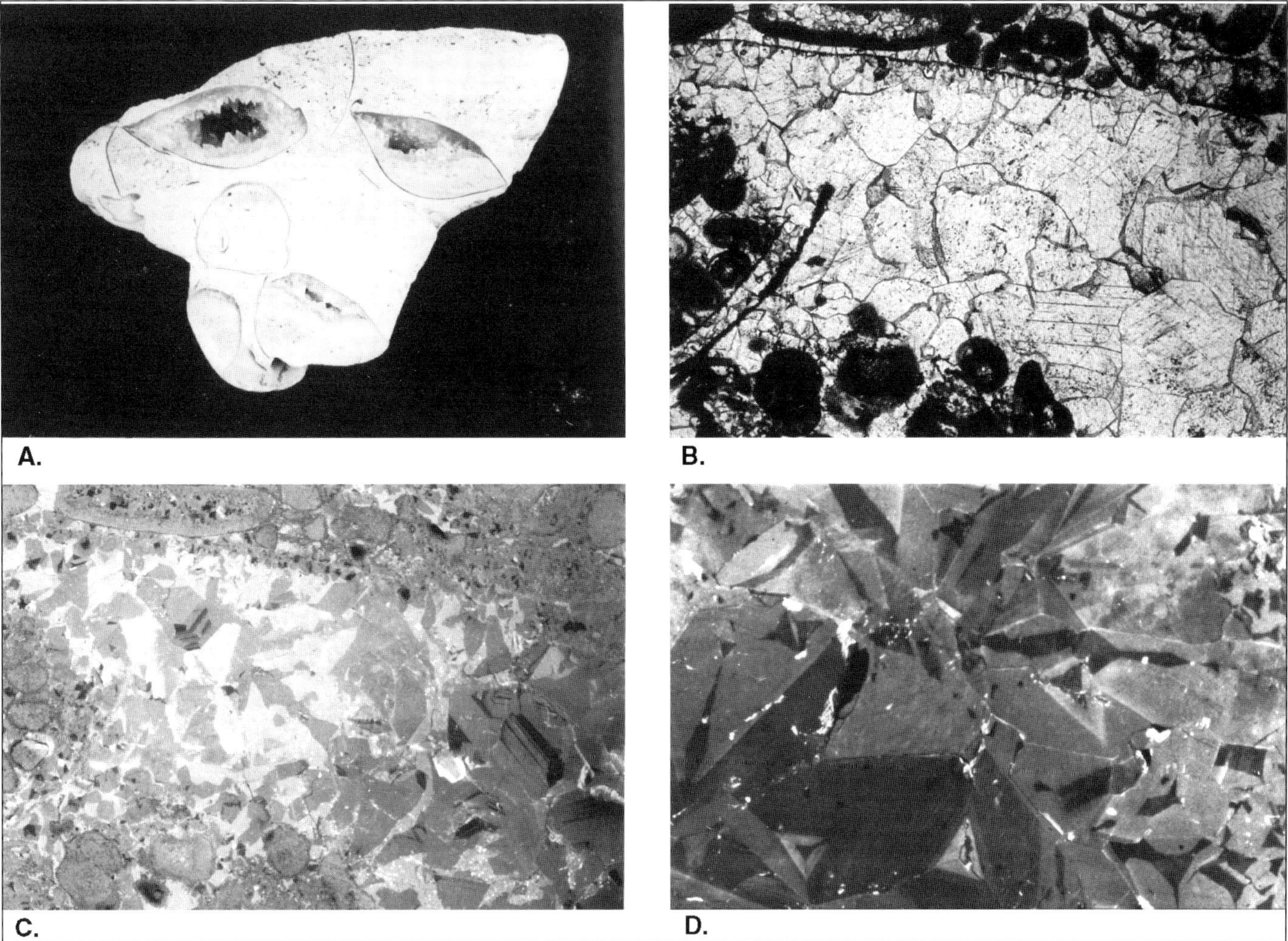

Figure 6. Texture and petrography of the WLF spars. (A) Geopetal spar partially filling brachiopod (*Epithyris* sp.) body chambers in bioturbated wackestone-packstone, northeast Oxfordshire. Width of specimen = 9 cm. (B) Thin section photomicrograph of spar-cemented peloidal grainstone from Oxfordshire. Transmitted light, field of view = 2.5 mm. (C) Cathodoluminescence photomicrograph of sector and sector-specific zonation in Oxfordshire spar (same field of view as B). (D) Example of complex zoning from Northamptonshire spar crystals, incorporating sector, sector-specific and subtle concentric zones. Fine grained, moderately luminescent spar adjacent to the bivalve shell is an early cement. Cathodoluminescence photomicrograph, field of view = 2.5 mm.

range of spar values is relatively restricted and is depleted in both carbon and oxygen with respect to the Oxfordshire spars. $\delta^{18}O$ compositions are between –6.1 and –10.1‰ PDB and $\delta^{13}C$ values between –2.6 and +1.0‰ PDB, with a single outlying analysis of –11.2‰ $\delta^{18}O$ and –4.2‰ $\delta^{13}C$. The most negative $\delta^{18}O$ analyses are from latest-stage coarse grained crystals which occur in the middle of brachiopod body chambers and shelter cavities. These also have some of the most negative $\delta^{13}C$ values.

Stratigraphic Trends

No pronounced vertical trends in isotopic composition of the spars have been observed, with the exception of several adjacent localities in southwest Oxfordshire where a prominent cycle-top shale occurs within the exposed WLF. In each of these quarries a small but consistent shift towards more positive $\delta^{18}O$ and $\delta^{13}C$ values is seen from spars in the limestones over about 2 m on either side of the shale. This shift is accompanied by an enrichment of spar Fe and Mg contents towards the shale (Figure 8).

INTERPRETATION OF PETROGRAPHIC AND COMPOSITIONAL DATA

Zoning

The occurrence of a major generation of pore filling calcite cement throughout the WLF of the study area suggests that a regional paleohydrologic system may have been active at some time subsequent to the

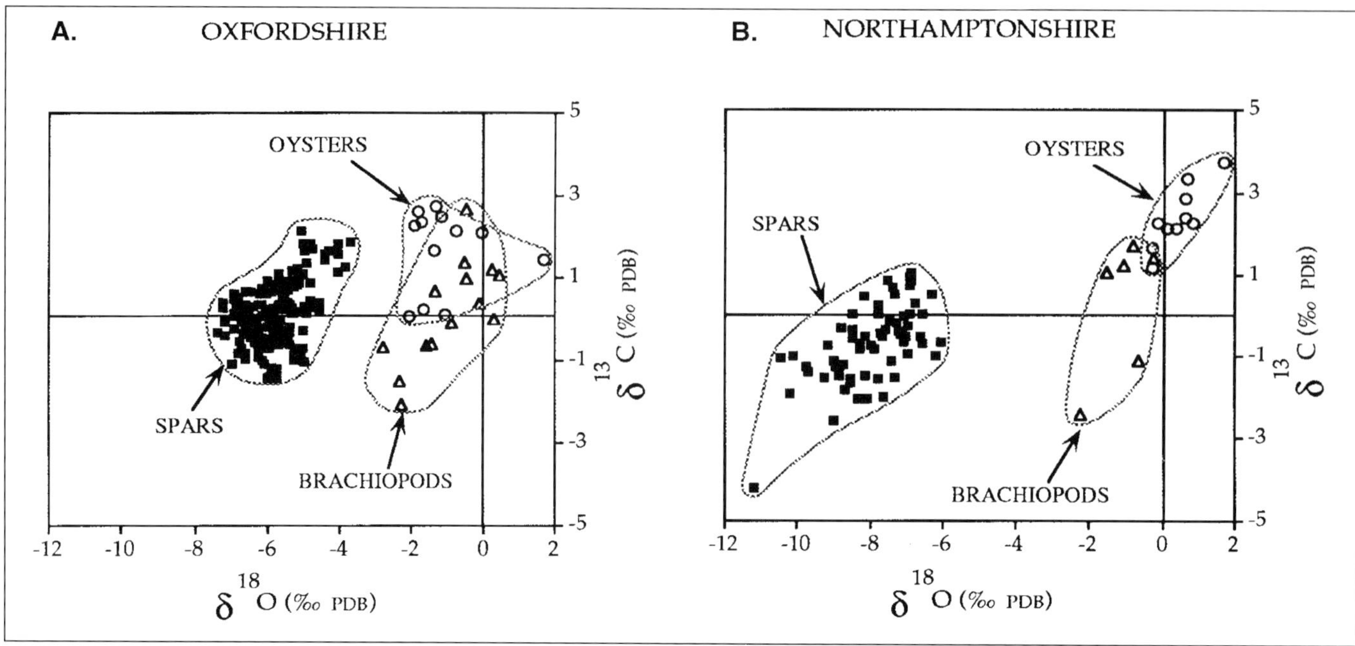

Figure 7. $\delta^{13}C$ vs. $\delta^{18}O$ cross-plots of spar and marine component compositions.

deposition of the WLF. Spatial trends in zoning patterns of calcite cements have been used in several studies to constrain pore fluid flow directions in subsurface paleoaquifers (e.g. Dorobek, 1987; Niemann and Read, 1988), but these rely on the presence of regionally mappable concentric zoning sequences. The lack of concentric zoning in the bulk of the pore filling spars from the Oxfordshire WLF, and its limited development in Northamptonshire, preclude such an approach. In contrast, the observations suggest that the pore fluid trace element composition was relatively invariant during the precipitation of much of the Oxfordshire cement and underwent only limited evolution during the precipitation of the Northamptonshire spars. The complex sector and sector-specific zoning gives no information about gross pore fluid evolution and is a product of trace element partitioning related to growth mechanisms operating on different crystal faces (Hendry and Marshall, 1991b).

Stable Isotopes

Oxfordshire

In Figure 9 the $\delta^{18}O$ range of the Oxfordshire spar cements is overlain on an equilibrium fractionation diagram which relates calcite $\delta^{18}O$ PDB values to temperature for a range of pore fluid compositions (relative to SMOW). Disequilibrium fractionation of stable isotopes within natural calcite spar crystals has recently been recognized (Dickson, 1991), but appears to be relatively small (i.e. < 1‰) for $\delta^{18}O$ and will have been averaged out by the bulk sampling of spar crystals employed in this study. The temperature range in which precipitation could have occurred can be constrained to lie between the minimum depositional paleotemperature and the local maximum burial temperature (Figure 5). Sea floor temperatures in the Bathonian are estimated to have been 20°–25°C, depending on the water depth (Hendry, 1990).

Pore waters involved in the cementation of shallow buried (< 1 km) sedimentary sequences are likely to have one (or some mixture) of three principal origins: depositional marine water, meteoric-derived groundwater, or compactional water from buried shales. The $\delta^{18}O$ analyses of the nonluminescent oyster calcites (and of the least-altered brachiopods) are considered to be a good indication of equilibrium precipitation from the depositional pore fluids. Applying the modified equation of Friedman and O'Neil (1977) to the analyses shown in Figure 7A, with the 20°–25°C paleotemperature range, yields a depositional water composition of –1 to +2.5‰ SMOW. This is, on average, more enriched in ^{18}O than the "standard" value of –1.2‰ taken for non-glacial sea water (Shackleton and Kennett, 1975). This may reflect the greater potential for evaporative enrichment in ^{18}O of the marine reservoir in shallow and relatively restricted platform settings as compared to open ocean basins. As the oysters tend to occur in faunally restricted WLF facies, their compositional range may be biased towards the most saline/isotopically enriched waters (Hendry, 1990). A general value of 0 ± 1‰ SMOW is therefore assumed for an average depositional pore fluid in Oxfordshire.

Figure 9 shows that the spars precipitated from pore fluids of between +1 and –6‰ SMOW. A synsedimentary, marine origin for the cements is incompatible with the paleotemperature estimates, as

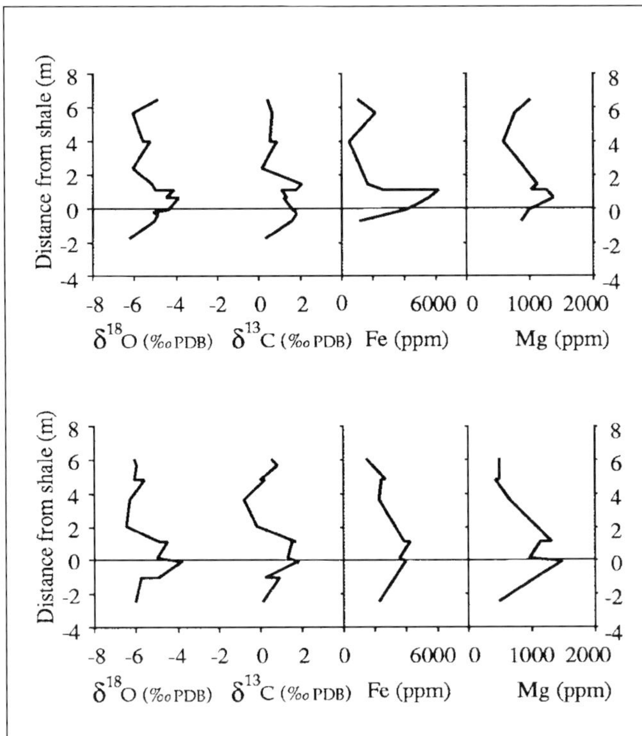

Figure 8. Stratigraphic trends of $\delta^{13}C$, $\delta^{18}O$, Fe and Mg in spars located in limestones adjacent to a cycle-top shale bed in the WLF of two quarries in southwest Oxfordshire.

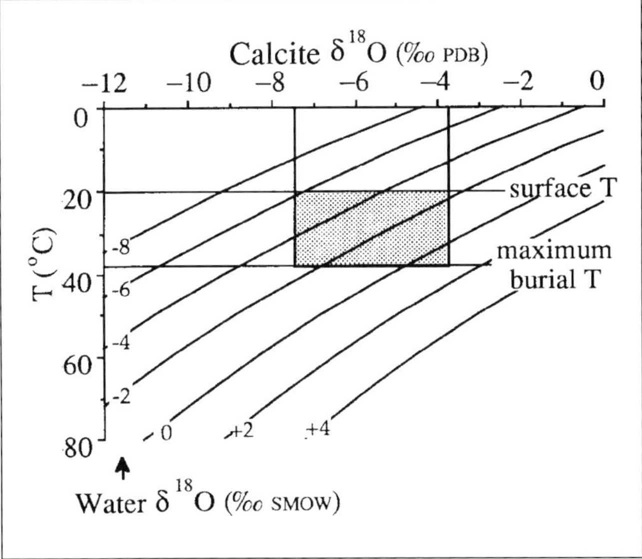

Figure 9. Relationship between the $\delta^{18}O$ of diagenetic pore fluids and calcite cements showing possible temperatures of precipitation of the WLF spars. Curves are constructed using the relationship $10^3 \ln \alpha_{calcite-water} = 2.78 \times 10^6/T^2 - 2.89$ (O'Neil et al., 1969). The vertical lines enclose the full range of sparry calcite $\delta^{18}O$ values from the Oxfordshire WLF. Estimated local marine water composition was 0±1‰ SMOW. The range of possible water compositions from which the spars could have been precipitated is defined by the curves which pass through the stippled box between the estimated depositional and maximum burial temperatures; in this case, about +1 to –6‰ SMOW. Only a single curve (δw = –3‰) can account for the entire range of spar compositions within the permitted temperature values.

even the most enriched value of –3.7‰ PDB could only have precipitated from depositional pore fluids of 0 ± 1‰ SMOW between 27°–37°C. Spar cementation in Oxfordshire therefore followed introduction of isotopically depleted meteoric water into the WLF from an external source. This is supported by consideration of $\delta^{13}C$ data. Occurrence of spars with negative $\delta^{13}C$ values (Figure 7A) indicates that isotopically negative carbonate was supplied to the pore fluid. Thermal degradation of organic matter produces isotopically negative carbon which can be incorporated into subsurface carbonate cements, but maximum burial depths in Oxfordshire were insufficient for thermal decarboxylation to have been locally important. Negative bicarbonate could theoretically have been transported in compactional fluids migrating up the limestones from deeply buried shales in the Wessex Basin, but if this were the case the most negative $\delta^{13}C$ values would be expected in cements closest to the basin center. In contrast, comparison of Figures 7A and 7B shows that the most negative spars are from Northamptonshire, which is located further from the basin center than Oxfordshire (Figures 1, 2). An alternative explanation for the negative $\delta^{13}C$ values in the spars is therefore required, and at shallow burial depths the most probable source of negative carbon for spar cementation is from dissolution of organically-derived CO_2 during meteoric water infiltration through the soil zone (Anderson and Arthur, 1983; Lohmann, 1988). Therefore, both carbon and oxygen isotopic data indicate that meteoric waters formed at least a component of the cementing pore fluid. The wholesale dissolution of skeletal aragonite provides supporting evidence for the involvement of meteoric water on a regional scale.

Estimating ancient meteoric water compositions is difficult as it depends on paleolatitude and paleotopography (e.g. coastal precipitation versus continental runoff) at the time of recharge, as well as the amount of water-rock interaction in the infiltration zone. Isotopic fractionation during evaporation and precipitation dictates that rainfall is depleted in ^{18}O compared with the coexisting marine reservoir. This difference may be only a few‰ in oceanic settings (Anderson and Arthur, 1983; Lohmann, 1988), but the presence of elevated land masses surrounding the study area in the Mesozoic means that significantly more depleted meteoric waters may have entered the limestones. Based on the estimated surface and maxi-

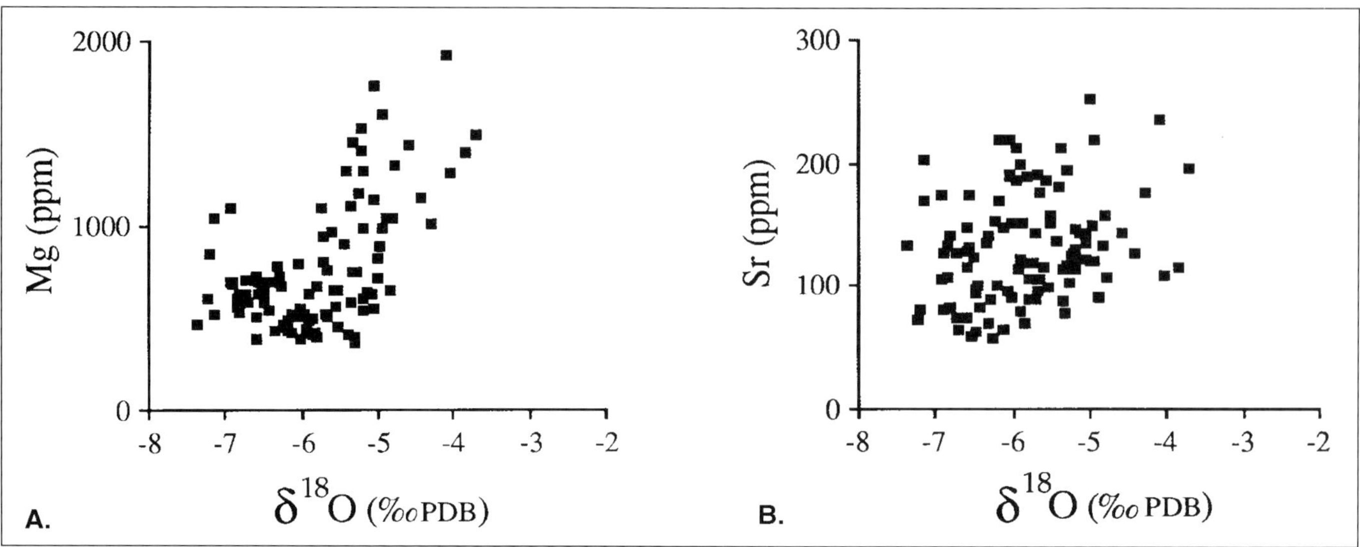

Figure 10. Scatter plots of: A. $\delta^{18}O$ vs. Mg, and B. $\delta^{18}O$ vs. Sr for the Oxfordshire spars.

mum burial temperatures, Figure 9 demonstrates that the most negative $\delta^{18}O$ pore fluid which can have been involved in precipitation of the Oxfordshire spars (at 20°C) is a water of –6‰ SMOW. This value is comparable with present day precipitation in subtropical continental regions (Yurtsever, 1975) and similar to values assumed for Bajocian meteoric water in the East Midlands (Emery et al., 1988); it will therefore serve as a working estimate which can be tested when the origins of the spars have been determined.

From Figure 9 it is apparent that only the most isotopically negative spars could have been precipitated from the estimated meteoric water composition of –6‰. Therefore the range of pore fluids which precipitated all of the Oxfordshire spars must have been between depositional marine and estimated meteoric $\delta^{18}O$ values. Three possibilities are examined below:

(1) The meteoric water was considerably more enriched in ^{18}O, with the most negative spars precipitated at temperatures > 20°C;

(2) the meteoric water composition was about –6‰ SMOW as estimated but $\delta^{18}O$ values were altered by water-rock interaction prior to spar cementation; or

(3) that mixing of meteoric and marine or compactional fluids took place prior to or during cementation.

The first option requires that meteoric water was derived from coastal precipitation, with minimal altitude-related isotopic fractionation (cf. de Wet, 1987). Figure 9 shows that a single pore fluid of –3‰ SMOW could have produced all of the spars, *if* cementation continued from near-surface temperatures until maximum burial.

Progressive water-rock interaction during meteoric diagenesis involves dissolution of metastable marine carbonates by undersaturated recharging waters, and calcite precipitation further downdip when supersaturation is attained. This process can produce characteristic compositional patterns in cement precipitates with respect to both timing of precipitation and distance from recharge ("inverted J" trends; Meyers and Lohmann, 1985; Lohmann, 1988). These arise from the small concentration of the total dissolved carbon in the water compared with the "rock" (reactive marine carbonate) carbon reservoir, coupled with an overwhelming dominance of oxygen in the fluid phase. As a result, meteoric cement precipitated in a carbonate aquifer should show relatively invariant $\delta^{18}O$ compositions, coupled with a spatial trend of progressively increasing $\delta^{13}C$ compositions (e.g. Saller and Moore, 1991). Negative soil zone carbon values typify cements located nearest recharge, but down-aquifer cement $\delta^{13}C$ values rapidly approach those of the dissolving marine carbonates. Additionally, as the abundance of metastable components decreases, undersaturated fluids can migrate further whilst experiencing less water-rock interaction (Allan and Matthews, 1982). Therefore, as an aquifer evolves, the $\delta^{13}C$ of the water at a fixed point would be expected to become more negative with time. Cements precipitating in such a situation might be zoned with respect to $\delta^{13}C$.

Considerably more water-rock interaction is required before the $\delta^{18}O$ values of cements precipitated from the water start to approach that of the marine carbonate (Meyers, 1989; Banner and Hanson, 1990) but in the distal portions of aquifers the $\delta^{18}O$ composition of the water may begin to become more enriched. At such low water/rock ratios the $\delta^{13}C$ of dissolved carbon is invariant, representing dynamic equilibrium with the metastable marine carbonates. Therefore the $\delta^{13}C$ in any cement precipitated should be constant and positive. Mg and Sr would also be expected to increase in these distal cements, in response to cumulative aragonite dissolution and Mg-calcite stabilization (Lohmann, 1988).

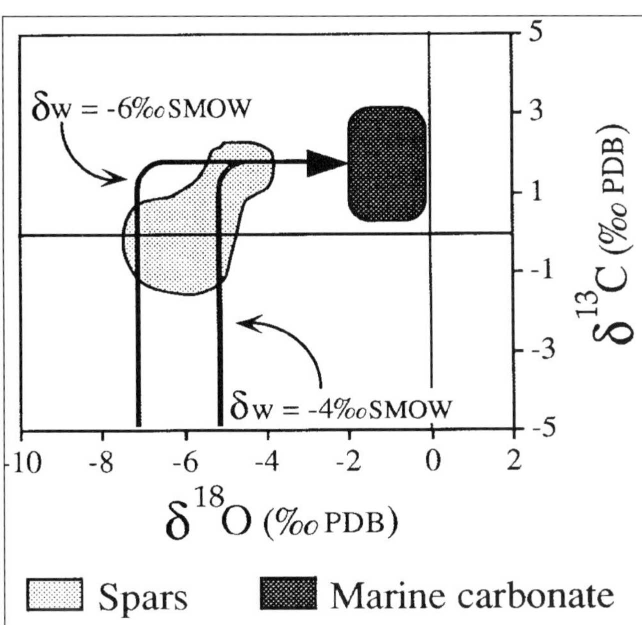

Figure 11. Predicted trends of calcite stable isotope compositions arising from progressive water-rock interaction, a temperature of 20°C and initial meteoric water compositions of –4‰ and –6‰ SMOW. With precipitation during increasing water rock interaction, cement $\delta^{13}C$ values should approach marine values before $\delta^{18}O$ values change appreciably. Compositional field of Oxfordshire spars is overlain for comparison.

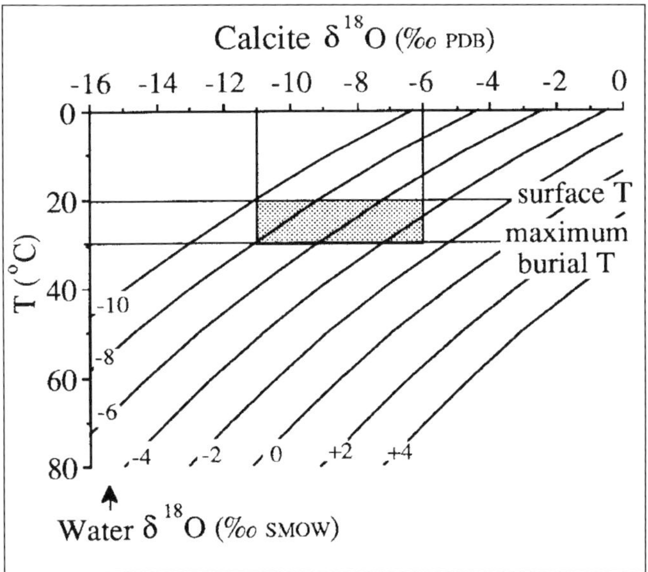

Figure 12. Relationship between the $\delta^{18}O$ of diagenetic pore fluids and calcite cements showing possible temperatures of precipitation of the WLF spars. The vertical lines enclose the full range of sparry calcite $\delta^{18}O$ values from the Northamptonshire WLF and the stippled box represents the range of fluid compositions corresponding to precipitation of those spars within the maximum range of burial depths experienced. Estimated local depositional water composition was +2 ±1‰ SMOW. The curves are constructed using the relationship $10^3 \ln \alpha_{calcite-water} = 2.78 \times 10^6/T^2 - 2.89$ (O'Neil et al., 1969). See text for explanation.

Minimal information on the temporal evolution of the diagenetic water composition is available from the single generation of WLF spars. Nevertheless Figure 7A shows that the spars with the most positive $\delta^{18}O$ values tend to have the most positive $\delta^{13}C$ compositions. These spars also have the highest Mg contents (Figure 10A), as would be expected for precipitation at low water/rock ratios. Despite this, there are problems in attributing the range in spar compositions to water-rock interaction alone. Firstly the trend in increasing Mg contents with enrichment in spar $\delta^{18}O$ compositions is not clearly mimicked by Sr (Figure 10B). Secondly the presence of spars with negative $\delta^{13}C$ values over a relatively wide range of $\delta^{18}O$ compositions (Figure 7A) is incompatible with the water-rock interaction model, where calcite $\delta^{13}C$ values are predicted to increase to marine values *before* any appreciable change in $\delta^{18}O$ away from the "meteoric water line" (representing equilibrium precipitation from the original water value). This is demonstrated in Figure 11.

From the foregoing arguments it is apparent that a meteoric pore fluid of –6‰ SMOW could only have produced the full range of spar compositions if progressive water-rock interaction was coupled with some mixing with a marine water. Mixing of marine and meteoric fluids in carbonate aquifers is generally assumed to result in undersaturation with respect to calcite and therefore dissolution rather than precipitation (Lohmann, 1988). Nevertheless, pore fluid mixing offers the simplest explanation for the compositions of the Oxfordshire spars. Before investigating the possibility further it is worth examining the data from Northamptonshire.

Northamptonshire

The $\delta^{18}O$ values of the calcite spar cements from the Northamptonshire WLF are overlain on an equilibrium fractionation diagram in Figure 12. Once again the boundary conditions of maximum estimated burial temperature and estimated surface paleotemperature are included. Depositional pore fluid compositions calculated from the oyster and *Kallirhynchia* sp. analyses of Figure 7B range from +1 to +4‰ SMOW for 20°–25°C. These values reflect a generally more restricted setting than in Oxfordshire, in accordance with the greater distance of Northamptonshire from the open marine Wessex Basin. A value of +2 ±1‰ SMOW is therefore assumed for the average depositional pore fluids in the Northamptonshire region.

From Figure 12 it can be seen that:

(1) the Northamptonshire spars precipitated from a pore fluid of between –3 and –10‰ SMOW, and therefore

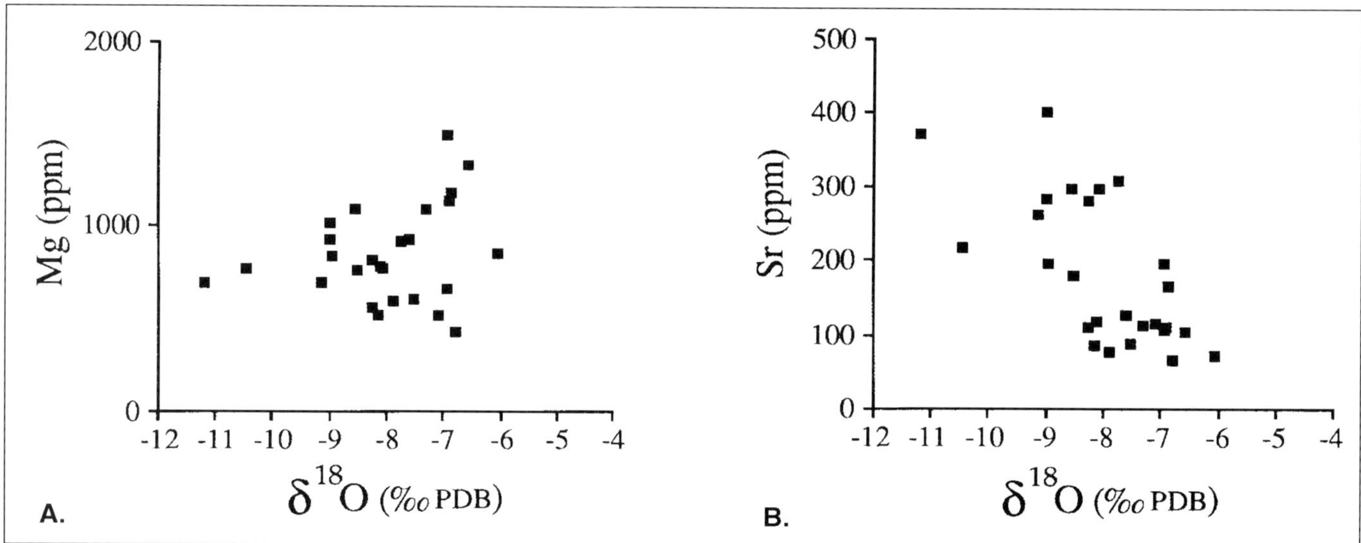

Figure 13. Scatter plots of: (A) $\delta^{18}O$ vs. Mg, and (B) $\delta^{18}O$ vs. Sr for the Northamptonshire spars.

isotopically negative waters must have entered the formation prior to or during spar precipitation,

(2) the most isotopically negative spars must have been derived from a pore fluid more depleted in ^{18}O than that which precipitated any of the Oxfordshire spars, and

(3) no single pore fluid composition can account for the full range of spar compositions.

Point (2) is a consequence of the spar compositions being more depleted in ^{18}O than in Oxfordshire, *and* having a shallower maximum burial depth (Figure 5). This is a critical observation; the similar paragenesis and nature of the spar cementation throughout the study area suggests that the same diagenetic paleoaquifer extended through Oxfordshire and Northamptonshire. The $\delta^{18}O$ value of the meteoric recharge must have been at least as low as –8‰ SMOW to account for the most negative Northamptonshire spars. However, it has already been shown that the most negative water composition which could have precipitated any of the Oxfordshire spars was only –6‰ SMOW. It follows that all of the spars in Oxfordshire must have been precipitated from a mixture of the meteoric water with marine fluids. The possibility, discussed above, that the Oxfordshire spars were precipitated from a more ^{18}O-enriched meteoric water (i.e. –3‰ SMOW) can effectively be ruled out. It would require that cementation occurred in a completely separate paleoaquifer from the Northamptonshire spars, and it is difficult to envisage how such a large difference in meteoric recharge compositions (–3 vs –8‰ SMOW) could have arisen in a relatively small area.

Figure 7B shows that the $\delta^{13}C$ range of the Northamptonshire spars extends to more negative values than in Oxfordshire, and to considerably more negative values than the $\delta^{13}C$ compositions of primary skeletal calcite. This is compatible both with a relatively smaller degree of water-rock interaction and/or with less mixing of marine fluids prior to precipitation of the Northamptonshire spars, as compared to those from Oxfordshire (cf. Lohmann, 1988).

Trace Elements

The interpretation of trace element data in terms of sources of cement components will not be detailed in this chapter. Instead, emphasis will be on using these results to refine the paleohydrology of the WLF. The wide spread of the analyses indicated by the large standard deviations in Table 2 partly reflects bulk sampling of sector and sector-specific intracrystalline zones and partly records regional variations in average compositions. Although ranges of Sr and Mg contents are comparable between the two areas, it is clear that the most ferroan and manganoan compositions characterize the Northamptonshire spars.

Pervasive sector zoning in the spars renders it impossible to calculate precise chemical compositions of the diagenetic fluids from the ICPAES results (Hendry and Marshall, 1991b). However, the low Mg ($\leq$ 2000 ppm) and Sr ($\leq$ 500 ppm) contents of the spars are typical of precipitates from fluids of low ionic strength such as meteoric water (e.g. James and Choquette, 1984; Lohmann, 1988; Saller and Moore, 1991). The elevated Fe and Mn contents of the spars (up to approximately 9300 ppm and 400 ppm respectively), together with their coarse, equant morphology, dull luminescence, sector zoning, and stable isotopic characteristics, are more typical of "burial" cements (e.g. Scholle and Halley, 1985; Choquette and James, 1987; Emery, 1987; Hurley and Lohmann, 1989). Nevertheless, the correspondence of the most ferroan and manganoan spars with the more argillaceous facies of the WLF in Northamptonshire suggests that Fe^{2+} and Mn^{2+} were mobilized from clays on a local scale. Both cations are readily soluble under reducing conditions and their abundance in the spars merely indicates

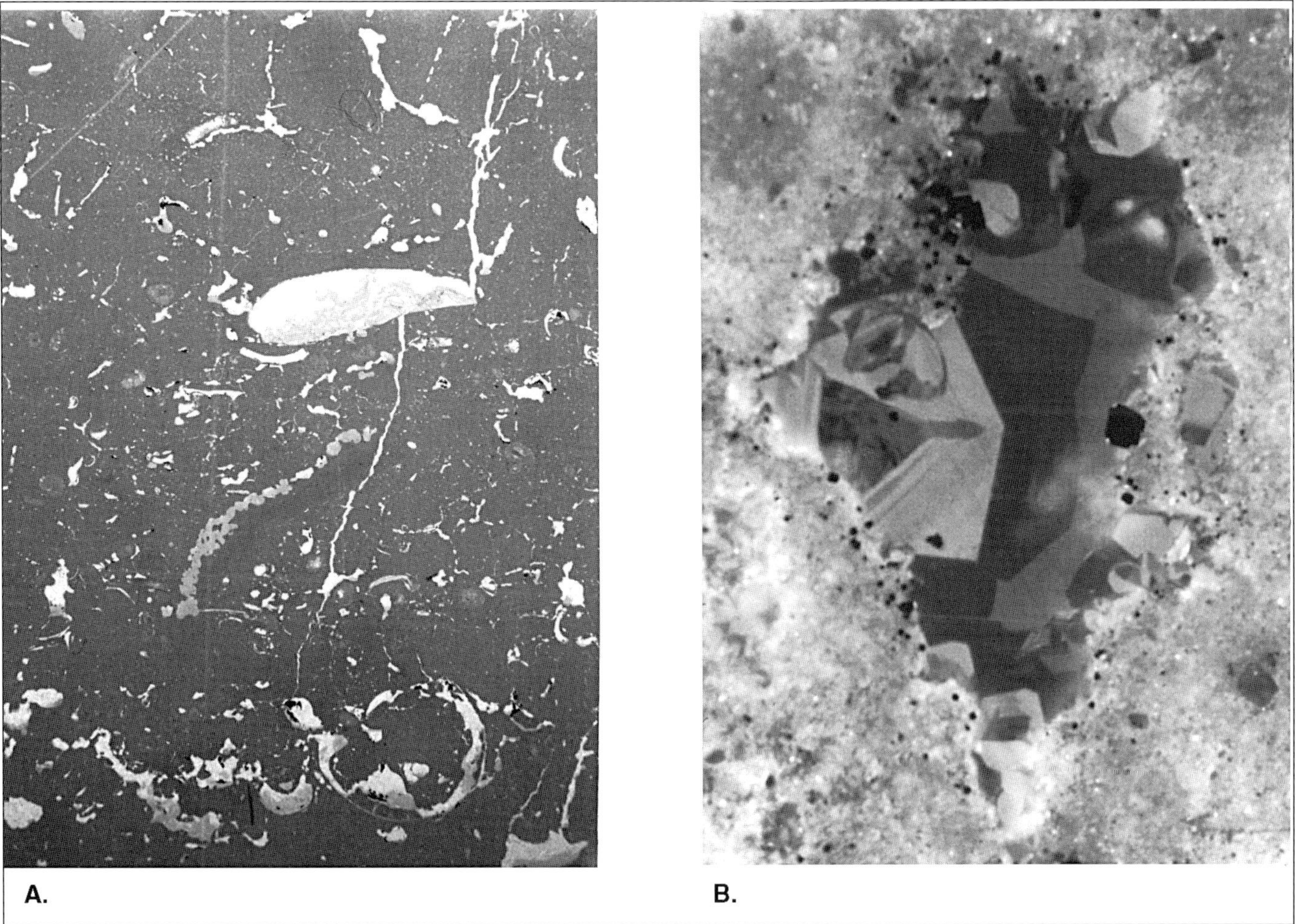

Figure 14. (A) Thin section photomicrograph of fenestral fabrics in an intraformational supratidal lime mudstone in the Oxfordshire WLF. Note the lack of evidence for aggressive dissolution, karstic features or vadose cements. Field of view = 1.8 cm. (B) Cathodoluminescence photomicrograph from a fenestral lime mudstone in the WLF of central Oxfordshire. Note the presence of pyrite lining the cavity, and its subsequent occlusion by sector-zoned dull luminescent spar. See text for discussion. Field of view = 1.2 mm.

that the pore fluids were anoxic during cementation.

The correlation of relatively high spar Mg contents and least-negative $\delta^{18}O$ compositions is poorly constrained in Northamptonshire, and elevated Sr contents characterize the cements most *depleted* in ^{18}O (Figure 13). This may reflect (1) a more complex Mg and Sr cation budget during cementation, or (2) less intense water-rock interaction than in Oxfordshire. In the latter case, the range of $\delta^{18}O$ values from the Northamptonshire spars must be a result of pore fluid mixing.

PALEOHYDROLOGY OF THE WHITE LIMESTONE FORMATION

Timing and Location of Meteoric Recharge in the WLF

The widespread occurrence and negative $\delta^{18}O$ values of the WLF spars show that a major meteoric paleoaquifer was established in the formation. Recognition of opportunities for meteoric recharge can provide constraints on timing and depth of spar cementation. This requires assessment of the tectonic, paleogeographic, climatic, and sedimentological history of the study area and surrounding regions.

Bathonian–Callovian

The regional development of the WLF spars mitigates against the involvement of a spatially-restricted synsedimentary meteoric lens such as would be developed beneath a temporarily emergent shoal or island (cf. Emery and Dickson, 1989). The lack of concentric zoning, an absence of any relationship of cementation textures to facies, the sheer volume of cement precipitated, and the persistence of the spar throughout all of the sampled vertical sequences, require that an active flow system channeled the diagenetic fluids through the entire thickness of the formation.

It is probable that little or no meteoric water was sourced from the peritidal horizons developed at the

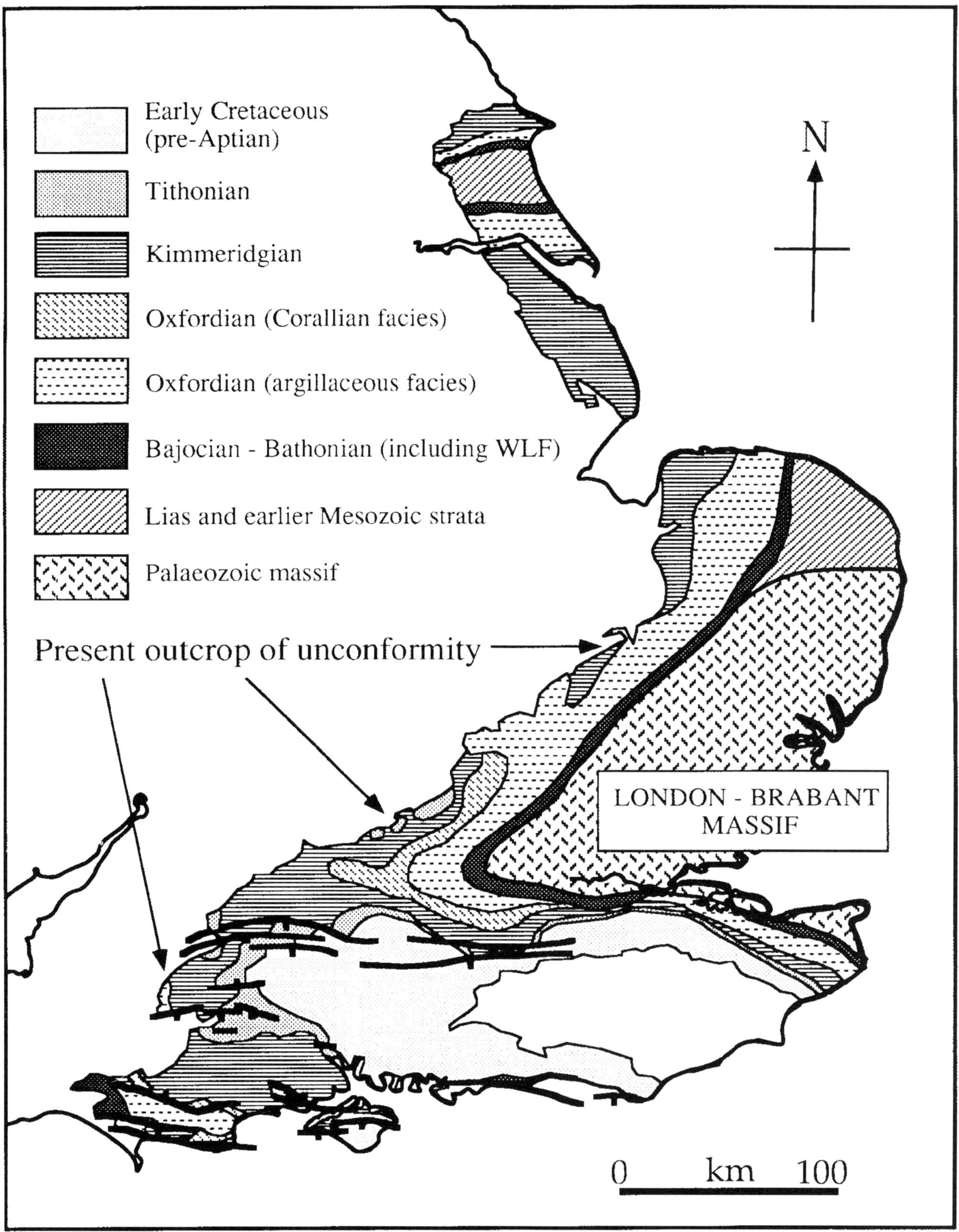

Figure 15. Geological map of subcrop beneath the late Cimmerian unconformity. Note the truncation of Bathonian strata onlapping the Paleozoic basement of the London-Brabant Massif. Stratigraphically higher Jurassic units were exposed considerably closer to the study area but are mainly clays, apart from isolated nearshore carbonates and sands in the Oxfordian (Corallian Formation). The unconformity is currently buried beneath Cretaceous and Tertiary strata. Adapted from Whittaker (1985).

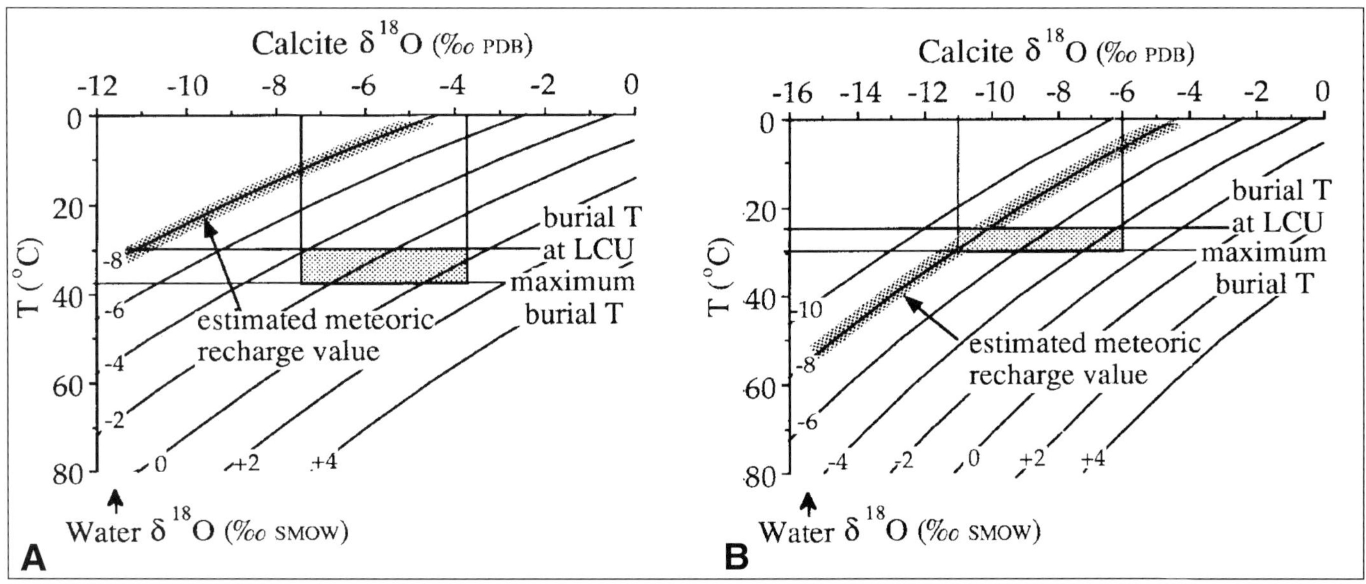

Figure 16. Modification of Figures 9 and 12 showing ranges of porefluid oxygen isotope compositions corresponding to precipitation of the WLF spars between burial depths at the late Cimmerian unconformity ("LCU") and maximum burial depths: (A) Oxfordshire, (B) Northamptonshire. The possible porefluid values are represented by the curves which intersect the stippled box in each case. The most ^{18}O-enriched possible Early Cretaceous meteoric water composition (-8‰ SMOW) is also highlighted for comparison (see text for discussion).

top of some of the sedimentary cycles within the formation. Although these represent temporary widespread emergence of the study area (Palmer and Jenkyns, 1975; Cripps, 1986), there was unlikely to have been significant freshwater recharge for two reasons. Firstly, there is no evidence for massive dissolution, karstic features or vadose zone diagenesis in the typically low permeability deposits. In contrast fenestral fabrics and shrinkage cracks in some of the micrite beds (Figure 14A) suggest desiccation in relatively arid conditions. Fenestrae contain pore-lining pyrite which suggests that they were filled with marine water prior to precipitation of the ubiquitous pore filling spar (Figure 14B). Secondly, although locally absent through contemporaneous erosion (Cripps, 1986), the peritidal horizons and overlying rootletted shales do not represent major long-lived unconformities, such as those associated with ancient meteoric recharge horizons (e.g. Longstaffe and Ayalon, 1987; Kaufman et al., 1988; Niemann and Read, 1988; Walkden and Williams, 1991). In a semi-arid climate and without associated sea level fall there would have been very little opportunity for meteoric diagenesis (Wright, 1991).

The uppermost WLF in the study area has been inferred by Cripps (1986) to represent two minor transgression-initiated shoaling cycles, both of which resulted in carbonate deposition over Oxfordshire but not significantly into Northamptonshire. Both cycles were terminated by coastal progradation, with supratidal carbonate marshes being developed in Oxfordshire whilst Northamptonshire remained emergent. Following the final episode of progradation, much of the study area experienced non-deposition and localized erosion; subaerial exposure may have endured for at least one ammonite zone (approximately 0.6 m.y.). No significant eustatic sea level fall or tectonic uplift was associated with this emergence (Cripps, 1986, and personal communication). Therefore, even though there may have been enough time for meteoric recharge prior to the next transgression, the low relief of the coastal plain and marsh environments would have provided a minimal hydrologic head for gravity drive of the water throughout the underlying strata. In addition, petrographic examination of the overlying Forest Marble Formation (Figure 3) reveals that it is cemented by very similar zoned calcite spars to those of the WLF.

The Forest Marble Formation is interpreted by Cripps (1986) as originating from deposition during a minor northeast directed transgression from the offshore Wessex shelf. This was followed by emergence and erosion of the uppermost part of the unit across much of the study area. The poor permeability of the clay rich Forest Marble facies in Northamptonshire, and the absence of a significant hydraulic head in the study area, both reduce the likelihood of major meteoric recharge at this time. Renewed transgression in the early Callovian inundated the study area. Marine conditions persisted through the Oxfordian and Kimmeridgian, allowing uninterrupted deposition of shales from the Wessex Basin across the study area and into the Anglo-Dutch Basin, precluding any possibility of meteoric recharge into the WLF.

Early Cretaceous

Shallowing associated with latest Jurassic eustatic sea level fall and initiation of tectonic uplift restricted the extent of uppermost Jurassic sedimentation and probably led to periodic emergence over the Central Midlands Platform. More importantly, continued regional shallowing combined with a major phase of uplift of pre-existing basement highs in the Valanginian–Barremian to cause exposure and erosion of the Middle and Upper Jurassic strata onlapping the flanks of the London-Brabant Massif (Chadwick, 1985b). Subsidence of the Wessex-Weald Basin and East Midlands Shelf continued during early stages of basement uplift, but slowed down in the Early Cretaceous. Consequently, paralic and continental sedimentation covered much of southern England by the Barremian (e.g. Sladen and Batten, 1984; Allen, 1975; Anderton et al., 1979).

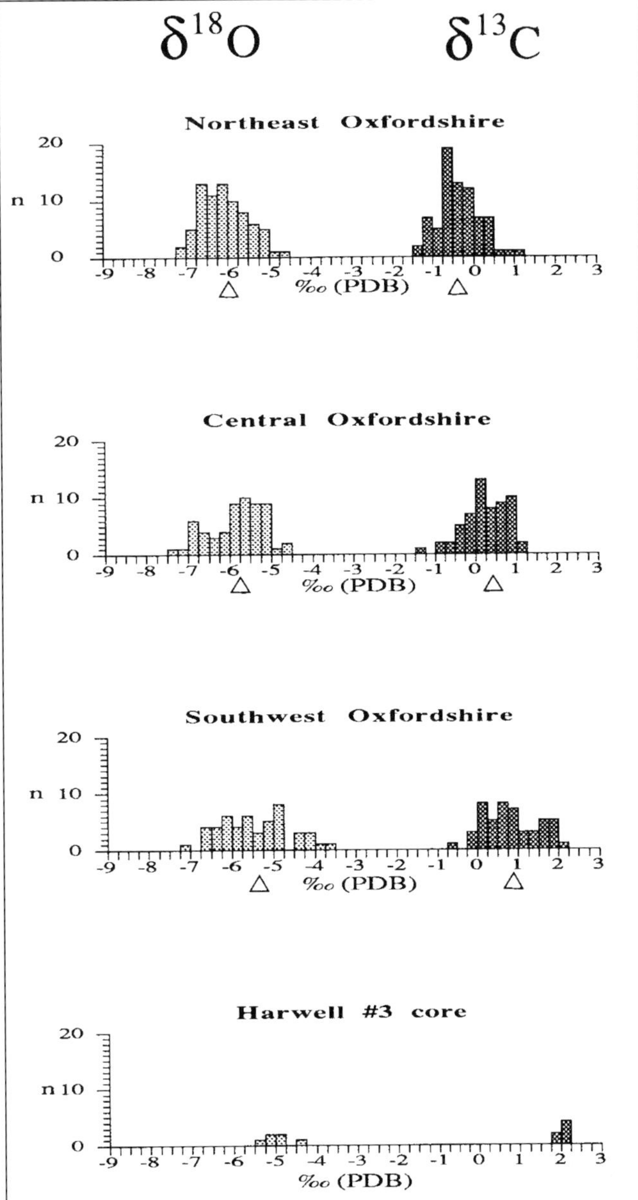

Figure 17. Histograms of $\delta^{18}O$ and $\delta^{13}C$ of the WLF spars in southwest, central and northeast Oxfordshire, and from Harwell borehole. Pointers indicate mean compositions for each area. Note that the spar analyses display unimodal normal distributions, but with a pronounced skew to more positive values in southwestern Oxfordshire. The Harwell borehole is situated in the northern part of the Wessex Basin; see Figures 1 and 2.

Early Cretaceous erosion in the study area itself was relatively minor and did not exhume Middle Jurassic deposits (Figure 5). However, an extensive compilation of subsurface data (Whittaker, 1985) has shown that Bathonian strata are truncated by the late Cimmerian unconformity precisely where they onlap the Paleozoic basement of the London-Brabant Massif, between about 50 and 80 km east of the study area (Figure 15). Meteoric recharge into the lateral equivalents to the WLF of the study area may therefore have occurred in the Early Cretaceous. The potential recharge zone, on the margins of the uplifted massif, would have been in a favorable position for receiving run-off from inland watersheds in addition to direct rainfall. The Bathonian strata are overstepped at the unconformity by Aptian deposits (Whittaker, 1985). This suggests that there might have been a maximum of 20 m.y. of exposure, assuming uplift at the end of the Tithonian.

Aptian–Recent

No further potential for channeling meteoric water through the WLF and for causing the extensive cementation is likely to have existed after the Aptian transgression, because much of central and southern England remained inundated by marine waters until regional Tertiary uplift. It is not possible to ascertain exactly when the Middle Jurassic limestone scarp finally became emergent, but opportunities probably existed for meteoric water infiltration from then until the present. However, the modern hydrology of the Middle Jurassic limestones in Central England is dominated by fissure flow in otherwise tightly cemented rocks (Morgan-Jones and Eggboro, 1981). The present day groundwater in the WLF at Harwell (the southernmost locality on Figure 2) has a $\delta^{18}O$ value of –7‰ SMOW (British Geological Survey, personal communication), which is not in isotopic equilibrium with the local spars. The active present day meteoric aquifer is limited in extent, and following Tertiary inversion of the London–Brabant Massif, no significant hydraulic head is likely to have existed to maintain an active regional paleoaquifer. Therefore the WLF spars appear unrelated to either the present day or earlier Tertiary hydrologies.

In summary the most favorable opportunity for regional scale meteoric water penetration into the WLF in the study area was during the early Lower

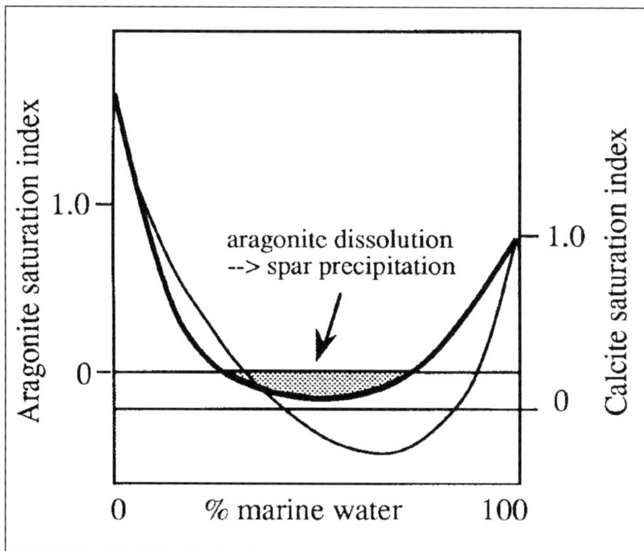

Figure 18. The effects of mixing supersaturated meteoric water and normal marine water (seawater) on net carbonate saturation levels. Heavy curve = hypothesized situation during precipitation of the WLF spars; light curve represents the case with a higher P_{CO_2} in the meteoric water component.

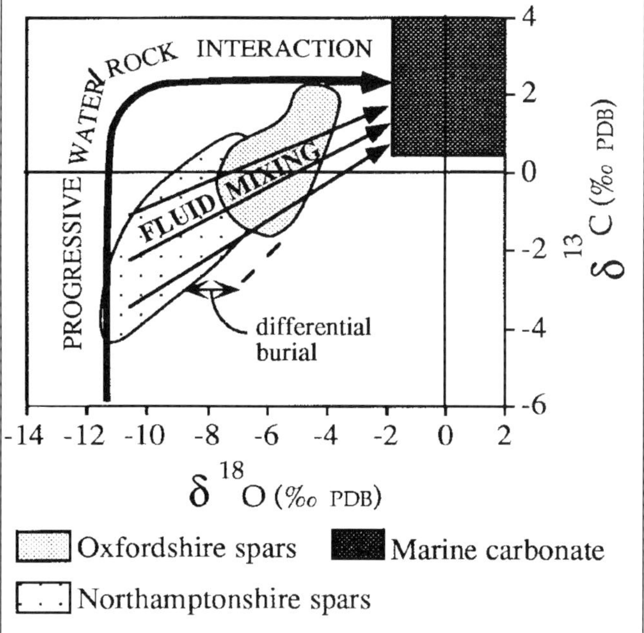

Figure 19. $\delta^{18}O$ vs. $\delta^{13}C$ cross-plot with showing ranges for the WLF spars from the whole study area. Arrows indicate trends of isotopic evolution in cements resulting from progressive water–rock interaction and from mixing of meteoric and basinal / interstitial (marine) waters. Offset of data fields for Northamptonshire and Oxfordshire represents differential burial of the paleoaquifer. See text for discussion.

Cretaceous, sourced from the late Cimmerian unconformity some 50–80 km distant. At this time the estimated burial depths of the WLF in Oxfordshire and Northamptonshire were approximately 260 m and 115 m, respectively (Figure 5).

Hydrologic Considerations

In a regional subsurface aquifer, the extent of gravity-driven meteoric water penetration depends upon the hydraulic head, the permeability of the strata, and the amount of interstitial water to be displaced. Meteoric water recharges an aquifer by infiltration and moves down a topographic gradient in the direction of maximum rate of decrease of gravitational potential energy. The fluid potential is proportional to the hydraulic head and also depends on the overburden pressure (see Freeze and Cherry, 1979). Mathematical modeling of simple isotropic systems discriminates local, intermediate and regional scales of flow (Toth, 1963). The former are dictated by variability in surface topographic relief whilst the latter are produced by a regional topographic slope with negligible local relief. The hydrology is more complex if variable permeability characteristics are incorporated into the models. Specifically, a high permeability horizontal layer at depth strongly promotes regional flow at the expense of local flow systems even in areas of pronounced surface topography (Freeze and Witherspoon, 1967). The effect is so important that if the permeability contrast is greater than two orders of magnitude, the hydrologic system can be approximated to horizontal flow in the aquifers and vertical seepage through the aquicludes (Freeze and Cherry, 1979; Bethke, 1989).

Middle–Upper Jurassic limestones outcropping on the flanks of the London–Brabant Massif in Early Cretaceous times would have supported regional groundwater flow systems because:

(1) They were sandwiched between considerably less permeable clays (Figure 3), very probably giving a permeability contrast of two or more orders of magnitude (cf. table 2.2 in Freeze and Cherry, 1979).

(2) There was probably an elevated recharge area on the uplifted flanks of the London-Brabant massif but negligible topography over the study area, which was covered by low-lying alluvial/coastal plain/shallow marine environments (cf. Anderton et al., 1979).

(3) The paleoaquifers were probably outcropping at the land surface and were dipping (at about 0.5°–1°) into the study area. The precise facies of the WLF on the flanks of the London-Brabant massif are not known, but even if the most permeable limestones were a short way downdip, vertical seepage through exposed overlying units will have focused meteoric waters into them (see above).

Under these circumstances the potential existed for

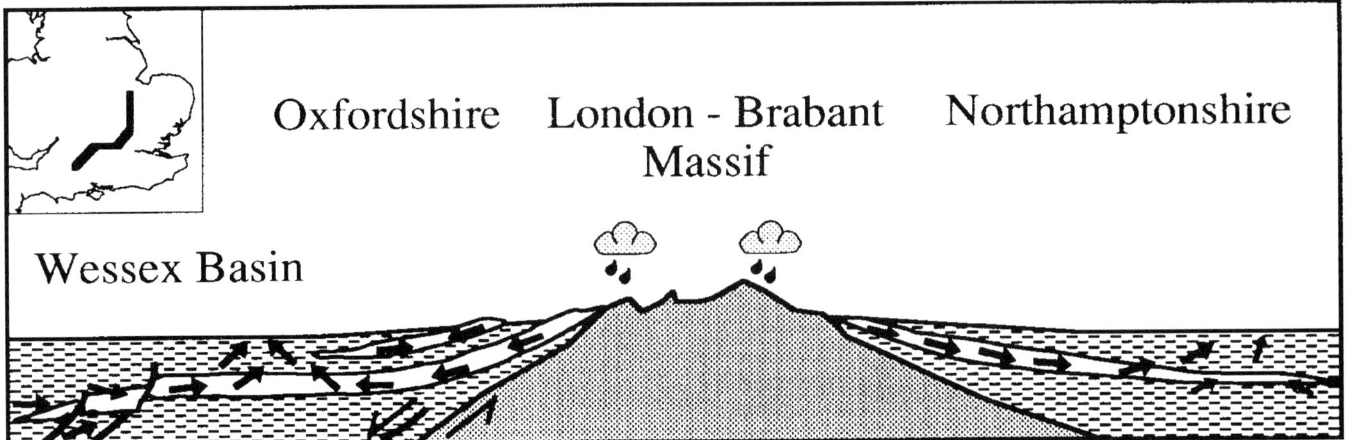

Figure 20. Schematic cross section of the study area showing Early Cretaceous paleohydrology of the Bathonian carbonates (which include the WLF) and the overlying Corallian Formation. Fluid movement through the surrounding shales (horizontal stippling) would have been by vertical seepage. Arrows indicate direction of fluid movement.

channeling meteoric groundwaters through the study area and out into the Wessex-Weald Basin, where calcite spars petrographically similar to those of the WLF occur in contiguous strata (Sellwood et al., 1989). Although these basinal cements were interpreted as having a burial origin, their published isotopic compositions are not incompatible with a significant component of meteoric water in the precipitating pore fluid.

Composition of the Diagenetic Pore Fluids

Now that the approximate timing of meteoric recharge into the WLF has been established, the regional variation in isotopic composition of pore fluids that precipitated the spars can be more tightly constrained. Equilibrium fractionation diagrams for the Oxfordshire and Northamptonshire spars are reproduced in Figures 16A and B with the minimum depth of precipitation corresponding to the lower Early Cretaceous burial depths in the respective areas. Figure 16A shows that the most negative pore fluid composition in the Oxfordshire area was about –4‰ SMOW, assuming precipitation was completed before maximum burial. The corresponding value in Northamptonshire (Figure 16B) is about –9‰ SMOW. Figure 16B also demonstrates that the meteoric recharge can not have been more enriched in ^{18}O than a $\delta^{18}O$ of –8‰ SMOW, to account for precipitation of the most negative spars within the maximum permissible burial depths.

Water-Rock Interaction

Figures 16A and B show that the full range of spar compositions in both parts of the study area could not have been derived from an invariant pore fluid. However, in Oxfordshire, the majority of the spars with $\delta^{18}O$ values more positive than –5‰ PDB are those sampled adjacent to a prominent and extensive shale bed in several quarries in the southwest of the area (Figure 8). The effect of this local variation on the overall distribution of spar compositions is emphasized in Figure 17 where the isotopic analyses from Oxfordshire are divided into four groups based on an east–west geographical subdivision of the area. The distribution of $\delta^{18}O$ values in southwest Oxfordshire is skewed towards ^{18}O-enriched compositions compared with both central and northeast Oxfordshire. The same skew can be seen for $\delta^{13}C$ values, but here it is superimposed upon a subtle regional trend in the mean values from southwest to northeast Oxfordshire. Bearing in mind that the recharge area lay to the east of the Oxfordshire outcrop, the regional isotopic trends are compatible with increasing water-rock interaction down-aquifer, and are particularly intense adjacent to the shale bed (cf. Lohmann, 1988). The reason for the pronounced shifts adjacent to the shale is uncertain, but it might have been involved in some compartmentalization of the aquifer and/or had considerably slower pore fluid flow adjacent to it. It is also possible that diffusion or compaction of relatively more ^{18}O-enriched waters from the shale locally influenced the overall pore fluid composition.

If the shale-related $\delta^{18}O$ analyses more positive than –5‰ PDB are put aside, Figure 16A shows that the majority of the Oxfordshire spars would have precipitated from a fluid of –2 to –4‰ SMOW during recharge from the late Cimmerian unconformity. Assuming that meteoric water entering the Northamptonshire and Oxfordshire WLF had an identical initial $\delta^{18}O$ value, at least as negative as –8‰ SMOW, considerable mixing of interstitial marine and/or basinal (shale-dewatering) fluids must have occurred *prior* to precipitation in Oxfordshire. Precipitation in Northamptonshire seems to have occurred

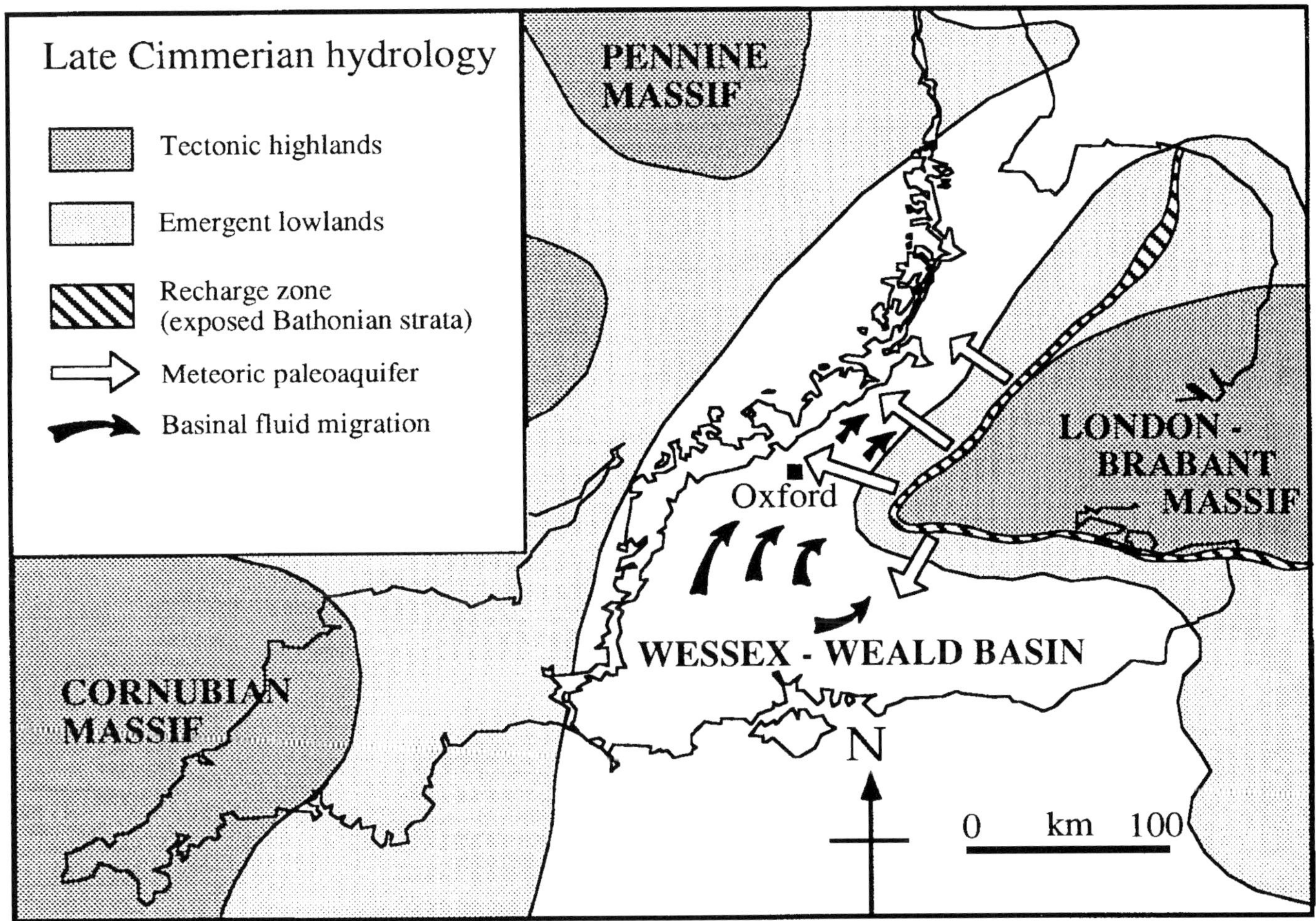

Figure 21. Plan view of WLF paleohydrology showing regional pore fluid flow directions during spar cementation throughout central England. Exposure and recharge of the Middle Jurassic limestones to the NW of the study area against the Pennine and Welsh Massifs cannot be evaluated owing to the lack of geological information, but remains a possibility. Paleogeography estimated from Allen (1975).

as the meteoric recharge mixed with, and finally displaced marine fluids out of the aquifer. The latest stage cements have the most depleted $\delta^{18}O$ values as well as negative $\delta^{13}C$ values, suggesting that meteoric fluids had flushed through the formation and that mineralogical stabilization had largely been completed. The slight depletion of ^{18}O values in the cements from the edge to center of some cavities in the Oxfordshire samples could likewise indicate a slightly greater meteoric component in the pore fluid with time, but might equally reflect decreased water-rock interaction as the metastable components were used up.

Pore Fluid Mixing And Controls On Cementation

Subsequent to uplift and subaerial exposure of the updip margin of an aquifer, some degree of mixing between the recharging meteoric waters and the interstitial waters is unavoidable as the latter are displaced and expelled through the overlying strata. The chemistry of the mixing zone determines whether or not precipitation will take place in it (Plummer, 1975; Wigley and Plummer, 1976). Changes in saturation during mixing reflect the fact that molalities and activities of individual ions are non-linear functions of end-member values, and result from the redistribution of carbonate species in response to changes in pH, P_{CO_2} and solution composition (Freeze and Cherry, 1979). P_{CO_2} and the ionic strength and saturation states of the end members are the most important inorganic controls on the saturation state of marine-meteoric water mixtures (e.g. Plummer, 1975; Budd, 1988). Plummer (1975) showed that if surface seawater is mixed with calcite-saturated groundwater at 25° C the proportion of sea water required to cause calcite undersaturation increases from $< 25\%$ at $P_{CO_2} = 10^{-4}$ atm to nearly all mixtures at $P_{CO_2} > 10^{-0.5}$ atm. However if groundwater is twofold saturated with respect to calcite, no mixtures are undersaturated at $P_{CO_2} < 10^{-1.8}$ atm although the overall saturation state is lowered with respect to the weighted average of the end members.

Modern shallow mixing zones in carbonates tend to be characterised by calcite dissolution and concomitant dolomitization (e.g. Smart and Whitaker, 1991), but these environments are commonly open to CO_2 exchange with the atmosphere. The WLF of the study area, buried beneath Upper Jurassic clays, will have been essentially closed with respect to CO_2. Any high P_{CO_2} values obtained during infiltration through the soil zone (James and Choquette, 1984) are likely to have been lowered through water-rock interaction in the upper, unconfined part of the aquifer. Additionally, dissolution of carbonate in the soil zone and updip portion of the paleoaquifer will have caused the water to rapidly exceed saturation with respect to calcite and probably approach saturation with respect to aragonite.

If the P_{CO_2} of the meteoric water component entering the Oxfordshire study area was relatively low, and it was supersaturated with respect to calcite from water-rock interaction updip, mixing with an interstitial marine fluid could have taken place without causing calcite undersaturation. Precipitation from a barely oversaturated mixed fluid would be compatible with the coarse, equant morphology of the spars and the lack of any petrographic evidence of skeletal calcite dissolution. However, it is quite likely that the mixing would have resulted in undersaturation with respect to *aragonite* (Figure 18). Regional aragonite dissolution occurred prior to or during calcite precipitation in the WLF, and the solubility contrast between aragonite and calcite provides an attractive driving force to sustain the single episode of spar precipitation at low supersaturations (James and Choquette, 1984; Budd, 1988). This removes the requirement for enormous throughputs of supersaturated meteoric water, as the extent of calcite production is only limited by either the amount of aragonite initially present or by kinetic barriers such as the inhibitory buildup of Mg^{2+} and Sr^{2+} in the pore fluid. Stabilization of mixed-mineralogy carbonate sediments cannot result in complete porosity occlusion, but in the buried WLF paleoaquifer additional carbonate may have been sourced from dissolution updip, as well as from local pressure dissolution seams within the formation. Primary porosity was probably also reduced by mechanical compaction, especially in micritic lithologies. However, the presence of unfilled porosity in relatively large voids (e.g. Figure 6A) probably indicates that cementation in the WLF was unable to continue once the aragonite had been exhausted.

Paleohydrologic Model for the White Limestone Formation

From the preceding discussions, a paleohydrologic model for spar cementation in the WLF can be proposed. The basis of the model is that precipitation of the spar was directly related to meteoric recharge at the late Cimmerian unconformity, into a confined aquifer incorporating the formation and bounded by Upper and Lower Jurassic shale aquitards. Figure 19 combines the stable isotopic data for spars over the whole of the study area. The large range of $\delta^{18}O$ and $\delta^{13}C$ values are considered to be the result of a combination of (1) mixing of meteoric and marine-like pore fluids prior to or during spar precipitation, and (2) progressive water-rock interaction altering the $\delta^{13}C$ of the dissolved bicarbonate towards marine values (Lohmann, 1988). Localised, intense water-rock interaction is probably recorded in the most isotopically positive spars. This model suggests that the $\delta^{18}O$ of the meteoric recharge can be represented by a meteoric calcite line of about –11‰ PDB, corresponding to a water composition of –8 to –9‰ SMOW. The $\delta^{13}C$ of the dissolved carbon in the meteoric water appears to have been roughly –1 to –4‰ PDB at the time of cementation. The amount of mixing was clearly greater in Oxfordshire than in Northamptonshire. This suggests that the formation was flushed with meteoric recharge more rapidly (relative to the interval of precipitation) in the northeastern area, and/or that Oxfordshire was near the limit of meteoric water penetration.

The degree to which meteoric recharge can flush interstitial pore fluids out of an aquifer depends upon its hydraulic head. In modern, confined coastal aquifers the fresh water potentiometric head can flush fresh waters tens to hundreds of kilometers offshore (Johnson, 1983). Chadwick (1985b) estimated from structural principles that over 300 m relief could have been generated on the London-Brabant Massif, coupled with an approximately 80–150 m eustatic drop in sea level from the Late Jurassic until the Aptian transgression. Uplift was thought to be at a maximum in the earliest Early Cretaceous, with a humid climate contributing to development of major active fluvial systems (Allen, 1975; Sladen and Batten, 1984). Even with a suitable hydraulic head, the necessity of expelling interstitial pore fluids through shales overlying the WLF will have served to increase the time for hydrodynamic equilibrium to be attained. It will also have promoted diffusive and dispersive mixing of interstitial and meteoric pore fluids.

Dorobek (1987) described Silurian meteoric cements in basinal limestones at least 150 km from inferred recharge in tectonically uplifted basin margin highlands, and buried beneath some 250 m of low permeability strata. He estimated that complete flushing of the aquifers would have taken no more than 3 m.y. The broadly similar scale of this situation to the WLF in earliest Cretaceous times suggests that late Cimmerian recharge would have been capable of rapidly extending through the paleoaquifer in the study area. However, the isotopic results show that the Oxfordshire WLF was not completely flushed with meteoric pore fluids by the end of the cementation event. Calculation of mixing proportions is hampered by the unknown total dissolved carbon contents of the meteoric and marine end-members (Lohmann, 1988), but estimation from Figure 19 suggests that as much as 50–75% of marine fluid was involved. This paradox might be explained by the relative proximity of Oxfordshire to the center of the Wessex Basin. Subsidence plots (Karner et al., 1987) indicate that the basin still would have been in the "immature" stage of hydrodynamic development in the Early Cretaceous (cf. Toth, 1980); its thick Jurassic marine shale sequence would still have been undergoing compactional dewatering. Cross-formational flow would have taken place channeling these fluids into the most permeable conduits to migrate towards the basin margins. Therefore, compactional waters of marine parentage could have been driven into the study area, and counteracted the penetration of gravity-driven meteoric recharge (cf. Hurst and Irwin, 1982). In fact some of the most positive spar $\delta^{18}O$ values are from Harwell borehole (Figure 17), significantly closer to the basin center than are the localities in the outcrop belt (Figure 2).

Dewatering of shales under a pressure head is most important during the first kilometer of burial, which is when almost half the available compactional water is released. Therefore the compactional pore fluid reservoir is limited and average flow rates are usually several orders of magnitude less than those of gravity-driven circulation (Bethke, 1986; Giles, 1987). Because of this, the downward limit of active meteoric circulation is usually thought to dictate the upper extent of the compactional flow regime (Galloway, 1984). However, active faulting on the margins of the Wessex-Weald Basin during late Cimmerian uplift might have provided episodic increases in basin dewatering, especially if the shales were overpressured (cf. Sibson et al., 1975). Compactional water would be channeled away from active faults through near-horizontal high permeability beds such as the WLF (Magara, 1976). Pulsed delivery of compactional fluids into the Oxfordshire WLF will have promoted mixing and therefore possibly acted as a trigger for aragonite dissolution and spar precipitation at an early stage of aquifer development. In contrast, the much greater distance of the Northamptonshire area from the basin margin would have removed it from the influence of the basinal waters, allowing more efficient meteoric flushing and gradual stabilization of the WLF.

A two-dimensional conceptual diagram of the late Cimmerian paleohydrology is presented in Figure 20 and a plan view shows the predicted directions of pore fluid movement in Figure 21. It is notable that the mixing of basinal and meteoric waters in Oxfordshire would be an oblique, rather than "head-on" system. This might account for the unusually thick mixing zone required from the small range of $\delta^{18}O$ values throughout much of the Oxfordshire area. Nelson and Read (1990) described a similar paleohydrologic reconstruction for regional late cementation in a Mississippian carbonate paleoaquifer. Ferroan spar cements were inferred to have been derived from gravity-driven meteoric fluids in updip portions of the paleoaquifer passing to basin-derived compactional fluids in downdip portions of the limestone some 200 km from the recharge area. Regional isotopic trends in the Mississippian cements are broadly similar to those in the WLF, with respect to the basin geometry.

Stratigraphic and Geographical Extent of the Paleohydrologic System

The WLF in Oxfordshire is immediately surrounded by lower and upper Bathonian carbonate strata (Figure 3), the whole package being sandwiched between shale aquitards. It is very likely that these adjacent carbonates were in hydrologic continuity with the WLF during spar cementation, and indeed a major generation of remarkably similar ferroan, sector zoned spar has been observed by the author in both. Late calcite spar cements in shallow marine carbonates of the Corallian Formation (Oxfordian, Upper Jurassic) in Oxfordshire also bear much petrographic and geochemical similarity to the WLF spars (cf. de Wet, 1987, phase 3 cements). The Corallian Formation spars have $\delta^{18}O$ compositions comparable with the most enriched WLF spars, but combined with negative $\delta^{13}C$ values. They are less abundant than the WLF spars, and post-date early aragonite dissolution and calcite cementation from shallow meteoric diagenesis. They occur in non fabric-selective vugs and fractures formed by prior dissolution of calcite.

The Corallian Formation is truncated and overstepped by lower Aptian deposits immediately to the east of the study area (Figure 15). It can therefore be assumed to have been recharged by the same late Cimmerian meteoric water, approximately −8‰ SMOW, that entered the WLF. The burial temperature of the Corallian Formation in the Early Cretaceous was about 26°C, suggesting precipitation of the Corallian spars from a pore fluid of about −1 to −2‰ SMOW. Negative $\delta^{13}C$ values of the spars suggest that the meteoric water component had undergone less water-rock interaction and contained a more isotopically depleted dissolved carbon reservoir than was the case in the WLF paleoaquifer. A diagram similar to Figure 19 can be constructed using the spar data in de Wet (1987), a temperature of 26°C, and an initial meteoric water $\delta^{18}O$ of −8‰ SMOW. If this is done, the mixing trend indicates about 40–50% meteoric pore fluid with a dissolved carbon $\delta^{13}C$ of about −6 to −10‰ PDB. Although this estimation is very rough (e.g. cementation might not have occurred until later Cretaceous burial), it suggests that mixing proportions were not dissimilar to those responsible for WLF cementation. The fact that antecedent calcite dissolution occurred in the Corallian paleoaquifer is probably a result of a higher P_{CO_2} in the meteoric waters. This could reflect reduced water-rock interaction in previously stabilized Corallian Formation carbonates, and greater soil development in their relatively closer and topographically lower recharge zone.

Bajocian carbonates in eastern England are also cemented by coarse, sector-zoned, ferroan calcite spars (Emery et al., 1988). Unlike the WLF and the Corallian Formation discussed above, these limestones were not exposed at the late Cimmerian unconformity, but were wholly enveloped by Middle Jurassic shales (Whitaker, 1985). By interpreting the $\delta^{18}O$ values of the spar cements in light of the subsidence and thermal history of the limestones, Emery et al. (1988) and Emery (1991) were able to show that cementation took place from burial fluids sourced either from adjacent shales and/or via faults from more distal subsurface reservoirs.

The paleohydrologic model constructed for the WLF necessitates cross-formational flow of meteoric water discharging from the limestone aquifers through Upper Jurassic shales. Hydrologic models (e.g. Freeze and Witherspoon, 1967) show that upward-directed discharge from subsurface aquifers can be spread over a wide area above and beyond the downdip pinchout (or mixing zone). Although shales are commonly thought of as aquicludes, a sufficient hydrologic head will eventually force fluids vertically through them (cf. Alexander et al., 1989). In this respect, additional evidence supporting the paleohydrologic model comes from several examples of late stage calcite cements in septarian concretions in the Oxfordian and Kimmeridgian shales of central England. These ferroan spars have negative $\delta^{18}O$ values which cannot be explained solely by burial, and seem to require a meteoric fluid input (Hudson, 1978; Astin and Scotchman, 1988). The most likely origin of this meteoric fluid is via cross-formational flow from underlying carbonate paleoaquifers.

In summary, available evidence from the WLF and surrounding formations strongly suggests that an Early Cretaceous paleohydrologic system extended from the margins of the London-Brabant Massif, and affected most of the Middle and Upper Jurassic strata in central England which were exposed at the late Cimmerian unconformity. Flow was concentrated in relatively thin carbonate and sandstone paleoaquifers such as the WLF, and vertical seepage was dispersed through the thicker argillaceous units. Mixing of meteoric waters with marine fluids expelled from compacting shales was most important adjacent to the Wessex Basin.

DISCUSSION

Unconformities and Cementation

Proximal marine facies associated with unconformities formed by basin margin tectonic uplift are particularly favorable for transmitting meteoric water into sedimentary basins (e.g. Grover and Read, 1983; Dutton and Land, 1985; Longstaffe and Ayalon, 1987; Niemann and Read, 1988; Walkden and Williams, 1991). In these situations the highest flow rates and fluxes will occur when permeable conduits, separated by aquitards, form aquifers dipping into the basin. Regional scale cementation in carbonate platforms can also occur from temporary widespread emergence in the absence of major tectonic uplift (e.g. Meyers and Lohmann, 1985; Kaufman et al., 1988; Horbury and Adams, 1989). In contrast, synsedimentary emergence of the WLF over much of the study area was not accompanied by meteoric cementation. This highlights the importance of understanding the different types of unconformity, and their genesis, in predicting cementation patterns (Wright, 1991). Regional cementation in the examples cited above was associated with significant eustatic sea level fall, providing prolonged exposure, an elevation head for deep penetration of surface meteoric waters, extensive karstification and cement production. However, the intraformational unconformities in the WLF were relatively short-lived and formed as a result of shallowing and progradation of coastal plain sediments from surrounding land masses *without* significant eustatic sea level fall. Consequently no hydraulic head existed for large scale inundation of the underlying sediment with meteoric waters. Emergence of the WLF took place in a semi-arid climate, but even with high rainfall rates cementation would have been minor without a hydraulic head. Comparison can be drawn with cycle-top peritidal carbonates in the Ordovician of the Appalachians. Grover and Read (1983) noted that these low permeability beds showed some evidence of subaerial exposure but relatively little vadose zone dissolution and no unequivocally meteoric cementation. They suggested that the sediments would have been waterlogged with marine to brackish waters for much of the time, and that active meteoric lenses would not have had a chance to develop. The presence of pyrite in the WLF examples (e.g. Figure 14B) suggests a similar scenario.

Development of Paleoaquifers and Controls on Cementation

Regional scale cementation from unconformity-sourced meteoric fluids in a number of North American Paleozoic platform carbonates exhibits distinctive compositional concentric zonation sequences (Dorobek, 1987; Niemann and Read, 1988). Mapping of these zonation patterns has allowed individual paleoaquifers to be identified and cements to be timed relative to specific unconformity recharge events (e.g., Kaufman et al., 1988). When meteoric waters flush into an aquifer a redox front is established at a depth where dissolved oxygen has been removed through oxidation of organic matter (e.g., Bishop and Lloyd, 1990). As Eh falls with distance down-aquifer the oxides and hydrated oxides of Mn become unstable before those of Fe (Barnaby and Rimstidt, 1989). Cement co-precipitates at successive positions down-aquifer may show a corresponding progression from no Mn and Fe (nonluminescent), to Mn but no Fe (brightly luminescent), to Mn and Fe (dully luminescent). The extent of redox interface migration into a confined aquifer depends upon the vigor and duration of flow. Topographic elevation of the recharge area, permeability and composition of the aquifer, duration of exposure, and the amount and seasonality of rainfall can all have major effects on the extent of oxidizing meteoric water penetration.

The maximum distance of the WLF study areas from their tectonically uplifted recharge zone was no more than 80 km, yet no early non-luminescent concentric zones occur in the WLF spars. The only bright luminescent early cement is related to sulfate reduction and pyrite formation and pre-dates the main episode of spar cementation (Hendry, 1990). Spar cementation therefore took place under fully reducing conditions. In contrast, Dorobek (1987) documented cements produced in oxidizing conditions over 100 km away from the tectonically uplifted recharge zone of Silurian paleoaquifers, and many other Paleozoic paleoaquifers contain cements precipitated from oxidizing pore fluids over comparably large distances (e.g. Grover and Read, 1983; Mussman et al., 1988; Niemann and Read, 1988). WLF spar petrography therefore suggests either that meteoric water penetration was slow and/or limited in extent, or that precipitation occurred at an early stage of aquifer evolution. As discussed above, the Early Cretaceous climate in Britain was probably relatively humid and the hydrologic head generated by uplift of the London-Brabant Massif should have been sufficient to generate a vigorous aquifer flow. Therefore the second option is preferred and is supported by the stable isotopic data which indicate cementation during mixing with and displacement of the interstitial pore fluids.

The stark contrast of the WLF cements with the Paleozoic examples cited above is most probably related to the different mineralogical compositions of the host sediments. The abundance and reactivity of metastable components will affect the width of the zone of mineralogical stabilization migrating down the aquifer, and thereby the extent to which the temporal and spatial isotopic and chemical trends are developed during precipitation. The Paleozoic carbonates were largely calcitic, and therefore precipitation would be expected to have been relatively slow, continuing as the organic matter in the host sediment was oxidized and the redox front migrated a long way downdip. Prolonged water-rock interaction during precipitation would have allowed a large range of evolving isotopic compositions and Eh conditions to be recorded in successive zones of pore filling cements. In the WLF, precipitation was driven by aragonite inversion. This rapid transformation (cf. Budd, 1988) would have happened at an early stage of aquifer development and in a single, relatively brief episode. Consequently only spatial water-rock interaction trends were recorded and it is likely that the redox front had not progressed far into the confined aquifer by this stage. At the other extreme, if mineral transformations are very slow such as in silicate diagenesis, the cementation front may not even be able to keep pace with erosion of the aquifer outcrop (cf. Bjørkum et al., this volume).

Implications for Porosity Prediction

Exposure and subaerial diagenesis of limestones can have varied and dramatic effects upon their porosity evolution both in the immediate locus of emergence and in lateral counterparts which remain buried. Recognizing the controls on cementation from different types of unconformities can be important for optimizing exploration strategies in carbonate successions (Wright, 1991; Esteban, 1991). As shown in this study, this requires an understanding of regional tectonics, paleoclimate, and basin development. Emergence of limestones is commonly assumed to result in an increase in porosity through karstification. Whilst this may be true in the immediate vicinity of the exposure unconformity, the WLF shows that pore-occluding cementation on a regional scale may be the result of localized emergence. Several important factors need to be considered when predicting the extent of unconformity-related cementation in carbonate paleoaquifers:

(1) the original composition of the carbonate (a function of its age and sedimentary facies);
(2) local tectonics and eustatic sea level histories during deposition of the carbonate;
(3) the burial and uplift history of the carbonate both in the basin center *and* on its margins;
(4) the regional geometry of the carbonate unit and its stratigraphic relationships with surrounding strata; and
(5) paleoclimates and paleogeographies during deposition of the carbonate and at times of subsequent unconformities at which it was exhumed or brought near to the surface.

The first two factors effectively dictate the proportion of aragonite to calcite in the paleoaquifer, as well as its initial porosity and permeability. For example, oolitic grainstones can be important reservoir rocks but their diagenetic potential is strongly influenced by the original mineralogy of the ooids, which has been recognized to change through geological time (Sandberg, 1983). Synsedimentary subaerial exposure of carbonate shoals or platform margins *as a result of a relative sea level fall* can lead to localized aragonite dissolution and calcite cementation if the climate is sufficiently humid. Again, this reduces the diagenetic potential of the rocks. The importance of (3) and (4) is in controlling the scale and nature of the paleoaquifer. In unconfined aquifers (for example where sea level fall has exposed large carbonate platforms) meteoric recharge is dispersed over a wide area. Flow patterns are dictated by local relief; net flow rates and consequent diagenetic alteration are generally lower than in confined aquifers such as the WLF where fluids are focused into the most permeable strata. The extent and rate of freshwater penetration down confined aquifers depends on the hydraulic head. This is controlled by differential burial of the aquifer, effective runoff (a function of net rainfall and local topography) and interaction with upward-migrating fluids expelled from thick basinal mud deposits. The duration of exposure is also important; unconformities produced from tectonic uplift of basin margins are likely to have recharged carbonate paleoaquifers for much longer than unconformities resulting from small-scale sea level changes (Wright, 1991).

Once meteoric waters enter an immature carbonate the flow initially disperses through intergranular porosity. Stabilization of any aragonite produces calcite cement in this primary porosity but also generates new secondary (moldic) porosity. Undersaturated waters recharging through the soil zone can continue to dissolve calcite and the resulting saturated pore fluids will flow down the hydraulic gradient away from the recharge zone. Once the carbonate they are flowing through is mineralogically stable no further cementation will take place until the waters encounter more metastable grains, or sufficient over saturation is achieved through (i) CO_2 degassing, (ii) temperature increase, or (iii) input of carbonate from pressure dissolution (Smart and Whittaker, 1991). In a confined aquifer the influence of (i) will be small, and at moderately shallow burial depths (ii) and (iii) are unlikely to cause major cementation. Thus Mussman et al. (1988) noted that meteoric cementation in the unconformity-sourced Mississippian Knox paleoaquifer was spatially restricted to where aragonite components of the limestone had been leached. It follows that stabilization of a mixed-mineralogy carbonate by early diagenesis reduces the potential for massive pore filling cementation in any subsequent meteoric paleoaquifer. For example, oolitic and bioclastic grainstones of the Great Oolite Group in the Weald Basin (south of the London-Brabant Massif) are lateral equivalents of the WLF but contain important hydrocarbon reservoir units in which primary porosity was not occluded by sparry calcite cement. These units also contain evidence of stabilization during synsedimentary meteoric and mixing zone diagenesis (Sellwood et al., 1987; McLimans and Videtich, 1989). It seems likely that the local early diagenetic removal of aragonite inhibited pore occlusion by later calcite spars.

It is still possible for regional cementation to take place in aquifers that are wholly calcitic, but a huge throughput of carbonate-charged water is required. In turn this necessitates a prolonged episode of exposure and an aerially extensive and well-vegetated recharge zone from which carbonate can be dissolved. For example, extensive cementation of Mississippian carbonates in eastern England took place in a regional paleoaquifer, following late Mississippian–Pennsylvanian tectonic uplift of the basin margins and concomitant eustatic sea level fall (Walkden and Williams, 1991). These sediments had already been stabilized by repeated synsedimentary emergence, but had not experienced significant pressure dissolution. Precipitation of the paleoaquifer cements was estimated as requiring an external supply of 500 km^3 of carbonate from karstification of the recharge areas, transported in up to 10^7 km^3 of water. Recharge into the Dinantian carbonates may have lasted for up to 35 m.y., and prolonged cementation during progressive burial of the aquifer is suggested by the wide range in cement $\delta^{18}O$ values and the presence of compositional concentric zonation. This can be contrasted with the WLF where aragonite was preserved during the synsedimentary emergence episodes. It was then available for redistribution in a relatively rapid episode of cementation as the Lower Cretaceous paleoaquifer was initiated.

CONCLUSIONS

Much of the Middle and Upper Jurassic sedimentary sequence in central England formed part of a basin-scale meteoric paleohydrologic system, recharged by marginal uplift and exposure at the late Cimmerian unconformity. Meteoric fluids were focused along the most permeable conduits, including the WLF paleoaquifer. Recharge was facilitated by outcrop of the paleoaquifer adjacent to a long-lived upland massif with well-developed drainage systems during predominantly humid climatic conditions. In contrast, Bajocian carbonates in eastern England were not exposed at this unconformity and experienced a different hydrologic regime more akin to their structural and stratigraphic setting (Emery, 1991). The paleohydrology of sedimentary units and the development of their host basins are intimately related and it is therefore important to understand diagenesis in the context of tectonic evolution. Porosity prediction in ancient carbonates requires a thorough assessment of their sedimentological facies and paleoenvironment, but also of the subsequent tectonic, stratigraphic and climatic history of the basin as a whole.

ACKNOWLEDGMENTS

This study forms part of the author's PhD research, undertaken at Liverpool University and supported by NERC. Liverpool University stable isotope laboratory is funded by NERC grants to J. D. Marshall. I am grateful to Jim Marshall for much useful discussion and guidance during the course of the research. Hilary Attenborough, Alison Warren, Walter Ritchie and Barry Fulton provided helpful technical back-up during the analytical work and preparation of the manuscript. Thanks also to Tony Dickson for kindly supplying the calcite ICPES standards. Careful reviews of the manuscript by Tony Adams, Dominic Emery, Andrew Horbury and Andrew Robinson proved most helpful and undoubtedly improved the quality of the finished product.

REFERENCES CITED

Alexander, J., J. H. Black and M. A Brightman, 1989, The role of low-permeability rocks in regional flow, *in* J. C. Goff and B. P. J. Williams, eds., Fluid Flow in Sedimentary Basins and Aquifers: Geological Society of London Special Publication 34, Oxford, Blackwell Scientific, p. 173-183.

Allan, J. R. and R. K Matthews, 1982, Isotope signatures associated with early meteoric diagenesis: Sedimentology, v. 29, p. 797-817.

Allen, P., 1975, Wealden of the Weald: a new model: Proceedings of the Geologists' Association, v. 86, p. 389-437.

Anderson, T. F. and M. A. Arthur , 1983, Stable isotopes of oxygen and carbon and their application to sedimentologic and palaeoenvironmental problems, *in* M. A. Arthur, T. F. Anderson, I. R. Kaplan, J. Veizer. and L.S. Land, eds., Stable Isotopes in Sedimentary Geology: Society of Economic Palaeontologists and Mineralogists Short Course 10, p. I.1-I.151.

Anderton, R., P.H. Bridges, M.R. Leeder and B.W. Sellwood, 1979, A Dynamic Stratigraphy of the British Isles: London, George, Allen and Unwin Ltd, 301p.

Andrews-Speed, C. P., E. R. Oxburgh and B. A. Cooper, 1984, Temperatures and depth-dependent heat flow in the Western North Sea: AAPG Bulletin, v. 68, p. 1764-1781.

Aplin, A. C., E. A. Warren and S. M. Grant, 1992, Mechanisms of quartz cementation in North Sea reservoir sands: constraints from fluid compositions: This volume.

Ashton, M., 1977, Stratigraphy and Carbonate Environments of the Lincolnshire Limestone Formation, Eastern England: PhD thesis, University of Hull, U.K.

Astin, T. R. and I. C. Scotchman, 1988, The diagenetic history of some septarian concretions from the Kimmeridge Clay, England: Sedimentology, v. 35, p. 349-368.

Banner J. L. and G. N. Hanson, 1990, Calculation of simultaneous isotopic and trace element variations during water-rock interaction with application to carbonate diagenesis: Geochimica et Cosmochimica Acta, v. 54, p. 3123-3137.

Barnaby, R. J. and J. D. Rimstidt, 1989, Redox conditions of calcite cementation interpreted from Mn and Fe contents of authigenic calcites: GSA Bulletin, v. 101, p. 795-804.

Bethke, C. M., 1986, Hydrologic constraints on the genesis of the Upper Mississippi Valley mineral district from Illinois Basin brines: Economic Geology, v. 81, p. 233-249.

Bethke, C. M., 1989, Modeling subsurface flow in sedimentary basins: Geologische Rundschau, v. 78, p. 129-154.

Bishop, P. K. and J. W. Lloyd, 1990, Chemical and isotopic evidence for hydrogeochemical processes occurring in the Lincolnshire Limestone: Journal of Hydrology, v. 121, p. 293-320.

Bjørkum, P. A., R. Knarud, and M. Bergen, 1993, How important is the late Cimmerian unconformity in controlling the formation of kaolinite in sandstones of the North Sea?–with examples from the Snorre Field: This volume.

Budd, D. A., 1988, Aragonite-to-calcite transformation during fresh-water diagenesis of carbonates: Insights from pore-water chemistry: GSA Bulletin, v. 100, p. 1260-1270.

Chadwick, R. A., 1985a, Permian, Mesozoic and Cenozoic structural evolution of England and Wales in relation to the principles of extension and inversion tectonics, *in* A. Whittaker, ed., Atlas of Onshore Sedimentary Basins in England and Wales: Post Carboniferous Tectonics and Stratigraphy: Glasgow, Blackie, p. 9-25.

Chadwick, R. A., 1985b, End Jurassic-Early Cretaceous sedimentation and subsidence, late Portlandian to Barremian, and the late Cimmerian unconformity, *in* A. Whittaker ed., Atlas of Onshore Sedimentary Basins in England and Wales: Post Carboniferous Tectonics and Stratigraphy: Glasgow, Blackie, p. 52-56.

Chadwick, R. A., 1985c, Cenozoic sedimentation, subsidence and tectonic inversion,*in* A. Whittaker ed., Atlas of Onshore Sedimentary Basins in England and Wales: Post Carboniferous Tectonics and Stratigraphy: Glasgow, Blackie, p. 61-63.

Choquette, P. W. and N. P. James, 1987, Diagenesis in limestones 3 –the deep burial environment: Geoscience Canada, v. 14, p. 3-35.

Cope, J. C. W., K. L. Duff, C. F. Parsons, H. S. Torrens, W. A. Wimbledon and J. K. Wright, 1980, A correlation of the Jurassic rocks in the British Isles Part Two: Middle and Upper Jurassic: Geological Society of London Special Report 15, 109 p.

Cripps, D., 1986, A Facies Analysis of the Upper Bathonian of Eastern England: PhD thesis, University of Aston, U.K.

de Wet, C. B., 1987, Deposition and diagenesis in an extensional basin: the Corallian Formation, Jurassic, near Oxford, England, *in* J. D. Marshall, ed., Diagenesis of Sedimentary Sequences: Geological Society of London Special Publication 36, Oxford, Blackwell Scientific, p. 339-354.

Dickson, J. A. D., 1991, Disequilibrium carbon and oxygen isotope variations in natural calcite: Nature, v. 353, p. 842-844.

Dorobek, S. L., 1987, Petrography, geochemistry, and origin of burial diagenetic facies, Siluro-Devonian Helderberg group, carbonate rocks, central Appalachians: AAPG Bulletin, v. 71, p. 492-514.

Dutton, S. P. and L. S. Land, 1985, Meteoric burial diagenesis of Pennsylvanian arkosic sandstones, Southwestern Anadarko Basin, Texas: AAPG Bulletin, v. 69, p. 22-38.

Emery, D., 1987, Trace-element source and mobility during limestone burial diagenesis–an example from the Middle Jurassic of eastern England *in* J. D. Marshall, ed., Diagenesis of Sedimentary Sequences: Geological Society of London Special Publication 36, Oxford, Blackwell Scientific, p. 201-217.

Emery, D., 1991, Patterns and rates of fluid flow during burial diagenesis of the Middle Jurassic Lincolnshire Limestone, eastern England: Marine and Petroleum Geology, v.8, p.442-451.

Emery, D. and J. A. D. Dickson, 1989, A syndepositional meteoric phreatic lens in the Middle Jurassic Lincolnshire Limestone, England, U.K: Sedimentary Geology, v. 65, p. 273-284.

Emery, D., J. D. Hudson, J. D. Marshall, and J. A. D. Dickson, 1988, The origin of late spar cements in the Lincolnshire Limestone, Jurassic of central England: Journal of the Geological Society of London, v. 145, p. 621-633.

Esteban, M., 1991, Palaeokarst: practical applications, *in* V. P. Wright, M. Esteban and P. L. Smart, eds, Palaeokarsts and Palaeokarstic Reservoirs: Reading, P.R.I.S. Occ. Publ. Series, 2, p. 89-119.

Freeze, R. A. and J. A. Cherry, 1979, Groundwater: New Jersey, Prentice-Hall Inc. 640 p.

Freeze, R. A. and P. A. Witherspoon, 1967, Theoretical analysis of regional groundwater flow: 2. Effect of water table configuration and subsurface permeability variation: Water Resources Research, v. 3, p. 623-634.

Friedman, I. and J. R. O'Neil, 1977, Compilation of stable isotope fractionation factors of geochemical interest, *in*. M. Fleischer, ed., Data of geochemistry: Geological Survey Professional Paper 440-KK, p. KK1-KK12 + Figures.

Galloway, W. E., 1984, Hydrologic regimes of sandstone diagenesis, *in* R. A. McDonald and R. C. Surdam eds., Clastic Diagenesis: AAPG Memoir 37, p 3-13.

Giles, M. R., 1987, Mass transfer and problems of secondary porosity creation in deeply buried hydrocarbon reservoirs: Marine and Petroleum Geology, v. 4, p. 188-204.

Grover, G. and J. F. Read, 1983, Paleoaquifer and deep burial related cements defined by regional cathodoluminescent patterns, Middle Ordovician carbonates, Virginia: AAPG Bulletin, v. 67, p. 1275-1303.

Haq, B. U., J. Hardenbol and P. R. Vail, 1987, Chronology of fluctuating sea levels since the Triassic: Science, v. 235, p. 1156-1167.

Hendry, J. P., 1990, Diagenetic studies of Bathonian carbonates from central and eastern England. PhD. Thesis, University of Liverpool, U.K.

Hendry, J. P. and J. D. Marshall, 1991a, Interpretation of complex non-concentric zoning fabrics in sparry calcite cements: Middle Jurassic limestones, central England: Sedimentary Geology, v. 75, p. 39-55.

Hendry, J. P. and J. D. Marshall, 1991b, Disequilibrium trace element partitioning in Jurassic sparry calcite cements: implications for crystal growth mechanisms during diagenesis: Journal of the Geological Society of London, v. 148, p. 835-848.

Horbury, A. D. and A. E. Adams, 1989, Meteoric phreatic diagenesis in cyclic late Dinantian carbonates, northwest England: Sedimentary Geology, v. 65, p. 319-344.

Hudson, J. D., 1978, Concretions, isotopes and the diagenetic history of the Oxford Clay, Jurassic, of central England: Sedimentology, v. 25, p. 339-369.

Hurley, N. F. and K. C. Lohmann, 1989, Diagenesis of Devonian reefal carbonates in the Oscar Range, Canning Basin, Western Australia: Journal of Sedimentary Petrology, v. 59, p. 127-146.

Hurst, A. and H. Irwin, 1982, Geological modelling of clay diagenesis in sandstones: Clay Minerals, v. 17, p. 3-22.

James, N. P. and P. W. Choquette, 1984, Diagenesis in limestones 9–the meteoric diagenetic environment: Geoscience Canada, v. 11, p. 161-194.

Johnson, R. H., 1983, The saltwater-freshwater interface in the Tertiary limestone aquifer, southeast Atlantic outer continental shelf of the USA: Journal of Hydrology, v. 61, p. 239-249.

Karner, G. D., S. D. Lake and J. F. Dewey, 1987, The thermal and mechanical development of the Wessex Basin, southern England *in* M. P. Coward, J. F. Dewey and P. L. Hancock, eds., Continental Extension Tectonics: Geological Society of London Special Publication 28, Oxford, Blackwell Scientific, p. 517-536.

Kaufman, J., H. S. Cander, L. D. Daniels and W. J. Meyers, 1988, Calcite cement stratigraphy and cementation history of the Burlington- Keokuk Formation, Mississippian, Illinois and Missouri. Journal of Sedimentary Petrology, v. 58, p. 312-326.

Lake, S. D. and G. D. Karner, 1987, The structure and evolution of the Wessex Basin, southern England:an example of inversion tectonics: Tectonophysics, v. 137, p. 347-378.

Lohmann, K. C., 1988, Geochemical patterns of meteoric diagenetic systems and their application to studies of palaeokarst, *in* N. P. James and P. W. Choquette, eds., Palaeokarst: Berlin, Springer-Verlag p. 58-80.

Longstaffe, F. J. and Ayalon, A., 1987, Oxygen-isotope studies of clastic diagenesis in the Lower Cretaceous Viking Formation, Alberta: implications for the role of meteoric water *in* J. D. Marshall, ed., Diagenesis of Sedimentary Sequences: Geological Society of London Special Publication 36, Oxford, Blackwell Scientific Publications, p. 277-296.

Magara, K., 1976, Water expulsion from clastic sediments during compaction–directions and volumes: Bulletin of the American Association of Petroleum Geologists, v. 60, p. 543-553.

McLimans, R. K. and P. E. Videtich, 1989, Diagenesis and burial history of Great Oolite Limestone, Southern England: AAPG Bulletin, v.73, p. 1195-1205.

Meyers, W. J., 1989, Trace element and isotope geochemistry of zoned calcite cements, Lake Valley Formation, Mississippian, New Mexico: insights from water-rock interaction modelling: Sedimentary Geology, v. 65, p. 355-370.

Meyers, W. J. and K. C. Lohmann 1985, Isotope geochemistry of regionally extensive calcite cement zones and marine components in Mississippian limestones, New Mexico *in* N. Schneiderman and P. M. Harris, eds., Carbonate Cements. SEPM Special Publication 36, p. 223-240.

Miller, J., 1988, Microscopical techniques I. Slices, slides, stains and peels *in* M. E. Tucker, ed., Techniques in Sedimentology: Oxford, Blackwell Scientific, p 86-107.

Morgan-Jones, M. and M. D. Eggboro, 1981, The hydrogeochemistry of the Jurassic limestones in Gloucestershire, England: Journal of Engineering Geology, v. 14, p. 25-39.

Mussman, W. J., I. P. Montanez and J. F. Read, 1988, Ordovician Knox paleokarst unconformity, Appalachians, *in* N. P. James and P. W. Choquette, eds., Palaeokarst: Berlin, Springer-Verlag p. 211-228.

Nelson, W. A. and J. F. Read, 1990, Updip to downdip cementation and dolomitization patterns in a Mississippian aquifer, Appalachians: Journal of Sedimentary Petrology, v. 60, p. 379-396.

Niemann, J. C. and J. F. Read, 1988, Regional cementation from unconformity-recharged aquifer and burial fluids, Mississippian Newman Limestone, Kentucky: Journal of Sedimentary Petrology, v. 58, p. 688-705.

O'Neil, J. R., R. N. Clayton and T. K. Mayeda, 1969, Oxygen isotope fractionation in divalent metal carbonates: Journal of Chemical Physics, v. 51, p.5547-5558.

Palmer, T. J., 1979, The Hampen Marly and White limestone formations: Florida-type carbonate lagoons in the Jurassic of Central England: Palaeontology, v. 22, p. 189-228.

Palmer, T. J. and H. C. Jenkyns, 1975, A carbonate island barrier from the Great Oolite, Middle Jurassic, of central England: Sedimentology, v. 22, p. 125-135.

Palmer, T. J., J. D. Hudson and M. A. Wilson, 1988, Palaeoecological evidence for early aragonite dissolution in ancient calcite seas: Nature, v. 335, p. 809-810.

Plummer, L. N., 1975, Mixing of sea water with calcium carbonate ground water: GSA Memoir 142, p. 219-236.

Richardson, S. W. and E. R. Oxburgh, 1979, The heat flow field in mainland U.K.: Nature, v. 282, p. 565-567.

Saller, A. H. and C. H. Moore, Jr., 1991, Geochemistry of meteoric calcite cements in some Pleistocene limestones: Sedimentology, v. 38, p.601-622.

Sandberg, P. A., 1983, An oscillating trend in Phanerozoic non-skeletal carbonate mineralogy: Nature, v. 305, p. 19-22.

Scholle, P. A. and R. B. Halley, 1985, Burial diagenesis: out of sight, out of mind, *in* N. Schneiderman and P. M. Harris, eds., Carbonate Cements: SEPM Special Publication 36, p. 309-334.

Sellwood, B. W., J. Scott, B. James, R. Evans and J. Marshall, 1987, Regional significance of 'dedolomitization' in Great Oolite reservoir facies of Southern England, *in* J. Brooks and K. Glennie, eds, Petroleum Geology of North West Europe: London, Graham and Trotman, p. 129-137.

Sellwood, B. W., T. J. Shepherd, M. R. Evans and B. James, 1989, Origin of late cements in oolitic reservoir facies: a fluid inclusion and isotopic study, Mid-Jurassic, southern England: Sedimentary Geology, v. 61, p. 223-237.

Shackleton, N. J. and J. P. Kennett, 1975, Palaeotemperature history of the Cenozoic and the initiation of Antarctic glaciation: Oxygen and carbon isotope analyses in DSDP sites 277, 279 and 281 *in* J. P. Kennett, R. E. Houtz, R. F. and others, eds., Initial Reports of the Deep Sea Drilling Project, v. 29, p. 743-755.

Sibson, R. H., J. McM. Moore and A. H. Rankin, 1975, Seismic pumping–hydrothermal fluid transport mechanism: Journal of the Geological Society of London, v. 131, p. 653-659.

Sladen, C. P. and D. J. Batten, 1984, Source-area environments of Late Jurassic and Early Cretaceous sediments in Southeast England: Proceedings of the Geologists' Association, v. 95, p. 149-163.

Smart, P. L.and F. Whitaker, 1991, Karst processes, hydrology and porosity evolution, *in* V. P. Wright, M. Esteban and P. L. Smart, eds, Palaeokarsts and Palaeokarstic Reservoirs: Reading, P.R.I.S. Occ. Publ. Series, 2, p. 1-55.

Toth, J., 1963, A theoretical analysis of groundwater flow in small drainage basins: Journal of Geophysical Research, v. 68, p. 4795-4812.

Toth. J., 1980, Cross-formational gravity-flow of ground water. A mechanism of the transport and accumulation of petroleum, the generalized hydraulic theory of petroleum migration *in* W. H. Roberts and R. J. Cordell, eds., Problems of Petroleum Migration: AAPG Studies in Geology 10, p. 121-168.

Walkden, G. M. and D. O. Williams, 1991, The diagenesis of the late Dinantian Derbyshire-East Midland carbonate shelf, central England: Sedimentology, v. 38, p. 643-670.

Whittaker, A.(ed.), 1985, An Atlas of Onshore Sedimentary Basins in England and Wales: Post Carboniferous Tectonics and Stratigraphy: Glasgow, Blackie, 71 p. and 28 enclosures.

Wigley, T. M. and L. N. Plummer, 1976, Mixing of carbonate waters: Geochimica et Cosmochimica Acta, v. 40, p. 989-995.

Wilson, N. P. and M. N. Luheshi, 1987, Thermal aspects of the East Midlands aquifer system, *in* J. C. Goff and B. P. J. Williams, eds., Fluid flow in sedimentary basins and aquifers; Geological Society of London Special Publication 34, Oxford, Blackwell Scientific, p. 157-169.

Wright, V. P., 1991, Palaeokarst: types, recognition, controls and associations, *in* V. P. Wright, M. Esteban and P. L. Smart, eds, Palaeokarsts and Palaeokarstic Reservoirs: Reading, P.R.I.S. Occ. Publ. Series, 2, p. 56-88.

Yurtsever, Y., 1975, Worldwide survey of stable isotopes in precipitation: Report of the Section for Isotope Hydrology, IAEA, 40 p.

Chapter 15

Examples From The Snorre Field

How Important Is the Late Cimmerian Unconformity in Controlling Formation of Kaolinite in Sandstones of the North Sea?

Per Arne Bjørkum
Geological Laboratory
Statoil
Stavanger, Norway

Ragnar Knarud and Morten Bergan
Saga Petroleum a.s.
Sandvika, Norway

ABSTRACT

Data on the content and distribution of kaolinite in the fluvial sandstones of the Statfjord Formation and the upper member of the Lunde Formation in the Snorre Field, northern North Sea, show no spatial relationship to the late Cimmerian unconformity. The reason is that the erosion rate during development of the unconformity was of the order of ten times greater than the propagation rate of the dissolution front in the sandstones caused by meteoric water flushing during subaerial exposure. This conclusion is consistent with similar observations from the nearby Gullfaks Field. It suggests that the diagenetic effect of the late Cimmerian unconformity on the underlying sandstones in the North Sea has been overestimated.

INTRODUCTION

The objective of this study is to examine whether diagenetic processes at the late Cimmerian unconformity influenced the kaolinite content in sandstones from the Snorre Field, northern North Sea. Sommer (1978), Hancock and Taylor (1978), Bjørlykke, et al. (1979), Bjørlykke (1983), Bjørlykke and Brendsdal (1986), Lønøy, et al. (1986), Bjørlykke, et al. (1989), Glasmann, et al. (1989), and others have concluded that the kaolinite in the Jurassic sandstones formed during meteoric water flushing. Meteoric water was introduced to the sandstones during or soon after deposition and during development of the late Cimmerian unconformity (Bjørlykke and Brendsdal, 1986). Sommer (1978), Hancock and Taylor (1978), Bjørlykke and Brensdal (1986) and Glasmann, et al. (1989) concluded that meteoric water flushing during late Cimmerian uplift and erosion may have resulted in kaolinitization of feldspars and mica in the underlying sandstones. The relative importance of synsedimentary meteoric water flushing and meteoric water flushing during development of the Late Cimmerian unconformity is, however, uncertain (Nedkvitne and Bjørlykke, 1992). Bjørkum, et al. (1990) concluded that the Late Cimmerian unconformity did not significantly

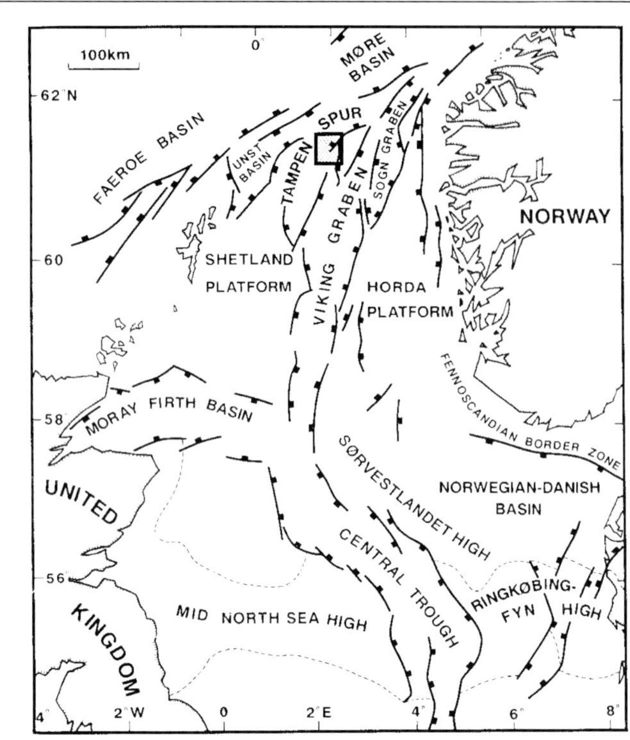

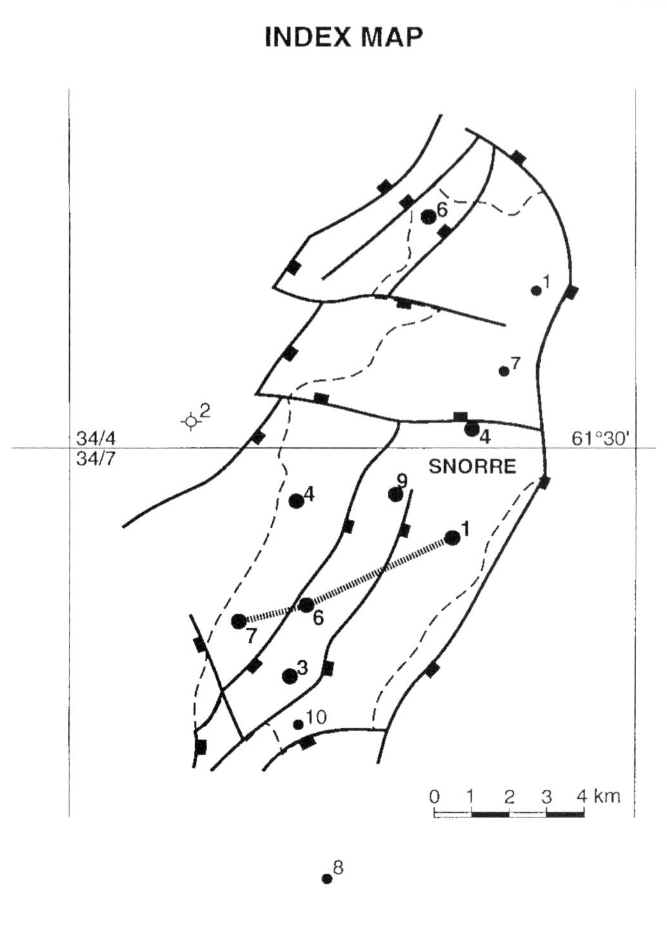

Figure 1. Geographic and structural map of the North Sea. Stippled rectangle shows the location of the Snorre Field. Index map shows Snorre Field, location of cross section on Figure 3, and well locations.

affect the kaolinite content of the Rannoch Formation of the Brent Group in the Gullfaks Field, pointing out that the average erosion rate during development of the Late Cimmerian unconformity was greater than the average propagation rate of the dissolution front in the sandstones being flushed by fresh water.

THE SNORRE FIELD

The Snorre Field is located at the Tampen Spur on the crest of an uplifted and rotated fault block at the north-western flank of the Viking Graben in the North Sea. An introduction to the Snorre Field is given by Hollander (1987) and Nystuen et al. (1989). A location map, general stratigraphic description and structural section of the area are shown in Figures 1, 2, and 3, respectively.

Depositional Setting

The reservoirs of the Snorre Field (Statfjord Formation and the upper member of Lunde Formation, Figures 1 and 2) consist of mobile fluvial channel belt sandstone bodies and, locally, isolated crevasse splay sandstone layers within flood plain inter-channel (mudstone) layers. The sandstone bodies and layers are interbedded with flood plain mudstones in sand/gross ratios of 0.3 to 0.8 in the various reservoir units. Deposition of the Norian-Rhaetian Lunde Formation and the Statfjord Formation occurred during the thermal cooling stage of the early Permian-Triassic rift phase (Badley et al., 1988; Steel and Ryseth, 1990) prior to late Jurassic rifting. A relative drop in sea level in the Early Rhaetian-Late Norian, after deposition of the middle member of the Lunde Formation, led to a retreat of the Triassic sea northward and to steepening of the depositional slope. As a result, sheet-like sandstone bodies were formed in flood channels of low-sinuosity ephemeral streams. In the Rhaetian, base level began to rise, and the high bedload stream systems that characterize the lower part of the upper member of the Lunde Formation (reservoir units D, E and F, Figure 3) were replaced by mixed-load streams of higher sinuosity in the upper part (reservoir unit B/C, Figure 3). This change in river types is marked by a transition from sheet-like sandstone bodies of relatively good lateral continuity to sinuous to ribbon-like sandstone bodies of more restricted lateral continuity.

The lower part of the Statfjord Formation is characterized by low sandstone content, whereas the upper part has a high content of extensive sheet-like sandstones with good reservoir quality. The lower part

Chronostratigraphy			Lithostratigraphy		
System	Series	Stage	Group	Formation	Member
Jurassic	Lower	Pliens-bachian	Dunlin	Amundsen	Calcareous
		Sinemurian		Statfjord	Upper
		Hettangian			
Triassic	Upper	Rhaetian			Lower
			Hegre Group	Lunde	Upper
		Norian			Middle
		Carnian			Lower
	Middle	Ladinian			
		Anisian		Lomvi	
	Lower	Scythian		Teist	Upper
					Lower

Figure 2. General stratigraphy for the Triassic and Lower Jurassic sequence, Tampen area in the North Sea.

represents a continuation of the depositional style that dominated during deposition of the uppermost part of the upper Lunde member. The Statfjord Formation sequence has an upward coarsening, progradational character (Nystuen et al., 1989). The sandstone facies in the upper part were deposited in low-sinuosity braided streams on an alluvial plain.

The Statfjord Formation was overlain by Lower Jurassic calcareous marine shales and siltstones of the Amundsen Formation of the Dunlin Group. These, as well as the upper part of the Dunlin Group and the Brent Group, were partly or completely eroded during the development of the late Cimmerian rift phase. The rifting resulted in block rotation and severe uplift, structuring, and possible subaerial exposure and erosion. The topseal in a down flank position is the Amundsen Formation while that in up flank positions is shales and marls of the Lower Cretaceous Cromer Knoll Group.

Structural Setting

The active stretching of the Cimmerian rifting phase (Ziegler, 1982) that is reflected in the depositional patterns of the Statfjord Formation continued throughout the Jurassic with a pronounced increase in activity at the boundary between the Early and Middle Jurassic. The rift processes and crustal thinning are also reflected in activation of growth faulting striking parallel to the graben axis, particularly during deposition of the upper part of the Jurassic Brent Group. The onset of the crustal cooling stage of the Cimmerian rifting resulted in rotation of crustal fault blocks in the Late Jurassic and Early Cretaceous. Over the Snorre Field, this led to uplift of the eastern side of the blocks, exposure, and erosion of Jurassic and Triassic rocks. During this inversion, up to 1500 m of sediments were eroded. The erosion took place during approximately 10–15 million years during the early Tithonian–Berriasian (Figure 4), suggesting an average erosion rate on the order of 100–150 m per m.y. in up flank positions.

MATERIALS AND METHODS

This study is based on mineralogic analysis of cores from the wells 34/7-1, -3, -4, -6, -7, -9, 34/4-4 and -6 (index map, Figure 1). The mineralogic compositions of 529 samples have been determined by X-ray diffraction (XRD), 150 thin sections have been studied using a transmitted light microscope, and 55 samples have been studied by scanning electron microscopy.

PETROGRAPHIC DESCRIPTION

The kaolinite in sandstones of the Statfjord Formation and upper member of the Lunde Formation occurs predominantly as kaolinite aggregates—pseudohexagonal, 5–30 mm, porefilling, kaolinite crystals with well-developed crystal surfaces. Vermicular aggregates and small quartz crystals are also commonly present. The highest relative amount of this type of kaolinite is observed in the coarse-grained sandstones. Partly dissolved feldspar is common in the sandstones in both formations. Micas are partly altered to kaolinite (expanded micas), but fresh micas may also be observed in the same thin section. Biotite is generally more altered than muscovite. Expanded mica edges have been interpreted to indicate that the mica has been transformed to kaolinite after deposition (Bjørlykke and Brendsdal, 1986; Nedkvitne and Bjørlykke, 1992). It should be stressed, however, that expanded mica edges are not sufficient evidence of post-depositional transformation because micas may undergo fresh-water diagenesis prior to deposition. The relative amounts of authigenic and detrital kaolinite have not been quantified because of difficulties involved in setting up objective criteria for their identification. For the purpose of this study it is, however, not necessary to know the relative amount of authigenic and detrital kaolinite.

The mineralogic compositions for each grain size in the Statfjord Formation and in the upper member of the Lunde Formation are shown in Figure 5. The plagioclase feldspar content is close to two times higher in the sandstones in the upper member of the Lunde Formation than in the sandstones of the Statfjord Formation. The kaolinite content increases with decreasing grain size in both formations. The increase in kaolinite content with decreasing grain size is more pronounced in the Statfjord Formation than in the upper member of the Lunde Formation.

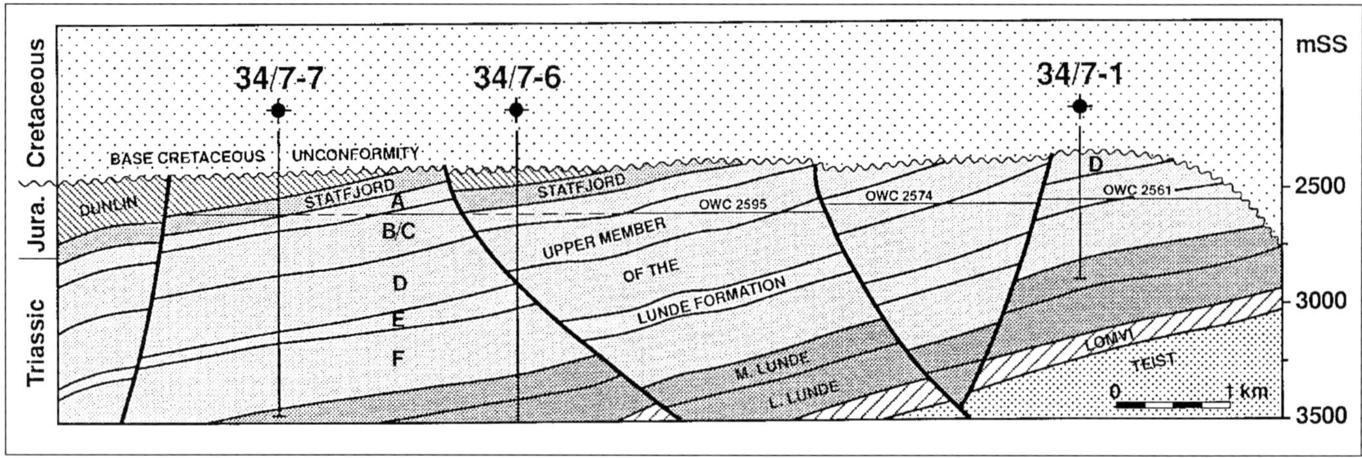

Figure 3. East-west cross section within block 34/7. See Figure 1 for location of section.

RESULTS AND DISCUSSION

Kaolinite Content and Distribution

The average kaolinite content for each grain size within both Statfjord Formation and the upper member of the Lunde Formation does not show any systematic variation relative to the late Cimmerian unconformity (Figure 6). The kaolinite content in the medium-grained sandstones in the reservoir unit D of the upper member of the Lunde Formation varies between 2 and 13% in well 34/7-1 (Figure 7) and between 2 and 12% in well 34/7-7 (Figure 8) and has a relatively uniform vertical distribution. These wells are located 7 km apart and have very different structural positions relative to the unconformity (Figure 3). In well 34/7-7, the analyzed section is separated from the unconformity by the uppermost part of the Lunde Formation, the Statfjord Formation, and by shales of the Amundsen Formation (a total of approximately 450 m vertical section). The same stratigraphic level in well 34/7-1 is truncated by the late Cimmerian unconformity. Sandstone connectivity is expected to be very good in unit D due to high sand/gross ratio (Nystuen et al., 1989). If the late Cimmerian unconformity were related to the formation of significant authigenic kaolinite in the Snorre Field, the kaolinite content in unit D in well 34/7-1 should be greater than in unit D in well 34/7-7. Also, we would expect to see an increase in kaolinite content towards the late Cimmerian unconformity in well 34/7-1. Such a relationship is not observed.

The kaolinite content in the sandstones in unit A of the upper member of the Lunde Formation is similar to the kaolinite content in the other reservoir units in the Lunde Formation (Figure 6). The sand/gross ratio in unit A in the upper member of the Lunde Formation is low (less than 0.2). Hence, sandstone connectivity is expected to be very low. It is therefore unlikely that the kaolinite in these sandstones was formed by meteoric water flushing during development of the late Cimmerian unconformity. If the kaolinite in the different units of the upper member of the Lunde Formation has the same origin, the late Cimmerian unconformity is unlikely to have had any significant influence on the kaolinite content.

Meteoric Water Flushing and the Unconformity

The diagenetic effect of meteoric water flushing during development of the late Cimmerian unconformity is dependent upon the propagation rate of the dissolution front in the sandstones relative to the erosion rate during development of the unconformity. Significant diagenetic effects from the meteoric water flushing are only to be expected if the propagation rate of the dissolution front exceeds the rate of erosion (Bjørkum et al., 1990). The sandstones in the Snorre Field contain between 20 and 45% feldspar, with plagioclase more abundant than K-feldspar (Figure 5). The experimentally determined dissolution rate constants for plagioclase (albite) and K-feldspar at 25° C and a pH of 5, are 3.9×10^{-12} mole m^{-2} s^{-1} and 0.56×10^{-12} mole m^{-2} s^{-1}, respectively (calculated from Busenberg and Clemency, 1976). Hence, the dissolution rate for albite is nearly one order of magnitude higher than that for K-feldspar. There is no petrographic evidence, however, suggesting that albitic plagioclase is preferentially altered compared with K-feldspar by such a magnitude. For simplification, we use the dissolution rate constant for K-feldspar for both K-feldspar and albite in the following calculations. This simplifies them and affects only the thickness of the dissolution zone, i.e. the zone where feldspars and micas are being dissolved and kaolinite precipitates. It does not affect the propagation rate of the dissolution front. We further assume that the average grain size is 0.5 mm; the feldspar concentration at the time of deposition is 25%; the average initial porosity is 33%; the partial pressure of CO_2 in the soil horizon is $10^{-2.5}$ atm (a typical value in mature

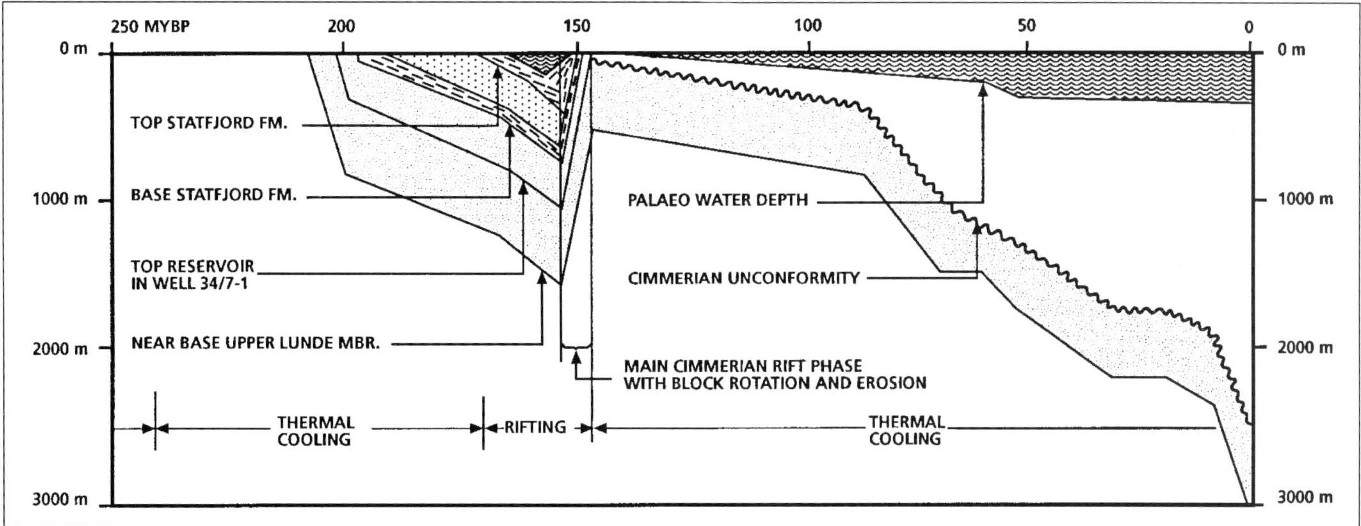

Figure 4. Burial history of well 34/7-1.

soils, around ten times the partial pressure of CO_2 in the atmosphere); and the average flow rate of meteoric water is 1 m yr^{-1} (see Bjørkum et al., 1990, for discussion of these parameters).

Based on the above assumptions, the thickness of the dissolution zone is calculated to be less than 1 m; the amount of kaolinite formed would be in the range of 10–20%; and the propagation rate of the dissolution front less than 10 m $m.y.^{-1}$. It follows that it would take of the order of 100 m.y. for the dissolution front to propagate through 1000 m of sandstone in the upper member of the Lunde Formation. The duration of meteoric water flushing in the late Kimmeridgian–Berriasian at the Snorre Field was approximately 10–15 m.y., suggesting that only about 100–150 m of sandstone could have been significantly affected by kaolinite formation.

The dissolution rates suggest that most of the mineral reactions caused by introduction of meteoric water will take place within a limited zone immediately below the late Cimmerian unconformity. If dissolution rate constants that are two orders of magnitude less are applied, the thickness of the dissolution zone is increased by two orders of magnitude: the zone where kaolinite is formed is of the order of 100 m thick. It is important to note however, that the dissolution rate does not affect the propagation rate of the dissolution front, only the thickness of the dissolution zone (Bjørkum et al., 1990). The diagenetic effect of the meteoric water introduced during development of the late Cimmerian unconformity is thus likely to be confined to a deeply weathered zone situated immediately below the unconformity, i.e. a few meters to a few tens of meters (Bjørkum et al., 1990; Emery et al., 1990). The distribution of kaolinite is therefore unlikely to give any information about the lateral continuity of the sandstone in reservoirs, as suggested by Nedkvitne and Bjørlykke (1992).

For a given flow rate, the amount of kaolinite that can form during meteoric water flushing is constrained by the amount of protons provided by CO_2 that has been introduced to the water in soil horizons. Nedkvitne and Bjørlykke (1992), Bjørlykke et al. (1992) and Bjørlykke and Aagaard (1992) argue that additional CO_2 may be provided by decomposition of organic matter along the flow path and that kaolinitization therefore will continue a long distance away from the meteoric water inlet area. It is reported by Bjørlykke (1984), however, that the amount of CO_2 released from organic matter in the subsurface is insufficient to explain large scale leaching of feldspar (and carbonate). Further, since most of the CO_2 in sedimentary basins is generated at temperatures higher than 50° C (Hunt, 1979), and not at the time when the sandstones were flushed by meteoric water, it is difficult to see how CO_2 added to the meteoric water along the flow path could give rise to a significant amount of kaolinite precipitation.

Bjørlykke and Aagaard (1992) and Bjørlykke et al. (1992) have suggested that kaolinite can form by hydrolysis of silicate minerals like feldspar and mica and does not require acid water, as protons can be provided by dissociation of water. Following this idea the following reaction can be set up:

K-feldspar Kaolinite

$$2KAlSi_3O_8 + 3H_2O = Al_2Si_2O_5(OH)_4 + 2K^+ + 4Si_2(aq) + 2OH^- \quad (1)$$

If the pH in the pore water were increased from 7 to between 10 and 11, this would produce between 10^{-3} to 10^{-4} mole l^{-1} H^+. Typical pH values for meteoric and ground waters are, however, between 6 and 9, suggesting that the maximum amount that can be produced is on the order of 10^{-5} mole l^{-1} H^+. This is one order of magnitude less than the amount of protons assumed to be provided by carbon dioxide introduced in the soil horizon at the inlet area. It is there-

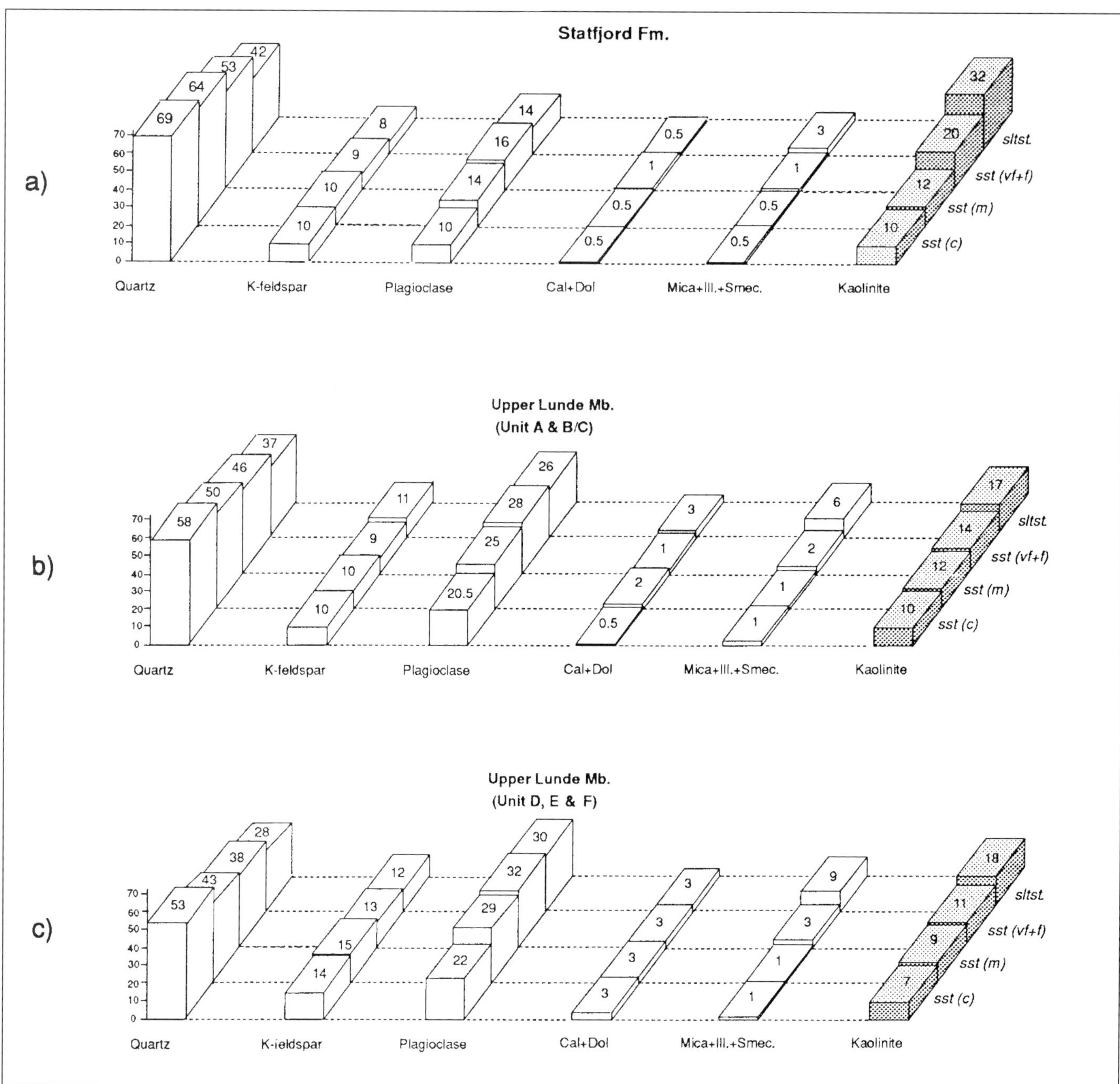

Figure 5. Bulk mineral composition for (a) the Statfjord Formation, (b) reservoir units A and B, C, and (c) D, E and F of the upper member of the Lunde Formation. The mineralogical composition is determined by X-ray diffraction.

fore unlikely that hydrolysis is an important source for protons during meteoric water flushing.

Erosion Rates and Preservation of Kaolinite

Erosion in the eastern and northern part of the Snorre Field has removed from 500 to 1500 m of Jurassic and Triassic strata on the crest of the rotated fault block during development of the late Cimmerian unconformity (i.e. during 10–15 m.y.). Hence, the average erosion rate in the Snorre Field area was in the range of 50–150 m m.y.$^{-1}$, between 5 and 15 times greater than the average propagation rate of the dissolution front ~10 m m.y.$^{-1}$ in the sediments flushed by meteoric water. Hence, sediments that were kaolinitized due to fresh water flushing during development of the late Cimmerian unconformity are not likely to be preserved in the sedimentary column in the Snorre Field. This conclusion is consistent with the results from the Gullfaks Field (Bjørkum et al., 1990). The results of these two studies justify a critical reevaluation of the assumption that the late Cimmerian unconformity did significantly affect the distribu-

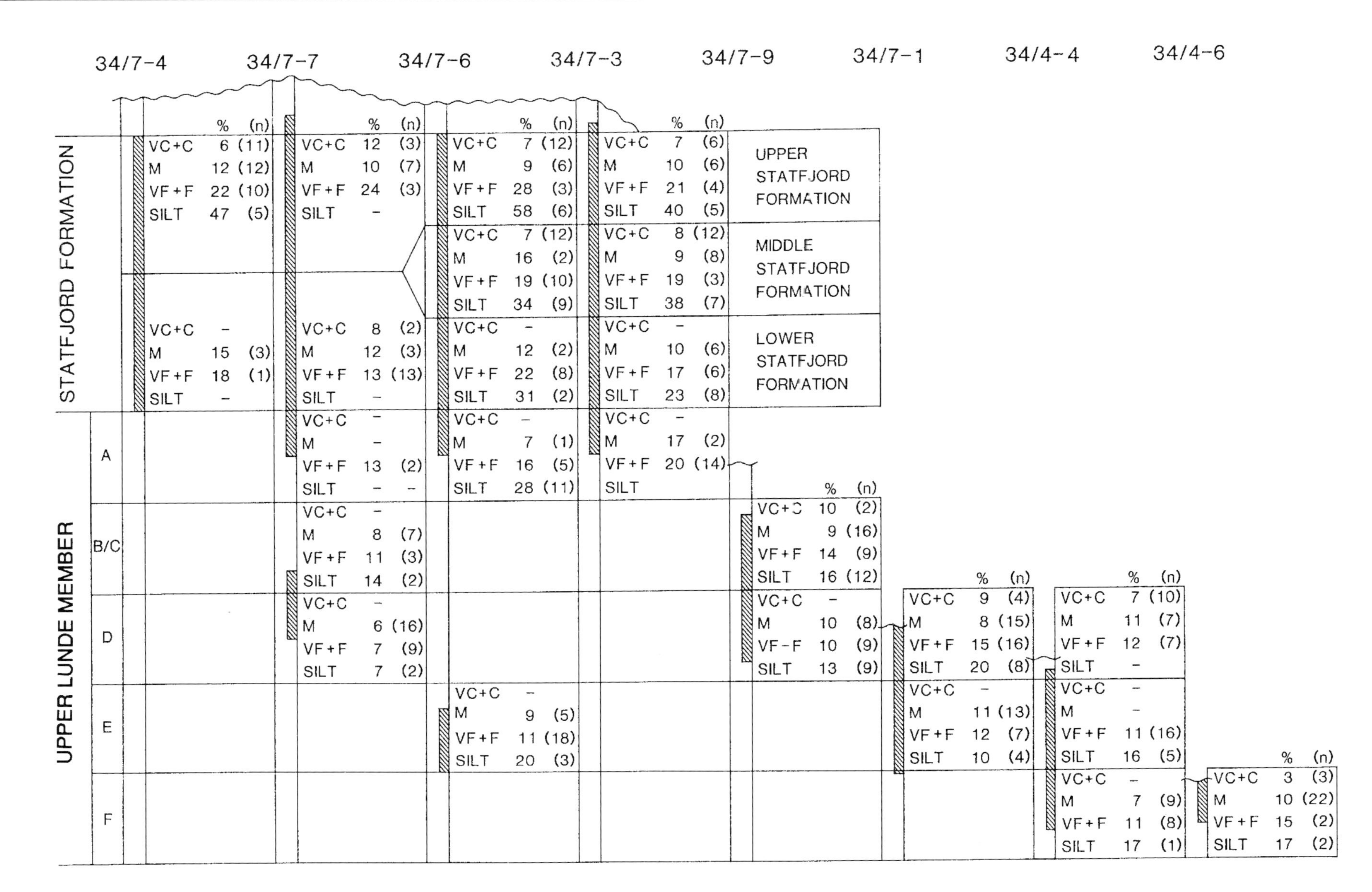

Figure 6. Bulk kaolinite content (%) for wells in the Snorre Field. Average values of bulk XRD analyses are given for each grain size fraction in various lithostratigraphic units cored. Silt = siltstone, VF+F = very fine- to fine-grained sandstone, M= medium-grained sandstone, VC+C=very coarse to coarse-grained sandstone; n is number of samples. Cored intervals are indicated by diagonally ruled vertical bars. The wavy line represents the late Cimmerian unconformity.

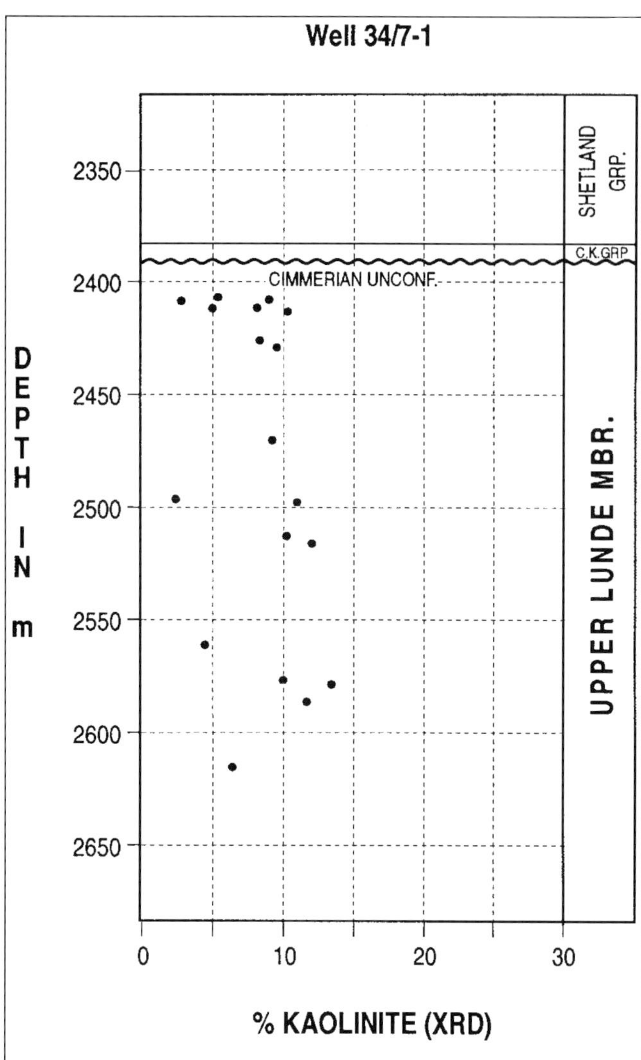

Figure 7. Kaolinite versus depth for medium-grained sandstones in reservoir unit D in well 34/7-1.

tion of kaolinite in the underlying sandstones situated in intrabasinal structural highs.

CONCLUSIONS

The kaolinite content for each grain size and facies in the Statfjord Formation and the upper member of the Lunde Formation in the Snorre Field has been compared for each reservoir unit in wells with different vertical and horizontal distances to the late Cimmerian unconformity. For a given grain size, the kaolinite content within a reservoir zone is similar throughout the field. This suggests that the unconformity had no significant influence on the origin and distribution of kaolinite in the Snorre Field. The conclusion is consistent with a study of the Gullfaks Field (Bjørkum et al., 1990). The explanation is probably that the rate of erosion during uplift, which is on the order of 50 to 150 m m.y.$^{-1}$, was an order of magnitude greater than the propagation rate of the dissolution front in the sandstones. Authigenic kaolinite that formed during development of the late Cimmerian unconformity is thus not likely to be preserved due to high erosion rates. Studies of both Snorre and Gullfaks fields therefore indicate that the late Cimmerian unconformity had little effect on the diagenesis of the underlying sandstones.

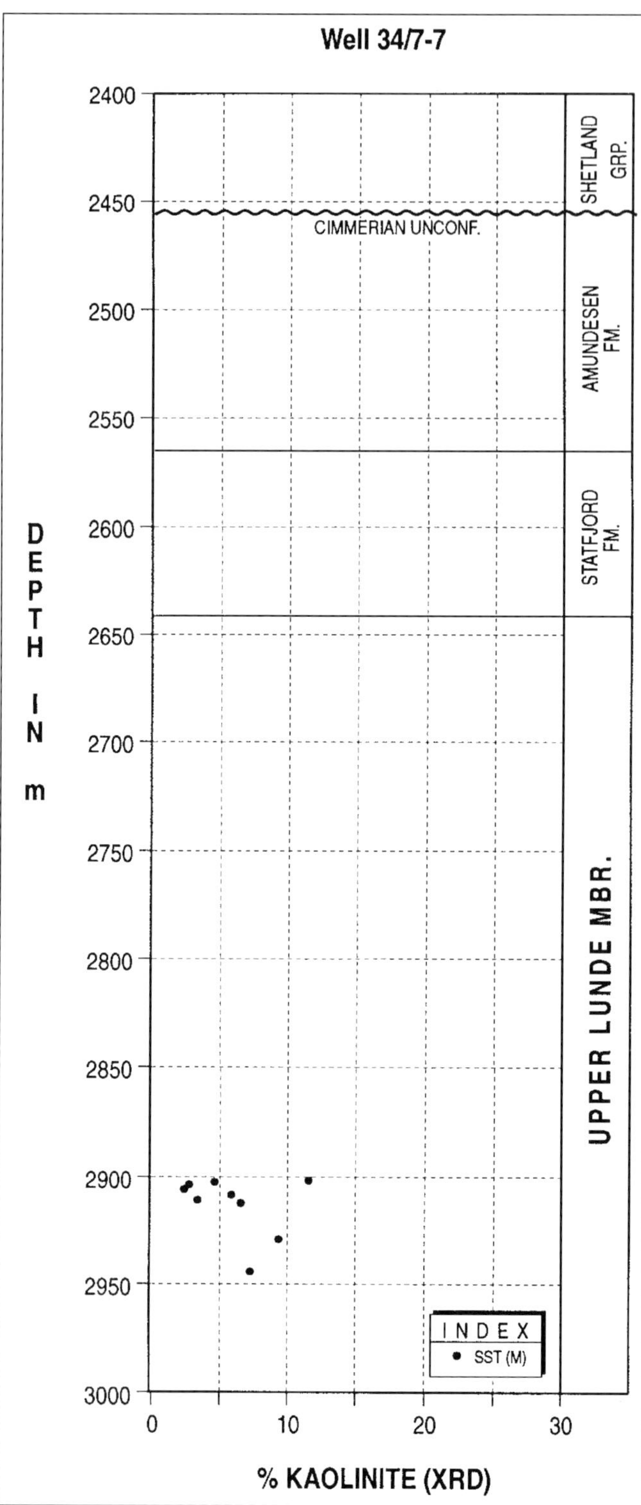

Figure 8. Kaolinite versus depth for medium-grained sandstones in reservoir unit D in well 34/7-7.

ACKNOWLEDGMENTS

This work was partly funded by the Norwegian Department of Energy and is part of the State Program for Improved Oil Recovery and reservoir technology (SPOR). XRD data and thin-sections were prepared by Bente Monsrud and Eva Solbrække, Saga Petroleum a.s. SEM analyses were performed by Tor Mellem, Saga Petroleum a.s. We thank Saga Petroleum a.s. and the license partners in PL 057 and PL 089 for permission to publish the results of these studies. Stephen Ehrenberg is thanked for constructive reviews of an earlier draft of this paper.

REFERENCES CITED

BADLEY, M.E., PRICE, J.D., RAMBECH DAHL, C. AND AGDESTEIN, T., 1988, The structural evolution of the northern Viking Graben and its bearing upon extension modes of basin formation: Journal of the Geological Society of London, v. 1456, p.455-472.

BJØRKUM, P.A., R. MJØS, O. WALDERHAUG, and HURST, 1990, The role of the late Cimmerian unconformity for the distribution of kaolinite in the Gullfaks Field, Northern North Sea: Sedimentology, v. 37, p. 395-406.

BJØRLYKKE, K., 1983, Diagenetic reactions in sandstones, A. Parker and B. W. Sellwood, eds., Sediment Diagenesis: A. Reidel Publishing Company, p.169-213.

BJØRLYKKE, K., 1984, Formation of secondary porosity: How important is it?, D.A. McDonald and R.C. Surdam, eds., Clastic Diagenesis: AAPG Memoir 37, p.277-286.

BJØRLYKKE, K. AND P. AAGAARD, 1992, Clay minerals in North Sea Sandstones, D.W. Houseknecht, and E.D. Pittman, eds., Origin, Diagenesis and Petrographics of Clay Minerals in Sandstones: SEPM Special Publication 47, p. 65–80.

BJØRLYKKE, K. AND A. BRENSDAL, 1986, Diagenesis of the Brent sandstones in the Statfjord Field, North Sea, D.L. Gautier, ed., Roles of organic matter in sediment diagenesis, SEPM Special Publication 38, p.157-167.

BJØRLYKKE, K., A. ELVERHØI, AND A.D. MALM, 1979, Diagenesis in the Mesozoic sandstones from Spitsbergen and the North Sea: A comparison: Geologische Rundschau, v.68, p.1151-1171.

BJØRLYKKE. K. NEDKVITNE, T., M. RAMM, AND G.C. SAIGAL, (1992), Diagenetic Processes in the Brent Group (Middle Jurassic) reservoirs of the North Sea—an Overview, Morten et al., eds., Geology of the Brent Group: Geological Society of London Special Publication 61, p. 263–287.

BJØRLYKKE, K., M. RAMM, AND G.C. SAIGAL, 1989, Sandstone diagenesis and porosity modification during basin evolution: Geologische Rundschau, v. 78, p.243-268.

BUSENBERG, E. AND C.V. CLEMENCY, 1976, The dissolution kinetics of feldspars at 25° C and at 1 atm CO_2 partial pressure: Geochimica et Cosmochimica Acta, v. 40, p.41-50.

EMERY, D., K.J. MYERS, AND R. YOUNG, 1990, Ancient subaerial exposure and freshwater leaching in sandstones: Geology, v.18, p. 1178-1181.

GLASMANN, J.R., R.A. CLARK, S. SLATER, N.A. BRIEDIS AND P.D. LUNDEGARD, 1989, Diagenesis and hydrocarbon accumulation, Brent Sandstone (Jurassic), Bergen High Area, North Sea: AAPG Bulletin, v.73, p. 1341-1361.

HANCOCK, N.J. AND A.M. TAYLOR, 1978, Clay mineral diagenesis and oil migration in the Middle Jurassic Brent Sand Formation: Journal of the Geological Society of London, v.135, p.69-72.

HOLLANDER, N. B., 1987, Snorre, in A.M. Spencer et al., eds., Geology of the Norwegian Oil and Gas Fields. Graham & Trotman, London, p. 307-318.

HUNT, J.M., 1979, Petroleum Geochemistry and Geology. W.H. Freeman and Company, San Francisco. pp.617.

LØNØY, A., J. AKSELSEN, AND K. RØNNING, 1986, Diagenesis of a deeply buried sandstone reservoir: Hild Field, northern North Sea: Clay Minerals, v. 21, p. 497-511.

NEDKVITNE, T. AND K. BJØRLYKKE, 1992, Secondary porosity in the Brent Group (Middle Jurassic), Huldra Field, North Sea—Implications for predicting lateral continuity of sandstones?: Journal of Sedimentary Petrology, v. 62, p. 23-34.

NYSTUEN, J.P., R. KNARUD, K. JORDE, AND K.O. STANLEY, 1989, Correlation of Triassic to Lower Jurassic sequences, Snorre Field and adjacent areas, northern North Sea, J.D. Collinson, ed., Correlation in Hydrocarbon Exploration: Norwegian Petroleum Society, Graham & Trotman, p. 273-290.

SOMMER, F., 1978, Diagenesis of sandstones in the Viking Graben: Journal of the Geological Society of London, v. 135, p. 63-67.

STEEL, R, AND RYSETH, A., 1990, The Triassic—Early Jurassic succession in the northern North Sea: megasequence stratigraphy and intra-Triassic tectonics, R.F.P. Hardman and J. Brooks, eds., Tectonic Events Responsible for Britain's oil and gas reservoirs: Geological Society Special Publication 55, p. 139-168.

ZIEGLER, P. A., 1982, Geological Atlas of Western and Central Europe. Shell International Petroleum Maatschappij B.V. and Elsevier, Amsterdam, p. 130.

Index